MARINE MANGANESE DEPOSITS

SIR JOHN MURRAY

1841—1914

Elsevier Oceanography Series, 15

MARINE MANGANESE DEPOSITS

edited by

G.P. GLASBY

*New Zealand Oceanographic Institute, Department of Scientific and Industrial Research,
Wellington, New Zealand*

ELSEVIER SCIENTIFIC PUBLISHING COMPANY
AMSTERDAM - OXFORD - NEW YORK

ELSEVIER SCIENTIFIC PUBLISHING COMPANY
335 Jan van Galenstraat
P.O. Box 211, Amsterdam, The Netherlands

Distributors for the United States and Canada:

ELSEVIER NORTH-HOLLAND INC.
52, Vanderbilt Avenue
New York, N.Y. 10017

Library of Congress Cataloging in Publication Data

Main entry under title:

Marine manganese deposits.

  (Elsevier oceanography series ; 15)
  Bibliography: p.
  Includes index.
  1. Manganese nodules. I. Glasby, G. P.
QE390.2.M35M37   553'.462   76-48895
 ISBN 0-444-41524-6

ISBN 0-444-41623-4 (series)
ISBN 0-444-41524-6 (vol. 15)

Printed in The Netherlands

FURTHER TITLES IN THIS SERIES

LIST OF CONTRIBUTORS

AMOS, A.F.
Lamont-Doherty Geological Observatory, Columbia University, Palisades, New York 10964, U.S.A.

AUBURN, F.M.
Faculty of Law, The University of Auckland, New Zealand.

BREWER, P.G.
Woods Hole Oceanographic Institution, Woods Hole, Massachusetts 02543, U.S.A.

BURNS, R.G.
Department of Earth and Planetary Sciences, Massachusetts Institute of Technology, Cambridge, Massachusetts 02139, U.S.A.

BURNS, V.M.
Department of Earth and Planetary Sciences, Massachusetts Institute of Technology, Cambridge, Massachusetts 02139, U.S.A.

CALVERT, S.E.
Institute of Oceanographic Sciences, Wormley, Godalming, Surrey, England.

CRONAN, D.S.
Department of Applied Geochemistry, Imperial College of Science and Technology, Prince Consort Road, London S.W.7 2BP, England.

FEWKES, R.H.
Department of Geology, Washington State University, Pullman, Washington 99163, U.S.A.

ELDERFIELD, H.
Department of Earth Sciences, The University, Leeds LS2 9JT, England.

FUERSTENAU, D.W.
Department of Materials Science and Engineering, University of California, Berkeley, California 94720, U.S.A.

GARSIDE, C.
Lamont-Doherty Geological Observatory, Columbia University, Palisades, New York 10964, U.S.A.
and The City University of New York.

GLASBY, G.P.
New Zealand Oceanographic Institute, Department of Scientific and Industrial Research, Wellington, New Zealand.

HAN, K.N.
Department of Chemical Engineering, Monash University, Clayton, Australia.

JENKYNS, H.C.
Department of Geological Sciences, The University, Durham, England.

KU, T.L.
Department of Geological Sciences, University of Southern California, University Park, Los Angeles, California 90007, U.S.A.

MALONE, T.C.
Lamont-Doherty Geological Observatory, Columbia University, Palisades, New York 10964, U.S.A.
and The City University of New York.

MERO, J.L.
Ocean Resources, Inc., La Jolla, California 92037, U.S.A.

MEYLAN, M.A.
Department of Oceanography, University of Hawaii, Honolulu, Hawaii 96822, U.S.A.

MURRAY, J.W.
Department of Oceanography, University of Washington, Seattle, Washington 98195, U.S.A.

PAUL, A.Z.
Lamont-Doherty Geological Observatory, Columbia University, Palisades, New York 10964, U.S.A.
and The City University of New York.

PRICE, N.B.
Grant Institute of Geology, University of Edinburgh, Edinburgh, Scotland.

RAAB, W.J.
Kennecott Exploration, Inc., 2300 West 1700 South, Salt Lake City, Utah 84104, U.S.A.

ROELS, O.A.
Lamont-Doherty Geological Observatory, Columbia University, Palisades, New York 10964, U.S.A.
and The City University of New York.

SOREM, R.K.
Department of Geology, Washington State University, Pullman, Washington 99163, U.S.A.

ACKNOWLEDGEMENTS

As editor, I would like to express my sincere thanks to all the contributors to this book for their tremendous efforts in producing manuscripts, to Mr J.W. Brodie for his encouragement in this project, to Dr J.S. Tooms who introduced me to the subject of manganese nodules and who supervised my Ph.D. on that topic, to my colleagues with whom I have collaborated in the field of manganese nodule research over the years, and to Miss E.M. Bardsley who prepared the index.

Authors would also like to thank various individuals for their help in producing their chapters. These are listed by chapter.

Chapter 1. Dr D.R. Stoddart kindly permitted quotation from the letters of J.Y. Buchanan written aboard H.M.S. *Challenger*.

Chapter 3. The following are thanked for permission to use unpublished data: D.Z. Piper for rare earth contents of concretions from Lake Shebandowan, E. Suess for data on the composition of manganese carbonate from the Baltic Sea and for confirming the occurrence of MnS in the same sediments, and R. Rossmann and E. Callender for information on the composition of concretions from Lake Michigan. J.W. Murray kindly supplied preprints of papers on the adsorptive capacities of precipitated manganese oxides. P.G. Brewer and F. Culkin provided critical discussion of some of the concepts discussed here.

Chapter 4. Drs M.G. Audley-Charles and R.A. Gulbrandsen for supplying photographs of their material, and Dr R.W. O'B. Knox for reading the manuscript.

Chapter 5. The section by W.J.R. is published by permission of Kennecott Copper Corporation which supported the research. Discussions with D. Felix and C.E. Schatz of Kennecott's Ocean Exploration Division and D. Norton contributed significantly to the content of this chapter. Thanks are also due to our laboratory personnel who performed the many analytical tasks. E.J. Mahaffey first observed the artificial nodule formation and performed all analytical work for that project.

Chapter 6. The authors are grateful for the encouragement and financial support of their research in the past few years from the National Science Foundation and Washington State University. It is impossible to acknowledge properly the stimulation received from discussions and meetings with many co-workers in this field but it is much appreciated. Apologies are presented if ideas have inadvertently been presented here as being those of the authors which in fact were first suggested by someone else. These and any other shortcomings in this chapter are regretted and are solely the responsibility of the authors. Special thanks are due to Allan R. Foster for the preparation of many of the specimens studied here and particularly for many enlightening discussions. In addition, the counsel of Professor Arthur Cohen, Director of the Washington State University Electron Microscope Laboratory, and his generosity in providing the SEM for our use is much appreciated. Professor

Charles Knowles, University of Idaho, is also thanked for making electron microprobe analyses possible. The authors gratefully acknowledge the support of the U.S. National Science Foundation in much of their research.

Chapter 7. This chapter could not have been written without the assistance of a large number of people. The following people not only generously gave us helpful comments, but also provided valuable reference specimens: Prof. Clifford Frondel, Dr Rudolf Giovanoli, Dr Michael Fleischer, Prof. Douglas Fuerstenau, Prof. M. Nambu, Dr Stanley Margolis, Dr David Horn, Prof. Howard Jaffe, and Dr Maury Morgenstein. Prof. Owen Bricker and Prof. Ronald Sorem kindly loaned us reference X-ray patterns of certain manganese oxides. The following people kindly gave us their time and helpful discussions: Dr M.H. Hey, Dr A.L. Mackay, Dr R.M. McKenzie, Dr Gale Hubred, Prof. Martin Buerger, Prof. P.B. Moore, Dr Ursula Marvin, Dr Bruce Brown, and Dr Heinz Nau. Some exploratory experimental work was carried out by Dr David Kohlstedt and Mr Windsor Sung. Results of their investigations will be reported elsewhere. A special acknowledgement is due to Ms. Roxanne Regan for patiently helping to organize this manuscript and in typing the various drafts. The study was supported by a grant from the NSF/ IDOE Inter-University Ferromanganese Program.

Chapter 8. Thomas O'Neil assisted in the measurements of the nodule densities. Financial support was supplied through the National Science Foundation (Grant DES72-01557 (Oceanography Section)) and through the Seabed Assessment Program of the Office for the International Decade of Ocean Exploration (IDO75-12960).

Chapter 9. The current research in experimental (Table 9-IX) and theoretical (Table 9-X) geochemistry is funded by the Natural Environment Research Council, grant GR3/2180.

Chapter 10. The chapter benefitted from many fruitful discussions with Drs R.G. Burns, M.L. Bender, G. Thompson, D.W. Spencer, W. Stumm, J.J. Morgan and D.Z. Piper. The work was sponsored by NSF grants GA-22292 and DES74-14362.

Chapter 13. Some of the research cited in this chapter was supported by contracts from the U.S. Department of Commerce, Environmental Research Laboratories (03-6-022-27; 03-3-022-144; 03-6-022-35105; 03-6-022-35106) and by the U.S. Energy Research and Development Administration (AT[11-1]2185). Lamont-Doherty Geological Observatory contribution No. 2401; City University Institute of Marine and Atmospheric Sciences contribution No. 79.

G.P.G.

CONTENTS

CHAPTER 1

HISTORICAL INTRODUCTION

G. P. GLASBY

One of the earliest records, if not the earliest, of iron-manganese concretions relates to the form of soil nodules appropriately known as "buckshot gravel" and is given by Liechhardt (1847; quoted in Bryan, 1952) in his book "Overland Expedition from Morton Bay to Port Essington". "Swarms of whistling ducks occupied large ponds in the creek, but our shot was all used, and the small iron-pebbles which we used as a substitute were not heavy enough to kill even a duck." Such are the beginnings of science.

The discovery of marine manganese nodules begins with the voyage of the H.M.S. *Challenger*, 1872—1876. Deep-sea nodules were recovered for the first time on 18 February 1873 approximately 300 km southwest of the island of Ferro in the Canary Group (Murray and Renard, 1891). The results of this cruise were unique in as much as they were to dominate thinking on manganese nodules for over 80 years. Large quantities of nodules displaying a wide range of morphologies and internal structures were recovered from the Atlantic, Indian and Pacific oceans (Fig. 1-1). The nodules were shown to consist of concentric bands of ferromanganese oxides around such diverse nuclei as pumice, coral, phosphorite nodules, volcanic ash, palagonite, sharks' teeth and glacial erratics. A slow growth rate for the nodules was established and some of the nodules were shown to have broken in situ and subsequently accreted manganese around the broken surfaces. Great diversity in appearance in the nodules was noted but generally nodules from a single site were similar in appearance and differed in size, form and internal appearance from those at another station; "so much so that now, after a detailed study of the collections, it is usually possible for us to state at sight from which *Challenger* station any particular nodule had been produced". In many cases, the external form of the nodule depended on the shape of the nucleus and was often complicated by the incorporation of multiple nuclei into the nodule. The complexity of formation of the nodules is well illustrated by reference to samples from Station 281 (22°21'S 150°17'W, 4,360 m) in the South Pacific. Here, an ash shower fall appears to have covered nodules growing in red clay. The resultant slabs have subsequently been covered on the upper surface by further manganese oxides to give a complex depositional history (Fig. 1-2).

The most extensive deposits of manganese nodules were found in the deeper-water regions of the Pacific and Indian oceans in association with

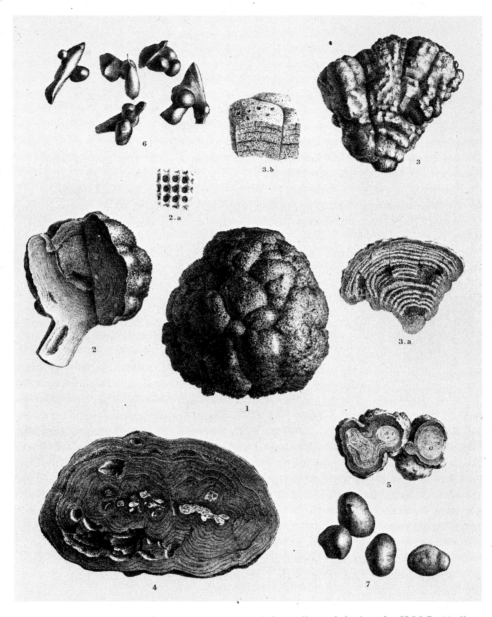

Fig. 1-1. Line drawing of manganese-iron nodules collected during the H.M.S. *Challenger* expedition (1872—1876) (Murray and Renard, 1891, Plate II).

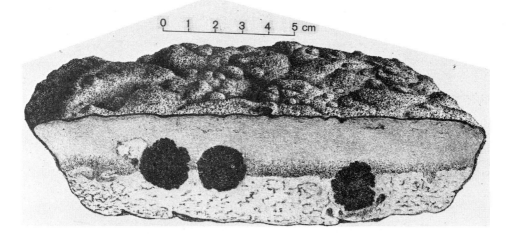

Fig. 1-2. "One of several large slabs dredged among the nodules from the South Pacific, in section, and showing part of the upper surface. About the middle of the section there is a dark line which appears to represent the upper surface of an old sea-bottom, with manganese nodules embedded or partially embedded in the clay. A fall of ashes would appear to have taken place, covering the floor of the ocean in some places to the depth of an inch. The coarser particles lie immediately on the clay, and contain much black mica, then follow layers of finer and finer particles. Subsequently the bottom was apparently, after consolidation, rent by cracks, and layers of manganese were deposited over the upper surface and down the cracks, binding the whole into a compact mass. Station 281; 2,385 fathoms. South Pacific." (Murray and Renard, 1891, Plate IV, fig. 3.)

sharks' teeth, earbones of cetaceans, cosmic spherules and dark chocolate-coloured clays (all indicators of low sedimentation rates). By contrast, the occurrence of manganese nodules appeared to be more limited in the Atlantic Ocean and restricted mainly to the vicinity of volcanic islands, although there were indications of an approach to the chocolate-coloured clays of the Pacific in the deeper waters about 20°N 50°W.

Chemical analyses revealed the major components of the nodules to be manganese and iron oxides with appreciable amounts of silica, alumina, lime, magnesia and water. Analytical sensitivities at that time were such that the minor elements, copper, nickel and cobalt, were more frequently referred to as being present in "trace" amounts. These analyses indicate that the nodules could be "classed along with the impure variety of manganese known as wad or bog manganese ore".

Basically, four major hypotheses of nodule formation were proposed as a result of the *Challenger* work.

(1) "The manganese of the nodules is chiefly derived from the decomposition of the more basic volcanic rocks and minerals with which the nodules are nearly always associated in deep-sea deposits. The manganese and iron of these rocks and minerals are at first transformed into carbonates, and subsequently into oxides, which on depositing from

solution in the watery ooze, take a concretionary form around various kinds of nuclei."
(2) "They are formed under the reducing influence of organic matters on the sulphates of sea-water, sulphides being produced and subsequently oxidised."
(3) "They arise from the precipitation of manganese contained in the waters of submarine springs at the bottom of the ocean."
(4) "They are formed from the compounds of manganese dissolved in sea-water in the form of bicarbonates, and transformed at the surface of the sea into oxides, which are precipitated in a permanent form on the bottom of the ocean."

In addition, Lockyer (1888) proposed a cosmic or meteoric origin for the nodules based on the similarity of the spectra of manganese nodules and meteorites, and Buchanan wrote in a letter to his father dated 15 March 1874 (D. R. Stoddart, personal communication, 1976): "I believe it (manganese) is formed by a beast; this my zoological colleagues will not admit and naturally so, as it would then be their duty to find it."

With the exception of (2), each of these hypotheses has been refined and discussed by modern proponents, although Murray and Renard (1891) themselves accepted the first interpretation.

The central character in the study of the *Challenger* nodule collection was John (later Sir John) Murray (see Frontispiece) who took over responsibility for editing the *Challenger* volumes on the death of the expedition leader, Sir Charles Wyville Thomson, in 1882. His work with the Rev. A. F. Renard, Professor of Geology and Mineralogy at the University of Ghent, was to result in the definitive account of the deep-sea deposits collected during the expedition (Murray and Renard, 1891). The discovery of nodules on the sea floor was a surprise to the scientists, Thomson (1873) describing the dredge haul of 7 March 1873 as containing "a number of very peculiar black oval bodies about one inch long". Thomson initially thought that these deposits were fossils but subsequent examination by J. Y. Buchanan, the expedition chemist, revealed that they were composed "of almost pure manganese peroxide". Buchanan must also take credit as being the first person to recognize the commercial possibilities of manganese nodules. In the previously quoted letter to his father he wrote: "Manganese is a mineral of great commercial importance and it is one of the principal substances used in the manufacture of bleaching powder and although of course the bottom of the sea at present could never be made a paying source of supply, its occurrence there with the certainty of having been formed there may turn out to be an important fact in geology."

Dredging was the main method used for the recovery of nodules during the expedition and considerable quantities (up to 80 litres) were recovered in a single haul (Fig. 1-3). After the *Challenger* expedition, Buchanan made a number of summer cruises between 1878—1882 on his yacht *Mallard*, in which he discovered manganese nodules on the bed of Loch Fyne (Buchanan 1878, 1891). These deposits and others from western Scotland were subsequently studied by Murray in conjunction with Robert Irvine aboard his yacht *Medusa*, in the years 1884—1892 (Murray and Irvine, 1894).

Fig. 1-3. Sifting the contents of the dredge, H.M.S. *Challenger* expedition, 1872—1876
(Tizard et al., 1885, p. 191). The largest quantities of nodules were recovered at stations
281 (22°21'S 150°17'W, 4,360 m) and 285 (32°36'S 137°43'W, 4,345 m) in the South
Pacific. In each case, the associated sediment was dark chocolate-coloured clay containing
abundant sharks' teeth.

For a modern scientist, one of the features of the *Challenger* report is the
excellence of the descriptive accounts of nodules and of the drawings of
nodules. The quality of curation has been such that even today the
Challenger collection of the nodules at the British Museum remains a unique
and valuable reference collection for scientists in this field. *Challenger*
nodules were recently discovered during renovations of the Redpath
Museum, McGill University (Stevenson and Stevenson, 1970). Apart from
extensive dehydration due to storage in hot, dry conditions, the nodules

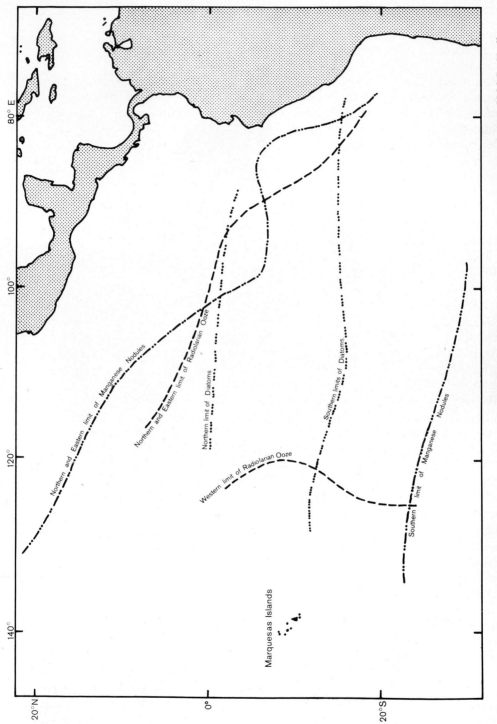

Fig. 1-4. Schematic diagram showing the distribution of manganese nodules in the Eastern Pacific as revealed by H.M.S. *Challenger*, 1872—1876, and U.S. Fish Commission Steamer *Albatross*, 1904—1905 (Agassiz, 1906, Plate 3).

remained in excellent condition. Interestingly, oxidation at the surface of the nodules has led to the formation of a thin superficial coating of birnessite (cf. Chapter 7).

Following the *Challenger* expedition, work on nodules was sporadic. More extensive collections of nodules from the Pacific were made during the *Albatross* expeditions of 1899—1900 and 1904—1905 and the limits of nodule distribution in the equatorial Pacific were mapped (Agassiz, 1902, 1906; Murray and Lee, 1909). In particular, the east—west trending zone of high nodule concentration lying off the west coast of the United States between the latitudes 6°30′N and 20°N was recognized (Fig. 1-4) (see also Dietz, 1955; Horn et al., 1972a, c). Manganese nodules were also recovered during the first *Valdivia* expedition of 1898 (Chun, 1908). Following this, few studies of deep-sea nodules were made until the cessation of the second world war, although nodules were collected during the *Carnegie* expedition of 1928—1929 (Revelle, 1944) and the *John Murray* expedition of 1933—1934 (Wiseman, 1937) and the manganese contents of sediments collected during the *Meteor* expedition of 1925—1927 were determined (Correns, 1937). The radium content of nodules was reported by Joly (1908), Piggot (1933, 1944) and Pettersson (1943) and a review of the mechanisms of nodule formation presented by Pettersson (1945).

As previously mentioned, shallow water marine concretions were first discovered in 1878 in Loch Fyne, Scotland, by Buchanan and were subsequently collected in Lochs Goil and Striven and on the Skelmorlie Bank in the Clyde Estuary by Murray and Irvine. They were discovered in the Black Sea during the Russian expeditions of 1890 and 1891 and were known in the northern Russian seas before the turn of the century. In the case of the Loch Fyne nodules, the factors controlling their morphology and the relationship between manganese oxides and carbonates were established at this time. Buchanan (1891) also established that littoral nodules from Loch Fyne contain higher contents of manganese and lower contents of copper, nickel and cobalt than oceanic nodules and that manganese is less highly oxidized in the littoral nodules than in their deep-sea counterparts. One interesting suggestion was that manganese in the Loch Fyne nodules was derived from industrial effluent; one firm alone having discharged 56,000 tons of manganese chloride into the river Clyde during the years 1818—1846 (Murray and Renard, 1891). This hypothesis has not been upheld by subsequent findings. The occurrence of lake concretions in the Northern Hemisphere (Sweden, Finland, the Soviet Union, North America) has also been known for a considerable time, in many cases since the end of the last century or before, although nodules from the English lake district were not discovered until the 1950's (Gorham and Swaine, 1965). A fuller account of early discoveries of the shallow marine and lacustrine nodules is given in Chapter 3. Fossil nodules were recognized in the Eastern Alps of Europe, Barbados, Timor and the Soviet Union late in the last century or early in this

(Chapter 4). Although lake and shallow marine nodules continued to receive some attention during the inter-war years, particularly in the Soviet Union, Scandinavia and North America, interest in deep-sea nodules seemed to wane during the early part of this century and has only been revived in the post-war years.

Following World War II, an extensive collection of deep-sea sediment cores was acquired as a result of the 1947—1948 Swedish Deep-Sea Expedition. Geochemical investigations of 15 of these cores showed a marked inter-relationship between Mn, Ni and Co which was attributed to the scavenging of Ni and Co by the manganese oxides (Landergren, 1964). Goldberg (1954) also put forward a colloidal scavenging hypothesis of nodule formation and showed that the incorporation of trace elements into manganese nodules could be explained in terms of the scavenging of the elements from sea water by manganese and iron oxides. The importance of ferric oxide in catalysing the oxidation of divalent manganous ions in sea-water to the tetravalent state was also emphasized by Goldberg and Arrhenius (1958). Pettersson and Rotschi (1952) also suggested on the basis of chemical data from the Swedish Deep-Sea Expedition that the nickel content of pelagic sediments is derived in part from a cosmic source. This conclusion was, however, subsequently disputed by Smales and Wiseman (1955) on the grounds of lack of similarity in the nickel/cobalt/copper ratios of deep-sea sediments and meteorites (see, however, Öpik, 1956).

It was not until 1965, however, that coherent hypotheses of nodule formation began to appear. In that year, Mero for the first time collated data on the regional variation of nodule composition throughout the Pacific and Manheim postulated the influence of diagenetic processes on the formation of nodules from shallow water, continental margin environments. Since that time, there has been a considerable expansion in the literature on manganese nodules and they have been the subject of such esoteric studies as the search for transuranic elements (Otgonsuren et al., 1969), electric monopoles (Fleischer et al., 1968) and cosmic spherules (Finkelman, 1970, 1972; Jewab 1970, 1971) and have even appeared in the fictional literature (Innes, 1965). Current ideas accept that any hypothesis of nodule formation must be multifaceted, i.e., a number of possible mechanisms may contribute to manganese deposition and trace element uptake any one of which may be dominant in a given situation.

The realization that manganese nodules may be a potential ore resource led the International Decade of Ocean Exploration (I.D.O.E.) to sponsor a conference "Ferromanganese Deposits on the Ocean Floor", at the Lamont-Doherty Geological Observatory in January 1972. This served to co-ordinate all existing data on nodules (Horn, 1972) and as a considerable stimulus to the further study of the geological, economic, legal, environmental and technological problems associated with the development of an incipient nodule industry. At the present time, extensive studies of the deep-sea floor are being carried out by a number of nations, particularly in the equatorial North Pacific but also in the South Pacific and Indian oceans, in order to determine optimum areas for nodules mining. So far, the main countries involved are the United States, Japan, West Germany and France. In November 1972, the *Hughes Glomar Explorer* was launched as the what was universally taken to be the first nodule mining ship (Anonymous, 1974). However, subsequent investigations revealed that this vessel was in fact more concerned with the recovery of a sunken Russian submarine (Anonymous, 1975). In spite of this, there is the prospect of development of an extensive deep-sea mining industry and Deepsea Ventures have already requested the U.S. Secretary of State for diplomatic protection for the development of a nodules mining site in the equatorial North Pacific (Deepsea Ventures, Inc., 1974). Because any nodules mining venture is likely to take place in international waters, the legal regime under which it takes place is clearly of major interest to many nations and has been the subject of extensive debate during the 1974 Law of the Sea Conference in Caracas and the 1975 Conference in Geneva. The next decade is therefore likely to see a continuing interest in marine manganese nodules as an economic resource.

CHAPTER 2

DEEP-SEA NODULES: DISTRIBUTION AND GEOCHEMISTRY

D. S. CRONAN

INTRODUCTION

Deep-sea ferromanganese oxide deposits have received increasing attention in recent years. From being considered an interesting curiosity two or three decades ago, they have grown in scientific importance and some have also become a major potential source of several economically valuable elements such as manganese, nickel, copper, cobalt, and lead. Until recently, however, many investigations on these deposits were empirical in nature. Only in the last few years have efforts been made to relate their composition to the nature of their environment of deposition, and thus to attempt to predict what type of nodules might occur in any given area. Regional variations in nodule composition are related in large part to differences between the environments in which the nodules form. It is evident therefore that an ability to characterize the composition of nodules in terms of these environments will greatly aid in the location of new deposits of economic importance. In this chapter, the distribution and geochemistry of nodules is discussed on this basis.

DISTRIBUTION

Data on the distribution of nodules both at the surface and within the sediment column have accrued from a number of sources. Until about 25 years ago, almost all of such information was obtained from dredging. Menard (1964), for example, concluded that there were high concentrations of nodules over much of the southeastern Pacific, because they had been reported in most of the dredgings in the area. With the advent of deep-sea coring, another tool became available for the estimation of nodule distribution, particularly that beneath the sediment surface. Mero (1965a) has discussed techniques whereby the number and size of nodules recovered in sediment cores taken in given areas can be used to estimate nodule concentration in those areas. In addition, Skornyakova and Zenkevitch (1961) have reported the estimation of surface nodule concentrations using a deep-sea grab.

Underwater photography is the tool most commonly used at present in estimating the distribution of nodules on the ocean floor. Modern stereo cameras can provide accurate information on the concentration of nodules in any given area, but samples are also necessary if accurate tonnage estimates are to be made. In recent years the use of underwater television has become increasingly important, particularly in exploration for economic deposits of nodules (Kaufman and Siapno, 1972). However, most of the information currently available on the worldwide distribution of nodules is based on bottom photographs, cores, and grabs (Skornyakova and Andrushchenko, 1970; Ewing et al., 1971), and these data provide the basis for the following discussion.

Before describing the worldwide distribution of nodules, it is of importance to consider some of the factors determining this distribution. One of the more important of these is the rate of accumulation of the sediments associated with the nodules. In general, low sedimentation rates lead to high concentrations of nodules at the sediment surface (Ewing et al., 1971; Horn et al., 1972a). Thus the highest concentrations of nodules usually occur in red clay or siliceous ooze areas where sedimentation rates are low, often in the order of $1-3$ mm/10^3 years or less (Hays et al., 1969; Opdyke and Foster, 1970). Abundant nodules are not, however, restricted to such areas. High concentrations can also occur where sedimentation is inhibited as a result of strong current scour, as for example under the Antarctic Circum-Polar Current between Antarctica and Australia (Watkins and Kennett, 1971, 1972: Payne and Conolly, 1972). Furthermore, high concentrations of nodules or thick encrustations can also occur on seamounts where sedimentation is limited as a result of current action.

A second important factor determining the abundance and distribution of nodules and encrustations in the oceans is the length of time that has been available for their accumulation. This is likely to be largely related to sea-floor spreading rates. For example, young rocks in the Median Valley of the Mid-Atlantic Ridge have negligible or thin coatings of ferromanganese oxides, whereas the older rocks on either side of the valley commonly have moderate to thick coatings (Aumento, 1969). Furthermore, lateral traverses across both the Mid-Atlantic and Carlsberg ridges have shown that the ferromanganese oxide encrustations increase in thickness as the sea floor gets older away from the ridge crest (Aumento, 1969; Glasby, 1970), indicating that the volume of ferromanganese oxides deposited is time-dependent.

A third important factor determining nodule distribution is the availability of suitable nuclei around which the ferromanganese oxide phases can accrete. Almost all nodules nucleate around a foreign object, usually a fragment of altered volcanic rock such as palagonite or pumice, although organic remains like sharks' teeth and whales' earbones are also common. The distribution of volcanic nuclei in the oceans is obviously related to variations in the incidence of submarine volcanism, and might explain the commonly recorded relatively high concentrations of nodules in volcanic areas (Murray and Renard, 1891; Bonatti and Nayudu, 1965; and others). In addition, the nature of the volcanic nuclei themselves might also influence the rate at which the ferromanganese oxides accrete. Those nuclei which are easily weathered may better catalyze the oxidation of manganese than do those which alter only slowly.

Also before considering the worldwide distribution of nodules, it is important to note that they can be highly variable in concentration over quite small areas of the sea floor, and thus ocean-wide nodule distribution charts show only average concentrations. For example, nodules can vary from covering 75% or more of the sea floor to almost zero coverage within a few hundreds of metres (Kaufman and Siapno, 1972) and their size and shape can also vary over the same distance. The patchyness of nodule distribution on a small scale must be related to local environmental factors at the site of deposition. For example, scour caused by bottom currents possibly channelled by small-scale topographic variations might locally prevent the accumulation of sediments, thus leading to localized high nodule concentrations at the sediment surface. Local topographic controls such as were described by Moore and Heath (1966) where nodules were preferentially concentrated on the slopes of abyssal hills, might also be of importance in determining small-scale variations in nodule distribution. Other factors causing such local variations could include the distribution of suitable nuclei, as appears to be the case on parts of the Carlsberg Ridge (Glasby, 1973c) and local variations in the pattern of sedimentation. Bottom photographs illustrating sediment erosion due to current action, for example, have been published by many authors (Mero, 1965a; Hawkins, 1969; Heezen and Hollister, 1971; and others). Such erosion might cause previously buried nodules to be exposed at the surface with deposition of the reworked sediment in adjacent areas. In addition, the alternate burial and exposure of concretions by migrating waves of sediment could also lead to irregularities in their distribution. During submersible operations over manganese deposits on the Blake Plateau, Hawkins (1969) observed pavement-like structures in the troughs of sand waves, but manganese slabs near the wave crests. He also observed changes in the nature of the ferromanganese oxide deposits with changes in slope, nodules grading into large blocks with a slope increase from 10 to 20 degrees. The latter may have represented the remnants of a previously continuous pavement, possibly fractured by scouring and undercutting. Other variations in the form of the deposits were also noted.

North Pacific Ocean

Nodule distribution in the Pacific is better known than in the other oceans. High concentrations occur in an east—west band in the North Pacific between about 6°30′N and 20°N, which stretches from near Central America to the Mariana Trench and varies in depth from about 3,200 to 5,900 m (Ewing et al., 1971; Horn et al., 1972a; Fig. 2-1). The limits of the area seem to be determined by increases in the rate of sedimentation towards the margin of the Pacific and toward the Equator. To the north nodule growth is inhibited by increases in terrigenous and biogenic sedimentation, to the east by turbidite and hemipelagic deposition, and to the south by the equatorial zone of rapid carbonate accumulation. In the northern portion of the area of high nodule concentration, the sediments are predominantly red clays with sedimentation rates of around 1 mm/1,000 years (Skornyakova and Andrushchenko, 1970; Opdyke and Foster, 1970). By contrast, in the south, immediately to the north of the carbonate zone, they consist largely of siliceous oozes where sedimentation rates are around 3 mm/1,000 years (Hays et al., 1969).

The distribution of nodules within this zone of high concentration is by no means uniform. Particularly in the west, the increasing frequency of seamounts and other topographic features which can disturb the general pattern of sedimentation restrict high nodule concentrations to deep intermontane basins (Ewing et al., 1971). The seamounts themselves though often have thick manganese encrustations or irregular shaped nodules (Hamilton, 1956; Londsdale et al., 1972). However, Scripps Institution of Oceanography cores from the western Pacific near 170°W show there to be a large area bounded by the Mid-Pacific Mountains, the Line Islands, the Marshall Islands, and the equatorial zone of high productivity, where nodules are particularly abundant (Cronan, 1967). Throughout most of this area, depths are in excess of 4,500 m.

Although nodules are much less common in the marginal areas of the North Pacific than in the central basin, together with encrustations they are locally abundant in some continental borderland areas. Thick crusts have been dredged from many localities in the Southern Borderland Seamount Province (Krause, 1961) and in the vicinity of the Mendocino Escarpment, and nodules have been found to be locally numerous in some linear basins off Baja California. The latter are associated with rapidly accumulating sediments and are thought to form as a result of the diagenetic remobilisation of buried manganese.

South Pacific Ocean

Nodules are more irregularly distributed in the South Pacific than in the North, possibly as a result of its greater topographic and sedimentological diversity. According to Menard (1964), Skornyakova and Andrushchenko (1970), Goodell et al. (1971), Glasby et al. (1974), and Glasby (1976), some of the highest concentrations of nodules in the South Pacific occur in the triangular Southwestern Pacific Basin bounded by the Austral Islands, in Tonga—Kermadec Trench, and the Pacific Antarctic Ridge (Fig. 2-2).

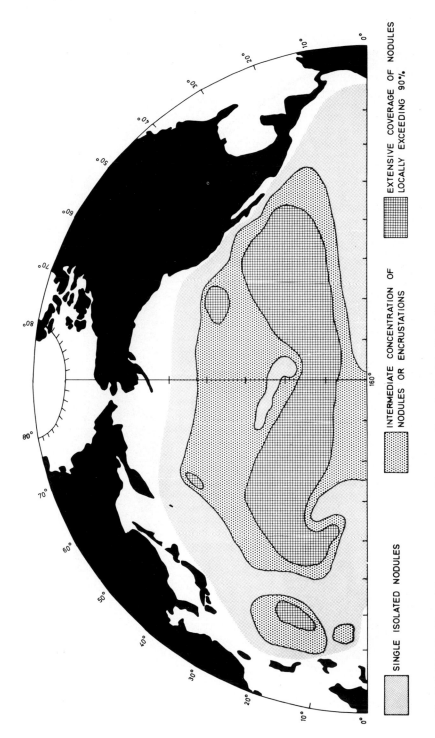

Fig. 2-1. Distribution of ferromanganese oxide concretions in the North Pacific.

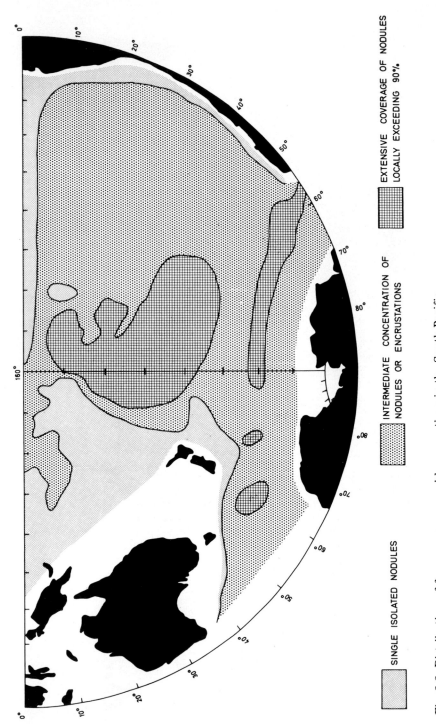

Fig. 2-2. Distribution of ferromanganese oxide concretions in the South Pacific.

Goodell et al. (1971) have reported one nodule deposit more than 350,000 km^2 in area on the eastern margin of this basin. The Southwestern Pacific Basin is an area of slow (0.3—0.5 mm/10^3 years, Skornyakova and Andrushchenko, 1970) deep-water sedimentation, where the deposits consist principally of very fine red to dark chocolate brown clays containing high concentrations of authigenic minerals (Arrhenius, 1963). As in the North Pacific, the slow rate of sedimentation favours nodule growth.

Another area in the South Pacific where nodule concentrations are high lies between 10° and 19°S and extends from about 132° to 162°W (Horn et al., 1972a; Fig. 2-2). It includes the Manihiki Plateau, the Society Islands, Tahiti and the Tuamotu Archipelago. Nodules occur on the flanks of topographic elevations and in intermontane basins throughout the area. Deep-water nodules are here associated with red clays, as they are elsewhere in the Pacific, whereas those from shallower depths such as the east flank of the Manihiki Plateau are associated with carbonates (Horn et al., 1972a).

The distribution of nodules over the East Pacific Rise and Nazca Plate is imperfectly known. Menard (1964) points out that abundant nodules have been recovered in dredge hauls, but bottom photographs have shown a sparse nodule distribution (see also Glasby, 1976). Menard suggested that the nodules may be buried too deep to be recorded in photographs but shallow enough to be recovered in dredges. Nodules have been recovered from the tops of cores and in dredges taken near South America between the Carnegie and Nazca Ridges, and have been found to be abundant in cores from the southern portion of the East Pacific Rise (Cronan, 1967). However, detailed information on the distribution of nodules in these areas is not available.

By contrast, Goodell et al., (1971) have reported an almost continuous belt of nodules and encrustations up to 550 km wide lying beneath the Antarctic Convergence near approximately 60°S and extending from 50° to 170°W (Fig. 2-2). This area is one of high velocity bottom currents, and the concretions consist of pavements, crusts, nodules and disseminated manganese oxides. Finally, extensive deposits of nodules have been reported to the south of Australia in the vicinity of Tasmania and from Campbell Plateau and Macquarie Ridge (Summerhayes, 1967) near New Zealand.

North Atlantic Ocean

Nodule distribution in the Atlantic (Figs. 2-3, 2-4) is more limited than in the Pacific, most likely as a result of the differing patterns of sedimentation in the two oceans. The Atlantic receives a much greater amount of terrigenous detritus relative to its size than does the Pacific, and sedimentation rates are often too high to permit the development of extensive nodule deposits. Much of the marginal North Atlantic is floored by turbidites and contains few ferromanganese oxide deposits. Isolated samples are not uncommon, but there is a general lack of the large nodule fields

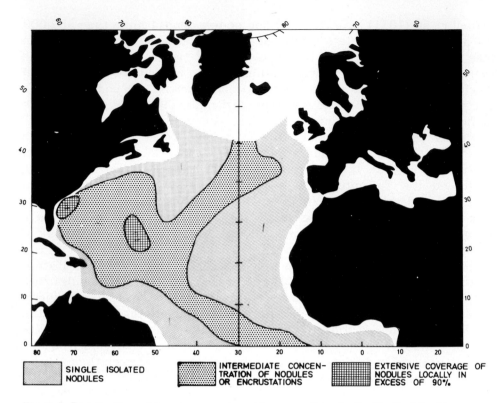

SINGLE ISOLATED NODULES

INTERMEDIATE CONCEN-
TRATION OF NODULES
OR ENCRUSTATIONS

EXTENSIVE COVERAGE OF
NODULES LOCALLY IN
EXCESS OF 90%

Fig. 2-3. Distribution of ferromanganese oxide concretions in the North Atlantic.

which characterize the North Pacific, particularly in the eastern part of the basin where some of the turbidite-floored abyssal plains extend up to the flanks of the Mid-Atlantic Ridge.

Another feature inhibiting nodule growth in the North Atlantic is its general shallowness compared with the Pacific. Much of the sea floor lies above the carbonate compensation depth, particularly over the Mid-Atlantic Ridge which occupies a large portion of the ocean (Biscay et al., 1976). Rapid biogenic sedimentation probably inhibits nodule growth in such areas, just as does terrigenous sedimentation in the marginal areas of the ocean.

The principal areas where ferromanganese oxides do occur in the North Atlantic, are those where sedimentation is inhibited for some reason. On the Blake Plateau at approximately 30°N 78°W, for example, the strong bottom currents associated with the Gulf Stream prevent sediment accumulation and enable nodules and encrustation to grow to a large size (Hawkins, 1969; Manheim, 1972). Similarly, current action over other topographic elevations such as the Kelvin Seamounts promotes nodule growth (Horn et al., 1972a). Nodules are rare on the Mid-Atlantic Ridge, but encrustations are common in exposed areas where carbonate accumulation is prevented either by

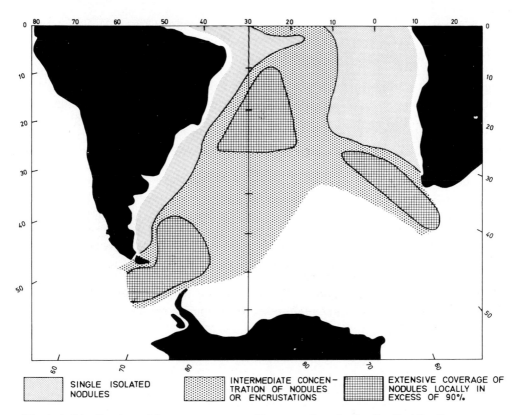

Fig. 2-4. Distribution of ferromanganese oxide concretions in the South Atlantic.

current action or rugged topography. Aumento (1969) has described extensive deposits of encrustations from the Mid-Atlantic Ridge near 45°N, and similar deposits have been obtained from the Kings Trough area at about 43°N 20°W, and further south. Almost the only samples of ferromanganese oxides that have been recovered in the eastern North Atlantic are encrustations from exposed rocks on seamounts and other topographic elevations (Matthews, 1962; Cronan, 1972a; Scott et al., 1972a). However, a deep-water area has been reported in the western North Atlantic at about 20°N 60°W which is below the compensation depth, beyond the reach of significant amounts of continental detritus, and where nodules are fairly abundant (Horn et al., 1972a).

South Atlantic Ocean

As in the North Atlantic, nodule distribution in the South Atlantic is largely dependent on the overall pattern of sedimentation. Nodules occur in greatest abundance within broad zones between the Mid-Atlantic Ridge and the abyssal plains adjacent to the continental margins (Fig. 2-4). Sedimentation rates are low in these areas relative to elsewhere in the basin, and it is largely this factor which promotes nodule growth. However, the data of Ewing et al. (1971), Horn et al. (1972a) and Cronan (1972a) show a greater frequency of nodule occurrences in the western South Atlantic than in the eastern portion of the basin, possibly indicating sedimentation differences between the two. Horn et al. (1972a) have suggested that these might result from a freer flow of bottom water in the western South Atlantic than in the east, largely because the Walvis Ridge acts as a barrier to northwards flowing bottom water in the latter. These authors also consider that the relatively high nodule concentrations to the south of the Walvis Ridge could be related to this bottom water flow.

Relatively few nodules and encrustations have been obtained from the Mid-Atlantic Ridge in the South Atlantic, but this may be due to the infrequency of dredging expeditions in the area rather than to their limited occurrence. Large concentrations of nodules and encrustations have been reported by Goodell et al. (1971) in the Drake Passage between South America and Antarctica, and in the Scotia Sea. Like nodules from similar latitudes in the Pacific, their abundance is probably related to high bottom current velocities.

Indian Ocean

Data on the distribution of nodules in the Indian Ocean have been summarized by Bezrukov and Andrushchenko (1972). Using previously published data (Bezrukov, 1963; Laughton, 1967; Cronan and Tooms, 1969; McKelvey and Wang, 1969), together with new information, these authors published a map showing the distribution of nodules throughout the ocean as a whole (Fig. 2-5). They found them to be more common in the basins than in the elevated areas, with highest concentrations occurring in the depressions far removed from land on either side of the 90° East Ridge. Inexplicably high concentrations were also found at the foot of some submarine ridges, occasionally in the vicinity of major fault zones. Nodules were also reported in the southern part of the ocean, principally from depressions, and encrustations were found to be common on the Carlsberg and Mid-Indian Ocean ridges. No nodules were recorded on continental slopes, or on abyssal plains adjacent to the continents such as are common in the Arabian Sea and Bay of Bengal. It is evident therefore that in their general distribution, Indian Ocean nodules show similarities to those in the Pacific and Atlantic oceans. They are most abundant in well-oxidized areas of slow sedimentation far removed from land, and are either scarce or absent in near continental areas where terrigenous sedimentation rates are high. Exposed rock outcrops are covered with encrustations rather than with nodules, and each can vary considerably in abundance over short distances.

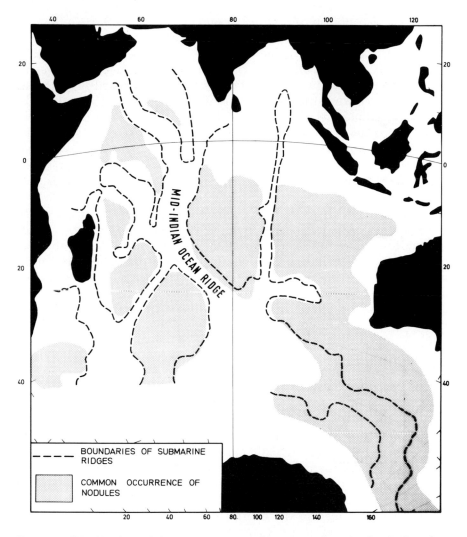

Fig. 2-5. Distribution of ferromanganese oxide concretions in the Indian Ocean (after Bezrukov and Andrushchenko, 1972).

Sub-surface nodules

In addition to their widespread distribution at the sediment surface, manganese nodules are also abundant within the sediments. Buried nodules have been found in all oceans (Menard, 1964; Mero, 1965; Cronan, 1967; Smith et al., 1968; Skornyakova and Andrushchenko, 1970; Goodell et al., 1971), but the bulk of the available information on their distribution is from the Pacific.

Examining the distribution of nodules in one hundred and thirteen Pacific sediment cores collected by the Scripps Institution of Oceanography, Cronan (1967) concluded that nodules in the upper 2—3 m of the sediment column were approximately equal in number to those at the surface. This was in general agreement with the work of previous authors. Menard (1964) had found in gravity cores from the Pacific that the concentration of nodules in the first metre of sediment was approximately half that found at the surface, and Bender et al. (1966) had found the ratio of nodules at the surface to those in the first metre of 48 gravity cores to be 1.7, slightly lower than Menard's figure. The surface to depth ratio of nodules in the upper metre of cores examined by Cronan (1967) was 1.8. Beneath the surface layer, the nodules were fairly evenly distributed throughout the cores.

The distribution of nodules in approximately 50 Lamont-Doherty cores from the Pacific and Atlantic oceans has been investigated by Horn et al. (1972b). They found the top 3 m of sediment to contain about one-third as many nodules as at the surface, a considerably lower proportion than found by Cronan (1967) in Pacific cores alone. This would suggest that the relative proportion of buried to surface nodules in the top few metres of Atlantic sediments is considerably lower than that in the Pacific, which would agree with the higher sedimentation rates occurring in the Atlantic mentioned previously.

Data on the distribution of buried nodules have also been presented by several other workers. Goodell et al. (1971) reported nodules 17 m below the surface in sediment 3.4 million years old in the South Pacific, but found most buried nodules to be concentrated in the first metre of sediments. The bulk of those obtained were in sediments younger than the Brunhes—Mahyuama boundary, suggesting a possible higher rate of ferromanganese oxide deposition or lower rate of sedimentation in this area in the past few hundred thousand years than in previous epochs. That the distribution of buried nodules in cores may not be entirely random has also been reported by Skornyakova and Andrushchenko (1970). Buried concretions recovered in large-diameter cores taken in the central Pacific during the 34th cruise of the *Vityaz* were found in some cases to be concentrated into distinct horizons. These may represent ancient erosion surfaces; or they possibly suggest that there have been epochs of extensive ferromanganese oxide deposition in the past in this area, separated by periods of little or no deposition.

Until recently, buried nodules were only known in the top few metres of sediments. However, as a result of the deep coring carried out since 1968 by the Deep-Sea Drilling Project, they have been found at considerably greater depths also. Many of the buried nodules in DSDP cores have probably fallen from the sediment surface as a result of the drilling (McManus et al., 1970). However, at least one nodule recovered at a depth of 131 m in DSDP 162 during Leg 16 of the project in the eastern equatorial Pacific has probably formed in-situ. It was enclosed in an undisturbed mottle of bleached sediment which contained the same microfauna as the surrounding darker sediment, and is unlikely therefore to have fallen from the surface.

With few exceptions, buried nodules described to date in the upper few metres of sediments are very similar in nature to their surface counterparts. Cronan and Tooms (1969) found few chemical differences between surface and buried nodules in Pacific sediments and concluded that compositional variations between nodules in single cores were never so large that the composition of the buried nodules was atypical of the average composition of surface nodules from the region in which they were found. Similarly, Goodell et al. (1971) found that buried nodules in Southern Ocean sediments were similar in most respects to their surface counterparts. However, the deeply buried nodule in DSDP 162 mentioned previously was compositionally distinct from nodules higher in the core suggesting that significant variations in nodule composition can occur between nodules formed at the surface and deep within the sediment.

ENVIRONMENTS OF FERROMANGANESE OXIDE DEPOSITION

Although manganese nodules are abundant throughout the oceans, it is evident from the previous section that they are more common in some environments than in others. Attempts can be made to classify them on the basis of these environments. For example, Price and Calvert (1970) divided Pacific nodules into abyssal, seamount, marginal, and elevated marginal varieties, illustrating important chemical and mineralogical differences between deposits forming under different conditions. In this section, an attempt is made to classify nodules from all three major oceans according to their depositional environment. Seven environments have been differentiated. These are, oceanic seamounts, plateaux, active mid-ocean ridges, inactive ridges, continental borderlands, marginal topographic elevations, and the deep-ocean floor between about 2,000 and 6,000 m. These contain nodules formed by all the known processes of nodule formation, namely simple precipitation from seawater, from hydrothermal solutions, weathering of volcanic rocks and diagenetic processes (Goldberg and Arrhenius, 1958; Zelenov, 1964; Cronan, 1967; Cronan and Tooms, 1969; Glasby, 1970). In all, a total of over 280 samples have been investigated (Fig. 2-6). Chemical and mineralogical data have been obtained from Mero (1965a), Manheim (1965), Barnes (1967b), Cronan (1967, 1972a, and unpublished data), Cronan and Tooms (1967a, 1969), Willis (1970), and Horn et al. (1972c). Average abundances of elements in nodules from all oceans are listed in Table 2-I, and average concentrations of Mn, Fe, Ni, Co and Cu in each of the seven classes are shown in Table 2-II.*

---

*See Note added in proof: p. 44.

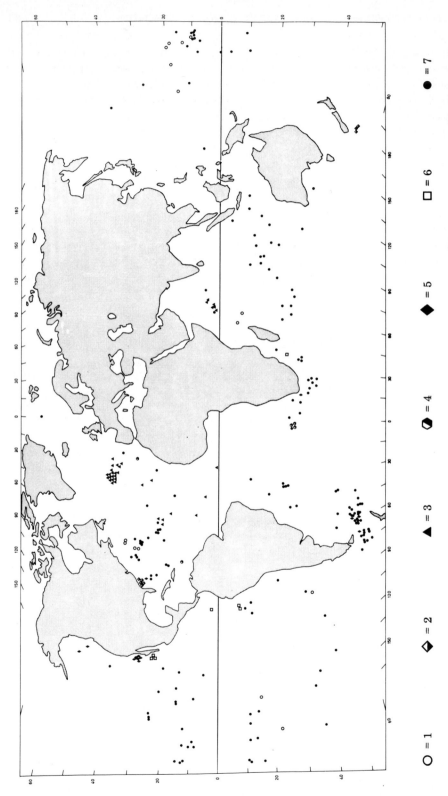

Fig. 2-6. Locations of ferromanganese oxide concretions classified according to environment of deposition: *1* = oceanic seamount ridge concretions; *2* = plateau concretions; *3* = active mid-ocean concretions; *4* = inactive ridge concretions; *5* = marginal seamount and bank concretions; *6* = continental borderland concretions; *7* = abyssal concretions.

○ = 1   ◇ = 2   ▲ = 3   ⬡ = 4   ◆ = 5   □ = 6   ● = 7

TABLE 2-I

Average abundances of elements in manganese nodules and other ferromanganese oxide deposits from the World Ocean (wt. %) (From Cronan, 1976)

Elements	World average	Elements	World average
B	0.0277	Sr	0.0825
Na	1.9409	Y	0.031
Mg	1.8234	Zr	0.0648
Al	3.0981	Mo	0.0412
Si	8.624	Pd	$0.553 \cdot 10^{-6}$
P	0.2244	Ag	0.0006
K	0.6427	Cd	0.00079
Ca	2.5348	Sn	0.00027
Sc	0.00097	Te	0.0050
Ti	0.6424	Ba	0.2012
V	0.0558	La	0.016
Cr	0.0014	Yb	0.0031
Mn	16.174	W	0.006
Fe	15.608	Ir	$0.935 \cdot 10^{-6}$
Co	0.2987	Au	$0.248 \cdot 10^{-6}$
Ni	0.4888	Hg	$0.50 \cdot 10^{-4}$
Cu	0.2561	Tl	0.0129
Zn	0.0710	Pb	0.0867
Ga	0.001	Bi	0.0008

TABLE 2-II

Average abundances of Mn, Fe, Ni, Co and Cu in manganese nodules and encrustations from different environments (wt. %)

	Sea-mounts	Plateaux	Active ridges	Other ridges	Continental borderlands	Marginal seamounts and banks	Abyssal nodules
Mn	14.62	17.17	15.51	19.74	38.69	15.65	16.78
Fe	15.81	11.81	19.15	20.08	1.34	19.32	17.27
Ni	0.351	0.641	0.306	0.336	0.121	0.296	0.540
Co	1.15	0.347	0.400	0.570	0.011	0.419	0.256
Cu	0.058	0.087	0.081	0.052	0.082	0.078	0.370
Mn/Fe	0.92	1.53	0.80	0.98	28.8	0.81	0.97
Depth (m)	1,872	945	2,870	1,678	3,547	1,694	4,460

Table 2-II shows that nodules from some environments are compositionally more distinctive than are those from others. For example, seamount nodules are characterized by a high cobalt content, continental borderland varieties by a high Mn/Fe ratio, and abyssal nodules by a high Cu content. By contrast there is little in Table 2-II to distinguish active ridge nodules from inactive ridge varieties other than the Mn/Fe ratio, nor is there much to distinguish marginal seamount and bank nodules from either. For convenience, each group of nodules will be considered separately.

Oceanic seamounts

Oceanic seamount nodules (Fig. 2-6) are distinguished by much higher than average cobalt concentrations, lower than average Ni and Cu, and close to average values of Mn and Fe. Mineralogically, they are characterized by an abundance of δMnO_2, the more highly oxidized of the two most common minerals in nodules (Barnes, 1967a; Cronan and Tooms, 1969).

The cause of cobalt enrichment in seamount nodules has been the subject of considerable discussion. It has been ascribed both to the oxidation of Co^{2+} to insoluble Co^{3+} under the highly oxidizing conditions associated with oceanic seamounts (Goldberg, 1963a; Barnes, 1967b; Cronan, 1967; Cronan and Tooms, 1969), and to local supplies of Co from the alteration of the seamount basalts (Burns, 1965). However, the latter alternative cannot fully explain the enrichment of Co in oceanic seamount nodules but not in those from other basaltic topographic elevations such as mid-ocean ridges and marginal seamounts.

One of the principal factors determining nodule composition is the degree of oxidation (redox potential) at the sediment—water interface (Cronan and Tooms, 1969; Glasby, 1972a). Redox potentials in marine sediments are dependent on a number of factors, one of which is the rate of sediment accumulation (Baas Becking et al., 1960; Price and Calvert, 1970), which partly determines the amount of organic material present. Slow accumulation rates tend to lead to high redox potentials, partly due to the oxidation of the organic matter before burial. The high current velocities which can occur over oceanic seamounts may inhibit the accumulation of sediment, and together with rapid rates of oxygen advection, may lead to an Eh sufficiently high for Co oxidation and precipitation to take place.

The lack of Co enrichment in nodules from other topographic elevations may result from higher sedimentation rates leading to a lower Eh in these environments. For example, marginal seamounts, because of their location near to land, probably receive greater amounts of sediment than do oceanic seamounts (Price and Calvert, 1970). Similarly, active oceanic ridges which are topographically more complex than seamounts might also receive greater amounts of sediment than the latter, partly by entrapment, and some of it possibly of local derivation.

A second factor of possible importance in the enrichment of Co in oceanic seamount nodules is their rate of growth, and thus the length of time available for the removal of Co from seawater. Marginal seamount and active mid-ocean ridge nodules may receive local supplies of elements from the continents and submarine volcanic sources respectively, leading to relatively high growth rates. By contrast, because of their isolation from both continental and volcanic sources of elements, nodules on inactive oceanic seamounts may accumulate more slowly, and thus be able to scavenge Co from seawater over a longer period.

In summary therefore, the high Co content of oceanic seamount nodules may be a function of several factors. These include a high degree of oxygenation of bottom waters, low rates of sediment accumulation leading to high redox potentials and slow nodule growth rates.

Comparing the composition of seamount nodules from each of the three major oceans (Table 2-III), it is evident that the high overall Co content largely reflects that in Pacific seamount nodules. The latter are higher in Co than are those from either the Atlantic or Indian oceans. If sedimentation rates do influence the abundance of Co in nodules, this would be in accord with the lower rates of sedimentation in the pelagic Pacific than elsewhere. The differences in the Mn/Fe ratio shown in Table 2-III might also reflect the influence of sedimentation rates. Other than where diagenetic remobilisation of manganese occurs as in continental borderland areas, increasing terrigenous sedimentation tends to decrease the Mn/Fe ratio in nodules (Manheim, 1965).

TABLE 2-III

Composition of seamount nodules from different oceans (wt. %)

Elements	Pacific	Indian	Atlantic
Mn	15.34	14.37	13.06
Fe	12.84	14.59	23.12
Ni	0.41	0.37	0.20
Co	1.50	0.91	0.46
Cu	0.062	0.060	0.047
Mn/Fe	1.20	0.98	0.56
Depth (m)	1,584	2,850	2,030

Nickel and copper are more evenly distributed among seamount nodules from the different oceans. Copper is uniformly low in all three oceans, and nickel is also below average (Table 2-II). These observations would suggest that the factors determining the concentrations of these elements in seamount nodules do not vary as greatly on a worldwide basis as do those determining Co and the Mn/Fe ratio.

Little can be said about variations in the composition of seamount nodules within each of the oceans as the sample distribution is so limited (Fig. 2-6). In the Pacific, about half the samples analyzed came from the Mid-Pacific Mountains, and these all contain Co values between 0.8 and 1.5%. However, values of up to 2.5% Co occur in some South Pacific seamount samples from the vicinity of the Tuamotu Archipelago and the Austral Islands (Cronan and Tooms, 1969). Associated with these high Co concentrations are often high values of Pb (Table 2-IV). Lead concentrations are considerably above average in Pacific seamount nodules, up to 0.5% in those containing the most Co. The reasons for this enrichment may be similar to those determining the enrichment of Co. Goldberg (1965) and Barnes (1967b) have suggested that under high redox potentials, Pb^{2+} may be oxidized to insoluble PbO_2, which may enter into solid solution with MnO_2. Barium and titanium are also often enriched in Pacific seamount nodules (Table 2-IV), but the causes are not entirely clear. Chromium is almost uniformly low in the deposits, except when they contain significant amounts of volcanic detritus. Mo and V are fairly similar to their Pacific average (Table 2-IV).

TABLE 2-IV

Pb, Ba, Mo, V, Cr and Ti in Pacific nodules (wt. %)

	Pb	Ba	Mo	V	Cr	Ti
Pacific seamounts*	0.261	0.503	0.046	0.060	0.001	1.06
Pacific average**	0.084	0.276	0.044	0.053	0.013	0.674

  *   Data from Cronan and Tooms (1969).
**   Data from Cronan (1976).

Only two seamount nodules from the Indian Ocean have been examined in this study, and obviously cannot be considered to be representative of such nodules in this ocean as a whole. However, their contents of Pb, Ba, Mo, V, Cr and Ti are similar to the average of these elements in Pacific seamount nodules (Cronan and Tooms, 1969), suggesting their incorporation under similar conditions. Four seamount samples were examined from the Atlantic Ocean, all from the northwestern part of the basin. In only one of the samples was Co above 0.8% and was as low as 0.066% in one other. The Mn/Fe ratio in the four samples varied from 0.34 to 0.9.

Inter-element relationships in seamount nodules (Table 2-V) differ considerably from such relationships in nodules as a whole (Cronan, 1969). The usual association between Mn and Ni is not significant in these samples, nor is there any correlation between Mn and Cu. Instead, there is a significant positive correlation between Mn and Co. Finally, the negative

TABLE 2-V

Inter-element relationships in oceanic seamount nodules and encrustations ($n = 15$)

	Mn	Fe	Ni	Co	Cu	Depth (m)
Mn	1.00					
Fe	−0.49	1.00				
Ni	0.43	−0.78	1.00			
Co	0.66	−0.73	0.64	1.00		
Cu	−0.12	−0.36	0.31	0.08	1.00	
Depth	−0.54	0.37	−0.31	−0.74	0.01	1.00

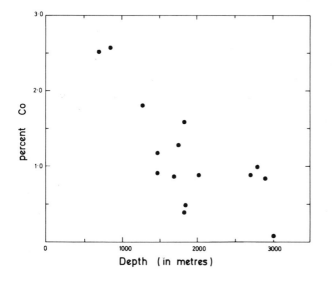

Fig. 2-7. Relationship between cobalt and depth in oceanic seamount concretions.

correlation found between Co and depth in Pacific and Indian Ocean nodules as a whole (Cronan and Tooms, 1969) is even more pronounced in seamount samples (Fig. 2-7), even though they were obtained over a much more limited depth range. This relationship would seem to confirm the common observation that, excluding continental margin areas, the highest cobalt values in nodules tend to occur on the most elavated seamounts.

Plateau nodules

The plateau nodules discussed in this chapter come from only two localities, the Blake Plateau in the northwestern Atlantic Ocean, and the Campbell Plateau in the southwestern Pacific. Taken together, nodules from these two areas do not differ markedly in Mn, Ni and Co from the world average, but are low in Cu and Fe (Table 2-II). Although they occur within the same depth range as oceanic seamount nodules, they are compositionally distinct from them.

The high Mn/Fe ratio in plateau nodules (Table 2-II) is of interest, especially as the bulk of the samples discussed here are from the Blake Plateau, and Atlantic nodules as a whole have an Mn/Fe ratio of less than unity (Cronan, 1972a). The reason for the high ratio may be related to the strong currents which sweep the Plateau and largely inhibit the accumulation of sediment. This would allow manganese to precipitate from seawater, relatively undiluted by Fe-bearing detritus. In this respect, Blake Plateau nodules may be similar to nodules from abyssal areas of minimal terrigenous sedimentation.

Comparison of the composition of Blake and Campbell Plateau nodules (Table 2-VI) reveals no major differences other than in their Mn and Fe contents. The low values of these elements in the Campbell Plateau nodules may result from the inclusion of phosphorite debris in the samples analyzed (Summerhayes, 1967). Summerhayes considered that volcanic influences were important in the Campbell Plateau deposits, but subsequent study has shown that such influences are probably of minor importance in view of their close similarity to Blake Plateau deposits where there is no volcanic activity (Glasby and Summerhayes, 1975). The high current velocities coupled with minimal terrigenous sedimentation in each area strongly suggest that the nodules grow by precipitation of Mn and associated elements from seawater and not as a result of submarine volcanic processes.

TABLE 2-VI

Composition of manganese nodules from the Blake and Campbell Plateaux (wt. %)

	Mn	Fe	Ni	Co	Cu	Mn/Fe
Blake Plateau	18.36	13.54	0.623	0.345	0.092	1.35
Campbell Plateau	14.00	7.18	0.689	0.351	0.076	1.94

The low Co contents of the plateau nodules relative to those from oceanic seamounts at similar depth, could perhaps be a result of their formation under lower redox conditions than occur in the seamount environment. The common occurrence of todorokite in Blake and Campbell Plateau nodules

(Manheim, 1965; Glasby, 1972a) compared with the more highly oxidized δMnO_2 in seamount nodules (Cronan and Tooms, 1969) would tend to support this view.

Both Cu and Co show positive correlations with Mn in plateau nodules (Table 2-VII) and there is a weaker correlation between Mn and Ni. None of the minor elements investigated show any correlations with iron.

TABLE 2-VII

Inter-element relationships in plateau nodules ($n = 11$)

	Mn	Fe	Ni	Co	Cu	Depth (m)
Mn	1.00					
Fe	0.08	1.00				
Ni	0.60	−0.22	1.00			
Co	0.69	−0.27	0.52	1.00		
Cu	0.68	0.14	0.82	0.56	1.00	
Depth	−0.36	−0.69	0.13	−0.19	−0.35	1.00

Active mid-ocean ridges

Active ridge nodules, or more commonly encrustations, are characterized by a low Mn/Fe ratio, low copper contents, slightly higher than average Co and slightly lower than average Ni (Table 2-II). The mineralogy of these deposits has not been established in any detail, mainly because the bulk of the samples analysed show only weak and diffuse powder patterns which are difficult to interpret. However, on the basis of the few samples which have been analysed successfully, concretions from active ridges would seem to contain δMnO_2 as their principal mineral, with subordinate amounts of todorokite (Cronan, 1967; Glasby, 1972a; Scott et al., 1972a).

The cause of the low Mn/Fe ratio in active ridge concretions could be related to a number of factors. Firstly, as many such deposits lie on outcrops of volcanic rock, they are not able to receive manganese by upward diffusion through interstitial waters as can nodules in sediment-covered areas. However, probably more important than this in causing Fe enrichment in these deposits are local volcanic sources of iron from the ridges themselves. On the basis of pH and Eh considerations (Krauskopf, 1957), the bulk of the iron entering the oceans from the continents should precipitate before the manganese, indicating a need for local oceanic sources of iron to account for its enrichment in mid-ocean ridge concretions. Such iron could be derived from at least two sources, weathering of Fe-bearing volcanics, and/or hydrothermal activity. Bertine (1972) has suggested from studies on weathered tholeiites from the Lau Basin, southwestern Pacific, that the bulk of the iron in sediments from this area is derived from breakdown of the basalts. In addition, Corliss (1971) has found the interiors of Mid-Atlantic Ridge basalt flows to be depleted in Fe and several other elements relative to the chilled margins of the flows, indicating that certain elements are selectively leached from the flows, and could contribute to nodule and encrustation formation.

Further evidence for a volcanic source of iron in active ridge areas has been provided by Scott et al. (1972a). These workers found that the older portion of a manganese encrustation from the Atlantis Fracture Zone in the North Atlantic had a Mn/Fe ratio (0.66) very similar to many present-day encrustations from the Mid-Atlantic Ridge. By contrast, the upper portion of the encrustation had a higher Mn/Fe ratio (1.06), more similar to that of modern encrustations from inactive topographic elevations. They suggested that the similarity between present-day Mid-Atlantic Ridge crusts and the older portion of the Atlantis Fracture Zone crust was due to each being similarly affected by volcanic activity, and that the change in the Mn/Fe ratio between the upper and lower portions of the latter resulted from the cessation of volcanic activity about 11 million years ago in the area of the Atlantis Fracture Zone where the sample occurred.

Comparison of Mid-Atlantic Ridge concretions with those from the Carlsberg Ridge (Table 2-VIII) reveals few compositional differences. Nodules from the Pacific—Antarctic Ridge have been discussed by Goodell et al. (1971). They were found to be similar to the active ridge nodules described in this chapter in that they contain an iron-dominated suite of elements including major amounts of Ti, V, Sr, Co, Ba and Zn.

TABLE 2-VIII

Composition of nodules and encrustations from the Mid-Atlantic and Carlsberg ridges (wt. %)

	Mn	Fe	Ni	Co	Cu	Depth (m)
Mid-Atlantic Ridge	15.33	20.47	0.242	0.414	0.076	2,671
Carlsberg Ridge*	16.46	14.62	0.506	0.270	0.131	3,763

* Includes data of Glasby (1970).

Some of the compositional variations which do occur between active ridge concretions from different areas could result, in part, from sampling differences. Almost all the samples from the Carlsberg Ridge, for example, were taken in an area (4c) (cf. Laughton, 1967) on the flanks of the ridge, whereas those from the Mid-Atlantic Ridge were taken at many widely spaced localities (Fig. 2-6). One possible factor affecting the differences in the iron content between deposits from the two areas could be their relative nearness to potential volcanic sources of iron. If such sources do contribute to nodule formation on active ridges, the highest iron contents would be expected in the youngest deposits closest to the spreading centres. As the encrustations move away from the ridge crest due to sea-floor spreading, they would receive a decreasing proportion of their constituents from volcanic sources, and an increasing proportion from normal seawater, and would change in composition towards that typical of nodules formed in non-volcanic areas. Aumento (1969) has found encrustations to thicken away from the Mid-Atlantic Ridge crest, indicating a continuing contribution of metals from seawater as the encrustations move away from local volcanic influences, and the work of Scott et al. (1972a) has indicated a decrease in the Fe/Mn ratio of such deposits as time passes. On this basis, the high Fe/Mn ratio of the Mid-Atlantic Ridge samples compared with those from the Carlsberg Ridge (Table 2-VIII) could be a result of many of the former having been obtained from the crestal area of the ridge in contrast to the latter which were almost all collected about 280 km from the spreading centre. Similarly, the relatively high Ni and Cu contents of Carlsberg Ridge nodules (Table 2-VIII) could also reflect a relatively greater contribution of these elements from normal seawater.

Nickel, copper, and, to a lesser extent, cobalt, all show positive correlations with manganese in active ridge concretions (Table 2-IX). None of the minor elements follow iron. As mentioned, Corliss (1971) has suggested that both iron and manganese can be supplied to the ocean floor at ridge crests as a result of hydrothermal leaching of basalts. However, balance calculations by Boström et al., (1972) have indicated that in order to supply all the Mn in ridge crest sediments, considerably more basalt has to be leached than is needed to supply the iron. These workers explain this by assuming that similar amounts of basalt supply both elements, but that iron may be selectively precipitated within the rocks. An alternative possibility is that the bulk of both the hydrothermal iron and manganese are precipitated

TABLE 2-IX

Inter-element relationships in active ridge nodules and encrustations (n = 54)

	Mn	Fe	Ni	Co	Cu	Depth (m)
Mn	1.00					
Fe	−0.47	1.00				
Ni	0.64	−0.75	1.00			
Co	0.32	0.10	−0.23	1.00		
Cu	0.55	−0.53	0.69	−0.38	1.00	
Depth	0.24	−0.36	0.44	−0.13	0.46	1.00

on the sea floor at the ridge crest, and that the "excess manganese" is precipitated from seawater by the catalytic action of the iron oxides. If this is the case, the association of Ni, Cu and Co with Mn in active ridge encrustations would suggest that a considerable proportion of these elements may also be supplied by normal seawater, and not by hydrothermal processes.

Inactive ridges

The few nodules from inactive ridges investigated in this study all come from the Atlantic Ocean (Fig. 2-6) and show no outstanding compositional peculiarities. Their Mn/Fe ratio and Ni contents are very close to the world average, similar to that found in Atlantic seamount nodules, and Cu is low (Table 2-II).

The principal differences between inactive and active ridge nodules lie in their higher Mn/Fe ratio, and higher Co contents (Table 2-II). The differences in the Mn/Fe ratio can probably best be ascribed to a volcanic source for iron on the active ridges, but not on the inactive ones. The higher than average Co content of the nodules might result from processes similar to those operating on oceanic seamounts.

Marginal seamounts and banks

The marginal seamount and bank nodules discussed in this chapter are from the Pacific and Atlantic oceans (Fig. 2-6). They contain near average concentrations of Mn, lower than average Ni and Cu, and higher than average Co and Fe. Apart from the active ridge deposits, their Mn/Fe ratio is the lowest encountered.

The principal differences between marginal and oceanic seamount nodules lie in their lower Co and higher Fe contents. Each could reflect differences between the two environments of deposition. The low Co content might result from lower redox potentials in marginal than in open ocean areas due to higher sedimentation rates and burial of organic matter, and the high iron content from proximity to continental sources of iron. Manheim (1965) has shown that iron increases relative to manganese in nodules with increasing continental influence, and suggested that this trend reflects the greater amount of releasable Fe relative to Mn available in continental materials. Mero (1965a) has also noted occurrences of iron-rich nodules bordering the continents in some areas of the Pacific, and concluded that they resulted from the differential precipitation of continentally derived Fe relative to Mn. The similarity in the Mn/Fe ratio in marginal seamount and mid-ocean ridge nodules could thus be due to local sources of Fe in each, continental on the one hand and volcanic on the other.

Comparison of marginal Atlantic nodules with those from the Pacific

TABLE 2-X

Composition of marginal seamount nodules and encrustations from the Pacific and Atlantic oceans (wt. %)

	Mn	Fe	Ni	Co	Cu	Depth (m)
Atlantic	14.87	23.61	0.281	0.339	0.074	2395
Pacific	16.34	15.50	0.308	0.489	0.081	1071

(Table 2-X) reveals few differences other than in their iron contents. The Mn/Fe ratio of the Atlantic samples is less than unity, in contrast to that in those from the Pacific. As most of the latter were taken in the Southern Borderland Seamount Province off Mexico, adjacent to an arid region in which there are few rivers, this difference might be due to low concentrations of fluvially transported iron in the area. The Mn/Fe ratio is much lower off the coast of Washington and British Columbia where the remaining Pacific samples were obtained (Fig. 2-6).

Continental borderlands

Continental borderland and nearshore nodules are discussed in a separate chapter of this volume, and thus will only be mentioned briefly here. Only seven samples were included in the present study, most from the western continental borderland of the Americas (Fig. 2—6), but from deeper waters than the marginal seamount and bank nodules from the same areas. Mineralogically, they consist principally of todorokite (Glasby, 1972a) and their outstanding chemical feature is their high Mn/Fe ratio.

Abyssal nodules

Abyssal nodules comprise the largest group of ferromanganese oxides discussed in this chapter. Relative to nodules from most other environments examined, they contain normal concentrations of Mn and Fe, higher than average Ni and Cu, and lower than average Co. Mineralogically, they consist principally of todorokite with subordinate amounts of birnessite and δMnO_2 (Barnes, 1967a; Cronan and Tooms, 1969; Glasby 1972a).

The average minor element composition of abyssal nodules may, in part, reflect their mineralogical composition. As abyssal nodules are principally composed of todorokite and as this mineral appears to concentrate Ni and Cu (Cronan and Tooms, 1969), the average enrichment of these elements in abyssal nodules relative to the other groups might be explained.

Comparison of the average compositions of abyssal nodules from different depth ranges (Table 2-XI) reveals some interesting differences. Copper, for example, increases by a factor of almost six between 2,000 and 6,000 m.

TABLE 2-XI

Composition of abyssal nodules from different depth ranges (wt. %)

Range (m)	Mn	Fe	Ni	Co	Cu
2,—3,000	18.90	18.13	0.506	0.344	0.108
3,—4,000	14.36	22.47	0.301	0.281	0.131
4,—5,000	17.02	16.95	0.588	0.230	0.342
5,—6,000	17.36	14.36	0.616	0.263	0.627

Excluding the 2,000—3,000 m interval, Ni also increases with depth, whereas Co, by contrast, tends to decrease slightly. Neither manganese nor iron show any consistent relationship to depth.

The causes of these variations with depth are somewhat problematic. The cobalt variations are probably too small to be of significance, although they do reflect the trend mentioned previously towards increasing Co concentrations in nodules from shallow open ocean environments. By contrast, the Ni and Cu variations are too large to be spurious, and may reflect mineralogical variations in the nodules. Other than in continental margin environments, available data indicate that todorokite-rich nodules tend to increase in abundance with increasing depth in the Pacific and Atlantic oceans, and this may explain the general increase in Ni and Cu with depth. The unusually high Ni content of nodules in the 2,000—3,000 m depth range is largely a result of the inclusion of three samples containing over 1% Ni in a total sample population of twelve. One of these nodules came from the South Atlantic, and the other two from the western Indian Ocean off South Africa. Willis (1970) has found that todorokite often occurs in relatively shallow pelagic concretions in these areas.

Comparison of the average composition of abyssal nodules from each of the three major oceans (Table 2-XII) shows most elements to vary quite considerably. Iron is lowest in the Pacific, higher in the Indian Ocean, and highest in the Atlantic. By contrast, the Mn/Fe ratio, Ni, Cu and to a lesser extent Co, behave oppositely. Manganese is high in the Pacific, but almost the same concentration in the other two oceans.

TABLE 2-XII

Average composition of abyssal nodules and encrustations from the Pacific, Indian and Atlantic oceans (wt. %)

	Mn	Fe	Ni	Co	Cu	Mn/Fe	Depth (m)
Pacific	19.27	11.79	0.846	0.290	0.706	1.63	4,794
Indian	15.25	13.35	0.534	0.247	0.295	1.14	4,567
Atlantic	15.46	23.01	0.308	0.234	0.141	0.67	4,162

Variations in the Mn/Fe ratio between abyssal nodules from the three oceans may be partly related to ocean-to-ocean differences in the rate of detrital and biogenous sedimentation. Rates are low over much of the abyssal Pacific leading to the oxidation of organic matter at the sediment surface, but are much higher in the Atlantic. Where such rates are sufficiently high to lead to a reducing environment below the sediment water interface, due to the preservation of organic matter as in some near shore areas, manganese reduction and high Mn/Fe ratios in surface nodules may result. However, in the absence of rates sufficiently high to cause such diagenetic remobilization, high detrital sedimentation rates as in the Atlantic could lead to low Mn/Fe ratios in view of the greater amount of releasable Fe relative to Mn in the products of continental weathering.

Regional variations in nodule composition within each of the three major oceans have been discussed by many workers (Mero, 1965a; Barnes, 1967a; Cronan, 1967; Price and Calvert, 1970; Skornyakova and Andrushchenko, 1970; Goodell et al., 1971; Cronan, 1972a, c, 1975) and thus need not be considered in detail here. Most of these studies have included data on nodules from all environments, and not just on those from abyssal areas. However, Price and Calvert (1970) examined regional variations in various ratios among Mn, Fe, Ni, Co, and Cu in abyssal nodules alone from the Pacific, and concluded that certain regularities existed. The regional distribution of Mn/Fe, for example, showed highest values in the eastern Pacific, decreasing westwards to lowest values in the southwestern and part of the northwestern Pacific. However, the data of Goodell et al. (1971) suggest that the Mn/Fe distribution may be somewhat more complex, as they found Mn to exceed Fe in nodules from the Southwestern Pacific Basin. Regularities were also found in Fe/Co and Mn/Ni ratios, nickel, for example, being enriched in nodules from a large area of the eastern Pacific.

The analyses of abyssal Pacific nodules discussed in this chapter are largely the same as those used by Price and Calvert (1970), namely the data of Mero (1965a), Cronan and Tooms (1969) and others, and the few samples not common to both data sets do not alter the distribution patterns significantly. Correlation coefficients (Table 2-XIII) confirm the Mn—Ni relationship, and also illustrate a strong correlation between Fe and Co. The latter is of interest in view of the seeming lack of such a correlation in nodules as a whole. Cronan (1969) found no significant correlation between Fe and Co in Pacific and Indian Ocean nodules from all environments, but a strong negative correlation between Co and depth. The absence of the Co-rich seamount nodules from the present data has removed the latter correlation, resulting in a stronger correlation between Fe and Co. It is evident therefore from these data, and those presented elsewhere in this chapter, that inter-element relationships in nodules vary depending on the sample populations chosen for analysis. Nodules from some environments often show different element associations than do those from other environments.

TABLE 2-XIII

Inter-element relationships in abyssal Pacific nodules (n = 63)

	Mn	Fe	Ni	Co	Cu	Depth (m)
Mn	1.00					
Fe	−0.51	1.00				
Ni	−0.54	−0.58	1.00			
Co	−0.34	0.61	−0.37	1.00		
Cu	0.14	−0.23	0.29	−0.05	1.00	
Depth	−0.13	−0.01	−0.11	0.13	0.14	1.00

TABLE 2-XIV

Inter-element relationships in abyssal Atlantic nodules (n = 82)

	Mn	Fe	Ni	Co	Cu	Depth (m)
Mn	1.00					
Fe	−0.84	1.00				
Ni	0.78	−0.71	1.00			
Co	0.33	−0.28	0.12	1.00		
Cu	0.51	−0.52	0.83	−0.06	1.00	
Depth	0.05	−0.10	0.07	−0.25	0.34	1.00

TABLE 2-XV

Inter-element relationships in abyssal Indian Ocean nodules (n = 32)

	Mn	Fe	Ni	Co	Cu	Depth (m)
Mn	1.00					
Fe	−0.29	1.00				
Ni	0.78	−0.52	1.00			
Co	−0.16	0.56	−0.31	1.00		
Cu	0.22	−0.51	0.49	−0.43	1.00	
Depth	−0.04	−0.35	0.06	−0.49	0.57	1.00

Regional variations in the composition of abyssal Atlantic nodules are less marked than are those in the Pacific. Manganese is greater than 15% in the basins on either side of the Mid-Atlantic Ridge, with the highest values of all occurring off South Africa. By contrast, it is low in the southwestern Atlantic, often less than 10% in the area of the Drake Passage and Scotia Sea. Iron behaves the reverse of manganese, being up to 40% in the area between South America and Antarctica, and generally more than 20% in the tropics. It is low in the basins on either side of the Mid-Atlantic Ridge (Cronan, 1972a, c). Ni, Co and Cu all follow manganese (Table 2-XIV).

Data on abyssal nodules from the Indian Ocean are relatively sparse, but those available indicate manganese enrichment in the deep-water areas to the east of the Carlsberg and Mid-Indian ridges, although it is also fairly high in some of the basin areas of the western Indian Ocean. Iron is generally lower than manganese in abyssal Indian Ocean nodules, including those from most of the eastern portion of the basin, and is also relatively low in those from the depressions in the west (Cronan, 1967). Willis (1970) has examined nodules from the Natal Basin, to the southwest of Madagascar, and has also found the Mn/Fe ratio to be generally greater than unity. As in the Pacific, Ni follows Mn, and Co follows Fe (Table 2-XV). Nickel and copper are most abundant in the eastern Indian Ocean, but also reach high values in some of the basins in the west. Cobalt, by contrast, and Pb too, are generally low in most of the basin areas (Cronan, 1967). These observations are supported by the more recent data of Bezrukov and Andrushchenko (1972). Although these authors only published analyses of two nodules from the west of the Mid-Indian Ridge system, these samples were significantly higher in Fe and lower in Mn, Ni and Cu than were those to the east. However, no marked fractionation of these elements was found across the 90°E ridge in the eastern Indian Ocean.

Factors causing regional variations in the composition of abyssal nodules are complex. The Mn/Fe ratio may be influenced by sedimentation rates and the burial of organic matter (Price and Calvert, 1970) and by proximity to continental or volcanic sources of elements (Cronan and Tooms, 1969). High sedimentation rates leading to low redox potentials and diagenetic remobilization of Mn near marginal areas and under the equatorial zone of high productivity may lead to the high Mn/Fe ratios common in some nodules from these areas. In addition, relatively high Mn/Fe ratios also occur in nodules from areas of very slow sedimentation (Horn et al., 1972c) such as the pelagic tropical North Pacific, possibly as a result of the slow precipitation of Mn from seawater in the absence of local diluent sources of iron. However, where terrigenous sedimentation rates are intermediate between these two extremes, as, for example, in the northern part of the North Pacific and over much of the Atlantic, the Mn/Fe ratio is relatively low, possibly as a result of the availability of continentally derived iron for incorporation into the nodules.

As might be expected, the influence of submarine volcanism on regional variations in the composition of nodules is greatest in the vicinity of active mid-ocean ridges, and this has already been dealt with. The influence of hydrothermal activity on nodule composition is likely to decrease in importance away from volcanic centres, but the slow alteration of submarine volcanic rocks in old inactive areas could contribute elements to nodules and encrustations over long periods of time.

The minor elements in abyssal nodules may also be affected by the proximity of local sources of elements, but environmental factors are perhaps of greater importance in determining their distribution. The significance of depth as an environmental parameter controlling regional variations in nodule composition has been the subject of some confusion. That both the mineralogy of nodules and their content of elements such as Co, Cu and Ni varies with depth in open ocean nodules has been shown by several workers (Menard, 1964; Barnes, 1967a; Cronan, 1967; this chapter). However, as pointed out by Glasby (1970), when shallow-water nearshore nodules are included in the computations, the relationships often break down. This means, of course, that depth per se is unlikely to significantly influence the composition of nodules, but does suggest that some factor which varies in a general way with depth in the open ocean could be doing so. The most likely such factor is the redox potential of the environment of deposition. This is at maximum in the elevated areas of the open ocean such as seamounts leading to the formation of δMnO_2-rich nodules, but lower in the basins and in continental margin areas resulting in the formation of nodules rich in todorokite. As Ni and Cu are enriched in the latter, regional variations in their contents in abyssal nodules will vary with regional variations in the content of todorokite and these in turn will depend on the redox potentials of the environments of deposition.

Even though mineralogical suitability dependent on redox potential is a probable pre-requisite for Ni and Cu enrichment in nodules, it alone cannot lead to the very high contents of these elements in some todorokite-rich nodules of potential economic value from the North Pacific, as todorokite-rich nodules from elsewhere are lower in Ni and Cu. The enrichment of these elements in the northern equatorial Pacific nodules may be the result of both biological productivity and the nature of the substrate. The Ni- and Cu-rich deposits occur in an east—west belt under the northern portion of the equatorial zone of high productivity, and are underlain by siliceous ooze (Horn et al., 1972c). According to Greenslate et al. (1973), planktonic organisms may extract metals from the surface waters in this zone, and, after death, transport them to the sea floor. Here, largely because of depth-induced dissolution of the organisms (the Ni- and Cu-rich nodule zone lies at or beneath the carbonate compensation depth), the metals scavenged by the organisms from the surface waters are liberated into the bottom and interstitial waters where they can be incorporated into growing manganese nodules. The high overall biological productivity in the zone would ensure a continuous flux of elements to the sea floor, although the intensity of the flux might vary regionally and with time. Different organisms could have differing capacities for concentrating metals from seawater, and thus the distribution of different planktonic species may influence the variable transportation of metals to the sea floor.

The nature of the nodule substrate, the siliceous ooze, may be of importance in facilitating the incorporation of elements into manganese nodules at the sediment surface. Because of the loose porous nature and high interstitial water content of this sediment (Horn et al., 1972c), Ni and Cu may be able to diffuse through its interstitial waters to remain enriched at the sediment surface where they could be incorporated into the forming ferromanganese oxides. In this way, these elements could be supplied to the nodules not only by the dissolution of calcareous organisms at or near the sedimentation surface, but also by the dissolution of siliceous organisms after burial. The observation of Greenslate et al. (1973) that the siliceous oozes themselves are relatively depleted in Ni and Cu could support this suggestion.

RARE ELEMENTS

One of the principal characteristics of deep-sea nodules is their enrichment in many rare elements which are normally present in the earth's crust in low concentrations. Data on these elements are often so sparse that their distribution in nodules from different environments cannot generally be discussed in the same way as can that of some of the more frequently determined elements. Average abundances of some of these elements are given in Table 2-I. Relative to their crustal abundances, Mo and Tl are enriched in nodules by a factor of more than 100. Ag, Ir and Pb between 50 and 100; B, Zn, Cd, Yb, W and Bi between 10 and 50; and V, Y, Zr, Ba, La and Hg between 3 and 10. Other rare elements which have been determined in deep-sea nodules include the rare earths (Goldberg et al., 1963; Ehrlich, 1968; Piper, 1972; Glasby, 1973a); Sn (Smith and Burton, 1972); Te (Lakin et al., 1963); Ga, Ge, Sr, Sn (Riley and Sinhaseni, 1958); Li, Be, S, Cl, Ge, As, Se, Br, Rb, Nb, and I (Cronan and Thomas, 1972); and As, Br, Rb, Nb, In, Sn, Sb, I, Cs, Hf, Os and Au (Glasby, 1973a).

There are several factors affecting rare element enrichments in nodules. Among these can be included the adsorptive and crystallochemical characteristics of the ferromanganese oxides themselves, the rate of supply of the elements concerned to the marine environment, their availability at the reaction site, and the nature of the environment of deposition.

Largely due to the very fine grain sizes and high surface areas of the ferromanganese oxide phases, adsorption is an important mechanism in enriching several elements in nodules. Enrichments of Bi, Cd, W, Ag, and Hg may be related to this process (Krauskopf, 1956). However, adsorption alone is unlikely to account for the occasional very high concentrations of some minor elements. The possibility of substitution of some elements into the lattices of the ferromanganese oxide minerals themselves has already been discussed and this may account for, among others, the high average Ni and Cu concentrations in todorokite-rich nodules. Some of the inter-element relationships in nodules are also probably due, in part, to direct substitution of one element for another in the ferromanganese oxide phases.

The relationships of rare elements in nodules to possible sources of the elements concerned has received a considerable amount of attention. In particular, submarine volcanism has been held responsible by several authors for higher than normal concentrations of some elements in nodules from volcanic areas. For example, Harriss (1968) suggested that mercury is enriched in nodules from volcanic regions as a result of submarine volcanic or hydrothermal processes; but this process cannot be universally operative as nodules from the Carlsberg Ridge with strong volcanic associations were found to be very low in mercury. Another suggested source influence on the rare element content of manganese nodules is that of meteoritic material. Harriss et al. (1968) found that some noble metals were characterized by an homogeneous distribution both within and between nodules, together with a lack of correlation with other elements present, and concluded that they might be extraterrestrial in origin. However, Glasby (1973a) considered that the behaviour of these metals was more likely to result from their uniform adsorption from seawater.

The importance of the nature of the environment of deposition on the rare-element content of manganese nodules must not be underestimated. This is perhaps most clearly shown in the contrast between shallow-water continental margin and lacustrine nodules, and deep-sea varieties (Chapter 3). For example, Glasby (1973a) has found a relatively uniform depletion of the elements As, Y, Zr, Nb, Mo, Cd, In, Sn, Sb, REE, Tl, Pb, Bi and possibly also Ag, Hf, Os and Au, in continental margin and nearshore nodules relative to their content in deep-sea varieties. Similarly, Harriss et al. (1968) found that Au, Pd and Ir are lower in freshwater concretions than in their deep-sea counterparts. By contrast, alkali metal and halide ions such as Rb, Sr, Cs, Br and I, which might not be expected to be incorporated into ferromanganese oxide phases, are present in similar concentration in both continental margin and deep-sea nodules (Glasby, 1973a). However, the halides are depleted in lacustrine concretions (Cronan and Thomas, 1972), probably as a result of compositional differences between marine and lake waters. Compositional differences between nodules from different deep-sea environments are rarely as large as between deep-sea and continental margin nodules, but nevertheless can also be related to environmental controls. The enrichment of Co and Pb in seamount nodules as a possible result of high redox potentials has already been discussed.

One of the more extensively studied groups of minor elements in nodules is the rare earths. Goldberg et al. (1963) considered that an observed enrichment of the heavy rare earths in marine deposits relative to their crustal abundances might be due to the increasing stability of the heavier rare-earth complexes with ligands derived from seawater. By contrast, Ehrlich (1968) believed that detrital material forms the principal source of REE in nodules, and that the elements are concentrated into the nodule structure partly by occlusion of rare-earth bearing detrital phases, and partly by their surface transfer. However, Glasby (1973a) has suggested that a comparison of the rare-earth abundance pattern in nodules relative to that in seawater, with solubility product data of the trivalent rare-earth hydroxides, indicates a similarity in the variation of these functions with atomic number for all elements except La and Ce. As no similarity was observed between the rare-earth distribution in nodules and in sediments, Glasby concluded that the rare earths could be incorporated into manganese nodules by the direct precipitation of the rare-earth hydroxides from seawater. Piper (1972) has also presented evidence that REE are precipitated from normal seawater, and considers that their distribution is influenced by the mineralogy of the manganese phases.

SUMMARY AND CONCLUSIONS

Manganese nodules are common on the deep-ocean floor, but vary considerably in abundance and composition from place to place. They reach their greatest concentrations in deep-water areas at or below the calcium carbonate compensation depth, where detrital sedimentation is at a minimum and overall sedimentation rates are very low. Extensive deposits are also associated with areas of non-deposition or sediment erosion. The localized distribution of nodules within areas of overall high concentrations can be related to a number of factors which include sea-floor topography, sediment distribution and distribution and abundance of potential nuclei around which the nodules can accrete.

Large-scale compositional variations between manganese nodules can be related to their differing environments of deposition, with distinctive compositional varieties occurring, for example, in abyssal areas, continental borderlands and on seamounts. The factors which determine environmentally differentiated compositional variations between nodules include seawater composition; biological productivity; the nature of the substrate; and the degree of oxidation of the environment of deposition by its influence, firstly on the oxidation states of metals incorporated into the deposits, and, secondly, on the mineralogy of the nodules themselves which partially determines their receptiveness to different ionic species. Regional variations in nodule composition throughout the World Ocean result from the interaction of these factors, a fuller understanding of which would facilitate the location of the most economically valuable varieties of these deposits.

*NOTE ADDED IN PROOF

The selection of data for such a study poses some problems. Not all of the published analyses of manganese nodules and encrustations are presented in the same manner. An even more fundamental problem is the heterogeneity of individual nodules and encrustations. It has been shown that some elements can vary several-fold in nodules from the same site and even within individual concretions. Many of the reported discrepancies between analyses of different nodules from the same site can probably be accounted for on this basis, as could discrepancies between analyses of different subsamples of the same nodule or encrustation. In order to minimize this problem, the bulk of the data used here is drawn from a limited number of large groups of analyses. None the less, the problem should always be borne in mind when comparing nodule and encrustation analyses from more than one source.

CHAPTER 3

SHALLOW WATER,CONTINENTAL MARGIN AND LACUSTRINE NODULES: DISTRIBUTION AND GEOCHEMISTRY

S. E. CALVERT and N. B. PRICE

INTRODUCTION

Ferromanganese nodules, concretions, crusts and coatings on rock and mineral particles are widespread sedimentary features in many nearshore marine and lacustrine environments. It has become clear in recent years that these sediment components provide important clues to the behaviour of a wide range of elements in Recent sediments (Sevast'yanov and Volkov, 1967a, b; Calvert and Price, 1972; Duchart et al., 1973; Callender, 1973; Price, 1976) and comparisons between the composition of nearshore and open-oceanic ferromanganese nodules have helped in understanding the factors controlling the wide compositional variations of these phases in the marine environment (Price, 1967; Price and Calvert, 1970).

It is now ten years since Manheim (1965) drew attention to the widespread occurrence of ferromanganese concretions in shallow waters. In this classic publication, Manheim provided an excellent summary of the known distributions and compositions of concretions from a very large, and a very obscure, literature. He was able to contrast shallow-water and oceanic concretions in terms of their bulk compositions and suggested that these different forms of ferromanganese oxides form in quite different ways.

Since the work of Manheim, several detailed descriptions of the geochemistry and mineralogy of ferromanganese concretions in shallow water environments have appeared, together with a considerable amount of information on the compositions of sediment pore waters. Moreover, the discovery or re-examination of several occurrences of concretions from lacustrine environments, also mentioned by Manheim, has attracted some attention.

In this chapter, the new analytical data on shallow-water ferromanganese deposits is reviewed and a discussion of the geochemistry of these sediment components is presented. The distribution and composition of lacustrine deposits, together with some aspects of the ferromanganese component of soils, are also discussed in view of the similarity of these forms to marine concretions. Information contained in a wider literature on the behaviour of the iron, manganese and related elements in surface environments is also included where relevant.

MARINE SHALLOW-WATER FERROMANGANESE CONCRETIONS

Distribution

The distribution of ferromanganese deposits in several shallow-water areas, notably the Baltic, Barents and Kara seas, has been documented by Manheim (1965, figs. 5 and 6). In view of the patchy distribution of concretions in many areas and the variability of bottom sediment type in these areas, further refinement of the data of Manheim cannot be made. Rather, emphasis will be placed here on the relationship between the distribution, morphology and composition of the concretions and the nature of the sediment in or on which they occur.

Ferromanganese concretions are widely distributed in the Gulfs of Bothnia, Riga and Finland and in the central Baltic Sea, in water depths ranging from 15 to 270 m (Gripenberg, 1934; Winterhalter, 1966; Gorshkova, 1967; Shterenberg, 1971; Varentsov, 1973). They occur in highest concentration in the Gulf of Bothnia where the surface sediments, consisting of muds, sands and pebbly sands, are entirely oxidized. Here, approximately 10—20% of the Gulf floor is covered by concretions with an average concentration of 0.5 kg m^{-2} to 5 kg m^{-2} (Winterhalter, 1966).

In the Gulfs of Finland and Riga and in the central Baltic, concretions are not found in the central, deeper-water areas where fine-grained, reducing sediment occurs. They are found, however, in shallower water around the margins of depressions or central deeps where the sediment is coarser-grained and oxidized (Manheim, 1965, fig. 6; Gorshkova, 1967; Shterenberg, 1971; Varentsov, 1974). No estimates of surface concentrations are available for these areas.

Knowledge of the distribution of concretions in the northern Russian seas has been available for some considerable time (Nordenskiöld, 1881, pp. 185—186), and a considerable amount of information on their compositions is now available (Samoilov and Titov, 1922; Gorskova, 1931, 1957, 1966, 1967; Klenova and Pakhomova, 1940; Klenova, 1960). In the Barents Sea, concretions are found in water depths ranging from 200 to 1,000 m, mainly in the northern part where the sediments are highly oxidized, sandy clays (Klenova, 1960). Concentration estimates are not available. In the White Sea, concretions are apparently very numerous and are confined to shallow water areas with sandy sediments (Gorshkova, 1931). In the Kara Sea, concretions are not abundant, and occur in waters shallower than 200 m in predominantly sandy muds (Gorshkova, 1957).

Ferromanganese nodules were discovered in the Black Sea during the Russian expeditions of 1890 and 1891 (Murray, 1900). Descriptions were provided by Samoilov and Titov (1922) and they have been re-examined by Sevast'yanov and Volkov (1966; 1967), Georgescu and Lupan (1971), Georgescu and Nistor (1973). Hirst (1974) and Manheim and Chan (1974) discussed the distribution in relation to the geochemistry of sediments and pore waters. The concretions are found in the oxygenated zone of the Black Sea off the Crimean coast and off the river Danube in water depths ranging from 40 to 80 m. They occur as a thin (0.5—1.0 cm) surface layer on brown, clayey silts containing abundant empty *Modiola* shells, and concentrations reach 2.15 m^{-2} (Sevast'yanov and Volkov, 1967).

Ferromanganese concretions were discovered in Loch Fyne, Scotland, in 1878 (Buchanan, 1878) shortly after the oceanic nodules were discovered by the *Challenger* expedition (Thomson, 1874). They were subsequently collected in lochs Goil and Striven and on the Skelmorlie Bank in the Clyde Estuary (Murray and Irvine, 1894; see Calvert and Price, 1970a, for locations). The concretions occur in water depths ranging from 20 to 210 m and are generally found either within muds or on muddy sediments containing pebbles and boulders. The surface sediments in the lochs are brown in colour while the subsurface sediments, below a few millimetres to a few tens of centimetres, are grey, reduced muds (Calvert and Price, 1970a).

Ferromanganese concretions are known to occur at a single site in Jervis Inlet, British Columbia (Grill et al., 1968a; MacDonald and Murray, 1969). They are found in water depths of between 338 and 366 m where the sediments are brown to greyish-brown silty clays with a surface veneer of cobbles, pebbles and sands. The subsurface sediments, below 1 cm depth, are olive-gray sands and silts containing H$_2$S.

Other isolated occurrences of ferromanganese concretions are known in shallow-water areas, such as the Gulf of California (Mero, 1965a) and Izu Bank in the Japanese Archipelago (Niino, 1955) but adequate supporting information on the associated sediment is lacking.

Morphology and size

The morphology and size of shallow marine ferromanganese concretions are highly variable and appear to be characteristic for a given area and a given sediment type. The forms are:

(1) Spherical nodules, generally 1—2 cm in diameter, known in Finnish and Swedish publications as "pea ore" (Fig. 3-1A). Such nodules have concentric internal structures (Fig. 3-1B) and are generally devoid of a nucleus. The surfaces are generally quite smooth, although mammillated surfaces are found on larger examples.

(2) Flat concretions, generally 2—10 cm in diameter, known in Finnish and Swedish publications as "penny ore" (Fig. 3-1D). These concretions have clearly marked growth rings and are concentrically layered internally. They are generally mammillated and convex on the upper surface and smooth or pitted and concave on the lower surface.

(3) Ring or girdle-shaped growths around nuclei, usually pebbles and boulders (Fig. 3-1C). They have the same general morphology as the penny ores and merely represent a variant of this growth form.

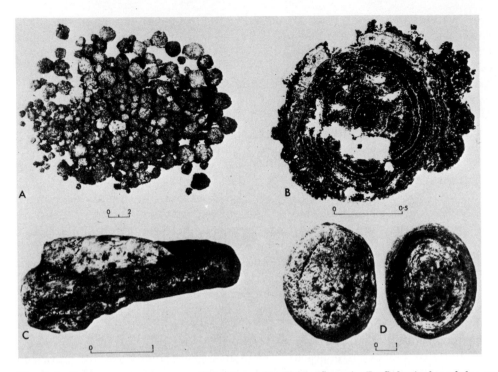

Fig. 3-1. Ferromanganese concretions from the Baltic Sea. A, B. Spherical nodules or "pea ores". C. Ferromanganiferous girdle around a pebble. D. Flat "penny ore" showing upper (left) and lower (right) sides. All scales in cm. From Winterhalter, 1966.

(4) Flat slabs, crusts and sheets usually a few millimetres to 1 cm thick. They are rather impure, containing a large proportion of detrital material and are very friable. Broken fragments may become recemented to form conglomeratic slabs.

Spherical nodules are invariably found in fine-grained sediments where growth has occurred on all surfaces. They are characteristically found, therefore, in the deeper-water areas of enclosed basins, such as lochs Goil and Fyne (Murray and Irvine, 1895; Calvert and Price, 1970a) and in the Gulf of Bothnia (Gripenberg, 1934; Winterhalter, 1966). In Loch Fyne, the nodules are distributed throughout the upper 10—15 cm of the sediment and occur only sporadically at the sediment surface.

Penny ores and ring-shaped concretions are invariably found at the sediment surface in shallow-water areas or wherever the sediment is coarse-grained. They are the most widespread growth form in the Baltic Sea and the Black Sea and on areas of the open continental shelf, such as the Barents and Kara Seas.

The relatively thin, flat concretions are known only from the Gulf of Bothnia (Winterhalter, 1966). They represent sediment impregnated with ferromanganese oxides and are common in the southern half of the gulf where the sediments are sandy.

Mineralogy

An extensive literature on the mineralogy of ferromanganese nodules has shown that they contain several different manganese oxide phases, poorly ordered iron oxides and various detrital impurities. A detailed account of this subject, together with a discussion of the problems of terminology involved, is given by Burns and Burns (cf. Chapter 7).

For the purposes of this chapter, the manganese phases represented by X-ray diffraction spacings of 10 Å and 2.4 Å are recognized. The 10 Å form is referred to as manganite, following Buser and Grütter (1956) while the 2.4 Å form is referred to as δMnO_2. Another form of manganite recognized by Buser and Grütter, with a basal spacing of 7 Å, is considered by Burns and Brown (1972) to represent a 101 plane of the 10 Å manganite and not a separate phase or a partially collapsed structure of the manganite.

Ferromanganese concretions from shallow-water environments appear to contain 10 Å manganite (Manheim, 1965; Cronan, 1967; Grill et al., 1968, Calvert and Price, 1970a; Glasby, 1972a). X-ray diffraction data for two such ferromanganese concretions are given in Table 3-I. Data on the X-ray diffraction characteristics of concretions from the Barents, Kara, White and Black seas are unfortunately lacking.

As pointed out by Buser and Grütter (1956) and Grütter and Buser (1957), the various forms of manganese oxides in nodules have different

TABLE 3-I

X-ray diffraction data for shallow marine ferromanganese concretions

Baltic Sea (Manheim, 1965)		Loch Fyne (Scotland)	
$d(Å)$	I	$d(Å)$	I
9.7	100	9.6	100
7.2	60	7.02	32
4.85	24	4.79	54
4.56	6	4.47	23
3.58	6	2.43	29
2.45		2.40	45
Band	96	2.34	
Band		2.19	32
2.23	3	2.02	27
2.06	5	1.99	23
2.00	7	2.80	32
1.42	21	1.44	18

oxidation grades, δMnO_2 being more highly oxidized than 10 Å manganite. The manganese oxides in nearshore environments, on this evidence, are therefore less oxidized than open oceanic forms, as pointed out by Buchanan (1891). Price and Calvert (1970) have used the mineralogy of nearshore and oceanic ferromanganese nodules to describe regional variations in their oxidation grades throughout the Pacific Ocean.

The nature of the iron phases in ferromanganese nodules has remained obscure. Cronan (1967) and Glasby (1972a) could not detect any crystalline iron phase in a large suite of nodule samples using conventional X-ray diffraction techniques, whereas Winterhalter (1966) and Varentsov (1973) have claimed that goethite can be detected in some nodules and crusts from the Baltic.

Chemical composition

Major elements

The composition of ferromanganese concretions from shallow marine environments is highly variable. Representative analyses are given in Table 3-II. There appear to be two sources of this variability, namely the different proportions of detrital aluminosilicate impurity in the concretions and the ratio of Mn/Fe.

The variable amount of detrital material incorporated into the concretions is shown by the variable amounts of Si, Al, Mg, K and Na in Table 3-II. A correction for the dilution of the ferromanganese oxide component can be made if the composition of the associated sediment is also known. Thus, it can be shown that Loch Fyne (Calvert and Price, 1970a) and Black Sea (Sevast'yanov and Volkov, 1967) concretions contain 25 and 27% by weight aluminosilicate impurity, respectively, assuming all Al is associated with this phase. Judging by the Si and Al contents of other shallow marine concretions (Table 3-II), and incidentally those from the deep sea (see analyses in Mero, 1965a), ferromanganese nodules generally appear to be contaminated to a roughly similar degree although the precise composition of the aluminosilicate component is highly variable.

The ratio of Mn to Fe in shallow-water concretions ranges from values of about 45 to < 0.1, different areas having concretions with roughly similar Mn/Fe values. For example, in the Black Sea, the ratio is always less than 0.77 (Sevast'yanov and Volkov, 1966), in the Baltic Sea the ratio is always less than 1.37 (Winterhalter, 1966) and in the Scottish loch concretions it is always greater than 2.56 (Murray and Irvine, 1895; Calvert and Price, 1970a).

Superimposed on these regional differences in the Mn/Fe ratio, there also appears to be a significant amount of variation in any one area. The range in the ratio for the Black Sea is 0.039 to 0.77 while in the Baltic Sea it is 0.007 to 1.37. This variability appears to be related to the morphology and the

TABLE 3-II

Chemical composition of ferromanganese concretions from shallow marine environments[1]

	1	2	3	4	5	6	7	8
Si	6.30	7.66	9.15	10.90	5.56	3.55	6.41	6.32
Al	2.02	1.54	1.82	1.89	1.65	0.64	1.71	2.28
Ti		0.13	0.28	0.29	0.10	0.10	0.11	0.21
Fe[*2]	14.67	22.47	19.68	22.78	26.54	18.20	6.16	3.92
Ca	1.48	1.22	1.70	1.36	4.45	10.26	1.32	5.57
Mg	0.72	0.58	0.21	0.43	1.04	0.90	1.82	1.87
K	0.07	0.75	1.43	1.45		0.41	0.99	1.03
Na	0.09	0.35	1.05	0.78		0.19	0.95	
P	1.46	0.69	1.24	0.69	1.14	0.84	0.34	0.35
Mn	28.20	14.03	13.54	9.90	6.79	14.10	32.76	30.19
S		0.08					0.07	
C		2.50	1.29	1.06	0.67			
CO$_2$	5.52	0.76	2.73	2.39	5.50	11.20	0.56	11.88
As					687			245
Ba		2500					2857	3090
Co		160	96	64	84	30	157	230
Cr		10	17	23	16	100	7	
Cu		48	9	17	37	10	67	17
Mo		130			18	30	231	55
Ni		750	35	47	281	100	314	77
Pb		38	9	24			nd	43
Rb								40
Sr								770
U		10						
V		150	68	98	186	10	157	
Y								28
Zn		80	113	135			23	60
Zr					42			55

[1] Major elements as wt.%, minor elements as ppm.
[2] Total Fe.

1: Barents Sea (from Samoilov and Titov, 1922). Single analysis of a flat concretion.
2: Baltic Sea (from Manheim, 1965). Composite analysis from the circum-Gotland region.
3: Gulf of Finland (Varentsov, 1973). Mean of 9 analyses.
4: Gulf of Riga (Varentsov, 1973). Mean of 19 analyses.
5: Black Sea (Sevast'yanov and Volkov, 1967; Sevast'yanov, 1967). Mean of: 15 analyses for Fe, Ti, Mn, P, Ni, Co, Cu, Mo, V and Cr; 8 analyses for Si, Al, Ca, Mg, C and CO$_2$; and 4 analyses for Zr.
6: Black Sea (Georgescu and Lupan, 1971). Single analysis of a nodule containing the least amount of shell material.
7: Jervis Inlet, British Columbia (Grill et al., 1968b). Mean of 2 analyses calculated on a total sample basis, except for S, Co, Cu, Pb, Ni, V and Zn, which are for HCl-soluble fractions only.
8: Loch Fyne, Scotland (Calvert and Price, 1970a). Mean of 2 analyses of composite nodule samples.

type of sediment associated with the concretions as is illustrated by the data from the Baltic (Winterhalter, 1966). Thus, spherical pea ores have Mn/Fe ratios lying mostly between 0.5 and 1.0 while slabs and flat concretions have ratios generally less than 0.2 (Fig. 3-2). Ring-shaped concretions, or penny ores, have intermediate values. As noted previously, pea ores occur in muddy sediments, penny ores and ring-shaped concretions occur on sandy and pebbly sediments, while slabs occur in areas where fragmented materials and boulders are abundant. Similar relationships between the Mn/Fe ratio and the morphology of concretions and the type of associated sediment are found in the other examples listed in Table 3-II.

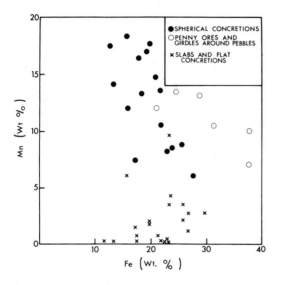

Fig. 3-2. Relationship between Mn and Fe contents of ferromanganese concretions from the Baltic Sea. From Winterhalter, 1966.

In addition to the high levels of Mn and Fe in ferromanganese concretions, it seems clear that to some extent Si, K, Mg, Ti and P are also enriched over their probable contributions from aluminosilicate impurities. These enrichments are deduced from comparisons between concretion and associated sediment compositions from data given by Sevast'yanov and Volkov (1967) and Calvert and Price (1970a) for the Black Sea and Loch Fyne, respectively. Moreover, in Loch Fyne, it appears that K is associated with the manganese phase, while Si, Ti and P are associated with the iron phase (Calvert and Price, 1970a). The association of Mg with either of the major phases could not be determined in this case because of the presence of a mixed manganese—calcium—magnesium carbonate (see p. 71).

High concentrations of P in shallow-water concretions have been reported by Samoilov and Titov (1922), Gorshkova (1931), Winterhalter (1966), Winterhalter and Siivola (1967), Sevast'yanov and Volkov (1967) and Calvert and Price (1970a). Phosphorus concentrations are well correlated with Fe in the concretions (Fig. 3-3). This close relationship has been explained by the adsorption of P by hydrous ferric oxides (Winterhalter and Siivola, 1967) or by the formation of a ferric phosphate in the iron-rich phase of the concretions (Sevast'yanov and Volkov, 1967).

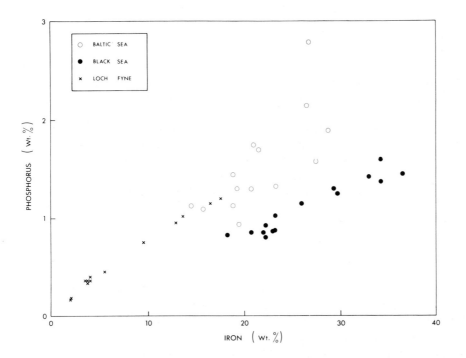

Fig. 3-3. Relationship between P and Fe contents of some shallow-marine ferromanganese concretions. Data for the Baltic Sea from Winterhalter and Siivola, 1967; for the Black Sea from Sevast'yanov and Volkov, 1966; for Loch Fyne from Calvert and Price, 1970a.

The enrichment of Ti in the Black Sea and Loch Fyne concretions suggests, following Goldberg and Arrhenius (1958), that the element is involved in authigenic oxide precipitation and is not simply a detrital element as suggested by Goldschmidt (1954). The correlation between Ti and Fe in the Loch Fyne concretions (Calvert and Price, 1970a) is also shown by oceanic manganese nodules (Goldberg, 1954).

The enrichment of K in the manganese phase of concretions from Loch Fyne is illustrated by comparing K/Rb ratios in the sediments and the various nodule phases. Thus, sediments have K/Rb ratio of 187 while the Mn-rich concretion cores have a ratio of 286. The presence of K in the lattice of the 10 Å manganite adequately explains this association in view of the reported compositions of this type of phase (Frondel et al., 1960a; Straczek et al., 1960). Likewise, the presence of Mg in the manganese phase is consistent with the same data and in any case the amount shown in Table 3-II (e.g. analyses 7 and 8) is too great to be associated solely with the aluminosilicate phase.

The presence of Ca in the oxide phases of the shallow marine concretions, which has been suggested for oceanic nodules (Brown, 1971), cannot be checked with the available data because of the presence of calcareous shell debris and manganoan calcite intimately associated with the concretions in the Black Sea and Loch Fyne, respectively. That a substantial amount of the Ca is present in these carbonate phases is shown by the CO_2 values (Table 3-II).

The organic-carbon content of shallow-water nodules, and its contrast with that in oceanic nodules, has been discussed by Manheim (1965). Carbon contents as high as 3.2% have been reported by Varentsov (1973). Associated sediments also generally have higher carbon contents than pelagic sediments due to a combination of high primary production and higher total sedimentation rates in nearshore areas. The adsorption of organic material onto dispersed oxides, which have very large surface areas, warrants examination (Goldberg, 1961a).

Minor elements

The concentrations of many minor elements in shallow marine ferromanganese concretions (Table 3-II) are much lower than those in oceanic nodules. This contrast has been discussed by Manheim (1965) and Price (1967). Nevertheless, before discussing this contrast, it can be shown that shallow-water concretions are enriched in minor elements compared with the associated sediments. Table 3-III shows available data for the Black Sea and Loch Fyne. On a strict comparison, As, Ba, Co, Mo, Ni, Pb, Sr and V are present in high concentrations in the concretions. This list can be extended by normalizing the values to Al (Table 3-IV) or Rb (Calvert and Price, 1970a); then Cu, Y, Zn and Zr are also enriched in the concretions. Only Cr shows no enrichment at all. Consequently, it can be concluded that shallow-water ferromanganese concretions are sites of co-precipitation and/or adsorption of minor and trace elements in the marine environment, just as oceanic nodules are in a more spectacular way (cf. Chapter 2).

Using analyses of manganese- and iron-rich phases of the concretions from Loch Fyne, Calvert and Price (1970a) showed that different groups of minor elements were preferentially enriched in one or other of the major oxide

TABLE 3-III

Major and minor element composition of ferromanganese concretions and associated sediments

	Black Sea[1]		Loch Fyne[2]	
	concretions	sediment	concretions	sediment
Si	5.56	17.83	6.32	22.60
Al	1.65	6.18	2.28	8.96
Ti	0.10	0.32	0.21	0.55
Fe	26.54	5.11	3.92	4.73
Ca	4.45	8.86	5.57	3.95
Mg	1.04	1.17	1.87	2.29
K			1.03	2.99
P	1.14	0.16	0.35	0.08
Mn	6.79	0.35	30.19	0.22
CO_2	5.50	9.62	11.88	4.02
As	687[3]	35[3]	245	15
Ba			3,090	590
Co	84	14	120	
Cr	16	46		
Cu	37	30	17	22
Mo	18	2	55	5
Ni	281	40	77	70
Pb			43	30
Rb			40	160
Sr			770	250
V	186	93		
Y			28	35
Zn			60	125
Zr	42	107	55	160

[1] Sevast'yanov and Volkov, 1967.
[2] Calvert and Price, 1970.
[3] Sevast'yanov, 1967.

phases. Thus, Ba, Co, Mo, Ni and Sr were enriched in the manganese phase while As, Pb, Y, Zn and Zr were associated with the iron phase. Although it is an oversimplification to accept that the principle of diadochy leads to statistical correlations between element pairs on the basis of chemical analytical data (Burns, 1973), it has nevertheless been suggested that some of these associations can be explained by substitution. For example, Ba and Sr can probably substitute for Ca and Mg (Levinson, 1960; Frondel et al., 1960a; Straczek et al., 1960) while Co, Mo and Ni may substitute for Mn^{2+} in the manganite lattice. Arsenic probably occurs as an adsorbed arsenate ion or replaces phosphate ions in a ferric phosphate (see p. 53). The high Y contents of the iron phase may be similarly due to its association with a ferric phosphate, as suggested by Fomina and Volkov (1969), or it may

TABLE 3-IV

Ratios of major and minor elements to Al in ferromanganese concretions and associated sediment*[1]

	Black Sea		Loch Fyne	
	concretions	sediment	concretions	sediment
Si	3.37	2.88	2.82	2.52
Ti	0.06	0.05	0.09	0.06
Fe	16.08	0.83	1.73	0.53
Ca	2.69	1.43	2.44	0.44
Mg	0.63	0.19	0.82	0.25
K			0.45	0..33
P	0.69	0.03	0.15	0.01
Mn	4.11	0.06	13.31	0.02
As	416	21	108	2
Ba			1358	65
Co	51	2		
Cr	10	7		
Cu	22	5	8	2
Mo	11	0.3	24	0.5
Ni	170	6	34	8
Pb			19	3
Sr			338	28
V	113	15		
Y			12	4
Zn			26	14
Zr	25	17	24	18

*[1] Data taken from Table 3—III. Minor element/Al ratios given $\cdot 10^4$.

actually be present as an yttrium phosphate, the theoretically stable solid phase in the marine environment (Sillén, 1961, Table 4).

Lead and zinc enrichments in the iron phases are not so readily explained. Lead is known to be associated with iron in some oceanic ferromanganese nodules (Calvert and Price, 1977); and Kee and Bloomfield (1961) have shown experimentally a considerable adsorption of Zn onto ferric hydroxides.

Finally, the enrichment of Zr in the iron phase of Loch Fyne concretions appears to be at variance with its supposed behaviour in the sedimentary cycle. It does apparently become associated with the authigenic fraction of marine sediments supporting the observations of Goldberg and Arrhenius (1958), Goldschmidt (1954) and Degenhardt (1957).

It is instructive to look at the order of enrichment of the different minor elements in shallow marine concretions with a view to considering the possible mechanisms of minor element uptake. For Loch Fyne, the order is, based on the data in Table 3-IV: As > Mo > Ba > Sr > Pb > Ni > Y > Cu > Zn > Zr. For the Black Sea the order is: Mo > Ni > Co > As > V > Cu > Zr > Cr. It is noteworthy that several elements that probably occur in anionic form in the marine environment, notably As, Mo and V (and also P), are among those elements most highly enriched in these two groups of concretions.

It should also be noted here that although the concentrations of many minor elements in shallow marine ferromanganese concretions are much lower than those in oceanic nodules, some elements are actually more concentrated in the shallow-water varieties. This applies particularly to As, Ba and P (see Table 3-II).

Rare earth elements

The yttrium and rare earth element (REE) concentrations in ferro-manganese concretions and associated sediments are shown in Table 3-V. In the Black Sea, the concentrations of total REE in the concretions are lower than in the associated sediments. This is in marked contrast to oceanic ferromanganese nodules where total REE concentrations are of the order 1,000—1,400 ppm compared with values of around 200 ppm in pelagic red clays (Piper, 1974).

The shale-normalized REE patterns (see Piper, 1974) of the Black Sea concretions are shown in Fig. 3-4. The main features are the lack of a Ce anomaly, an enrichment in Pr-Gd, with a marked Pr anomaly, relative to La-Ce and Er-Yb, and a depletion in Dy. Surprisingly, these patterns are more similar to those shown by deep oceanic nodules than to those recovered from shallow water (Piper, 1974).

Although the absolute abundance of the REE in the Black Sea concretions is lower than in the sediments, the shale-normalized patterns are quite similar (Fig. 3-4). Hence, there does not appear to be any preferential uptake of any one REE by the ferromanganese oxides, except perhaps a slight enrichment in Ce relative to La (Table 3-V). In addition, the Y content of the nodules does appear to be enriched, and as discussed previously, Fomina and Volkov (1969) suggest that this may be explained by its association with a ferric phosphate phase. The REE, on the other hand, appear to be present almost entirely in the occluded aluminosilicate debris in the concretions.

Total concentrations of REE in Loch Fyne concretions are similar to those in the Black Sea (Table 3-V). The shale-normalized REE patterns show both similarities and differences to those in Black Sea (Fig. 3-4); Pr is similarly enriched relative to the other REE, whereas Nd-Gd are depleted.

Glasby (1973a) has argued that the incorporation of REE into ferro-manganese nodules is by way of direct precipitation of the RE hydroxides from sea water rather than by the occlusion of detrital materials. While this

TABLE 3-V

Yttrium and rare earth contents of ferromanganese concretions and associated sediments from shallow marine environments[1]

	1	2	3	4
La	6.2	17.0	18.2	5.85
Ce	12.0	28.2	29.7	11.36
Pr	2.5	6.7	7.0	2.62
Nd	8.2	12.5	12.0	6.01
Sm	2.5	2.7	3.7	0.96
Gd } Eu }	2.5	2.7	3.7	0.89 0.19
Tb } Y }	14.7	12.7	11.0	0.24
Cy	.5	.5	.4	0.94
Ho				0.08
Er	.7	.8	.7	0.30
Yb	.6	.8	.7	
Ce/La	1.93	1.66	1.63	1.94

[1] All values in ppm.

1: Black Sea concretions, mean of 4 samples (Fomina and Volkov, 1969).
2: Black Sea surface oxidized sediment, mean of 4 samples (Fomina and Volkov, 1969).
3: Black Sea subsurface reduced sediment, mean of 4 samples (Fomina and Volkov, 1969).
4: Loch Fyne concretions (Glasby, 1973a).

may be true for oceanic nodules, as also argued by Piper (1974), where REE concentrations are often several orders of magnitude greater than in pelagic sediments, it is probably not so important in shallow-water concretions. Moreover, the patterns obtained for the Black Sea concretions and sediments (Fig. 3-4) support this latter contention.

Rates of accretion of shallow marine ferromanganese concretions

Estimates of 0.01 to 1.0 mm year^{-1} have been obtained by Manheim (1965) for the accretion rate of shallow marine ferromanganese concretions. These values were obtained from considerations of minimum permissible growth rates in areas where post-Pleistocene sedimentation has been established for a known time period and from the observations that the nodules would not survive burial below the uppermost oxidized sediment layer.

Allen (1960) obtained an estimate of the deposition rate of Mn on living mollusc shells from the Clyde Estuary by measuring the thickness of

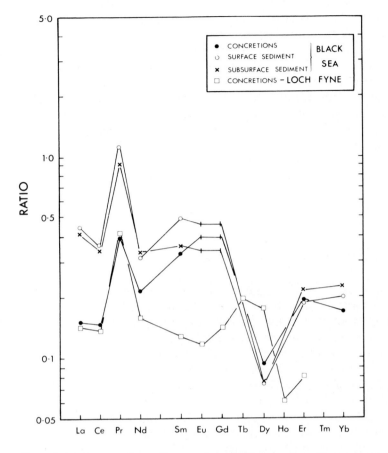

Fig. 3-4. Rare-earth patterns for shallow-marine ferromanganese concretions and sediments. Data taken from Fomina and Volkov, 1969, for the Black Sea, and from Glasby, 1973a, for Loch Fyne and normalised to average shale according to the method of Piper, 1974.

manganese crusts on shells of known age. Deposition rates of 1 mm in 29—60 years were obtained. Calvert and Price (1970) suggested that if these rates could be applied to the manganese nodules in Loch Fyne in the Clyde Estuary, the nodules would be 500 to 1,200 years old.

Radiometrically determined growth rates of Loch Fyne and Jervis Inlet ferromanganese nodules have been obtained by Ku and Glasby (1972). Using ^{230}Th and ^{231}Pa methods, they obtained accretion rates of about $0.3 \cdot 10^{-3}$ mm year^{-1} for both areas. Although these rates are considerably greater than those found in deep-sea nodules, of the order 1 to $40 \cdot 10^{-6}$ mm year^{-1} (Bender et al., 1966; Barnes and Dymond, 1967; Ku and Broecker, 1969), they are also much lower than rates estimated by Manheim (1965) for shallow marine concretions. These rates give maximum ages for the concretions of 10,000—12,000 years. In the case of Loch Fyne, the area of concretion formation was almost certainly under ice during the period (Sissons, 1967) when sea level was 50 m lower than at present. Radiocarbon dates of organic matter in the sediments from the concretion site (S. E. Calvert and N. B. Price, unpublished data) yield sediment accumulation rates of ~ 0.08 mm year^{-1}. This means that the concretions would be buried below the upper oxidized zone in the sediments (~ 10 cm thick) in 1,250 years. Therefore, it seems reasonable to conclude that the accretion rate of the Loch Fyne concretions is actually higher than $0.3 \cdot 10^{-3}$ mm year^{-1} as given by Ku and Glasby (1972) on the basis of their maximal concretion ages. The rate may be nearer $1 \cdot 10^{-2}$ mm year^{-1} on the basis of the Recent sedimentation rates and the fact that the concretions would not survive burial (see discussion on diagenesis, p. 82).

LACUSTRINE FERROMANGANESE CONCRETIONS

The occurrence of iron- and manganese-bearing concretions in lakes in northern temperate latitudes has been known for a considerable time. The so-called "bog ores" and "lake ores", essentially iron-rich ferromanganese concretions, have been mined for at least 2,000 years and used as the raw material for the smelting industry in Sweden (Naumann, 1922). A very extensive literature on their occurrence, composition and mode of formation exists in Sweden, Finland and the Soviet Union, much of which is given by Naumann (1922) and Manheim (1965).

More recently, the compositions of lacustrine ferromanganese concretions from North America have been investigated after a long period of neglect following their discovery in the latter part of the last century (see Kindle, 1932). In this case, ferromanganese concretions were also used in the smelting industry for the production of cast iron and ferromanganese (Vogt, 1906; Harder, 1910).

Distribution

Ferromanganese concretions are widely distributed in the lakes of Karelia, the Kola Peninsula, Finland, southern Sweden and southern Norway. They occur in areas where disseminated deposits of iron and manganese, "bog manganese", are also abundant in marshes, water courses, bogs and shallow ponds. The formation of *bog lakes* requires abundant precipitation throughout the year, high humidity, low soil temperatures, reduced evaporation and run-off and abundant plant growth. These particular physiographic and climatic conditions are found in a very large area around the margins of the Pleistocene ice sheet. The water bodies are generally fairly shallow, the surrounding areas being wholly or largely composed of peat deposits and which with time become filled with bog vegetation.

In southern Sweden, the distribution of lake concretions and bog iron and manganese deposits is shown by Naumann (1930, fig. 34). They appear to be confined to areas where the bedrock is primary silicate rock or where glacial outwash sands and gravels are common and where the land escaped the first post-Pleistocene marine incursion (the Yoldia Sea, 9,600 yrs BP).

In North America, the early discoveries of lacustrine concretions have been discussed by Moore (1910) and Kindle (1932, 1935 and 1936). When first discovered in some Nova Scotian lakes, flat concretions were positively identified as crude aboriginal pottery (Honeyman, 1880).

The best documented occurrences of ferromanganese concretions are to be found in Nova Scotia (Kindle, 1932, 1935, 1936; Beals, 1966; Harriss and Troup, 1969, 1970; Terasmae, 1971) and in the Great Lakes region (Rossmann and Callender, 1969, 1970; Cronan and Thomas, 1970, 1972; Calender, 1970; Damiani et al., 1973; Moore et al., 1973; Mothersill and Shegelski, 1973; Sly and Thomas, 1974) (Fig. 3-5). Other isolated occurrences are known in Loughborough Lake, Ontario (Kindle, 1932), Mosque Lake, Ontario (Harriss and Troup, 1969, 1970; Terasmae, 1971), Lake Schebandowan, Ontario (Carpenter et al., 1972), Trout Lake,

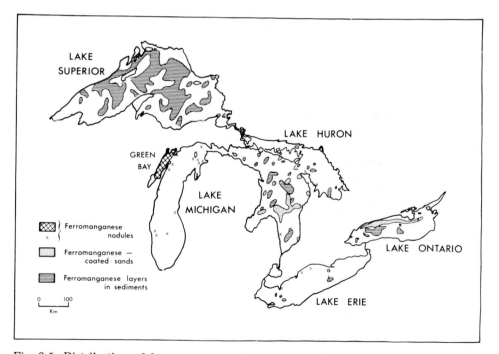

Fig. 3-5. Distribution of ferromanganese deposits in the Great Lakes (modified from Sly and Thomas, 1974). Ferromanganese layers occur at a depth of 2—5 cm in the sediments. Only the occurrences of nodules are indicated in Lake Michigan.

Wisconsin (Twenhofel et al., 1945), Lake Ossipee, New Hampshire (Kindle, 1935), Lake George, New York (Schoettle and Friedman, 1971), Oneida Lake, New York (Gillette, 1961; Dean, 1970; Dean et al., 1973; Ghosh and Dean, 1974), Chatauqua Lake, New York (Clute and Grant, 1974), Lake Champlain (Johnson, 1969) and the Minnesota lakes (Zumberge, 1952). This list is by no means exhaustive and there is every reason to suppose that it will be extended in view of the recent interest in the economic potential of the deposits.

The occurrence of iron-rich micro-concretions in the Recent sediments from Lake Malawi has been described by Müller and Förstner (1973). The concretions occur in the peripheral parts of the lake shallower than the anoxic hypolimnion, and contain poorly crystalline iron oxide, nontronite and vivianite.

Lacustrine ferromanganese concretions appear to be associated with a far more restricted range of sediment types than their shallow-marine counterparts. Most commonly, they are found on gravels and coarse-grained sands. This appears to be the case in all the lakes in Nova Scotia examined by Kindle (1932, 1935, 1936), and Harriss and Troup (1969), in Lake Ontario (Cronan and Thomas, 1970), and in the majority of the lakes in Karelia (Strakhov, 1966; Shterenberg et al, 1966; Varentsov, 1972a, b). In Swedish lakes and in Lake Michigan, where the information is much more extensive than elsewhere, concretions are found on well-sorted sands and sandy muds but not on finer-grained sediments (Ljunggren, 1953; Rossmann, 1973).

Morphology and size

The morphology of lacustrine ferromanganese concretions, judging by some of the older descriptions, would appear to cover a very wide range. In fact, local names have been applied to different varieties of the basic forms of concretions, often by ore miners of the last century. Thus, Bohnenerze and Russkugeln are forms of spherical nodules (Figs. 3-6, C and D) differing only in size, the range being 1—4 cm in diameter (Naumann, 1930). Gelderze, Kuchenerze, Schilderz and Kantenerz are all apparently different forms of flat or disc-shaped concretions (Fig. 3-6, A and B). The size range here is approximately 1—15 cm in diameter. In addition, Naumann (1922) illustrates a third type of concretion, namely tube- or pipe-shaped objects which are probably associated with root systems of aquatic vegetation in some Swedish lakes.

In the North American lakes, the concretions are either spherical or disc-shaped. Kindle (1932) refers to the flattened variety as "pancake-shaped" and illustrates several examples from Ship Harbour Lake, Nova Scotia having convex and concave sides (Fig. 3-6). Some discs also have a rock nucleus and would correspond to the girdle-shaped concretions commonly found in the Baltic Sea (Winterhalter, 1966).

Lacustrine concretions appear to have delicately mammillated surfaces, caused to some extent by the presence of sand impurities. They are also internally laminated on a fine scale, the laminae being composed of alternating iron- and manganese-rich bands (Harriss and Troup, 1969).

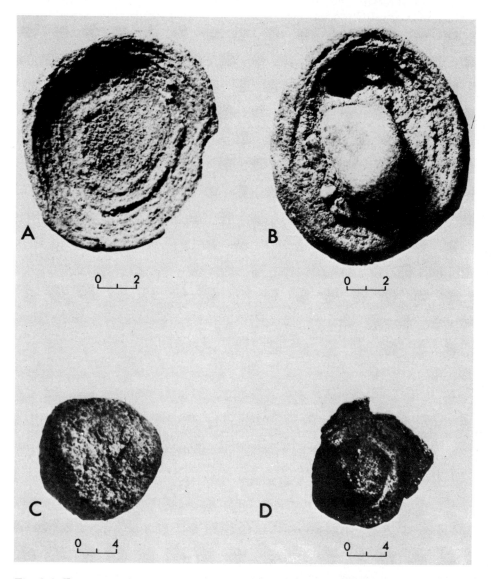

Fig. 3-6. Ferromanganese concretions from lakes. A. Flat "penny ore". B. Girdle around a pebble. C and D. Spherical nodules or "pea ores". Scales in cm. A and B from Kindle, 1935; C and D. from Naumann, 1930.

Mineralogy

Data on the mineralogical composition of lacustrine ferromanganese concretions are scanty, most reports providing only a brief indication of the probable phases present. The most extensive study is that of Ljunggren (1955b) who investigated a suite of Swedish concretions by differential thermal and X-ray diffraction methods. The phases identified were δMnO_2, having reflections at 2.44, 2.20 and 1.41 Å, in manganiferous concretions, and goethite in iron-rich concretions.

In Green Bay, Lake Michigan, Rossmann (1973) reports the presence of todorokite (10 Å manganite), birnessite (7.25—7.56 Å), psilomelane and goethite. Psilomelane is also reported from concretions from Lake Pinnus-Yarvi, Karelia (Shterenberg et al., 1966). Concretions from some areas of Lake Ontario contain goethite (Damiani et al., 1973). In Lake George, New York, concretions contain goethite with one manganese-rich sample containing what is described as "birnessite (δMnO_2)" having reflections at 7.27, 2.44 and 1.41 Å (Schoettle and Friedman, 1971). Bowser et al. (1970) also report goethite and birnessite (δMnO_2) in nodules from Wisconsin and Michigan lakes.

In contrast, concretions from Grand Lake and Ship Harbour Lake, Nova Scotia, Mosque Lake, Ontario (Harriss and Troup, 1969) and parts of Lake Ontario (Cronan and Thomas, 1970) contain no crystalline Mn or Fe phases detectable by conventional X-ray diffraction techniques. These concretions are described as amorphous.

Mössbauer spectra of concretions from Lake Michigan, Lake Ontario and Lake Shebandowan, Ontario are very similar to spectra of marine ferromanganese nodules, and show that only a few percent of the total iron can be present as Fe^{2+} (Carpenter and Wakeham, 1973).

Chemical composition

The chemical composition of lacustrine ferromanganese concretions is summarised in Table 3-VI. Complete or composite analyses are available only for the Karelian, Finnish and Swedish lakes. On the basis of these analyses, the concretions contain terrigenous detritus with a high Si/Al ratio probably indicating admixed sand-grade material.

Major elements
The data in Table 3-VI show that the Mn and Fe contents of lacustrine ferromanganese concretions are highly variable. Certain lakes contain both varieties. The total range in the Mn/Fe ratio is from approximately 0.05 to 4.6, the lowest values coming from Swedish and Finnish concretions (Ljunggren, 1955a) and the highest values coming from the Great Lakes (Rossmann, 1973).

The variation in the Mn/Fe ratio and its relationship to associated sediment type are well illustrated by the data on Green Bay, Lake Michigan presented by Rossmann (1973). Nodules collected from sandy sediments have ratios < 1, whereas concretions associated with sandy muds have much more variable ratios, ranging up to 4.6 (Fig. 3-7; Table 3-VII). This appears to be a similar relationship to that observed for the Baltic Sea by Winterhalter (1966) and illustrated in Fig. 3-2.

Apart from Mn and Fe, only P appears to be enriched, relative to Al, in lacustrine concretions, over the likely contributions from aluminosilicate impurities. A high degree of correlation between P and Fe is shown by the data on the compositions of Swedish lake concretions given by Naumann (1922).

Minor elements

The wide variations in the major element compositions shown in Table 3-VI are also reflected in the minor element data. Concretions from different lakes have, for example, Ba and Zn contents which range over an order of magnitude while Cu and Pb contents range over 2 orders of magnitude. Part of this variability may be due to the variety of analytical methods used and also to the manner of reporting analyses, many workers being concerned with the composition of the acid-soluble fractions rather than the entire samples.

The concentrations of most of the minor elements in Table 3-VI are similar to those in shallow marine concretions (Table 3-II). In Green Bay, Lake Michigan, however, Ba and Zn contents are present in higher concentration than marine forms. The high Ba values are readily explained by the presence of psilomelane in these concretions (Rossmann, 1973) whose composition is given as $BaMn^{IV}{}_8O_{16}(OH)_4$ (Palache et al., 1944). Judging by the analyses of pure psilomelane given by Palache et al. (1944) and the Ba contents of the Green Bay concretions, it can be shown that they contain approximately 16% psilomelane at a maximum if all the Ba is held in this mineral. Concentrations of Ba in some Swedish lake concretions (Ljunggren, 1955; Manheim, 1965) and oxidate crusts from the English lakes (Gorham and Swaine, 1965) are similar to those in shallow marine concretions.

Apart from Ba, enrichments of minor elements in lacustrine concretions cannot be investigated in the absence of complete chemical analyses of the concretions and their associated sediments.

Associations of minor elements with the Mn or Fe phases of lacustrine concretions can be seen in a very general way from the data in Table 3-VI. Thus, Co, Ni and Zn appear to be present in higher concentrations in Mn-rich concretions and this is supported by a study of inter-element correlations in Lake Ontario (Cronan and Thomas, 1972).

TABLE 3-VI

Chemical composition of lacustrine ferromanganese concretions[*1]

	1	2	3	4	5	6	7
Si	7.85	6.86					
Al	1.32	1.11					
Ti	0.09	0.08	0.30				
Fe[*2]	35.63	38.15	15.14	22.97	19.95	20.76	17.2
Ca	1.21	0.37		1.18		1.17	
Mg	0.45	0.23		0.35		0.25	
Na	0.08		0.64	0.49		0.03	
K	0.17	0.06	1.62	0.63		0.14	
P	0.29	0.22	0.28				
Mn	4.73	2.86	7.25	6.94	8.75	9.15	33.8
S	0.03	0.03	0.26				
C	1.4	1.5	0.76	1.05			
CO_2	0.28			1.57			
As					136	519	
Ba	1,000		2,912		8,115	10,326	
Cd	10				4		
Ce					161		
Co	80	130	34	198	69	116	198
Cr	10	10	34			24	
Cu	40		12	55	9	26	12
La			29		51		
Mo	30	50	10			36	
Ni	40	40	26	240		239	272
Pb	27		2,551				25
Rb			72				
Sr	300		59			148	
V	10	10	58	1	1		
Y			34				
Zn	50		1,112	263	205	324	1,633
Zr			69				

[*1] Major elements as wt.%, minor elements as ppm.
[*2] Total Fe.

   1: Swedish lakes, estimated average composition (from Manheim, 1965).

   2: Karelian—Finnish lakes, estimated average composition (from Manheim, 1965).

   3: Oxidate crusts from lakes Windermere and Ullswater (from Gorham and Swaine, 1965). Mean of 8 analyses, except for Cu(6) and Mo(4).

   4: Green Bay, Lake Michigan (from Rossmann and Callender, 1969). Mean of 23 analyses of HNO_3—H_2O_2 extracts.

   5: Green Bay, Lake Michigan (from Edgington and Callender, 1970). Mean of 6 analyses of total samples.

   6: Green Bay, Lake Michigan (from Rossmann, 1973). Mean of 52 analyses of HCl—H_2O_2 extracts.

8	9	10	11	12
				4.89
				1.15
				0.44
16.0	40.0	20.0	33.52	23.33
		1.3		0.97
		0.9		0.06
		0.08		0.33
		0.24		0.33
				0.001
27.0	15.7	20.5	3.57	21.94
				1.07
				2.39
				10,300
222	135	305	220	110
				2
8	10	90	1,314	4
136	95	725	702	12
21	24			4
				6
511	250	460	1,177	181

7: Grand Lake, Nova Scotia (from Harriss and Troup, 1969). Mean of 14 analyses for Mn and Fe and 12 analyses for the minor elements.

8: Ship Harbour Lake, Nova Scotia (from Harriss and Troup, 1969). Mean of 5 analyses.

9: Mosque Lake, Ontario (Harriss and Troup, 1969). Mean of 2 analyses for Mn and Fe. One analysis only for minor elements.

10: Lake Ontario (from Cronan and Thomas, 1970). Mean values from single nodule site; number of analyses not specified. Values represent compositions of acid-leached (50% hot HCl) fractions.

11: Lake George, New York (Schoettle and Friedman, 1971). Mean of 7 analyses. Analytical methods not given.

12: Eningi—Lampi Lake, Central Karelia (Varentsov, 1972a). Mean of 8 analyses.

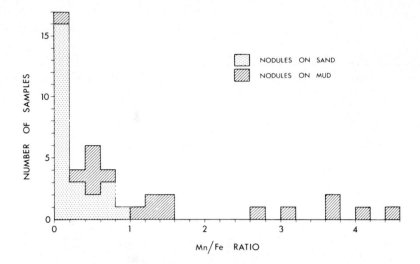

Fig. 3-7. Mn/Fe ratios of ferromanganese concretions occurring on sands or muds in Green Bay, Lake Michigan. Data from Rossmann, 1973.

TABLE 3-VII

Chemical composition of ferromanganese concretions from Green Bay, Michigan, associated with sandy and muddy sediments*[1]

	Concretions on sands*[2]	Concretions on muds*[3]
Mn	5.0	14.4
Fe	28.6	10.6
Ca	0.83	1.75
Mg	0.14	0.37
K	0.09	0.21
Na	0.03	0.04
As	596	288
Ba	7,626	13,760
Co	98	127
Cr	26	23
Cu	16	35
Mo	25	48
Ni	99	456
Sr	114	196
Zn	252	429

*[1] Major elements in wt.%, minor elements in ppm. Data from Rossmann, 1973.
*[2] Mean of 25 samples.
*[3] Mean of 17 samples.

TABLE 3-VIII

Rare earth element composition of ferromanganese concretion from Lake Shebandowan, Ontario[*1]

	ppm	Normalized to shale[*2]
La	18.5	0.45
Ce	38.0	0.46
Nd	16.5	0.44
Sm	3.2	0.43
Eu	0.68	0.42
Tb	0.55	0.45
Yb	1.48	0.42
Lu	0.26	0.43

[*1] Unpublished data of D. Z. Piper, University of Washington, Seattle, Wash.
[*2] See Piper, 1974.

In the concretions from Green Bay, Lake Michigan (Rossmann, 1973), Ba, Co, Cu, Ni, Mo, Sr and Zn are all positively correlated with Mn while As and Cr are correlated with Fe. These correlations, except for Pb, Zn and Cu, are consistent with those found for many shallow marine concretions (see Calvert and Price, 1970). Barium and Sr may be present in the 10 Å manganite lattice whereas the other cations may substitute for Mn^{2+}. The correlation between As and Fe is precisely that found for shallow marine nodules while the association of Cr with Fe is probably due to their association in iron-bearing terrigenous detritus.

Rare earth elements

The rare earth element (REE) composition of lacustrine ferromanganese concretions has not been reported. Table 3-VIII presents an analysis of a concretion from Lake Shebandowan, Ontario, kindly provided by D. Z. Piper. Total REE content is low and the Ce/La ratio is approximately the same as that in marine sediments (Haskin and Gehl, 1962; Piper, 1974) and shales (Haskin and Haskin, 1966). The shale normalized pattern (Table 3-VIII) is featureless, there being no enrichments or depletions among the series.

Accretion rates

Limits on the growth rates of lacustrine ferromanganese concretions are given by the length of time the lakes have been ice-free. Since all occurrences described so far are in northern temperate latitudes, this time period is

probably not more than 10,000 years, the end of the Gothi-glacial period
(Charlesworth, 1957). For a 1 cm diameter concretion, this gives a minimum
accretion rate of $5 \cdot 10^{-4}$ mm year^{-1}. This may be contrasted with an estimate
of 0.1—1.5 mm year^{-1} by Harriss and Troup (1969), for concretions in lakes
in Nova Scotia and Ontario, assuming that the fine laminations in the
concretions represent annual growth increments.

Such high accretion rates are consistent with the notion held by the
Scandinavian bog ore miners (malmfiskarens) that after a given lake has been
cleared of concretions, mining could be resumed after 30—50 years as a new
"crop" of concretions would have formed. Naumann (1922) discounts this
belief as an exaggeration of the possible growth rates and the actual rates
obviously lie somewhere between these wide extremes.

Krishnaswami and Moore (1973) have reported estimates of accretion
rates of concretions from Lake Alstern, Sweden and Oneida Lake, New York
using the ^{226}Ra method. The rates obtained are $1.4 \cdot 10^{-3}$ mm year^{-1} and
$2.6 \cdot 10^{-3}$ mm year^{-1}, respectively. The authors point out that these rates are
surprisingly low in view of the geological occurrence of the concretions. As
in the case of the shallow marine and oceanic nodules, a mechanism or
mechanisms for maintaining concretions at, or close to, the surface of the
sediment, which accumulates at a higher rate, is required if these accretion
rates are accepted.

MANGANESE CARBONATES IN SHALLOW MARINE AND LACUSTRINE SEDI-
MENTS

Mixed manganese—calcium carbonates have recently been found in several
shallow marine and lacustrine environments, and are of interest because of
their frequent occurrence together with ferromanganese oxides. The locations and compositions of these phases are given in Table 3-IX. The
carbonates in the Baltic Sea were identified on the basis of the compositions
of total sediments or of sediment leachates and the relationship between Ca
and CO_2 contents of the total sediments. This latter approach, combined
with X-ray diffraction techniques, has also revealed the presence of
manganoan carbonate in Oslo Fjord, but a specific composition is not
available (Doff, 1970).

Shterenberg et al. (1968) obtained sufficient material for a detailed X-ray
diffraction analysis of a manganese carbonate from ferromanganese nodules
in the Gulf of Riga. The composition of the phase was deduced using the
d-spacing of the strongest line and the compositional diagram of Mn, Mg and
Ca carbonates given by Goldsmith and Graf (1960). Note that this method
can provide only a compositional range, and in this particular case the
composition given by Shterenberg et al. (1968) does not correspond with the
X-ray data given.

TABLE 3-IX

Composition of manganoan carbonates in recent sediments

Locality	Composition	Source
A. Marine:		
Baltic Sea	$(Mn_{70}Ca_{30})CO_3^-$	Manheim, 1961a
	$(Mn_{60}Ca_{32}Mg_8)CO_3$	
Baltic Sea	$(Mn_{56.8}Ca_{25.5}Mg_{9.7}Fe_{8.0})CO_3$	Hartmann 1964
	$(Mn_{90}Ca_{10})CO_3$	E. Suess, pers. comm., 1975
Eastern Pacific	$(Mn_{50-80}Ca_{50-20})CO_3$	Lynn and Bonatti, 1965
Gulf of Riga	$(Mn_{40}Ca_{25}Mg_{35})CO_3$	Shterenberg et al., 1968
Loch Fyne	$(Mn_{47.7}Ca_{45.1}Mg_{7.2})CO_3$	Calvert and Price, 1970a
B. Lacustrine:		
Lake Pinnus-Yarvi, Karelia	$(Mn_{50.6}Ca_{45.3}Fe_{4.0})CO_3^-$	Shterenberg et al., 1966
	$(Mn_{34.2}Ca_{28.9}Fe_{37.1})CO_3$	
Green Bay, Lake Michigan	$(Mn_{64}Ca_{32}Fe_4)CO_3^-$	Callender, 1973
	$(Mn_{81}Ca_{16}Fe_3)CO_3$	

The composition of the manganoan carbonate from Loch Fyne is based on a complete chemical analysis of a large, reasonably pure concretion (Calvert and Price, 1970a). X-ray diffraction data confirm this composition. The carbonate appears to occur preferentially in the coarser-grained horizons of the sediments in the loch, as pure concretions and as the cement in aggregates of ferromanganese concretions. There has evidently been some migration and precipitation of carbonate into more porous, organic-free horizons in the sediments.

Two examples of manganoan carbonate from lake sediments, shown in Table 3-IX, occur in intimate association with ferromanganese nodules, and chemical and X-ray diffraction analyses are available. It is noteworthy that the lacustrine carbonates contain no Mg, but do contain Fe, in contrast to most of the marine examples. Damiani et al. (1973) report the presence of rhodochrosite in concretions from Lake Ontario with supporting electron and X-ray diffraction data, but chemical analyses are not available to check the presence of Mn in the lattice.

The formation of manganoan carbonate in reduced sediments from the eastern Pacific was used by Lynn and Bonatti (1965) as confirmation of the presence of manganese in the lower valency state in such sediments. The presence of carbonate phases having similar compositions in reduced basin sediments of the Baltic Sea and Oslo Fjord confirms this association. Presumably the carbonate is precipitated where interstitial Mn^{2+} and $CO_3{}^{2-}$ activities are relatively high; judging by the high dissolved Mn concentrations observed in many shallow marine sediments (see Duchart et al., 1973), it would seem that the carbonate activity is the critical control (see also Robbins and Callender, 1975).

The intimate association of manganoan carbonate with ferromanganese nodules, in the Gulf of Riga, Loch Fyne, Lake Pinnus-Yarvi and Lake Michigan, demonstrates that completely reduced conditions, that is with reduced sulphur species present, are not required for the formation of this material. Rather, conditions close to the oxidized/reduced boundary in the sediments, where a supply of Mn^{2+} from below is available (see p. 83), are occasionally adequate for the precipitation of concretionary manganoan carbonate. The mechanism involved warrants further study.

GEOCHEMISTRY OF MINOR ELEMENT ENRICHMENTS AND DEPLETIONS IN SHALLOW MARINE AND LACUSTRINE FERROMANGANESE CONCRETIONS

All types of ferromanganese concretions, including oceanic, shallow marine and lacustrine varieties, appear to have higher concentrations of minor elements compared with terrigenous sediments. Some of these relative enrichments have been noted on pp. 54 and 65. However, the degree of enrichment of minor elements in ferromanganese concretions is different for shallow marine and oceanic forms. The much lower concentrations of some elements in the shallow marine concretions compared with oceanic nodules have been discussed by Manheim (1965) and Price (1967). By contrast, the somewhat higher concentrations of some other minor elements in shallow marine concretions, mentioned on p. 57, have not previously been discussed.

In the discussion which follows, two somewhat separate problems will be examined, namely the problem of enrichments of minor elements in marine and lacustrine ferromanganese concretions, which involves the adsorptive behaviour of manganese and iron oxides with respect to cations and anions, and the significantly lower concentrations of some minor elements in shallow marine and lacustrine concretions compared with oceanic nodules.

Adsorption of minor elements by manganese and iron oxides

The ability of both ferric and manganese oxides and/or hydroxides to co-precipitate efficiently cations and anions from aqueous solution is well known. An extensive literature is available on the adsorptive capacities of ferric oxides (e.g., Kurbatov et al., 1951; Duval and Kurbatov, 1952; Dyck, 1968; Hingston et al., 1968a, b, 1970; Tewari et al., 1972) and manganese oxides (e.g., Kozawa, 1959; Morgan and Stumm, 1964; Grasselly and Hetenyi, 1968; Murray et al., 1968; Posselt et al., 1968a, b; Tewari et al., 1972; Loganathan and Burau, 1973; Anderson et al., 1973; Varentsov and Pronina, 1973; Murray, 1975a, b). These properties are used moreover in several analytical schemes for the pre-concentration of a wide range of trace elements from sea water, including Mo and W (Ishibashi et al., 1953), As, Co, Cr, Mo, Ni, Se and W (Ishibashi et al., 1962), Se (Chau and Riley, 1965), V (Chan and Riley, 1966b), Mo (Chan and Riley, 1966a), Cr (Chuecas and Riley, 1966) and Co (Fukai, 1968).

The co-precipitation or scavenging of minor elements from sea water onto ferromanganese oxides was discussed by Goldberg (1954) and the different behaviour of the iron and manganese oxide components of oceanic nodules was deduced by examining correlations between minor elements and iron and manganese. Subsequently, it has become abundantly clear that the behaviour of trace and minor elements is intimately linked to the cycling of manganese and iron oxides in soils (Jenne, 1968) and in natural waters (Stumm and Morgan, 1970).

Hydrous oxides of manganese and iron carry surface charges which vary in sign and in intensity with the pH of the medium. The potential-determining ions are principally H^+ and OH^- ions because of the acid-base equilibria involved in the hydrolysis of aquo-metal ions (Morgan and Stumm, 1965; Atkinson et al., 1967). Solid metal hydroxides and oxides exhibit ampho-teric behaviour and show strong tendencies to interact specifically with both anions and cations, anionic adsorption taking place where the metal oxide is negatively charged and cations being adsorbed where the metal oxide is negatively charged. The properties of manganese oxides have been of particular interest in view of their industrial importance and because of their use in waste-water treatment (Posselt et al., 1968b). Stumm and Morgan (1964, 1970) and Morgan and Stumm (1965) provide extensive accounts of these properties and distinguish between surface complex formation (ligand exchange) for anionic adsorption and ion exchange for cationic adsorption.

As noted previously, the charges on hydrous oxides vary with the pH of the medium. Over a pH range of about 1.5 to about 7.0, manganese and iron oxides are both negatively and positively charged. At low pH, the charge is positive and with increasing pH the charge decreases, becomes zero and then is increasingly negative. The pH value where the surfaces are uncharged is referred to as the isoelectric point or the pH of zero point of change (pH_o or pH_{zpc}). Isoelectric points of hydrous manganese and iron oxides have been summarized by Parks (1965). More recent determinations of the isoelectric points of well-characterized synthetic manganese oxides have been provided by Healy et al. (1966) and Murray (1974) and are shown in Table 3-X. The isoelectric points for δMnO_2 and Mn(II) manganite, and for hydrous MnO_2, the phases present in marine and lacustrine ferromanganese concretions, are quite low and lie outside the range of environmental pH (see Baas Becking et al., 1960). Therefore, on the basis of this evidence, and without a determination of the charge of natural hydrous manganese oxides, it may be assumed that the manganese oxide component of nodules and concretions is electro-negative.

TABLE 3-X

Isoelectric points of hydrous oxides of manganese and iron

Compound	Isoelectric point	Source
A. Manganese oxides:		
δMnO_2	1.5±0.5	Healy et al., 1966
Mn^{2+}-manganite	1.8±0.5	Healy et al., 1966
"Hydrous manganese dioxide"	2.25	Murray, 1974
αMnO_2	4.5±0.5	Healy et al., 1966
γMnO_2	5.5±0.3	Healy et al., 1966
βMnO_2	7.3±0.2	Healy et al., 1966
B. Ferric oxides:		
$\alpha FeOOH$ (synthetic)	4.2—8.6	Van Schuylenborg and Arens, 1950; Jones, 1957; Parks and de Bruyn, 1962; Parks, 1965; Hingston et al., 1967, 1968, 1972
$\alpha FeOOH$ (natural)	3.2	Parks, 1965
$\alpha Fe_2 O_3$ (synthetic)	1.9—9.04 8.45—9.27	Parks, 1965 Atkinson et al., 1967
$\alpha Fe_2 O_3$ (natural)	5.4—6.9	Johansen and Buchanan, 1957; Parks, 1965
Ferric oxides from soils	4.5—7.0	Sumner and Reeve, 1966; Jones, 1957.

The isoelectric points of hydrous ferric oxides, as summarized by Parks (1965) and including some more recent determinations, are also shown in Table 3-X. In this case, isoelectric points of both synthetic and natural oxides are available. For natural $\alpha FeOOH$ (goethite) and $\alpha Fe_2 O_3$ (haematite), the isoelectric points are much higher than the probable values of natural manganese oxides, and there appears to be a clear relationship between the crystallinity and the isoelectric point for these oxides (Van Schuylenborgh and Arens, 1950). The isoelectric points for the iron oxides in soils appear to cluster around pH 5.5 to 7.0 (Jones, 1957; Sumner and Reeve, 1966; Hingston et al., 1967, 1968a, b, 1972).

The wide range of isoelectric points for hydrous ferric oxides, and the high values frequently produced in synthetic products may explain the conclusion of Goldberg (1954) that ferric hydroxide in sea water carries a positive charge. This conclusion was based on the work of Harvey (1938) who prepared a synthetic ferric hydroxide sol having a positive charge at pH 7.4. However, he did not measure the charge on natural ferric hydroxide particles. At normal sea water pH values, natural ferric oxides and hydroxides are probably negatively charged and this is supported by the electrophoretic mobility measurements of iron particles precipitated in sea water over a range of salinities by Aston and Chester (1973). Hence, both the iron and manganese oxide components of ferromanganese oxide nodules and concretions carry negative charges at normal sea water pH values, that is around pH 8.1. However, the isoelectric points of the two oxides are quite different, that of the iron oxides being much closer to pH 8.1 and well within the range of possible environmental pH values. This means that the iron oxides carry a much weaker surface charge than manganese oxides at these pH values. The charges will vary with variable environmental pH values, and for some of the environments considered here, they can be considerable. Thus, while surface ocean water has pH values in the range 8.05 to 8.25 (Postma, 1964), subsurface pH values can fall to around pH 7.4 in the oxygen minimum (Park, 1968). In nearshore environments, especially where hydrographic restrictions lead to extensive oxygen consumption, for example in the Baltic Sea, values may be as low as pH 7.0 (Manheim, 1961a, b).

In lake waters, the total range of pH values is very large, from less than pH 1.0 in volcanic lakes in Japan to values in excess of pH 12.0 in closed alkaline lakes. In Swedish lakes, where ferromanganese concretions occur, pH values are found in the range 6.2 to 6.6 (Ljunggren, 1953). Even lower values are found in bog pools; values of 3.2 to 3.8 are known in some Russian bog pools (Shadowsky, 1923) and values of pH 3.3 have been recorded in pools in Mud Lake, northern Michigan (Jewell and Brown, 1929). The cause of such low pH values in these environments has been ascribed both to the presence of $H_2 SO_4$, produced by the oxidation of FeS_2 and by base exchange in peat (Hutchinson, 1957), and to the presence of acidic yellow organic substances (Shapiro, 1957).

In nearshore marine sediments, the pH is frequently lower than in the overlying water. For example, values of pH 6.8 have been reported in the Baltic Sea (Manheim, 1961a) and a range of pH 7.3—8.1 is reported for the pore waters of sediments off Southern California (Brooke et al., 1968). In lakes, pH values may be somewhat lower, e.g., pH 6.5 in a coastal pond in Massachusetts (Emery, 1969). Measurements of sediments in bog lakes, which have lower pH values in the waters, are not available.

On the basis of this meagre information, we conclude that the pH of waters and sediments in shallow marine and lacustrine environments of interest here varies between 3.0 and 8.0. This range does not include the isoelectric points of synthetic manganese dioxides most like those occurring in ferromanganese concretions, but does include the isoelectric points of several different synthetic and natural iron oxides and hydroxides. The adsorptive behaviour of the latter group of oxides is therefore probably quite different in oceanic and shallow marine/lacustrine environments. In oceanic waters, the iron oxides are electronegative whereas in shallow marine environments they are either very close to their isoelectric points or are electro-positive.

The importance of the charges on the constituent oxides of ferro-manganese concretions is somewhat modified by the complex mixtures of the two components. In such a mixture of oxides, the overall charge is the result of the net effect of the charges on the component oxides (Parks, 1965). Hence, in manganiferous concretions (Mn/Fe > 1.0), the manganese phase would probably dominate the charge characteristics at pH values below, but close to, the isoelectric point of the iron oxide because the charge would be greater on the manganese oxide, in turn because the difference between the isoelectric point and the actual pH is greater. On the other hand, in ferruginous concretions (e.g., in the Baltic, Black and Kara seas and in Swedish, Finnish and Karelian lakes), the iron oxide component may well dominate the charge characteristics so that the overall charge would be electro-positive below the isoelectric point of the oxide.

Data on the adsorptive capacities of manganese and iron oxides and hydroxides are provided by experimental work on synthetic oxides and on the extensive information on the composition and minor element association of the oxide fractions of soils and lake sediments. Thus, a wide variety of cations is adsorbed on synthetic hydrous manganese oxides; this includes Ba^{2+}, Ca^{3+}, Li^+, K^+, Sr^{2+} and U^{4+} (Chakravarti and Dhar, 1927), Zn^{2+} (Kozawa, 1959), Ag^+, Ba^{2+}, Ca^{2+}, Mg^{2+}, Nd^{3+} and Sr^{2+} (Posselt et al., 1968a), Ba^{2+}, Ca^{2+}, Co^{2+}, K^+, Li^+, Na^+, Ni^{2+} (Murray et al., 1968), Ag^+ (Anderson et al., 1973), Ca^{2+}, Co^{2+}, K^+, Na^+, and Zn^{2+} (Loganathan and Burau, 1973), and Ba^{2+}, Ca^{2+}, Co^{2+}, Mg^{2+}, Mn^{2+}, Ni^{2+}, Sr^{2+} and Zn^{2+} (Murray, 1975a). In all these experiments, cation adsorption increased with increasing pH over a range of pH 2.0—9.0, that is increasing adsorption above the isoelectric point.

Experimental work on synthetic hydrous ferric oxides has shown marked differences in the adsorption of cations at pH values above and below the isoelectric point. Thus, Co^{2+} and Ba^{2+} are adsorbed only at pH values greater than 6.0—6.5 (Kurbatov et al., 1951; Duval and Kurbatov, 1952), Cu^{2+} is adsorbed only at pH values greater than 5.5 (Hem and Skougstad, 1960) and Ag^+ is adsorbed in increasing quantities as the pH of a hydrous ferric oxide precipitate is increased from 4.0 to 8.0 (Dyck, 1968). In contrast, anion adsorption on ferric oxides decreases as the pH is raised up to and through the isoelectric point. Thus, Tanaka (in Morgan and Stumm, 1965) shows the tendency for Co^{2+} and Sr^{2+} to be more effectively adsorbed onto synthetic ferric hydroxide at pH values above 5.0 and 7.0, respectively, whereas $MoO_4{}^{2-}$, $PO_4{}^{3-}$, $WO_4{}^{2-}$ and $VO_4{}^{3-}$ are only co-precipitated effectively at pH values lower than 5.0, 11.0, 8.0 and 10.0, respectively. Kim and Zeitlin (1969) have also shown that the co-precipitation of Mo by precipitated ferric hydroxide in sea water is markedly dependent on the pH, the yield decreasing from pH 4.0 to 8.5. The hydroxide was shown to be positively charged between pH 4.0 and 7.5 but the charge density decreased with increasing pH.

Extensive data on the adsorption of a wide range of anions onto ferric hydroxides have been provided by soil scientists concerned with the availability of trace elements to plants. Thus, Cl^-, F^-, $AsO_4{}^{3-}$, $PO_4{}^{3-}$, $MoO_4{}^{2-}$, $SeO_3{}^{2-}$, and $SO_4{}^{2-}$ are all efficiently co-precipitated by ferric hydroxide at pH values lower than the isoelectric point (Jones, 1957); Riesenauer et al., 1962; Hingston et al., 1968a, b; 1970, 1972). The adsorption isotherms differ for undissociated, partially dissociated and completely dissociated acids because of the formation of a coordination complex on the surface oxide which requires the presence of protons either already present at pH values lower than the isoelectric point or derived from a weak acid. The experimental work is confirmed by the association of P (Bauwin and Tyner, 1957a, b; Saunders, 1965), Cl (Sumner and Reeve, 1966), Mo (Jones, 1957; Riesenauer et al., 1962; Smith and Leeper, 1969) and Si (Jones and Handreck, 1963; Beckwith and Reeve, 1963; Miller, 1967; McKeague and Cline, 1963a, b) with the iron-oxide component of soils of a wide range of types. Kirkman (1973a, b) has shown that most of the dispersed iron oxide in various New Zealand soils, readily removed by treatment with sodium carbonate and sodium dithionite, is present as fine-grained coatings on mineral particles and is most abundant in the B horizons of the soils, the zone of oxide accumulation (see Buckman and Brady, 1960). Note that the pH values in podzolic soils, where most Mn and Fe remobilization and precipitation occurs, are quite low; ranges are given as 3.48—4.80 (Buckman and Brady, 1960) and 3.9—5.5 (Swaine and Mitchell, 1960). This oxide fraction probably represents the material responsible for the extensive anion adsorption reported by other workers. In addition, Present (1971) has shown that As and Sb occur in higher concentration in the dispersed Fe-rich B horizons of some Canadian podzolic soils, whereas Ag, Cu, Pb, Zn and Zn contents are unrelated to the Fe contents.

The role of specific adsorption (i.e., the formation of chemical bonds between an ion and an adsorbent) of anions on ferric hydroxide also becomes important from the work of Hingston et al. (1967, 1970) and Breeuwsma and Lyklema (1973). They point out that some anions, e.g. F^-, SO_4^{2-} and PO_4^{3-}, have a specific affinity for the surface of ferric hydroxide and that they adsorb to a much greater degree than can be predicted from their concentrations in solution. These anions therefore enter the coordination layer by ligand exchange (see Morgan and Stumm, 1965). For strong acids which are fully dissociated, this exchange occurs only when the oxide surface is positively charged. However, in the presence of a weak acid, the anion can be adsorbed by the oxide surface provided a sufficiently large energy is available to dissociate the acid. Under these circumstances, adsorption in excess of the positive charge, and indeed on negatively charged surfaces, is possible. Adsorption maxima for different weak acids are therefore found where the proportions of undissociated acid and the anion are similar, or in other words at the pK value of the acid (Hingston et al., 1967). Thus, silicate shows maximal adsorption on goethite at pH 9.2, while phosphate adsorption envelopes show breaks at pH 7.0 and 11.5 (see Breeuwsma and Lyklem, 1973, for another interpretation).

Specific adsorption of cations on oxides is also well recognized (Murray et al., 1968; James and Healy, 1972a, b, c). For example, Morgan and Stumm (1964) and Stumm et al., (1970) have shown that the adsorption of cations by hydrous manganese dioxide is capable of reversing the surface charge, which takes place by the replacement of a proton by a divalent cation. This indicates that specific adsorption and not just electrostatic attraction is involved. Murray (1975a) has also demonstrated that there are marked selectivity sequences for adsorption of alkaline earth and transition metal cations onto hydrous MnO_2, namely $Ba > Sr > Ca > Mg$ and $Co > Mn > Zn > Ni$ over the pH range 2—8, and that there is measurable adsorption of these cations at the isoelectric point of the oxide.* Estimated specific adsorption potentials for the transition metals are large compared with the possible electrostatic contribution to the free energy of adsorption. They appear to be similar to the specific adsorption potentials for TiO_2 obtained by James and Healy (1972a, b, c).

James and Healy (1972a, b, c) explain the large specific adsorption potentials for TiO_2 as being due to the possession of a high dielectric constant (78.5) which means that the coulombic and chemical interactions, as opposed to the solvation energy, dominate the free energy of adsorption. Murray (1974) estimates that the dielectric constant for hydrous MnO_2 is

---

*Note that the isoelectric point and the zero point of charge do not occur at the same pH in the presence of specific adsorption.

approximately 32 and therefore has a much greater specific adsorption than say SiO_2 (which has a similar zero point of charge) with a dielectric constant of 4.5.

The role of metal oxides in the behaviour of minor nutrient elements in lakes has been extensively documented. Einsele (1938), Einsele and Vetter (1938), Pearsall and Mortimer (1939), and Mortimer (1941, 1942, 1971) have shown that P, Si, Fe and Mn abundancies in lake waters are all closely related to the seasonal oxidation/reduction cycle in lakes, the concentrations of these substances increasing in the lake water during periods of oxygen depletion and disappearing during oxygenation due to the formation of an oxidized microzone at the sediment surface. The distribution of the different fractions of P in lake sediments, its availability to the plankton, the fate of fertilizers and its bearing on the productivity of lakes, has been of particular interest (MacPherson et al., 1958; Harter, 1968; Frink, 1969a, b). Williams et al., (1971) have shown that the amounts of Fe and inorganic P in the sediments of several North American lakes can be chemically extracted in constant proportions, leading them to conclude that the P is present in a "short-range order Fe—inorganic P complex". The complex contains 5—10 atoms of Fe for each atom of P, this ratio exceeding that found in naturally occurring Fe phosphates such as strengite ($FePO_4 . 2H_2O$) and vivianite ($Fe_3 (PO_4)_2 . 8H_2O$). Experiments with lake muds show that phosphate is most effectively removed from solution in the pH range 5.0—6.5, a result similar to that obtained with precipitated ferric hydroxide (Ohle, 1953).

In addition to P, several other anions are also controlled by the oxidation/reduction cycle in lakes, involving their adsorption and desorption onto ferric hydroxides. They include AsO_4^{3-} (Kanamori, 1965), I^- (Sugawara et al., 1958), SO_4^{2-} (Koyama and Sugawara, 1951) and H_4SiO_4 (Kato, 1969). These relationships can be reproduced in the laboratory by using precipitated ferric hydroxide as an adsorbent, the substance being released to the water when the hydroxide is redissolved.

The information on the cation and anion exchange properties of manganese and iron oxides, and the differences in adsorptive behaviours of the two oxide components in ferromanganese concretions over the environmental pH range, provide a basis for understanding the different degrees of enrichment of minor elements in the concretions. We do not, unfortunately, have any data on the surface charge characteristics of naturally occurring manganese and iron oxides with which we can check this hypothesis.

As previously pointed out, several elements are present in shallow marine and lacustrine concretions in higher concentration than in oceanic nodules, and inter-element correlation data (see Calvert and Price, 1970a and unpublished; Callender, 1973) show that these (apart from Ba) and some additional elements are associated with the Fe-phase in shallow marine and lacustrine concretions. The list includes As, P, Ti, V, Y, and Zr. These elements occur in sea water either as anions or as hydroxy complexes (Sillén,

1961; Riley and Chester, 1971, Table 4-I). On the other hand, elements occurring as cations or as positively charged ion pairs (Ba, Co, Cu, Ni, Pb and Zn) are associated with the Mn-rich phase of the concretions. From these observations, it can be inferred that the charge characteristics of the manganese and particularly the iron oxide phase of the concentrations may be of fundamental importance in controlling the adsorption of minor elements from sea and lake waters, as argued previously by Goldberg (1954).

Cation adsorption on highly negatively charged manganese oxides and anion adsorption on weakly positively charged iron oxides under different environmental conditions may provide an explanation for the contrasting element enrichments discussed here. That specific adsorption of both cations and anions is also clearly involved is apparent from the large body of experimental work on synthetic systems. This may in fact explain the apparently anomalous behaviour of some anions (e.g. Mo) with negatively charged MnO_2 and some details of the adsorption of silicate and phosphate on ferric hydroxides. Nevertheless, the importance of the charge of naturally occurring oxides, in view of the wealth of observational data on minor element associations with Mn or Fe, and the degree of specific adsorption of cations and anions on these oxides, warrants further study.

Minor element depletions in shallow marine and lacustrine ferromanganese concretions

In view of the evidence for the enrichment of some minor elements in shallow marine and lacustrine concretions, both with respect to associated sediments and oceanic nodules, attention is now turned to the relative depletions of some minor elements, principally cations, compared with oceanic forms. Such depletions may be produced by a number of processes, including: (a) stabilization of the minor elements as insoluble phases (mainly sulphides) in subsurface reducing sediments (see discussion on diagenesis); (b) stabilization of minor elements as metal-organic complexes in shallow sediment pore waters; and (c) a lower effective co-precipitation of minor elements with ferromanganese oxides because of a higher accretion rate of concretions in shallow marine and lacustrine environments.

Manheim (1965) suggested that the explanation of the higher Co and Ni content of Baltic nodules compared with Cu, Pb and Zn may be due to the formation of sulphides of the latter group of elements within reducing sediments and their consequent immobility. Available data on the concentrations of dissolved trace metals in reduced sub-surface sediment pore waters (Duchart et al., 1973), show, however, that Cu, Zn and Pb concentrations are generally higher than those of Co and Ni. Moreover, Price (1967) has pointed out that sedimentary iron sulphides contain much more Co and Ni than Cu, Pb and Zn (see Minguzzi and Talluni, 1951) and this is easily understood on crystal chemical grounds (Mohr, 1959). It is unlikely therefore that sulphide precipitation in sediments is responsible for the distribution of minor elements in nearshore manganese nodules or consequently for the depletion of minor elements in these nodules compared with oceanic forms.

The influence of organic materials on the behaviour of minor metals in nearshore sediments has been discussed by Price (1967). He has drawn on the experimental work of Krauskopf (1956), Kee and Bloomfield (1961) and Swanson et al. (1966) and has pointed out that the transition metal and lead chelates produced by the reaction between dissolved metals and the products of degraded plant material are water-soluble rather than colloidal and that they are stable in air. The chelates so produced have theoretical stabilities which increase in the order Mn > Fe > Co > Ni > Cu, Zn > Pb (Irvine and Williams, 1953) while their ease of co-precipitation with ferric oxides decreases in the order Fe, Mn, Zn < Co, Ni, Cu and Pb (Kee and Bloomfield, 1961). Therefore metal—organic complexes, produced in nearshore sediments, having higher concentrations of labile organic material compared with oceanic sediments, may co-precipitate with surface ferromanganese oxides to varying extents, producing a general depletion in minor elements in these precipitates compared with those from the deep sea and also producing different degrees of depletion. Thus, Mo and Zn do not appear to be depleted to the same degree as Co and Ni in nearshore nodules if a comparison is made with Co, Ni, Mo and Zn contents of oceanic nodules from the Atlantic or the Pacific (cf. Chapter 2). The behaviour of Zn is entirely in keeping with the work of Kee and Bloomfield if organic association is involved. Similarly, the association of Mo with organic materials has been pointed out by Krauskopf (1956) and its ready co-precipitation with manganese oxides is well known.

The significantly higher accretion rates of nearshore ferromanganese nodules compared with those from the deep sea (Ku and Glasby, 1972) may also have an effect in causing minor metal depletion (cf. Chapter 2). This could only be an important effect if the rate of minor metal uptake was relatively constant compared with overall growth rates of manganese and iron oxide layers. In view of the varying degrees of minor metal depletion and enrichment discussed above, this effect would appear to be minimal. However, if the variable accretion rates also produce manganese and/or iron oxide phases of different crystallinities, as demonstrated experimentally by Bricker (1965), there could be a control on minor element content of the oxides because of the different degrees of substitution of elements in products of variable structural order.

FORMATION OF SHALLOW MARINE AND LACUSTRINE FERROMANGANESE CONCRETIONS

The mode of formation and the geochemistry of shallow marine and lacustrine ferromanganese concretions have been examined from many

different viewpoints in recent publications. It is now generally accepted that
these concretions, in contrast to open oceanic nodules, are formed as a
consequence of diagenetic reactions and recycling of metals in the pore
waters of the sediments, a mechanism first formally stated by Manheim
(1965). The role of volcanism, which features prominently in many
discussions of oceanic nodules, is not involved here as is evident from the
geological setting of the areas where the concretions occur. The role of
diagenesis in the formation of ferromanganese concretions and the problem
of the role, if any, of micro-organisms in the precipitation of manganese and
iron is discussed here.

Diagenesis in Recent sediments

Diagenetic processes affect the composition of Recent sediments in a
number of ways. In the first place, surface and subsurface sediments are
frequently quite different in composition, particularly with respect to Mn
and Fe. This has been known for some considerable time by Russian workers
(e.g., Gorshkova, 1931, 1957, 1966; Klenova, 1936a, 1938, 1960; Bruevich,
1938; Trofimov, 1939, Klenova and Pakhomova, 1940; Belov et al., 1968).
Thus, surface sediments in the North Russian seas and the Black Sea are
yellow to brown, highly oxidized sands and muds containing substantial
amounts of dispersed Mn and Fe as oxides and hydroxides, while the
subsurface sediments are greenish grey to black, reduced sediments containing very little Mn and non-silicate Fe as FeS_2 (see also Turner, 1971).

Secondly, the pore waters of many shallow water and/or nearshore
sediments contain high concentrations of dissolved metals (Debyser and
Rouge, 1956; Hartmann, 1964; Sevast'yanov and Volkov, 1967; Presley et
al., 1967, 1972; Brooks et al., 1968; Duchart et al., 1973) which are released
from buried surface oxides and which migrate to the sediment surface during
sedimentation and compaction. The interactions of elements released during
diagenesis lead to the formation of a distinctive suite of authigenic mineral
phases in both the oxidized and the reduced sediment, these mineral suites
reflecting the interstitial physicochemical environment rather than the
physiographic or hydrographic environment.

Diagenetic reactions in Recent sediments are controlled by the rate of
sedimentation, the supply of reacting materials and of organic matter, and
the rate of burial of these materials in the sediments. Under conditions of
rapid deposition, that is in shallow water and/or nearshore areas, relatively
labile organic material may be rapidly buried. This material supports a rich
bacterial flora which exhausts the interstitial molecular oxygen supply.
Kanwisher (1962) has shown, for instance, that in nearshore environments
molecular oxygen is present only within 1 cm of the sediment surface.
Within the anaerobic interstitial environment, nitrate and sulphate become
the electron acceptors for the oxidation of the organic matter. With the
reduction of sulphate by bacteria (Murray and Irvine, 1895; Miller, 1950;

Thode et al., 1951), sulphide species are produced and the redox potential becomes negative (Kaplan and Rittenberg, 1963). Under these conditions, transition elements present in their higher oxidation states are reduced to their lower oxidation states; Mn and Fe pass from solid tetravalent and trivalent forms, respectively, to soluble divalent forms.

Dissolved ferrous iron reacts with interstitial HS^- to form iron sulphides, a process that is well documented (Berner, 1964, 1967, 1970; Rickard, 1969; Roberts et al., 1969; Hallberg, 1972). The reaction of interstitial Mn^{2+} is not so well known. The occurrence of MnS in reducing sediments from the Baltic Sea has been described by Debyser (1961), and this has recently been confirmed by mineralogical and chemical analysis by E. Suess (personal communication, 1975) in the highly reducing sediments of Landsort Deep. The possible control of the concentration of Mn^{2+} by the formation of $MnCO_3$ was mentioned by Lynn and Bonatti (1965) and mixed manganese—calcium carbonates are now known to be common in some shallow-water environments (Calvert and Price, 1970a).

A particularly clear example of the control on the distribution of interstitial Mn^{2+} is discussed by Li et al. (1969) and Calvert and Price (1972). Concentration profiles frequently show subsurface maxima which are produced by a combination of solution of oxides close to the surface and precipitation of manganoan carbonate in subsurface sediments. This appears to be a quasi-steady state process governed by the recycling of Mn between solid and dissolved phases during continuing sedimentation.

The behaviour of minor metals during the burial and solution of surface oxidized sediments has been discussed by Duchart et al. (1973). Any metals associated with the surface oxides and hydroxides of Mn and Fe will be released upon burial and are therefore available for reaction or migration in the pore waters. Concentrations of several trace metals, notably Co, Cu, Ni, Pb and Zn are up to 2 orders of magnitude greater than in sea water (Presley et al., 1967, 1972; Brooks et al., 1968; Duchart et al., 1973). Such concentrations appear to be characteristic of sediments with oxidized surface and reduced subsurface layers whereas sediments which are entirely reduced (i.e., form in anoxic waters) or entirely oxidized have much lower pore water concentrations.

The precipitation of Mn and Fe oxides in surface sediments and the formation of concretionary ferromanganese oxides from expelled interstitial solutions, as stated by Manheim (1965), is a mechanism similar to, but more generally applicable than, the suggestion of Aarnio (1918) and Ljunggren (1953, 1955a) that lake concretions form by the precipitation of Mn and Fe from groundwaters migrating upwards in the porous lake sediments. Such a source of metals is also suggested by the morphologies of many concretions, flattened and girdle-shaped forms having detailed shapes that can only be produced from an underlying source (Murray and Irvine, 1894, figs. 1 and 4; Manheim, 1965, fig. 13).

A contrary view has recently been expressed by Varentsov (1973). He has pointed out, from material balance calculations, that the amounts of Mn and Fe derived from interstitial solutions of sediments in the Gulfs of Riga and Finland are insufficient to account for the quantities of oxides present in the surface sediments as nodules. The sedimentation rate in these areas is very low, not more than 20 cm of sediment accumulating since the last glaciation. Varentsov suggests that the metals are supplied almost entirely by land drainage, in solution, in organic complexes and in particulate form.

The supply of iron and manganese to shallow, temperate lakes from podzolic soils is discussed by Strakhov (1966). A considerable transport of iron is, of course, a feature of such soils (Blookfield, 1952; Buckman and Brady, 1960) and ferromanganese concretions are formed in the oxide accumulation layer (Winters, 1938; Drosdoff and Nikiforoff, 1940; Clark and Brydon, 1963; Taylor et al., 1964; Taylor and McKenzie, 1966; Sokolova and Polteva, 1968). Strakhov (1966) maintains that the conditions for podzolization, namely high rainfall together with the rapid accumulation of humus, is the prime requirement for the formation of "ore lakes" in high latitudes and lacustrine ferromanganese concretions are indeed virtually restricted to this physiographic setting.

A special case of lacustrine ferromanganese concretions is found in deep, stratified lakes where precipitation takes place around the lake margin at depths shallower than the anoxic zone. Such an example has been described by Müller and Förstner (1973) in Lake Malawi where nontronite, goethite and vivianite micro-concretions have formed from the large reservoir of dissolved Mn and Fe in the anoxic lake water. These metals have in fact been derived largely from thermal springs, as is also the case in other East African lakes (see Degens et al., 1973).

As previously discussed, some fractionation of Mn and Fe takes place during diagenesis (see also Krauskopf, 1957). Much more Fe appears to be held in reduced subsurface sediments, as the sulphide, compared with Mn so that the pore solutions become enriched in Mn relative to Fe. Surface oxides are similarly enriched (Strakhov, 1965; Sevast'yanov and Volkov, 1967; Cheney and Vredenburgh, 1968). Therefore, regardless of the composition of the original precipitated oxide materials, subsequent reactions within the sediment will lead to surface precipitates having higher Mn/Fe ratios. Ferromanganese concretions formed in rapidly accumulating nearshore areas and in lakes consequently have Mn/Fe ratios greater than those of the originally deposited dispersed Mn and Fe oxides. This explains to some extent the wide variability in the compositions of shallow marine and lacustrine concretions, from one area to another, and within any one area. In the Baltic Sea, Winterhalter (1966) has shown that concretions forming in muddy sediments, where diagenesis is much more important, have higher Mn/Fe ratios than concretions forming on coarser-grained sediment. The composition of the precipitate which forms directly from Baltic Sea water is probably most similar to that of the flat slabs having approximately 1% Mn and 21% Fe. Wherever post-depositional processes enhance the supply of Mn to the surface sediments, the resulting precipitates contain approximately 15% Mn and 17% Fe.

This reasoning also applies to marginal oceanic areas where sedimentation rates are relatively high. Smooth regional variations in the composition of ferromanganese nodules are more readily explained, therefore, by gradational variations in diagenetic and normal sedimentational processes rather than volcanism or other catastrophic events (Price and Calvert, 1970; Glasby et al., 1971; Glasby, 1973d).

The role of micro-organisms in the precipitation of manganese and iron

In addition to the precipitation of Mn and Fe by purely inorganic processes, other explanations involving the action of micro-organisms have been proposed. The discovery of small amounts of polynuclear organic materials in some marine manganese nodules led to the suggestion that micro-organisms may produce organic compounds within the nodules (Thomas and Blumer, 1964) and perhaps be responsible for the accretion of nodular Mn and Fe (Graham, 1959; Graham and Cooper, 1959). Sorokin (1972) has reported the presence of relatively large microfloral biomasses on ferromanganese nodules from the central Pacific, although no specifically Mn- and Fe-oxidizing bacteria could be found. He suggested that heterotrophic microflora utilise the organic component of organometallic complexes and thereby stimulate the oxidation of Mn^{2+} adsorbed on the surface of the nodules. While suggesting that biological activities could be involved in the precipitation of Mn and Fe on the sea floor, Sorokin (1972) nevertheless cautions against the acceptance of such agencies when compounds that can be readily oxidized inorganically are involved.

Experimental work by Ehrlich (1963, 1968) on oceanic nodules has led to the suggestion that bacteria isolated from the nodules (*Arthrobacter*) enhanced the adsorption of Mn^{2+} from culture media and promoted the oxidation of adsorbed Mn^{2+} to Mn^{4+}. Sorokin (1972) has pointed out that the experimental conditions used by Ehrlich are quite unnatural and that contamination could have been important. Moreover, although some microbiological control in the behaviour of Mn in laboratory cultures can be deduced, Ehrlich (1963) conceded that it remains to be determined whether these processes also occur in the sea. Subsequent work by Trimble and Ehrlich (1968, 1970) and Ehrlich (1971) has been concerned with demonstrating reduction of MnO_2 by nodule bacteria, the role of MnO_2-reductase and the ability of bacteria to oxidize Mn^{2+} under hydrostatic pressure. In the absence of incontrovertible evidence that bacteria are involved here, it is difficult to assess this information for systems which can be readily understood by purely physicochemical mechanisms (see W. Stumm, in Ehrlich, 1964).

Perfil'ev et al. (1965) have presented a great deal of evidence to support the contention that bacteria bring about the oxidation and deposition of Mn and Fe oxides in laboratory cultures of Karelian lake muds. The organisms principally concerned, *Metallogenium* and *Siderococcus*, are readily isolated from the oxide deposits and are well illustrated by Perfil'ev and his co-workers. It is concluded that the bacteria oxidize upward migrating Mn^{2+} and Fe^{2+} produced in the lower reduced sediment layers by the burial and solution of surface oxides. The oxides are precipitated in two distinct layers close to the sediment surface, a Mn-rich layer lying above a Fe-rich layer. The deposits are approximately 90% pure oxides, with Mn/Fe ratios generally in the range 1 - 4, and with most of the Mn present as Mn^{3+} (Shapiro, 1965). Concretionary oxides are not formed under the laboratory conditions used.

Once again, interpretation of the data is difficult since no direct evidence of bacterial precipitation is available. The critical test of conducting a sterile control experiment is not possible because the presence of sulphate-reducing bacteria is necessary in order to produce negative redox potentials suitable for the production of Mn^{2+} and Fe^{2+} ions and their supply to the sediment surface (see Gabe and Gal'perina, 1965).

CHAPTER 4

FOSSIL NODULES

H. C. JENKYNS

INTRODUCTION

The discovery of large quantities of ferromanganese nodules on the deep-sea floor during the H.M.S. *Challenger* expedition (1872—1876) aroused considerable scientific interest and ensured their niche in the annals of post-Victorian science (cf. Chapter 1). That such nodules can become entombed in sediments was proved much later by their occurrence in deep-sea cores (cf. Cronan and Tooms, 1967b; Goodell et al., 1971; Horn et al., 1972b; Margolis, 1975). Yet successful burial may be only the first step towards survival as part of the sedimentary record.

Geologists, particularly Europeans, were not slow to recognize fossil equivalents of oceanic concretions (e.g. Gümbel, 1878, in the Eastern Alps; Jukes-Browne and Harrison, 1892, in Barbados; Molengraaf, 1915, 1922 in Timor). To Gümbel (1861), in fact, must go the credit of describing fossil ferromanganese nodules before they were discovered in the Recent; in describing Lower Jurassic pelagic limestones from the Eastern Alps, Gümbel mentioned that iron and manganese were frequently concentrated as veins, plasterings, and nodular concretions. After Buchanan (1878, 1891) published an account of littoral* manganese nodules from the waters off the western coast of Scotland, fossil analogues were recognized in Russia (Sokolow, 1901). Curiously, recognition of, or interest in, fossil nodules faded during the course of the 20th century and only recently has interest in ancient occurrences been revived.

Equally curious is the literature on sedimentary manganese ores: two volumes of the proceedings of the 1956 International Geological Congress in Mexico are devoted to manganese deposits, yet scarcely a reference is made to Recent iron-manganese accumulations. But clearly at least some of the sedimentary ore deposits described are related in genesis to Recent ferromanganese deposits.

Identifiable ferromanganese nodules do appear to be relatively rare in the sedimentary record. Most Recent nodules occur in ocean basins. The "new global tectonics" requires that most ocean crust, and its sedimentary cover, is eventually consumed at a subduction zone (Isacks et al., 1968; Vine and

---

*"Littoral" is used here in the general sense of "near shore".

Hess, 1970). Clearly, the preservation potential of oceanic ferromanganese nodules is poor. Only when slivers of oceanic crust and/or their sedimentary cover are emplaced on land during continent—continent or continent—island arc collision do such nodules stand some hope of preservation. We may look therefore in certain mountain belts for red clays or radiolarian cherts to find ferromanganese nodules of oceanic provenance.

In order that manganese nodules be preserved in the sedimentary record they should be formed in a setting whose ultimate basement comprises continental crust. Littoral manganese nodules thus stand some chance of survival, if diagenesis permits. But the most favourable setting for the preservation of nodules is an essentially "oceanic" situation developed on continental crust: certain ancient pelagic sediments fulfilled this requirement. Such sediments were apparently deposited in seas which had flooded extensive parts of continents, into which negligible terrigenous material was transported, and where depositional rates were low.

In the following account, the major occurrences of fossil ferromanganese oxide nodules in sedimentary strata on land are described. They were apparently formed in three main environments: (1) deep-sea; (2) continental-margin seamount; (3) littoral (Table 4-I). It is difficult to evaluate the vast literature on sedimentary manganese ores, and in most cases description is restricted to those deposits which have been specifically recognized as equivalent to Recent ferromanganese nodules.

DEEP-SEA ENVIRONMENT

Cretaceous nodules (I)

Description and stratigraphic location. Ferromanganese nodules of Late Cretaceous age occur in a red clay matrix in the western part of the island of Timor, East Indies (Fig. 4-I). Their colour is black to brownish-black. The largest "have the size of lemons, the smallest are about equal in size to nuts" (Molengraaf, 1922). Some are "no larger than peas". Spherical or ellipsoidal form is common, although compound forms are more irregular in shape. The outer surfaces of the nodules possess tubercles. Nuclei commonly comprise radiolarian chert, around which concentric layers of ferromanganiferous material has precipitated. Radiolaria may also be included within the mineral layers themselves. Associated with the large ferromanganese nodules are micronodules (2—3 mm, El Wakeel and Riley, 1961).

The red clay that contains the ferromanganese nodules is, in fact, part of an included raft that floats in a clay olistostrome that was emplaced on Timor during the Miocene (Audley-Charles, 1972).

Mineralogy. Molengraaf (1922) did not present any mineralogical data on the Cretaceous Timor nodules. Audley-Charles (1965) commented that the

TABLE 4-I

Major occurrences of ferromanganese nodules in the sedimentary record on land, and suggested environmental settings

	Deep-sea	Continental-margin seamount	Littoral
Pleistocene			
Pliocene			
Miocene			
Oligocene	Sicily		Ukraine, U.S.S.R.
Eocene	Barbados — Washington State, U.S.A.		
Paleocene	Borneo	Western Alps, France	
Cretaceous	Timor — Alpine-Mediterr. region		
Jurassic	?Roti	Alpine-Mediterr. region	
Triassic		Alps, Balkans, Himalayas	
Permian			Montana, U.S.A.
Carboniferous		France, Germany, Austria	
Devonian			
Silurian			Israel
Ordovician			
Cambrian			Newfoundland
Precambrian			Botswana

Fig. 4-1. Concentrically laminated ferromanganese nodules in oceanic red clays of Late Cretaceous age. Near Niki Niki, western Timor. Length of rule visible is 46 cm. Photograph courtesy of M.G. Audley-Charles.

west Timor nodules contain about 20% clay, 20% chalcedonic silica and about 40% complex iron-manganese minerals.

Chemistry. The quantities of iron and manganese in these nodules are relatively constant (Mn circa 12.5%; Fe circa 21%). The Mn—Fe ratio is thus approximately 0.6. The silica and alumina contents are high (SiO_2 circa 30%; Al_2O_3 circa 14%). Ti reaches 1.3%; Ni, Cu, Co and Ba may reach several thousand ppm.

Palaeogeographic setting. The red clay that contains the ferromanganese nodules exhibits a quite extraordinary chemical and mineralogical resemblance to Recent deep-sea red clays (Molengraaf, 1922; El Wakeel and Riley, 1961). Quartz and feldspar crystals, fragments of volcanic glass and serpentinized material are present; ill-defined Radiolaria occur in small quantities; sharks' teeth are moderately abundant. There is no doubt therefore that these Upper Cretaceous red clays can be interpreted as deep-ocean sediments. They were presumably scraped off a descending lithospheric slab and embroiled in a chaotic olistostrome (Audley-Charles, 1972).

Jurassic (?) nodules

Description and stratigraphic location. Ferromanganese nodules of sup-posed Jurassic age occur on the island of Roti, to the southwest of Timor (Molengraaf, 1915). The matrix is siliceous, slightly calcareous clay, which is either red or white. The nodules are spheroidal and cake-shaped, and are of centimetre scale. Concentric structure is generally lacking, as are nuclei. Radiolaria occur dispersed throughout the nodules, and in some cases the nodules have grown out from the siliceous tests.

It seems that the red clays of Roti are also part of the same olistostrome that is present on Timor (Audley-Charles and Carter, 1972). The dating is thus suspect since the age was derived from a belemnite and radiolarian fauna in a limestone from "a complex of strata" (Molengraaf, 1915). This limestone could well be a separate block not indicative of the age of the main bulk of oceanic sediments.

Mineralogy. No mineralogical data are available for the nodules from Roti. The high Ba content (see below) might suggest the presence of psilomelane and barite.

Chemistry. Two nodules analysed from Roti yield Mn contents circa 44% and Fe contents circa 1%. The Mn/Fe ratio is thus around 40. Particularly noteworthy is the high concentration of barium, which assays at 8.2% and 10.5% (Molengraaf, 1915).

Palaeogeographic setting. From the description given by Molengraaf (1915) the nodules were apparently formed on a red clay substrate in a deep-sea setting, presumably floored by oceanic crust.

Cretaceous nodules (II)

Description and stratigraphic location. Ferromanganese nodules, either manganese-rich or iron-rich, occur in siliceous marls and radiolarian cherts of Cretaceous age in eastern Timor (Wai Bua Formation, Audley-Charles, 1965). The manganese-rich nodules are of centimetre scale and have nobbly surfaces with smooth black skins. They are hard and dense, and lack concentric banding. There is no trace of skeletal structure in the nodules.

The iron-rich nodules are larger, ranging up to 15 cm; they are dark rusty red in colour and possess well-developed concentric banding. A skeletal nucleus is recorded from one nodule.

Mineralogy. The manganese-rich nodules consist mainly of pyrolusite as fine-grained microcrystalline aggregates. The iron-rich varieties contain goethite, with pyrolusite, hausmannite and probably cryptomelane.

Chemistry. The Mn content of one manganese-rich nodule is 52%; the Fe content is 0.07%. The Mn/Fe ratio is therefore 743. The values of Cu and Ba are 4,500 ppm and 5,500 ppm, respectively: these are the highest values attained in any of the fossil nodules from Timor. A low silica content (1.1%)

also characterizes this nodule. The iron-rich nodule has a Mn content of 13.5% and a Fe content of 32%, giving a Mn/Fe ratio of approximately 0.4. In this nodule the trace-element abundances (Cu, Ni, Co) are not as great as in the Mn-rich nodule, although Ba attains 4,000 ppm. Si, Ti and Al, however, have roughly similar values in both manganese- and iron-rich types.

Palaeogeographic setting. The Wai Bua Formation apparently stratigraphically overlies shallow-marine facies (Audley-Charles, 1968). The ferromanganese nodules presumably formed in a deeply submerged continental margin setting, near or below the compensation depth for calcite, where silica-rich sediments could accumulate. The tectonic position of the Wai Bua Formation — para-autochthonous, according to Audley-Charles (1972) — would suggest that the basement was continental crust.

Oligocene—Miocene nodules

Description and stratigraphic location. Ferromanganese nodules of Late Oligocene to Early Miocene age occur in the so-called Numidian Flysch which outcrops in north-central Sicily (Alaimo et al., 1970). The nodules occur chaotically in a grey-brown clay matrix; they are of centimetre scale and built up of concentric shells.

Mineralogy. The nodules are composed of 7 and 10Å manganite, lithiophorite, goethite, haematite and quartz. The last three minerals are quantitatively the most important.

Chemistry. Microprobe analyses of the nodules reveal a disparity in the contents of Mn and Fe. Average Mn values of 40%, 27.7% and 4.2% with corresponding Fe values of 16.9%, 1.3% and 61.4% are recorded. Mn-rich and Fe-rich layers alternate. Ni is enriched (maximum value 1.75%) in those zones well endowed with Mn and poor in Fe. Ni, Zn and Cu are sorbed with Mn, and Co with Fe.

Palaeogeographic setting. The Numidian Flysch, represented by a kilometres-thick succession of redeposited clastic material and basinal clays, has been interpreted as a wedge of continental rise sediments deposited at the foot of the African platform (Wezel, 1970a, b). The basement is thus assumed to be continental crust depressed to considerable depths.

Other occurrences

Molengraaf (1915) records ferromanganese nodules from East Borneo in red clays and cherts identified, on radiolarian faunas, as probably Jurassic (Danau Formation, Molengraaf, 1909); this formation has, however, been ascribed to the Permo—Carboniferous by Van Bemmelen (1949). Nodules also occur in Paleocene/Lower Eocene radiolarian cherts, associated with altered basalts, in North Borneo (Stephens, 1956).

Sorem and Gunn (1967) have related certain features of the Eocene manganese deposits of the Olympic Peninsula, Washington State, U.S.A. to Recent ferromanganese nodules. Particularly relevant in this respect is the colloform layering exhibited by hausmannite. The manganese deposits are associated with pillow lavas (altered tholeiites), limestones, some of which are pelagic, shales and sandstones. According to Garrison (1973) at least part of the Eocene sequence of the Olympic Peninsula could represent a slice of the ocean floor, plus its sedimentary cover, inserted onto a continental margin (see also Glassley, 1974; Lyttle and Clarke, 1975).

Manganese deposits*, commonly nodular, may also be associated with the radiolarian cherts and umbers that overlie Alpine—Mediterranean ophiolite complexes (e.g., Geiger, 1948; Tromp, 1948; Burckhardt and Falini, 1956; Debenedetti, 1965; Ealey and Knox, 1975). Such radiolarites, usually dated as Late Jurassic or Early Cretaceous, were apparently laid down on Mesozoic ocean floor and emplaced on a continental margin during orogeny (e.g., Bernoulli and Jenkyns, 1974). Pyrolusite nodules (diameter 1—100 cm) occur in Campanian Fe—Mn-rich sediments overlying the pillow lavas of the Troödos Massif, Cyprus (Elderfield et al., 1972; Ealey and Knox, 1975); there is some doubt, however, as to whether these are true ferromanganese nodules (A. H. F. Robertson, personal communication, 1975).

In many "geosynclinal" areas there is an association of submarine volcanics, cherts, and iron-manganese accumulations whose depositional setting was either an ocean floor or a rifted and sunken continental margin. Many of these mineral deposits may be related in their genesis to Recent ferromanganese nodules.

Descriptions of the Eocene—Oligocene deep-sea sediments of Barbados are a little puzzling. According to Jukes-Browne and Harrison (1892), the red, pink, yellow, white or mottled siliceous earth "often exhibits hollow spaces (up to ½ inch in diameter) which appear to be the casts of small manganese nodules, the manganese having been dissolved away and the space being either left empty or filled with loose powdery earth". Elsewhere in the paper, they refer to the presence of manganese oxides as coatings on joint planes or as minute spherules. Probable underwater correlatives of the succession do contain nodules (Hurley, 1966). Molengraaf (1922), however, states that "no manganese nodules occur in the red clay of Barbados". The structural setting of these deep-sea deposits (Lohmann, 1973), their chemistry and mineralogy (El Wakeel, 1964) leaves no doubt that they were deposited on ocean crust and then emerged during the Late Tertiary.

Discussion

Fossil deep-sea nodules can be divided into two types: those that accumulated on ocean floor, and those that formed on a foundered

---

*Some of these are certainly metalliferous spreading-ridge deposits (Bonatti et al., 1976).

continental margin. True oceanic nodules are apparently represented by the occurrences in western Timor and Roti (Molengraaf, 1915, 1922; Audley-Charles, 1972). The Mn/Fe ratios of the Timor nodules are a little less than unity; there are high contents of silica and alumina, and trace elements are relatively abundant. This parallels the composition of some Recent nodules from red and brown clays in the Atlantic, Pacific and Indian oceans (Mero, 1965a; Cronan and Tooms, 1969). The nodule from the Jurassic (?) red clay on Roti, with its high Mn/Fe ratio, also has recent parallels in regions where red clays accumulate near continental borderlands (Mero, 1965a; Cronan and Tooms, 1969). In these areas diagenetic enrichment of manganese by remobilization and precipitation is supposedly important (Price and Calvert, 1970). Many of the iron-poor nodules off the Pacific coast of Central and South America are formed in areas where the sediments are reducing at depth, and are coloured greenish-grey (Lynn and Bonatti, 1965). Since the fossil nodule from Roti is embedded in clays that are still red, it seems unlikely that reducing conditions strongly influenced the genesis of this nodule, or the sediments would have been bleached. Iron oxides (i.e., haematite) are stable in weakly reducing conditions (Garrels and Christ, 1965), but the stability field is small. The high Mn content of this nodule may thus have been induced by direct precipitation and may not be a diagenetic product (see also Bender, 1971). This returns one to the idea of Mn-rich hydrothermal effusions as a source for Fe-poor nodules (e.g., Mero, 1965a; Moore and Vogt, 1976). The high Ba content may be relevant to this problem, since this element attains values of up to 3.5% in hydrothermal precipitates on the East Pacific Rise (Boström and Peterson, 1966; Boström et al., 1973).

The deep-sea nodules from ancient continental-margin settings (Cretaceous of eastern Timor, Oligocene—Miocene of north-central Sicily) are both characterized by a chemical composition that may be Fe-rich or Mn-rich. This is in accord with their assumed palaeogeographic setting. As mentioned above, iron-poor nodules are characteristic of continental borderlands; however, manganese-poor (and iron-rich) nodules also seem, in some cases, to be spatially related to land-masses (Mero, 1965a; Manheim, 1965). Close juxtaposition of iron-rich and manganese-rich deposits in Recent oceans is sometimes ascribed to submarine volcanism (e.g., Bonatti and Joensuu, 1966) but this does not seem applicable to either sets of nodules (Audley-Charles, 1965; Alaimo et al., 1970).

Recent continental-borderland nodules are not usually enriched in trace elements (Price and Calvert, 1970). The high content of Ni in certain of the Tertiary Sicilian nodules is thus anomalous; this enrichment may be related to retention of the element in 10Å manganite (cf. Cronan and Tooms, 1969). Apart from Ba, trace elements are not particularly enriched in the Cretaceous Wai Bua nodules, and this is in accord with a continental-margin setting (Audley-Charles, 1965, 1972).

Particularly difficult to understand is the mechanism by which continental-margin nodules survive diagenesis. Why is manganese not lost in burial? This problem is particularly acute with the Mn-rich nodules in the grey-brown Numidian Flysch, whose colour must reflect diagenetic reducing conditions. Have the iron-rich nodules lost manganese, whilst their iron-poor neighbours retained it or were further enriched? Difficult questions: this problem is discussed at more length below, with reference to ancient littoral nodules.

CONTINENTAL-MARGIN SEAMOUNT ENVIRONMENT

Devonian—Carboniferous nodules

Description and stratigraphic location. Ferromanganese nodules of Late Devonian age occur in red pelagic limestones of the Montagne Noire, Southern France, and the Rheinisches Schiefergebirge, West Germany (Tucker, 1971, 1973a, 1974). Bandel (1972, 1974) has described ferro-manganese encrustations from the Lower Devonian to Lower Carboniferous of the central Carnic Alps, Austria, where pelagic limestones are also found. Similar nodules occur in the Devonian of the Pyrenees (Perseil, 1968). The concretions are developed as encrustations around limestone clasts and skeletal fragments which may be intensely bored. The size of the nodules ranges from a few millimetres to a few centimetres. Colours are reddish-brown to black; occasionally there is concentric arrangement of the differentially coloured bands. Colloform structures are frequent (cf. Sorem, 1967). Encrusting Foraminifera are often found within the ferromanganese layers.

Mineralogy. Haematite and a trace of rhodochrosite have been identified by Tucker (1973a) in his material. Perseil (1968) records manganite, cryptomelane, todorokite and polianite in samples from the Montagne Noire and Pyrenees.

Chemistry. The chemical composition of the Devonian nodules is rather variable. The Fe content rises to a maximum of 31% with a corresponding Mn content of 5%. The Mn/Fe ratio varies from 0.1 to 0.45. Small quantities of Ni (generally less than 0.1%) have been detected (Tucker, 1973a). Perseil (1968) mentions traces of Ni, V and Ba.

Palaeogeographic setting. The pelagic limestone in which the Devonian and Carboniferous nodules occur is stratigraphically condensed (Tucker, 1973a, b), and was apparently deposited on topographic highs. Such highs were either block-faulted "reef" limestones, volcanic ridges or basement rises. Evidence from probable algal borings, the encrusting Foraminifera, red algae, and the stratigraphic context, suggest formation of the nodules no deeper than a few hundred metres. The general palaeotectonic setting of the

Devonian—Carboniferous of this part of Europe can be interpreted in terms of a block-faulted continental margin floored by continental crust (Burrett, 1972; Floyd, 1972).

Jurassic nodules

Description and stratigraphic location. Ferromanganese nodules and crusts of Jurassic age are widely distributed in the Alps, Sicily (Figs. 4-2, 4-3) and Carpatho—Balkan chain and other parts of the Mediterranean (Tethyan) region (e.g., Jenkyns, 1967, 1970a; Wendt, 1969a, 1970; Germann, 1971, 1972, and references therein). They are usually located in stratigraphically condensed red limestones. The nodules are carbonate-rich due to the presence of included micrite and vary in colour from black, through brown to reddish purple. They are commonly of centimetre scale. Certain of them exhibit well-defined concentric lamination; others are more massive. Colloform and pillar-like microfabrics are frequently present (Fig. 4-4; cf. Cronan and Tooms, 1968; Sorem and Foster, 1972c; Monty, 1973, on Recent

Fig. 4-2. Concentrically laminated ferromanganese nodules in grey pelagic limestone of Middle Jurassic age, Rocca Argenteria, western Sicily. Length of pen is 13.5 cm.

nodules). Nuclei, though occasionally absent, are usually constituted by shell fragments, limestone clasts, or (rarely) volcanic debris. Void spaces occur in some nodules. Traces of boring algae and/or fungi, encrusting foraminifers, serpulids and bryozoans occur in many of these Tethyan nodules (Wendt, 1969a, 1970; Jenkyns, 1971).

Mineralogy. All Jurassic nodules contain abundant calcite. Those from Sicily contain goethite, haematite and a manganese mineral tentatively identified as todorokite (Jenkyns, 1970a). The nodules described by Germann (1971) from the northern Calcareous Alps also contain goethite and haematite, with some pyrolusite. Determination of the manganese minerals is difficult, however, and much of the Mn-oxide/hydroxide is X-ray amorphous.

Chemistry. The chemistry of Jurassic Tethyan nodules is extraordinarily varied. In many cases Fe is present in greater quantities than Mn, but in both western Sicily and the northern Calcareous Alps Mn-rich, Fe-poor nodules occur (Jenkyns, 1970a; Germann, 1971). The Mn/Fe ratio rises to nearly 10 in some west Sicilian nodules, to around 17 in one example from the northern Calcareous Alps. Mn-contents of 5—40% and Fe-contents of 10—35% are typical for these Tethyan nodules.

Fig. 4-3. Bedding surface with large discoid ferromanganese nodule. Matrix is red pelagic limestone of Middle Jurassic age. Monte Inici, western Sicily, Italy. Length of hammer handle is 28 cm.

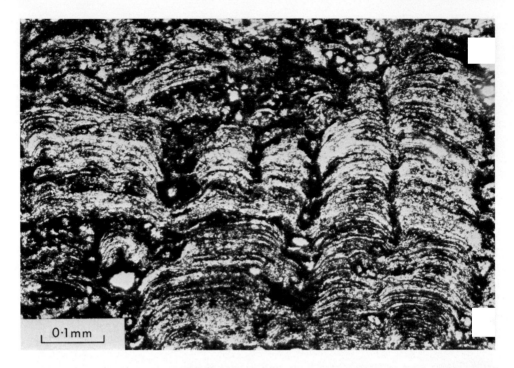

Fig. 4-4. Columnal or stromatolitic microfabric in ferromanganese nodule of Early to Middle Jurassic age. Thin section. Unken Syncline, northern Calcareous Alps, Austria.

Trace element abundances also vary widely; Ni, Co and Pb rise to over 0.1% in some of the Alpine nodules; Ni is generally more abundant than Co. V, from one analysis, assays at 260 ppm. Cu is always low (circa .04%). Electron microprobe analyses of the west Sicilian nodules show Ni, Co and Ti contents that can rise to or above 1%. In parts of some nodules Ba and Cr rise to 3.0% and 0.2%, respectively. V is also enriched (circa 0.1%): Cu is always low (circa 0.15%). Those nodules containing roughly balanced amounts of Mn and Fe are most enriched in trace elements.

Inter-element relationships vary from nodule to nodule. Germann (1971), on the basis of whole nodule analyses, noted a correlation between Ni and Mn and Co and Mn, except in those nodules which were particularly Fe-poor and Mn-rich. In the west Sicilian nodules, electron microprobe analysis gave the following common relationships: Ni, Ba with Mn, and Ti with Fe. Co was occasionally sorbed with Mn.

Palaeogeographic setting

The invariable location of Jurassic ferromanganese nodules in stratigraphically condensed limestones suggests that they were formed on topographic highs influenced by ambient currents (Jenkyns, 1967). The association of the nodules with stromatolites, encrusting Foraminifera, and probable algal borings suggests that formation took place no deeper than a few hundred metres (Jenkyns, 1971). Consideration of the palaeogeography of the whole of the Tethyan region during the Jurassic suggests that these topographic highs or seamounts were fault-bounded blocks of limestone undergoing little or no subsidence (Jenkyns and Torrens, 1971). The block-faulting and differential subsidence was related to extensional tectonics on the southern continental margin of the nascent Tethyan Ocean (Bernoulli and Jenkyns, 1974). The depositional setting was thus depressed continental crust where pelagic conditions were widespread.

Cretaceous—Paleocene nodules

Description and stratigraphic location. Black to brown carbonate-rich nodules occur in stratigraphically condensed pink pelagic limestones of Late Cretaceous/Paleocene age in the Briançonnais region, Western Alps (Bourbon, 1971a, b; Fig. 4-5). They average 5 cm in diameter, and are generally concentrically laminated, and contain cores of reworked limestone fragments or pieces of older nodules. Radial cracks traverse the laminated cortex. Colloform structures are common. Perhaps the most striking aspect of these nodules is the intimate association of encrusting Foraminifera with the laminated cortex (Fig. 4-6).

Mineralogy. The Paleocene nodules from the Briançonnais region contain phosphates, metallic oxides and sulphides, with calcite.

Chemistry. The two nodules analysed by Bourbon (1971a) show considerable variation in chemistry, with Mn contents of 2.4% and 3.9%, and Fe contents of 2.7% and 1.5%, respectively. Mn/Fe ratios are thus 0.9 and 2.6. Co and Ni are present in a few hundred ppm; Cu less than 20 ppm; V (from one analysis) 115 ppm. Perhaps most striking is the high Ba content (greater than 2%) of the more Mn-rich nodule.

Palaeogeographic setting. Much of the Briançonnais zone was a region of topographically high ground throughout the latter half of Mesozoic time (Lemoine, 1953) and pelagic conditions were widespread. The depositional environment of the Upper Cretaceous—Lower Tertiary ferromanganese nodules can be related to a current-swept submarine horst or seamount (Bourbon, 1971b) probably located on the northern continental margin of the Tethyan Ocean. The presence of encrusting Foraminifera might suggest formation of the nodules in relatively shallow water.

Other occurrences

The red limestones of the Alpine Jurassic, the Cretaceous—Paleocene of the Briançonnais, and the Devonian—Carboniferous of Europe have rough

Fig. 4-5. Ferromanganese nodule, with limestone core, in pink pelagic limestone of Paleocene age. Col de la Pisse, Briançonnais region, Western Alps, France. Diameter of coin is 1.7 cm.

facies parallels in certain Triassic deposits that occur in the Tethyan zone. Wendt (1969b, 1970, 1973) has described condensed limestones containing ferromaganese crusts and nodules from the northern Calcareous Alps, Balkans and Himalayas. These mineral deposits are also associated with sessile Foraminifera.

Discussion

The most striking fact about the shallow seamount nodules is that they contain encrusting Foraminifera, commonly *Nubecularia* and *Tolypammina*. Other faunal/floral elements — boring algae, fungi, bryozoans, serpulids — occur locally. This richness in indigenous fauna seems, therefore, to be a characteristic of those nodules formed on shallow topographic highs where sedimentation was slow and hard substrates were always available for colonization. The lack of descriptions of encrusting faunas in Recent ferromanganese nodules might at first suggest that most settings where such

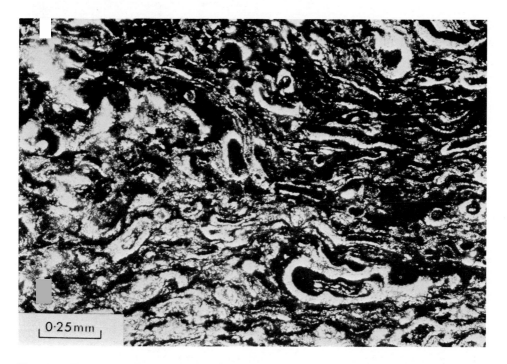

0·25mm

Fig. 4-6. Encrusting Foraminifera (*Nubecularia*) in ferromanganese nodule of Paleocene age. Thin section. Torrent du Grand Bois, Briançonnais region, Western Alps, France.

accumulations form today are considerably deeper than with these ancient occurrences; this, however, would be an oversimplification. Wendt (1974) has shown that encrusting Foraminifera can occur within Recent nodules taken from depths greater than 5 km. Serpulids, bryozoans, corals and sponges, however, seem to be more common on material from seamounts. Interesting in this respect is a nodule described by Niino (1955) obtained from off Japan. This nodule was dredged from a bank at a depth of 260 m; its surface was rough and partly covered by encrusting bryozoans and worm tubes. Foraminifera and sponge spicules were present within the nodule. Its chemistry was characterized by a low Fe content, a high Mn content (37%) and considerable quantities (11.7%) of calcium carbonate. This Recent nodule seems therefore to be directly comparable with certain of the Alpine and Sicilian examples. According to Wendt (1974) nodules encrusted with organisms are tolerably common on the Blake Plateau.

The presence of indigenous fauna and flora in these nodules raises the question of the role of organisms in precipitating iron and manganese from dilute solution in sea water. Foraminifera, fungi and bacteria can apparently all fulfill this function and have been suggested as important agents in the genesis of iron-manganese deposits (e.g., Harder, 1919; Thiel, 1925; Graham and Cooper, 1959; Ehrlich, 1963, 1966, 1968; Johnson and Stoke, 1966;

Krumbein, 1971). According to Krumbein (1971), some fungi show a preference for precipitation of manganese in concentric rings. This leads on to the origin of the colloform and pillar structures in both Recent and ancient nodules. Hofmann (1969) commented on the close resemblance of a ferromanganese pavement from San Pablo seamount to calcareous stromatolites. Monty (1973), taking this a step further, suggested that the columnar structures in Recent nodules *were* stromatolites, produced by the action of bacteria. Bacteria are known to be important in forming some Recent siliceous stromatolites (Walter et al., 1972) and are abundant in many Recent ferromanganese nodules in depths too great for algal growth (e.g., Ehrlich, 1963). The former presence of abundant bacteria in ancient seamount settings — given the faunal and floral association of the nodules — can readily be assumed.

Confirmatory evidence for a primary origin of the columnar structures in Jurassic nodules is suggested by the preferential siting of encrusting Foraminifera between the pillars (Wendt, 1969a). There is no doubt, however, that some colloform structures are produced during diagenesis since they are clearly replacive (Cronan and Tooms, 1968; Jenkyns, 1970b; Tucker, 1973a). Bourbon (1971a) has proposed that both primary "stromatolitic" and secondary inorganic structures may be developed within nodules, and it may be best to adopt this uneasy compromise.

The geochemistry of ancient seamount nodules, disregarding the carbonate, is comparable in a general way with that of Recent concretions formed on continental-margin banks (e.g., Blake Plateau) and oceanic guyots (Jenkyns, 1970a; Germann, 1971) (see also Price and Calvert, 1970). The element—depth relations of Cronan and Tooms (1969) are relevant in this respect. Ba was taken by these authors to be negatively correlated with depth: the high Ba content of some of the west Sicilian Jurassic nodules and Paleocene Briançonnais nodules thus suggests formation in shallow water. V is relatively enriched in the Tethyan Jurassic and Paleocene Briançonnais nodules, again indicative of moderate depths. Co is locally high in the Sicilian nodules, as in Ni, but since these elements are supposedly indicative of shallow and deep water, respectively, it is difficult to draw any sound conclusion. In the Jurassic nodules from the northern Calcareous Alps the higher concentrations of Ni over Co suggest depths in the order of 4 km; the Pb contents of these Alpine nodules, however, suggest formation in water shallower than 1 km (Germann, 1971). Clearly, this kind of speculation on geochemical data must be treated with discretion. The low Cu contents of all fossil seamount nodules is consistent with their formation in shallow-water zones. Todorokite, tentatively identified in the west Sicilian Jurassic nodules, is most commonly found in shallow marine settings (Glasby, 1972a).

Using the lifespan of foraminiferal colonies as an index, Wendt (1970) arrived at growth rates between 6 and 20 cm/1,000 years for Alpine and

Hungarian Jurassic nodules. Rates of deposition for these pelagic limestones can be estimated at a few millimetres per 1,000 years or less (calculated from data in Wendt, 1963, 1969a; Harland et al., 1964). Although rates of nodule growth and sedimentation were undoubtedly irregular, these nodules could clearly have maintained themselves at the sediment—water interface for long periods of time. The growth rates of these Jurassic nodules are comparable with those of some Recent littoral concretions (e.g., Manheim, 1965; Ku and Glasby, 1972), but considerably faster than those estimated for deep ocean or seamount nodules (cf. Bender et al., 1966; Barnes and Dymond, 1967; Somayajulu et al., 1971).

The limestone in which fossil seamount nodules occur is generally red; presumably, therefore, oxidizing conditions persisted after deposition, and migration of constituents should not have been great. The void spaces in some nodules, however, together with the presence of iron-manganese segregations in the enclosing limestone, can be taken as evidence for some post-depositional movement of soluble constituents (Jenkyns, 1970a; Tucker, 1973a).

LITTORAL ENVIRONMENT

Permian nodules

Description and stratigraphic location. Ferromanganese nodules of Permian age occur in one outcrop of argillaceous sandstone in the Park City Formation, near Dillon, Montana, U.S.A. (Gulbrandsen and Reeser, 1969). The bed containing the nodules is a little less than 1 m in thickness and is irregularly impregnated with manganese; it is in the upper part of this bed that the concretions are most abundant. The nodules are generally spherical, usually possess well-defined concentric lamination, and have irregular and pustulate outer layers (Fig. 4-7). They contain considerable amounts of fine sand. Most fall in the size range 2—4 cm; larger ones may be compound.

Mineralogy. The nodules are composed of two manganese minerals: chalcophanite and todorokite, plus goethite, fluorite, quartz, dolomite and clay minerals. Of the two manganese minerals chalcophanite is the most abundant.

Chemistry. The most striking factor in the chemistry of the Permian nodules is the high manganese—iron ratio. The two nodules analysed by Gulbrandsen and Reeser have Mn contents of 14.5% and 5.8% and Fe contents of 0.43% and 0.38% respectively. This yields Mn/Fe ratios of 34 and 15. Zn is particularly high (2.5 and 1.4%); Ni is also enriched (1.3 and 0.68%); so is As (0.04 and 0.06%). Co has values of 0.22% and 0.08%.

Palaeogeographic setting. Study of the sedimentary make-up of the Park City Formation suggests a shallow-marine environment of deposition in the

Fig. 4.7. Littoral ferromanganese nodules from the Park City Formation of Permian age. Light spots on nodules are remnants of rock matrix. Near Dillon, Montana, U.S.A. Photograph courtesy of R. A. Gulbrandsen.

stable interior of a large landmass. Cressman and Swanson (1964) considered that the sandy sediments of the Park City Formation were deposited at depths less than 50 m. The fauna comprises brachiopods, bryozoans, crinoids, bivalves and gastropods (Yochelson, 1968), which again suggests depths in a few tens of metres.

Oligocene nodules

Description and stratigraphic location. Ferromanganese nodules of Oligocene age occur in a sandy clay, which may be glauconitic and is often impregnated with manganese (Sokolow, 1901). The locality is Nikopol, which lies on the southern edge of the Ukrainian Shield, U.S.S.R. The bed containing the nodules generally varies between 1.5 and 3 m in thickness; the nodules are concentrated at the base. The nodules, of centimetre scale, are usually irregular in shape with knobbly outer surfaces; some possess well-defined concentric structure, others are homogenous. Small cavities occur in some nodules. Quartz grains can be included.

Mineralogy. According to Sokolow (1901), these Oligocene nodules are composed of pyrolusite. Varentsov (1964) also mentions psilomelane; furthermore, in discussing the Nikopol manganese formation in its widest development, he also records mixed ores including manganite, oxidized manganoan calcite and rhodochrosite.

Chemistry. The analyses quoted by Sokolow (1901) are limited. Most notable, however, are the high Mn and low Fe contents of the nodules. Mn values of 57.3, 52.5, 50.2, 43.8, 43.0% with corresponding Fe values of 0.3, 1.6, 0.7, 2.2 and 6% are typical. Mn/Fe ratios thus vary from 7.2 to 190. The analyses collected by Varentsov (1964) reveal a similar trend with Mn/Fe ratios in the region 8—10, 15—20 or higher. The Fe content rarely exceeds 3%. Some enrichment in minor elements (Ni, Co, Cr, V, Cu, Ba) is noted by Varentsov.

Palaeogeographic setting. The Oligocene sandy clays in which the ferromanganese nodules occur contain a fauna of cetacean remains, fish, crustacea, brachiopods, corals, and abundant bivalves and gastropods. Recognition of the bivalves as shallow-water forms leads Sokolow to postulate a littoral environment of deposition. Varentsov (1964), apparently unaware of Sokolow's work, presented a similar interpretation, suggesting formation of the "concretionary-lenticular, nodular, irregularly lumpy, variously pisolitic, oolitic and earthy ores" in shallow littoral areas of marine basins and lagoons.

Other occurrences

Ferromanganese concretions of probable Silurian age have been described by Bentor (1956) from Israel. Although not specifically interpreted as fossil nodules the description warrants their interpretation as such. Their matrix is a deep-red shale 0.5—1.5 m thick. They range in size from less than 1 mm to about 50 cm, many are kidney-shaped, with or without radial or concentric stucture. Their mineralogy is chiefly characterized by pyrolusite and psilomelane. The Mn content of three nodules is 48.7, 47.1 and 40.2%, with corresponding Fe values of 2.8, 1.8 and 2.9%. The Mn/Fe ratio is thus between 26 and 14. The trace metal contents are generally low, except for Cu (max. 1.0%) and Ba (max. 3.6%).

The depositional setting of these nodules was apparently in shallow lagoons (Bentor, 1956).

The description of Bentor's would also hold for several sedimentary manganese deposits, of differing ages, that occur in the United States. A glance at the compilation of Harder (1910) will confirm this. The rock matrix of such nodules, buttons and kidney-shaped ores is usually a sand or clay interpreted as a shallow-marine or lagoonal deposit. Analysis of the ores show high Mn and low Fe contents.

Although not oxide deposits, the manganese carbonate nodules that occur in the Lower Cambrian of southeastern Newfoundland (Dale, 1915) are worthy of mention. These nodules are generally set in a matrix of red or green shale; similar ore deposits and similar sediments occur in North Wales (e.g., Mohr, 1964; Glasby, 1974a). The Newfoundland nodules contain rhodochrosite, manganoan calcite, some manganese oxides, haematite, barite and phosphate; they assay high in Mn and low in Fe. Their stratigraphical proximity to algal stromatolites and association with shallow-water fauna indicates formation at no great depth. Dale (1915) specifically suggested "a more or less closed basin or coastal shoals or lagoons".

Ferromanganese nodules of Precambrian age occur in Botswana, southern Africa (Litherland and Malan, 1973). The matrix is an iron-stained mudstone whose maximum thickness is 2 m. The nodules exist as isolated forms about 1 cm in diameter or as intergrown aggregates up to 40 cm across. Mineralogically, they comprise nsutite and pyrolusite. Stratigraphically associated facies include mudstone, shale, sandstone and chert; manganiferous stromatolites occur locally. The sedimentary facies, and the stromatolites, can be taken as indicators of shallow-marine conditions.

Discussion

Two important questions are raised with reference to occurrences of ancient littoral nodules: firstly, the cause of the high manganese—iron ratio; secondly, the mechanism by which the nodules are prevented from dissolving during early diagenesis.

Sokolow (1901) was aware of the modern counterparts of the Oligocene nodules he described. In his paper he pointed to the Recent concretions described by Buchanan (1891) from Loch Fyne and Murray (1900) from the Black Sea. Dale (1915) drew attention to the Clyde Sea area, described by Murray and Irvine (1894), as a depositional model for the Cambrian nodules of Newfoundland; Glasby (1974a) used Loch Fyne and the Baltic as modern parallels for the depositional environment of the North Wales ores. To these occurrences of Recent shallow-marine nodules can be added those in the Barents, Kara and White seas (Manheim, 1965, and references therein; Winterhalter, 1966) and a British Columbia fjord (Grill et al., 1968a, b). In these settings the bottom sediments usually comprise mixtures of sand, silt and clay.

The best Recent analogues are the manganese-rich, iron-poor nodules of Loch Fyne (Buchanan, 1891; Calvert and Price, 1970a) and Jervis Inlet, British Columbia (Grill et al., 1968a, b). The average Mn content of the Loch Fyne nodules is 30%; Fe is around 4%; this gives a Mn/Fe ratio of 7.5. The nodules, however, are not compositionally homogenous. Those nodules from the British Columbia fjord have Mn and Fe contents of about 33% and 4.4% respectively: the Mn/Fe ratio is between 7 and 8. From the above it is apparent that Recent shallow-marine nodules forming in or on sand/clay substrates possess a strong chemical resemblance to their ancient counterparts. One can thus interpret the chemical make-up of the fossil nodules as primary. Loch Fyne seems to be a particularly apposite model for the Oligocene manganese deposits for in both cases associated manganese carbonates are developed (Varentsov, 1961; Calvert and Price, 1970a). Furthermore, the molluscan faunas in Loch Fyne are closely comparable to those in the Nikopol Formation. The greater age gap between the Loch Fyne sediments and the Permian deposits vitiates any possible comparison in this case. The presence of todorokite in the Permian nodules allows a comparison with those from Loch Fyne and Jervis Inlet.

High Mn/Fe ratios in shallow-marine nodules are usually ascribed to migration of manganese to the sediment-water interface, to form nodules, with retention of less mobile iron as sulphides deeper in the sediment column (Murray and Irvine, 1895; Lynn and Bonatti, 1965; Manheim, 1965; Cheney and Vredenburgh, 1968). Although this mechanism can explain high Mn/Fe ratios in ferromanganese nodules that are forming *now*, it fails to explain how such nodules could survive as part of the sedimentary record without loss of manganese. Manheim (1965) and Calvert and Price (1970a) assumed that littoral nodules would dissolve on burial as they passed into zones where reducing conditions prevailed. At the very least one would expect any survivors to have high Fe/Mn ratios. Why, then, do we have this considerable record of ancient shallow-marine nodules with a chemistry that suggests that little or no redistribution of elements has taken place?

The key to this problem may lie in the nature of the sediments deposited above the ferromanganese nodules. If these sediments were such that they rendered upward diffusion difficult or impossible, the concentration of dissolved manganese in interstitial waters may have built up to such a level that further dissolution of nodules was impossible. Thus, if some kind of sediment "seal" operated to close the system, littoral ferromanganese nodules might be assured of preservation. This hypothesis is difficult to evaluate; particularly so since the sediments that were laid down above the ferromanganese nodules may have been later removed by submarine erosion without leaving any stratigraphic record. In any case if we are to accept the high Mn/Fe ratios of these fossil nodules as due to diagenetic enrichment of manganese, then clearly this redistribution process must have been severely curtailed when the nodules were buried.

The minor element composition of the fossil littoral nodules probably reflects the nature of surrounding terrain. The As content of the Permian nodules is comparable to values found in the Loch Fyne samples; generally, however, these particular fossil nodules are more enriched in trace metals than their Recent counterparts.

The apparent absence of indigenous fauna and flora in these fossil (and Recent) shallow-water littoral nodules may be related to higher sedimentation rates than with the seamount examples described previously.

CONCLUSIONS

Fossil ferromanganese oxide nodules occur in rocks deposited in deep-sea, continental-margin seamount and littoral environments. The record of ferromanganese nodules ranges from Precambrian to Recent (Table 4-I). In general the chemistry of ancient nodules is comparable to that of Recent counterparts formed in similar settings. Ancient deep-sea marginal nodules are characterized by Mn-rich and Fe-rich species; true oceanic deep-sea nodules tend to contain roughly balanced amounts of Mn and Fe and are well endowed with trace elements. Seamount nodules are rather variable with respect to contents of Mn and Fe; if the host metals are present in comparable amounts trace elements are often abundant. These shallow-water seamount nodules invariably contain encrusting Foraminifera and other organisms and are characterized by "stromatolitic" microfabrics. Fossil littoral nodules are always Mn-rich and Fe-poor. The mineralogy of all fossil nodules is highly variable.

Certain stratified iron-manganese deposits are obviously related in their genesis to Recent manganese nodules. Fe-poor ferromanganese nodules from continental borderlands and littoral zones resemble, in their chemistry, some sedimentary ore bodies. Such nodules are often associated with manganese carbonates. Stromatolites also occur with certain littoral nodules.

The occurrence of oceanic ferromanganese nodules in ancient red clays or cherts necessarily has tectonic implications, since their very occurrence on land implies overthrusting of some oceanic lithosphere and its sedimentary cover. This suggests continent—continent collision or at least subduction of oceanic crust at some past geological epoch.

More generally, the presence of ferromanganese nodules in a sedimentary sequence points to slow sedimentation under oxidizing conditions. This in turn suggests the presence of stratigraphic lacunae, which may or may not be confirmed by palaeontological evidence.

CHAPTER 5

MORPHOLOGY

W. J. RAAB and M. A. MEYLAN

INTRODUCTION

Ferromanganese oxides are deposited in a variety of forms, and in numerous terrestrial, lacustrine, and marine environments. The forms that are common in the deep sea have been known for a century. Murray and Renard (1891), reporting on samples collected by the *Challenger* expedition (1872—1876), noted hydrates of manganese as "colouring matters, or as thin or thick coatings on shells, corals, sharks' teeth, bones, and fragments of rock". Minute grains of ferromanganese oxides, or micronodules, were also observed to be common constitutents of deep-sea sediments.

The basic categories of manganese deposition are: (1) stains; (2) agglutinations; (3) nodules; and (4) crusts. Other less common forms of occurrence are manganese dendrite-impregnated sediment, and manganese-lined worm tubes (Morgenstein, 1972). Stains are very thin deposits of manganese on some solid object such as a volcanic rock fragment or an outcropping rock. Agglutinations are clusters of discrete nuclei united by a thin (generally < 1 mm thick) encrustation of manganese. Thicker encrustations on single or multiple discrete nuclei are recognized as manganese nodules. Crusts are the relatively thick manganese depositions on submarine rock outcrops or on relatively large objects such as boulders or volcanic slabs. On the Blake Plateau, a crust of manganese oxide forms a pavement that may be continuous over an area of about 5,000 km^2 (Pratt and McFarlin, 1966). Manganese micronodules are a special case of the nodule category, being individual grains generally less than 1 mm in diameter, and often lacking a discernible nucleus. In additon to its encrusting forms, manganese also occurs as a replacement of material such as rock fragments and biological debris.

There is no accepted delineation between objects that are stained or very thinly encrusted with manganese, and objects that may be called manganese nodules. The thickness of the encrustation relative to the size of the object that it is accreted upon could be used in making such a distinction. A boulder or cobble with a 2 mm thick crust of manganese probably would not be considered a nodule, but a small pebble with such a thickness of manganese might be so considered.

Manganese nodules occur in an assortment of shapes. No widely used scheme of morphological classification exists, so individual investigators have resorted to use of general terms (spherical, ellipsoidal, discoidal, etc.) to describe the shapes of nodules from particular study areas. Murray and Renard (1891) recognized three morphological groups in the extensive nodule collection at *Challenger* Station 160: (1) more or less pyramidal or irregularly grape-shaped; (2) spheroidal or ellipsoidal; and (3) flattened, mammillated, and irregular in form. Goodell et al. (1971) grouped nodule shapes into: (1) spheroidal; (2) ellipsoidal; (3) tabular-discoidal; (4) polygonal; and (5) tubercular. The informal, but highly descriptive, classification of Heezen and Hollister (1971) includes cannonballs, potatoes, grapes, and slabs; Horn et al. (1973c) described hamburger-shaped nodules from the North Pacific. Raab (1972) observed that the majority of nodules from the mud-water interface in a North Pacific area were noticeably asymmetric, and could be described as oblate, discoid, or prolate. Spheroidal shapes were seen primarily in small nodules (less than 3 cm diameter) or in buried nodules recovered in cores.

Most nodule shapes would fit into the four main pebble shape classes of Zingg (1935): oblate (tabular or disk-shaped), equant, bladed, and prolate (rod shaped). The multinucleate nodules which resemble grape clusters and nodules formed around odd-shaped nuclei such as sharks' teeth, however, cannot often be easily placed in simple geometric categories.

The basic forms of manganese deposition reflect differences in thickness and mode of accretion of deposit, as well as differences in the character of the nucleating substrate. Grant (1967) devised a morphogenetic classification of Southern Ocean concretions based on the relative thickness of the oxides and single vs. multiple nuclei. "Nodules" have a single nucleus and relatively thick oxides. "Crusts" have a single nucleus and relatively thin oxides. "Botryoidals" have multiple nuclei and relatively thick oxides. "Agglomerates" have multiple nuclei and relatively thin oxides. The use of the terms "nodule" and "crust" as morphogenetic types is unfortunate because of their much wider use as more general terms. Goodell et al. (1971) noted that the ratio of the oxide crust thickness to the concretion minimum diameter for "crusts" and agglomerates" was generally less than 0.4, while for "nodules" and "botryoidals" the ratio was usually greater than 0.4. The measurements involved in determining the ratio are somewhat subjective, because the measurement of the thickness of the oxide crust is often difficult if the crust is not clearly differentiated from the subcrust or nucleus.

Goodell et al. (1971) also pointed out a general relationship between geometric nodule shape and the morphogenetic classes of Grant (1967). Spheroidal and ellipsoidal shapes are characteristic of the "nodule" morphogenetic class. Many "crusts", frequently somewhat flattened, display an ellipsoidal shape, although tabular-discoidal shapes are also typical. "Botryoidals" and "agglomerates" are usually tubercular, but quite often are irregularly ellipsoidal.

Manganese nodule facies can also be distinguished by using other combinations of morphological parameters. Meyer (1973) recognized six nodule facies in a part of the North Pacific on the basis of external shape, size, surface texture, symmetry of laminations, and relative tendency to occur intergrown, i.e., in a multinucleate, polylobate configuration (Table 5-I). Not assigned to this classification were a group of kidney- or half-moon-shaped concretions, and crusts. Meyer detected differences in metal content between the forms, with "B"-type nodules having the highest contents of Ni, Cu, and Mn and crusts the lowest.

During *Moana Wave* cruise Mn 74—01 to the northeastern equatorial Pacific, Meylan and Craig (see Meylan, 1974) devised a field classification of manganese nodules in order to characterize nodule types seen in different free-fall grab samples and at different stations. The nodules were classified

TABLE 5-I

Various aspects of the six nodule facies of Meyer (1973)

Facies	Size (cm)	Shape	Modifications	Surface	Intergrowths
B	3—15, avg. 5—6	Discoidal, often flattened on bottom	Irregular flat ellipsoidal, mushroom	May be botryoidal, often smooth on upper side, gritty on bottom, or with knobby band in equatorial zone	Very rare
E/S	2—6, avg. 3—4	Ellipsoidal or spheroidal	Discoidal	Normally smooth	Rare
Kr	0.5—5, avg. 1—2 when intergrown	Spheroidal	Elongated or ellipsoidal	Rough and gritty	Frequent
Kg	Intergrown 1.5—6, avg. 2.5	Spheroidal	Elongated or ellipsoidal	Very smooth	Frequent
SG	5—8, avg. 7	Very spheroidal	—	Smooth	None observed
G	Up to 25	Irregular spheroidal	—	Bumpy to cauliflower-like on upper surface, bottom gritty	—

on the basis of size, shape and surface texture. This information was expressed by a three-member symbol, the prefix indicating size, the central member indicating primary morphology, and the suffix indicating surface texture (Table 5-II). Using this field classification, Meylan and Craig (1975) determined that 29% of the 571 nodules collected during the cruise belonged to 5 classes, in descending order of frequency of occurrence: s[E]r, s[S]r, s[D]r, m[D]$_b^s$—r, and m[D]b. When only size and shape were considered, 72% of the nodules fell into 5 classes and 96% into 8 classes, in descending order: m[D], s[D], s[S] and s[E], s[P], m[P], m[E], and l[D]. Considering only shape and surface texture, 42% of the nodules fell into 5 classes, in descending order: [D]b and [D]$_b^s$—r, [E]r, [D]s, and [E]s—r. Compared to nodule collections from the Southwestern Pacific Basin and Pacific—Antarctic Ocean, there appears to be more diversity in the morphology of nodules collected at individual sites in the northeastern equatorial Pacific.

TABLE 5-II

Field classification of manganese nodules (from Meylan, 1974)

Prefix:	s = small = < 3 cm	nodule
	m = medium = 3—6 cm	size
	l = large = > 6 cm	(maximum diameter)
Primary morphology:	[S] = Spheroidal	
	[E] = Ellipsoidal	
	[D] = Discoidal (or tabular-discoidal form)	
	[P] = "Poly" (coalespheroidal or botryoidal form)	
	[B] = Biological (shape determined by tooth, vertebra, or bone nucleus)	
	[T] = Tabular	
	[F] = Faceted (polygonal form due to angular nucleus or fracturing)	
Suffix:	s = smooth (smooth or microgranular)	
	r = rough (granular or microbotryoidal)	surface texture
	b = botryoidal	

Examples: l[D]b = large discoidal nodule with botryoidal surface

m—l[E]$_{r\text{-}b}^{s}$ = medium to large ellipsoidal nodules with smooth tops, rough to botryoidal bottoms

Manganese nodules typically display a mammillated surface as a result of the botryoidal growth habit of encrusting ferromanganese oxides (Murray and Renard, 1891). The size of the mammillae (hemispherical protrusions) relative to concretion size varies greatly, as does the "relief" of the mammillae, i.e., the prominence above the general nodule surface. Usually at least two distinctly different sizes of mammillae occur together on a single nodule, with very small mammillae superimposed on larger ones (Goodell et al., 1971). At *Challenger* Station 160, Murray and Renard observed that the spherical or elliposoidal nodules were less mammillated than the irregular varieties.

The mammillarity of a concretion is not always uniform on every side; this fact was first reported by Murray and Renard (1891) and subsequently noted by Hubred (1970), Goodell et al. (1971) and Meyer (1973), among others. Non-uniformity is especially pronounced on many of the larger non-spheroidal nodules, especially tabular or discoidal forms. Differences in mammillarity are usually related to asymmetrical thicknesses of ferro-manganese oxides, the surface having thicker oxide accumulations displaying mammillae of higher relief and larger size (Goodell et al., 1971). However, Morgenstein (1972) concluded that, for Kauai Channel crusts, there was no correlation between crustal thickness and botryoidal size or shape.

Describing what they concluded to be typical deep-sea nodules collected at *Challenger* Station 248, Murray and Renard (1891) remarked that "two surfaces of these nodules present a marked difference of aspect; the inferior surface, which we believe to have rested in or on the clay ... is seen to be covered with an immense number of little rugosities, or rounded points, about 1 to 2 mm. in diameter, and the same in height. ... On the other, or superior, surface of the nodule, which appears to have projected above the surface of the clay, the asperities are not nearly so numerous, and the mammillae are smoother, larger, and less pronounced ... " Heightened mammillarity is favoured by low percentages of admixed fine detritus; large percentages of admixed coarse detritus depress mammillarity (Goodell et al., 1971).

The morphology of lacustrine and shallow-water marine ferromanganese concretions frequently differs from typical deep-sea nodule forms. Saucer-shaped crusts, which are almost unknown in deep-sea deposits (being reported only by Hubred, 1970), are found in the Baltic Sea and other nearshore areas (Manheim, 1965), and in several freshwater lakes (e.g., Kindle, 1932; Beals, 1966; Dean, 1970). The saucer-shaped concretions usually lie on the bottom with convex side upward. Another typical shallow-water or lacustrine morphology is the "skirted" concretion, where a rim of encrusting oxides encircles the nucleating object at or just above the sediment—water interface (e.g., Manheim, 1965; Terasmae, 1971).

Both the saucer-shaped and the skirted concretions often display a lack of oxides on the underside (e.g., Kindle, 1932; Manheim, 1965), the portion of the rock nucleus resting on the sediment being uncoated. Glasby (1973b) found a similar phenomenon on certain glacial erratics recovered from north of the Indian—Antarctic Ridge, the undersides being devoid of manganese. Morgenstein (1972) retrieved mushroom-shaped nodules from the Kauai Channel, partially buried uncoated pyroclastic fragments making up the "stem" of these nodules. Lacustrine saucer-shaped and skirted concretions are mainly horizontally banded, in contrast to deep-sea forms where accretion occurs more uniformly in all directions from the nucleus.

The nucleus of a manganese nodule is the solid object onto which ferromanganese oxides are initially precipitated, and it often determines the external form of a concretion. According to Murray and Renard (1891), "any solid body suffices for the support of the original and subsequent concretionary deposits. Basic and acid silicates — like pumice and glassy lapilli, almost always profoundly altered — are perhaps the most frequent nuclei, then follow teeth of sharks and other fish, otoliths, bones of Cetaceans, siliceous and calcareous sponges, and even agglomerations of the deposits in which casts of Foraminifera can be recognized". Other types of nuclei have been reported as well. On the Campbell Plateau, manganese nodules have formed around previously existing phosphorite nodules, and one nodule even had a nucleus consisting of small fragments of volcanic rock, mineral grains, zeolites, biogenic rocks, and bryozoan skeletons (Summerhayes, 1967). While most lacustrine nodules have formed around small rocks, some of the discoidal concretions of Lake George, New York, have tree bark or clay fragment nuclei (Schoettle and Friedman, 1971).

Nodules lacking apparent foreign nuclei have either formed around fragments of pre-existing nodules with indeterminate laminations or else the original nucleus has been replaced. Almost any object that serves as a nucleus can be replaced, given enough time. Volcanic rock fragments, which seem to be the most common nuclei of deep-sea concretions, are readily altered, especially if glassy. Klenova (1936b), cited in Manheim (1965), reported glacial erratics replaced by oxides. Even bones can be replaced by manganese. Nodules without a nucleus tend to be more compact, heavier, and break less easily than nodules with a nucleus (Murray and Renard, 1891).

Nodules collected at one station tend to resemble each other, and differ in size, form and internal structure from those collected at another station (Murray and Renard, 1891; Sorem and Foster, 1972a). This observation was extended for analytical purposes by Goodell et al. (1971), who defined a population as a group of concretions, all of which are characterized by similar nuclei and ferromanganese oxide accumulation geometry, and which were collected in a single dredge haul or at a single sediment core horizon. Multiple populations can occur in an area sampled by a single dredge haul, or even in the same core interval. Table 5-III summarizes the principal morphologic parameters of manganese nodules that can be used to distinguish different populations or facies.

The existence of varying forms of manganese accumulation at different localities must be attributable to several factors, including type of nucleation substrate available, the nature of the physical environment (e.g., bathymetric position, bottom current activity, sediment type and rate of supply, and activity of benthic organisms), and the nature of the chemical environment

TABLE 5-III

Principal morphologic parameters of manganese nodules

Parameter	Nature and extent of variability
(1) Size	Generally 0.5 to several metres. Concretions larger than about 20 cm in diameter generally assume the form of slabs.
(2) External shape	Numerous terms used. Most frequently are spheroidal ("peas" to "cannonballs"), ellipsoidal ("potatoes"), discoidal or tabular (including slabs), polygonal (faceted or irregular shape often due to shape of nucleus) and polylobate (intergrowths of spheroids or ellipsoids produce form resembling grape clusters). Discoidal or ellipsoidal nodules often flattened on one side. Lacustrine nodules may also be saucer-shaped or "skirted".
(3) Surface texture	Usually mammillated, but often smooth. Surfaces of larger mammillae frequently display smaller mammillae. Very small but prominent (high relief) mammillae give surface a gritty texture. Large, prominent mammillae can produce a polylobate appearance. Upper and lower surfaces of nodule may have different textures.
(4) Nature of ferro-manganese oxide crust	Nodule-like objects may have only a thin stain of ferromanganese oxides; manganese nodules may be composed almost completely of oxides. Typically deposited as concentric growth laminations. May also be partially a replacement of an easily altered nucleus. Oxide crust thickness often quite variable between upper and lower surfaces; asymmetry usually greater for larger nodules.
(5) Character of nucleus	Can be any solid surface. Often dictates external shape of nodule. If glassy volcanic in nature, may be rapidly replaced by ferromanganese oxides. Multinucleate nodules usually represent intergrowths of smaller spheroidal or ellipsoidal nodules.

(e.g., souce and rate of supply of elements). In the marine environment, the rocks most recently exposed to sea water tend to have the thinnest accumulations of manganese. Laughton (1967) and Glasby (1973a) examined basalts from the Carlsberg Ridge and found the crest to have the freshest rocks and thinnest manganese crusts, whereas basalt from the flanks was altered to a friable white clay and had thicker manganese crusts. Manheim (1965) observed that, for Gulf of Maine concretions, the extent and thickness of thin crusts varied directly with the dimensions of the host material.

Based on a study of volcanic sea-floor areas, Glasby (1973c) and Horn et al. (1973b) have concluded that the distribution and morphology of nodules is controlled by the availability of potential nucleating agents. Glasby assumed that in such areas nodules will form where the sediment surface is enriched in fragments of volcanic debris such as palagonite. He explained the discontinuous distribution of nodules on the western flanks of the Carlsberg Ridge as being due to either an initial heterogenous distribution of volcanic debris or the occurrence of topographic features capable of trapping downward-moving sediment.

Horn et al. (1973b) point out that there are three potential sources of nuclei in volcanic sea-floor areas: (1) basalt exposures along fracture zones; (2) basalt outcrops on steep slopes of seamounts (submarine volcanoes); and (3) ejecta of explosive phases of submarine volcanism. To explain the occurrence of nodules with volcanic nuclei found at great distances from nuclei sources, Horn et al. postulate that coarse-grained vesicular volcanogenic grains may be buoyant enough to allow widespread dispersal by strong currents. Manganese accumulation on ejecta of a single explosive phase may produce nodules of similar morphology; differences in morphology at a particular location could then be accounted for by assuming they were produced by different eruptions (Hubred, 1970).

Many manganese nodule deposits apparently have formed around consolidated ash layers that have been broken up, chemically weathered, and encrusted with manganese oxides. Menard (1960) described such an occurrence in the northeastern Pacific, where tabular manganese-encrusted phillipsite slabs decrease in thickness toward the west. Payne and Conolly (1972) depicted part of a nodular pavement in the Tasman Sea that has been broken up after formation, possibly by very strong bottom currents. Mero (1965a) postulated that the slab-like nodules of the Pacific were probably due to the coalescence of closely spaced groups of similar-diameter nodules. Aumento et al. (1968) described slabs of manganese pavement from San Pablo Seamount, slabs from the crest being more compact than those from the middle slopes.

Slabs can also form in non-volcanic terranes. Pratt and McFarlin (1966) recovered irregular to discoidal slabs from the Blake Plateau, with holes and indentations near the edges of the slabs.

Bathymetric position is another important control on nodule distribution and morphology. Shipek (1960) found that in the eastern Pacific the greatest concentration of manganese occurred at depths greater than 4,500 m on subdued abyssal hill surfaces. Moore and Heath (1966) and Moore (1970) discovered that the steepest slopes in the central equatorial Pacific were the sites of densest nodule coverage. Craig (1975) concluded that there is a significant increase in nodule occurrence by number and by weight in high topographic areas (abyssal hill tops and upper slopes) and an increase in the relative number of smaller nodules and patchiness of the deposits in low topographic areas (lower abyssal hill slopes and valleys between abyssal hills) (see also Margolis and Burns, 1976).

Laughton (1967) similarly reported that on the flanks of the Carlsberg Ridge nodules were most common on ridges where the slopes were high, whereas the lower slopes were devoid of them. Numerous reports document, by use of bottom photographs, the association of mid-ocean ridge or seamount slopes with manganese-encrusted volcanic flows. Hawkins (1969) noted that on the Blake Plateau, nodules graded toward a pavement at the base of a scarp.

On several scales, the intensity of ferromanganese encrustation is greatest at some central location, and decreases more-or-less gradually with increasing distance from the locus of maximum density. On the scale of the Pacific—Antarctic Ocean, Goodell (1965) discovered, from the study of thousands of bottom photographs, that beneath the Antarctic Convergence lay vast deposits of concretionary masses and ferromanganese drapes on rock outcrops. These thick deposits give way north and south to carpets of nodules, then to scattered nodules. The distribution of manganese in the Southern Ocean is at least partly due to the activity of strong bottom currents in the area of West Wind Drift, with glacial erratics serving as nuclei where volcanic nuclei are not available. Goodell (1965) pointed out the importance of bottom currents, finding it difficult to account for elongate or irregular-shaped nodules with multiple nuclei without postulating movement on the sea floor. Hollister and Heezen (1967) noted that manganese nodules north of the polar front become larger and more numerous from west to east.

On a smaller scale, such as on the Blake Plateau, an encrusting pavement of manganese grades laterally into manganese and phosphate nodules (Pratt and McFarlin, 1966). On a still smaller scale, a few square kilometres of abyssal hills in the equatorial Pacific, Andrews and Meylan (1972) noted wide variations in density of nodule coverage at a single camera station. Manganese pavements and submarine volcanic outcrops seemed to be typical of abyssal hill slopes, while variable nodule spacing was observed on hill tops and in valleys. Density of nodule coverage was generally greater in the valleys. Andrews and Meylan attributed the variations in nodule density to differences in proximity to volcanic outcrops and to variations in the local rate of sediment accumulation as affected by bottom currents.

Many questions regarding nodule morphology remain unanswered, and much basic research and thinking is still to be done. Probably the most important step to be taken is in the direction of agreement on the terminology and structuring of shape and surface feature classifications. Obviously, this is necessary for comparison of results between different workers and between different areas. When the nature and variability of morphologic parameters can be more or less uniformly described, the following important questions will be more easily answered: Why do nodules from different areas of the ocean typically differ in morphology? Why are lacustrine nodules often characterized by shapes not seen in deep-sea nodules? How does the mode of nodule accretion affect morphology? Why do individual nodules often display more than one type of surface texture? Are particular concretion shapes indicative of a certain range of elemental ratios or chemical composition? What effect does bathymetry have on nodule morphology and distribution? And what controls do bottom currents and benthic organisms exert on the form of nodules?

In the following section, one of the authors (W. J. Raab) describes in considerable detail the external morphology and internal structure of manganese nodules collected from one deep-sea area 4,000—5,000 m in the Pacific Ocean. While the observations were made on a relatively restricted suite of nodules from an area north of the equatorial carbonate zone in a region · of clay sedimentation, they demonstrate the processes controlling nodule morphology in general.

CENTRAL PACIFIC NODULES

This section forms a part of the larger study by Raab (1972) and Raab and Norton (1973), and presents some morphological and chemical characteristics of some Pacific deep-sea nodules as seen on a megascopic scale. Analytical and sampling techniques have previously been described by Raab (1972).

Shape of nodules

The shape of the nodules is variable. The majority of the nodules from the area studied could probably be described as oblate, discoid, and prolate with all gradations between these shapes; most nodules are, however, noticeably asymmetric. A common feature is a flattened curvature of one and an exaggerated curvature of the other surface of the nodule (Fig. 5-1). Truly symmetrical shapes are therefore quite rare among the nodules and nearly all nodules show some flattening.

The closest approach to truly spherical shapes have been found in the small nodules (less than 3 cm diameter) and in buried nodules recovered from sediment cores. It is possible that these small nodules were never exposed to the action of water but have remained buried throughout their growth history. That these nodules have been buried throughout their growth history is also suggested by their coarse gritty or goose bump-textured surface which is characteristic of the buried portion of surface nodules (see surface textures in the following section). The shape of the nodules is therefore controlled by their growth pattern which reflects the gross internal structures and the proximity of the nodule to the mud—water interface (this will also be further discussed in the next section).

Surface textures of nodules

As has already been pointed out, manganese nodules exhibit several kinds of surface textures. Textures commonly found in the study area are: (1)

Fig. 5-1. Common nodule shape. A. Flattened curvature on one and exaggerated curvature on the other surface. The flattened surface was the surface last exposed to seawater. B. Photograph shows two halves of the same nodule. The left half reveals that the internal structure is not controlling the external shape of the nodule. The nodule also reveals a pisolitic or knobby equatorial band. The surface with the exaggerated curvature was last exposed to seawater.

smooth — the nodule surface shows essentially no visible pattern (Fig. 5-2A at *a*). Often smooth surfaces also develop black lustrous patches; (2) gritty — the surface appears to be composed of sand-size and finer particles which are loosely cemented to the nodule (Fig. 5-2A at *b*). The base to which these particles are attached often has a texture which is best characterized as "goose bumps"; (3) "goose bumps" — a texture characterized by innumerable small welts with or without attached sand-size particles. This texture is found as a base for the gritty particles as well as on so-called β-breaks (breaks which happened at the sea floor and are healed). These welts may actually be individual grains of ferromanganese oxides which have been partially cemented to the main nodule surface or they may represent internal growth structures (p. 166, Fig 6-10A) of an accreting manganese oxide band; and (4) pisolitic or knobby-nodules which have developed a distinct equatorial zone often exhibit a band characterized by a knobby texture (Fig. 5-1B). This zone appears to be composed of fused grains larger than 2 mm, hence the name pisolitic. This pisolitic band may face in any direction, either towards the gritty or towards the smooth surface of the nodule; it is, however, always confined to the equatorial zone of the nodule. Some nodules are knobby altogether; individual knobs may be as large as 2 cm (Fig. 5-3).

Most frequently two kinds of textures, smooth and gritty, are exhibited by the same nodule. Those nodules which have developed a pronounced flattening also exhibit a pisolitic or knobby band near the equatorial rim created by the flattening. The reason for the development of these knobs at the equator is not known. The equatorial band does, however, mark the position to which the nodule was buried in the sediment (Fig. 5-4). The smooth texture is always associated with the nodule surface exposed to the water, whereas the gritty surface is always associated with surface exposed to the sediment. This association of the gritty surface with the mud has been demonstrated by the intimate adherence and entrapment of clay on the gritty nodule surface in many hundreds of nodules recovered by grab samplers, compared to only one nodule recovered where no clay was trapped in the irregularities of the gritty surface, but clay adhered to the smooth surface of that nodule. This nodule apparently had been turned over in the recent past; however, only enough time had elapsed to clean the nodule of adhering clay, but not enough time to destroy the gritty surface texture. That the smooth side faces toward the water has also been demonstrated by the nodules recovered in box cores (Fig. 5-5). The direction in which smooth and gritty surfaces face is also recorded by internal features and by the composition of nodule layers; all these features bear on the origin of nodules.

Internal fractures and break-up of nodules

Pacific Ocean nodules frequently exhibit two types of internal fractures: (1) radial or random fractures, and (2) concentric fractures (Figs. 5-6, 5-11).

Fig. 5-2. Surface textures. A. Smooth (last exposed to seawater) at *a*, gritty (last exposed to sediment) at *b*. The photograph also shows an α-break referred to later in the text. Note the black band of ferromanganese oxide at the bottom which held the nodule together until breakage after recovery. This fracture was completely open to the sediment on which the nodule rested. B. Detail of gritty surface texture. White patches represent clay adhering to the nodule indicating that this portion of the nodule was resting in the sediment.

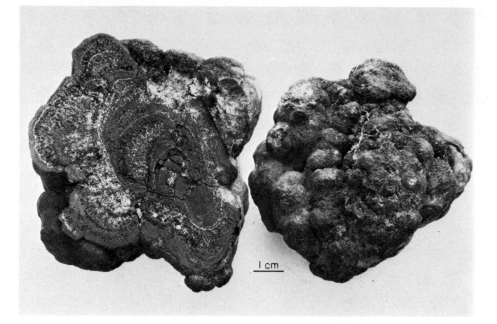

Fig. 5-3. Knobby surface texture. The individual knobs do not reflect the shape of the core fragment. The left portion of each half was last exposed to seawater.

Both types are generally filled with clay and are readily visible in cut nodules. Occasionally, the fractures are clay-free and are lined with an over-growth of ferromanganese oxides. The radial or random type appears to be an extension feature having the greatest separation toward the centre of the nodule and tapering toward the edges. Many of these fractures do not emerge at the nodule surface but some are open toward the side which faced the sediment at the time the nodule was collected. Many of the radial fractures show characteristics of shrinkage cracks; they have been interpreted to be the result of aging of older oxide material (Foster, 1970).

The concentric fractures are parallel to the zoning of the nodule. They are never continuous around the entire nodule and are frequently terminated by radial fractures (Figs. 5-6, A and B).

It is commonly observed that the nodules have old fracture surfaces (called β-breaks). This fracturing took place at the sea floor and the fracture surface is nearly always healed, i.e., the surface does not reveal the internal structure of the nodule but shows a thin crust of manganese material cover-ing the entire break surface (Fig. 5-7). The causes of nodule break-up have generally been attributed to benthic organisms or bottom currents; both agents may also serve to work against nodule burial by sediment, and maintain nodules at the mud-water interface (Menard, 1964). It appears

Fig. 5-4. Sea floor photograph showing the depth to which nodules are buried in the sediment. The equatorial rim of nodules coincides with the position to which the nodules are buried in sediment.

reasonable that both benthos and current could perform these tasks; however, either to prove or disprove their effectiveness is difficult.

Nodules usually break in two ways: (1) spalling of a layered segment (Fig. 5-8); and (2) breaking off of an irregular portion of the nodule involving the entire cross-section of the nodule (Fig. 5-9).

Breaks which transect the nodule layers more or less at right angles are much more common than peeling or spalling of layers. Spalled areas have been observed only in larger nodules (size 7 cm or larger in the longest dimension). Peeling or spalling appears to be more difficult even when clay-rich layers are present within the nodule. Spalling may take place where radial and concentric fractures intersect. It may require special conditions to remove a flake once it has been separated. Such an unusual condition could be obtained occasionally when a benthic organism chooses a nodule as a resting place (Fig. 5-10). The muscular action of this organism could also be responsible for the irregular break-up of nodules. However, the rarity with

Fig. 5-5. Nodules recovered in a box core. The photograph shows the extent to which nodules are buried in the sediment. The largest nodule just below the identification tag also reveals a knobby equatorial band and a smooth (most recently exposed to seawater) upper nodule surface.

which organisms resting on nodules have been photographed suggests that this mechanism is uncommon.

Nodule break-up may also be caused by internal features. The nodules which break during or after recovery always display considerable quantities of clay on the break surface. These breaks (called α-breaks) frequently appear to be essentially complete before recovery, i.e., formed at the mud— water interface. In nearly all cases on freshly broken surfaces, the fracture appears to be open toward the mud (and is filled with sediment) whereas toward the smooth (seawater) surface there is a crust of clay-free manganese material which apparently held the nodule together (Figs. 5-2 and 5-11). Only during handling of the nodule is this crust broken and the pre-existing internal break revealed.

The manganese nodules which show no external sign of fractures when cut or purposely broken often reveal internal fissures which are widest in the centre and terminate toward the margin of the nodule. These fissures are also lined with clay (Fig. 5-6). The clay apparently collected in the fissure after its formation, and therefore must have been precipitated from solutions circulating through the nodule. It has also been observed (Foster, 1970) that some of these fractures are lined with ferromanganese oxides (p. 159, Fig. 6-6A) which, like the clay, must have been precipitated after formation of the fissure. Both ferromanganese oxides (Foster, 1970) and clay could act as wedges and force the fractures apart. Once a fracture has broken the outer skin of the nodule in the direction of the sediment, break–up can then be facilitiated by differential forces acting on different parts of the nodule or, if the nodule should for some reason be turned over, currents could wash out the sediment and finally cause the break-up. The intersection of several internal fractures could also weaken a nodule to a sufficient extent that small exterior forces would break it. It is therefore conceivable that the break-up is generally initiated from the interior of the nodule but finalized by some external force such as a current, the benthos or slumping of sediments.

It has been noted that β-breaks are more common in the small nodules (largest dimension 6—7 cm) than in the large nodules (larger than 7 cm). In addition, many of the large nodules contain nodule fragments as cores; these cores are generally of the size and shape of the fragments from smaller nodules. On the other hand α-breaks appear to be more common in larger nodules. However, this latter observation may be related to the greater mass of the larger nodules wich would influence the stability of the nodule during recovery and handling.

As has already been pointed out, the shape of the large nodules is generally not controlled by the shape of the "core fragment" but mostly by the asymmetric growth in the sediment direction. Some of the large nodules, however, have apparent β-break surfaces. These surfaces are at times true β-breaks but are occasionally surfaces which reflect the shape of the core (Fig. 5-12). When an exceptionally large fragment with a long and sharp break surface serves as a core, the new overgrowth then often appears as an irregular flat surface which resembles a β-break.

Interior zoning of nodules

All the nodules studied show some zonal pattern. The general pattern of layers is produced by variations in the mineral content of the zones. Most generally alternations are produced by clay-rich and clay-poor ferromanganese layers, but compositional differences in the ferromanganese layers (as evidenced by differences in polishing hardness) also produce zonal variations.

Although the zoning appears to be concentric, individual bands are not necessarily continuous around the entire nodule circumference. Some bands completely terminate at or just beyond the equator of the nodule; termination is generally characterized by a sudden tapering of the band (Fig. 5-13).

Fig. 5-6. Internal fractures. A. Radial and concentric fractures. The radial fractures are thought to be shrinkage cracks. All are widest near the centre and rapidly terminate toward the margin of the nodule; only a few penetrate the outer nodule skin. White patches in the fractures represent clay which originally filled all fractures but was largely displaced by plastic (light grey). B. Radial and concentric fractures largely confined to the older nodule core. Surface last exposed to seawater is the upper section of the photograph. C. Surface of a radial fracture completely confined to the interior of the nodule. Note the black rim of clay-free ferromanganese oxide surrounding the clay-lined (light gray) fracture surface. Surface exposed to sediment is in the upper section of the photograph. Photographs A and B also show older nodule fragments serving as growth nuclei.

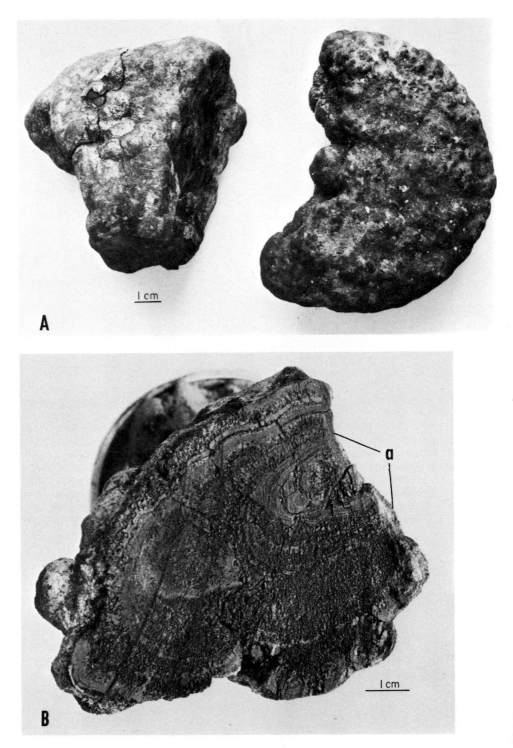

Fig. 5-7. Old or β-breaks. A. These nodules broke on the sea floor. B. and C. The fracture which truncates zones of both the older core fragment and the newer rind has a thin (1 mm) layer of ferromanganese oxides covering the break producing a sharp angular unconformity (at a).

Fig. 5-8. Partial break-up of a nodule caused by spalling. A. Partially detached flake; the frontal surface represents a healed β-break. B. Fragment completely separated from the nodule by radial and concentric fractures (at *a*); the cavity and the front of the nodule represent healed β-break surfaces. C. The nodule reveals a spalled central portion (at *a*, compared with the cavity in B above), a healed β-break at *b* and a clay-filled α-break at *c*.

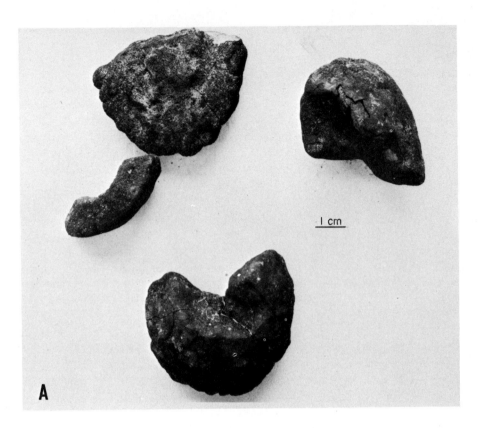

Fig. 5-9. Complete break-up of a nodule involving the entire cross section of the nodule. A. All breaks are old or β-breaks. These nodule fragments are of the size commonly found as cores in whole nodules. B α-break at *a* involving the entire cross-section of the nodule and a clay-covered β-break at *b*.

Fig. 5-10. Sea-floor photograph. The action of the foot of this bottom dweller could cause the removal of a spalled nodule section or the break-up of the entire nodule.

Fig. 5-11. Recent or α-breaks. A. This nodule broke after recovery. Note the continuous black band of ferromanganese oxide on the upper (most recently exposed to seawater) portion and the discontinuous black band on the lower portion. This fracture was partially open to the outside. B. Nodule broken after recovery. Note that the fracture was clay-filled except for the black band of ferromanganese oxide on the upper (exposed to seawater) surface which held the nodule together until after recovery.

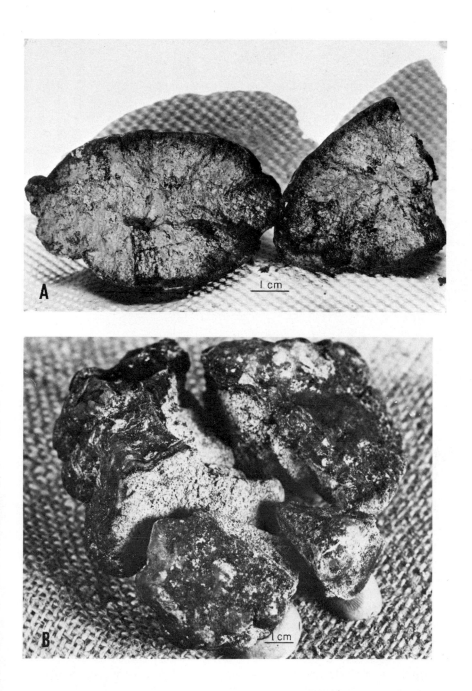

Fig. 5-12. False β-break. The shape of the nodule is controlled by the shape of the core fragment. The outer rind reveals continuous growth. Note also the radial and concentric fractures which are confined to the core fragment.

Other bands are continuous but show drastically different thicknesses. These continuous bands can generally be divided into two parts: (1) a thick portion which is essentially uniform in its width throughout one-half of the nodule, and (2) a tapered thin portion. The tapered thin portion starts near the equator of the nodule and rapidly tapers to a very thin lamina at the pole of the nodule (Fig. 5-13). Frequently the thick portion of a continuous band appears very clay-rich and ferromanganese oxide develops in a characteristic dendritic pattern, whereas the tapered counterpart on the opposite side of the equator visually appears essentially clay-free and often has a dense, hard appearance. Similar observations regarding tapered zones and changes of the appearance of the ferromanganese matter have also been made by Foster (1970) using the reflecting microscope.

Some zones, especially close to the core, are quite regular in thickness and even textured, suggesting a very uniform environment (possibly temporary burial) during that episode of growth (Fig. 5-14). These uniform bands are generally clay-rich and ferromanganese oxide forms as dendrites.

The cores of the manganese nodules from the study area are over-whelmingly composed of broken fragments of older nodules (Figs. 5-1B, 5-3, 5-6A and B, 5-12 and 5-13 A and B). Only in rare instances are nodule cores composed of clay or rock; animal remains (e.g. sharks' teeth or whales' earbones) are almost completely absent.

Core fragments range in size but generally do not exceed 3—4 cm (longest dimension). The fragments appear identical to small nodules which are characterized by abundant β-breaks. In a few instances the core fragment of the large nodules shows its own core fragment which once again is a portion of a still older nodule. In none of the core fragments have rock or clay cores been observed. It is conceivable that nodules with clay, rock, or tooth cores are the youngest type of nodule growing on the sea floor. Attachment of the ferromanganese oxide crust to this core could be relatively weak and break-up of these nodules may be more common than with nodules composed of oxide material throughout. The nodule fragment serving as core also is chemically different from the rind which forms the present outer portion of the nodule.

NODULE CHEMISTRY

In relation to morphological character, the smooth surface (most recently exposed to seawater) and the gritty, lumpy surface (most recently exposed to clay) are markedly different chemically from each other and from the interior of the nodule. Where tops and bottoms of the nodules can be recognized, the tops are high in Fe, Co and Pb, low in Cu, Ni, Mo, Zn and Mn; the bottoms show a reverse pattern being high in Cu, Ni, Mo Zn, and Mn and low in Fe, Co, and Pb (see Raab, 1972, pp. 36—37, Table 1). In nearly all these analyses the Al content, a measure of admixed clay, is essentially constant. The differences in metal concentrations are sometimes small but most frequently amount to more than 40—100%; these variations in metal content are quite reproducible from nodule to nodule and are independent from the admixed clay content.

In addition to this reversal in concentrations, the tops and bottoms of the nodules are zoned continuously around the external surface. Thus the bottom centre portion is highest in Cu, Ni, Mo, Zn, and Mn and decreases towards the equatorial rim; the lowest Cu, Ni, Mo, Zn, and Mn concen-trations are found at the top centre portion of nodules. Fe, Co, and Pb behave in a reverse pattern; they are highest at the top centre and lowest at the bottom centre with intermediate values near the equator (Fig. 5-15; see also Raab, 1972, pp. 36—37, Table 1).

Averaged values for the nodule cores also show them to be generally high in Cu, Ni, Mo, Zn and Mn and low in Fe, Co, and Pb as compared with the

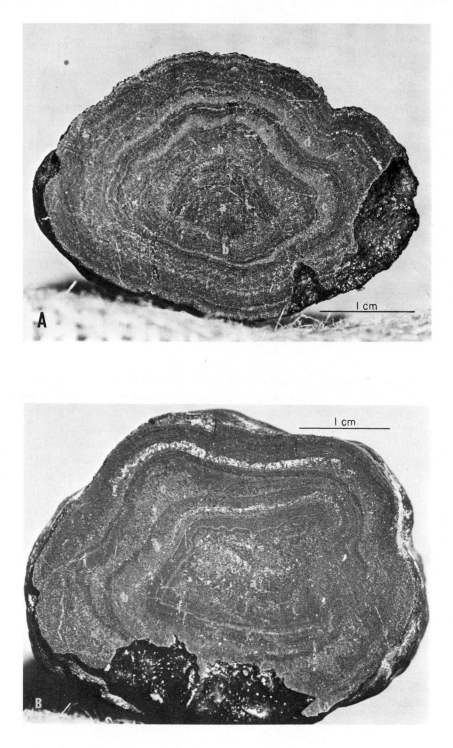

Fig. 5-13. Zoned nodules. A. Core fragment at *a* surrounded by a uniformly thick band at *b* suggesting a growth stage while completely buried in sediment. Also visible are sharply tapered bands at *c* and *d*. B. Core fragment at *a*, tapered bands at *b* and *c*. C. Core fragment at *a*, uniformly thick portion of a band at *b*, thickening of the band at the equator at *c* and wedging out at *d*.

Fig. 5-14. Nodule with a clay-rich zone at *a*. This band is quite regular in thickness suggesting a uniform environment of growth possibly during temporary burial in sediment. The left half of the nodule shows the sample locations 1-1 through 1-10 of Raab (1972, p. 40, Table 3).

Fig. 5-15. Chemical zoning study of a knobby nodule. A. Sample sites for upper smooth (most recently exposed to seawater) surface. B. Sample sites on lower gritty (most recently exposed to sediments) surface. See also Raab, 1972, pp. 36—37, Table 1.

rinds growing around them. The differences when compared with the outer nodule layer chemistry for Mn, Ni, Zn, Co, and Mo were less pronounced and more erratic, but were strong (generally greater than 30%) for Fe, Pb, and Co. The Fe, Co, and Pb deficiency in cores relative to rinds may be related to:

(1) A difference in solution composition which formed the original core nodules.

(2) Differences in chemical environments, i.e., undersaturation with respect to Fe, Co, and Pb and therefore selective removal of these elements.

(3) Insufficient time before break-up and incorporation in the present nodule to become saturated with respect to Fe, i.e., Fe-oxides may act as a cementing agent in nodules.

In contrast to the surface layers, individual layers within nodules tend to be more uniform in composition. In two nodules examined in detail, a single major layer was analyzed in great detail (Figs. 5-14 and 5-16). The layers were chosen because they showed marked thinning in one-half of the nodule. Visually, each layer appeared richer in clay than the next neighbouring layer; however, chemically the Al content was essentially the same as the average for the nodules. As previously shown by Raab (1972, p. 40, Table 3), the layers are essentially compositionally homogeneous except in the area of the equator. In the equatorial region where the layers are thickest, suggesting the most active precipitation of ferromanganese material, Al and Cu are higher than elsewhere in the layers. On a constant Al basis, i.e., equivalent clay basis, several other elements show a slight increase in the equatorial region. The composition of internal layers therefore contrasts markedly with the surface layer composition.

These compositional and morphological features suggest that:

(1) The external surface exposed to seawater is unstable with respect to Cu, Ni, Mo, Zn, and Mn and these elements are selectively removed from the nodule, implying that, in some portions of the oceans, seawater is undersatuarated with respect to these elements.

(2) That internal layers are isolated from this effect and persist essentially unchanged since they were precipitated.

That no compositional zonation can be recognized in internal bands also suggests that precipitation could have been very rapid, advancing at a rate greater than leaching. The thinning of the band on one side of the equator could thus reflect less supply of ferromanganese components in solution to the top of the nodule. If leaching of Cu, Ni, Mo, Zn, and Mn causes the depletion of these elements at the top of nodules, then older nodules should be depleted in these elements and enriched in Fe, Co, and Pb. It is conceivable that some high Fe-Co nodules reported from the vicinity of Tahiti may represent older, more leached nodules. The smooth upper surface of nodules previously described is also thought to be a product of this leaching, possibly augmented by some mechanical erosion.

Fig. 5-16. Compositional zoning of internal zones. Left half of the nodule shows the sample sites 2-1 through 2-10 analysed by Raab (1972, p. 40, Table 3).

EXPERIMENTAL NODULE GROWTH

Small iron-manganese concretions were formed in and on clay sediment after termination of a hydrometallurgical experiment (Fig. 5-17). For the experiment, nodules were ground to -80 mesh (particle size <0.18 mm) and allowed to dissolve in an aqueous solution of H_2SO_4 and SO_2. During the course of experimentation all the metal oxides dissolved but most of the clay contained in the nodules remained in suspension. After shelving of the sealed reaction vessel, the sediment settled and formed a layer about 2 cm thick. After about five months, brownish-black spherical aggregates, about 0.5—1.0 cm in diameter had grown on, partially submerged in and completely within the sediment. The nodules remained constant in size for an additional 40 months prior to their removal for analysis. Nodules formed on the sediment surface were matted aggregates of fine hair-like particles which appeared rigid while in the vessel. When removed from the vessel, the concretions proved to be X-ray amorphous gels with little strength to support their weight. Nodules formed within the sediment (Fig. 5-18) were composed of individual dendrites radiating outward from a common centre; much clay remained in the spaces between individual dendrites. Some clay-rich zones in deep-sea nodules have great textural simularities with these laboratory accretions. In ocean nodules these zones are composed of ferromanganese oxide dendrites with interstitial clay generally separating individual dendrites. Because these zones are clay-rich and often very uniform in thickness around the entire nodule they are believed to have formed during temporary burial of the entire nodule.

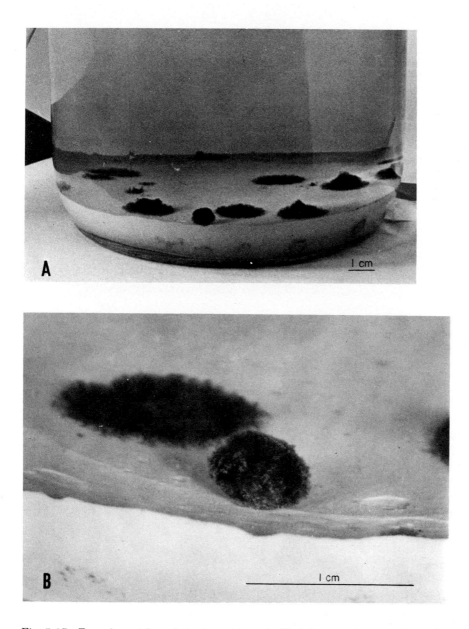

Fig. 5-17. Experimental nodule formation. A. Nodules growing on the surface of and within the sediment. B. Detail of a surface nodule.

Fig. 5-18. Experimental nodule formation. Dendritic growth of nodules within the sediment. Several of these dendrites are connected with the portions protruding above the surface of the clay.

The composition of the experimental solution and of the nodules formed (Table 5-IV) is of course not significant in terms of ocean nodule formation. However, the formation of nodules in and on sediment which contained particles smaller than 0.18 mm implies that nucleation sites need not be large and that spherical aggregates are formed regardless of the location of the nuclei either within the sediment, at the mud—water interface or on the sediment surface. Ocean nodules without a recognizable non-ferromanganese oxide core may therefore well have formed by spontaneous nucleation or nucleation on a clay-size particle.

A POSSIBLE MODE OF FORMATION OF MANGANESE NODULES

That the formation of nodules requires some special conditions is illustrated by:

(1) The discontinuous distribution of nodule fields.

(2) The chemical variations between nodule fields; some are Mn, Cu, Ni-rich, others are high in Fe and low in Mn, Cu, and Ni, etc., yet the underlying sediments are essentially constant in their trace metal composition (all nodules studied here were collected from the clay zone from approximately 5,000 m depth).

TABLE 5-IV

Experimental nodule formation; chemical composition of starting and end products

	% Mn	% Fe	% Cu	% Ni	% Co	% Mo	Eh	pH
Original deep-sea nodule used	31.1	5.55	1.23	1.44	0.24	0.074		
Solution at termination of hydrometallurgical experiment	1.24	0.26	0.05	0.07	0.01		0.557	2.50
Solution after formation of experimental nodule	1.22	0.04	0.05	0.05	0.01	<0.03	0.664	2.21
Experimental nodule	0.82	34.22	0.25	0.35	0.33	0.37		
Sediment after formation of experimental nodule	0.14	17.03	0.06	0.02	<0.01	0.28		

(3) Chemical gradients which could be ascribed to upward migration of trace metals due to diffusion cannot be demonstrated (Raab and Norton, 1973).

(4) Nodules accrete from the sediment side, but dissolve on the seawater side of the mud—water interface.

Because some components of manganese nodules dissolve in contact with seawater, it appears unlikely that nodules form from direct precipitation from seawater. Manganese flux due to the diagenetic expulsion of interstitial water or due to diffusion is inadequate to explain the formation of manganese nodules (Raab, 1972; Raab and Norton, 1973).

Intraplate igneous activity as that encountered by the Deep Sea Drilling Project in the Pacific is considered here to be intimately associated with nodule formation. Intrusive sills (with associated baked sediment contacts) could increase the temperature in the sediment colum causing: (1) increased solubility of metals in interstitial waters, thus effectively leaching nodule components from the sediment column (for experimental evidence see Raab, 1972); and (2) generate a convective fluid flow, thereby increasing the rate of fluid migration out of the sediments.

At the mud—water interface these warm solutions would spread along the interface similar in behaviour to the Red Sea brines and precipitation could take place both from ascending and overlying solutions, thus explaining the concentric layering of nodules and the uniformity of metal content in individual internal layers. Once the metal-rich brine layer had precipitated its metals or was dissipated, newly formed nodules would be exposed to normal seawater and the previously described process of chemical leaching of Cu, Ni, Mo, Zn, and Mn could become effective.

Intraplate intrusives thus provide the energy necessary to extract and transport metals from the sediment column to the mud—water interface and are suggested as being ultimately responsible for the formation of deep-sea manganese nodule deposits.

CHAPTER 6

INTERNAL CHARACTERISTICS

R. K. SOREM and R. H. FEWKES

INTRODUCTION

The concretionary and encrusting nature of modern marine manganese deposits has been recognized for many years. It is therefore somewhat surprising that the details of the layering of these nodules and crusts are still so poorly understood. Although Hamilton (1956) and Mero (1962) clearly showed the variety of nodular structures present, it was Cameron (1961) and Arrhenius (1963) who first showed the complexity of the layering by means of photomicrographs. Burns and Fuerstenau (1966) were among the first to show that chemical variations exist from one layer to the next in a single nodule, and Sorem (1967) emphasized the mineralogical and textural features, as well as large structural features which indicate non-uniformity of deposition, now known to be typical of individual nodules the world over.

In the past few years it has become clear that a thorough understanding of nodule origin and the many natural factors affecting the economic use of nodules requires much more data than are presently available. It is therefore fortunate that recently many workers in diverse disciplines have undertaken the detailed research necessary to obtain the needed information on physical features, mineralogy, and chemical composition of nodules. The outlook for new developments in this field in the next few years is indeed bright.

It is the purpose of this chapter to present information currently available on the internal features of manganese nodules, to offer genetic interpretations of these features, and to suggest new lines of research. To judge from the limited data available on ocean floor crusts rich in manganese and iron (see, for example, Aumento et al., 1968), much of what is learnt about nodules may be applied eventually to an understanding of the origin of these more continuous masses. A genetic relationship between these two is likely.

METHODS OF STUDY

Manganese nodules are complex heterogeneous three-dimensional objects, and a thorough understanding of the internal character of a nodule cannot be obtained by using any single research method. Not only must a variety of details be studied, but the scale of the details must cover a range from megascopic to sub-microscopic. Furthermore, interpretation of many internal characteristics must commonly depend upon knowledge of the modern external habit of the nodule, so a complete investigation of nodule structure must include an examination of the exterior as well as the interior. For these reasons, a thorough study requires the use of a variety of special equipment, including the ore microscope, the scanning electron microscope (SEM), electron probe microanalyzer, X-ray macroprobe spectrograph, X-ray diffraction apparatus, and a wide assortment of analytical techniques. Although a discussion of the technical operation of the various instruments is beyond the scope of this chapter, the general application of those methods most commonly used is described briefly below.

Ore microscope

The familiar concentric shell structure of nodules is commonly visible on broken or sawn nodule surfaces, but a good quality polished section is necessary if the microscopic details of the internal structure are to be studied. The structure of a whole nodule is best revealed by a complete section through the centre, and photographs of whole nodule sections form excellent "maps" with which to control all analytical work on nodules (Fig. 6-1)* (Sorem and Foster, 1972a). Many microscopes and other analytical instruments will not accept whole sections of large nodules, however, and parts of nodules must be sectioned separately for many studies.

Preparation of good polished sections of nodules is more difficult than with most common ores, but with experience it is relatively straightforward if modern mounting, impregnating, and diamond polishing methods are used (Sorem and Foster, 1972a). Generally the larger the section the more difficult and time-consuming the preparation. Ideally, a scratch-free section should result so that the specimen may be studied with the polarizing ore microscope at even the highest magnification. If a nodule contains non-opaque minerals of interest, a matching thin-section may be prepared. Polished thin-sections are particularly valuable for some work.

For best results, polished sections should be examined using vertical illumination, like that of the ore microscope, so that both physical relationships and optical properties may be determined. With a good instrument, objects as small as 0.1 micron in size can be resolved, but low to moderate magnification is generally used. Numerous properties may be observed to aid in mineral identification, including colour, reflectivity, bireflectance, and anisotropism, as well as physical properties like habit, cleavage, grain size, texture, hardness, and to a lesser extent chemical reactivity and magnetism.

---

*All photographs in this chapter are of polished sections using vertical illumination. Opaque minerals appear light, non-opaque minerals appear dark. Near outer edge of nodules is at top in all photomicrographs.

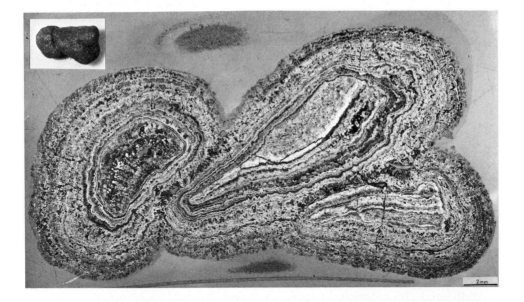

Fig. 6-1. Polished section. Photomacrograph showing whole nodule structure. Accretion has produced crustal zone of thin continuous shells (layers) around older diversely oriented nodule fragments. Reflectivity of different layers ranges widely and is dependent primarily upon mineralogy and texture. Layers prominent at this magnification are called "gross layers". Photograph courtesy of D. M. Banning. Nodule DH 7—4, East Pacific Ocean, depth 3,660m. (Inset: exterior view of nodule before sectioning.)

The investigation of polished sections of manganese nodules is particularly challenging because of the extremely small grain size and intricate mineral relationships, but the work is very fruitful. Anyone who undertakes this kind of research, however, should be sure that he has a modern ore microscope in good adjustment and an adequate background in its use. Reference to the text by Cameron (1961) would be helpful. Only good quality sections should be used.

Useful optical properties
In the study of nodule sections, the single most useful property in plane-polarized light is *reflectivity*. Most nodule minerals lack colour but differ in reflectivity, with resulting differences in appearance which range from white to dark gray. Fortunately the eye is sensitive to very small differences in reflectivity, so that adjacent mineral layers less than one micron thick which differ in reflectivity by only a few percent can commonly be distinguished. In the future, it is expected that quantitative measurements of reflectivity will be useful in estimating the approximate chemical composition of nodule minerals.

An important related property is *bireflectance*, which is characteristic of the crystalline oxide minerals in nodules and is accompanied by *anisotropism* between crossed nicols. Detection of these effects permits the instant distinction between crystalline and amorphous nodule minerals, since thus far no crystalline isotropic mineral has been found in nodules. The importance of this distinction cannot be overemphasized, for recognition of crystalline oxides is tantamount to identifying the nodule material richest in manganese, copper, and nickel. To our knowledge, no well-crystallized iron minerals have yet been found in marine nodules.

Other properties

The relative hardness of different minerals in nodule sections, as revealed by polishing behaviour and as detected by the pseudo-Becke line method, is also useful in identifying intricate intergrowths of fine-grained minerals. The scratch hardness (Talmage) is also of interest and should be estimated, along with an evaluation of tenacity, whenever a microsample is extracted for X-ray diffraction analysis.

Additional characteristics to aid in mineral identification are being sought. Among the techniques currently being explored are chemical etching, magnetic testing, and monochromatic photography. At present, however, we know of no available reference data usable on a microscopic scale.

Other nodule characteristics which have attracted interest in the past include specific surface area, which is uncommonly great (Weisz, 1968), specific gravity, and solubility (Raab, 1972). Although data on these properties may not be of general use for mineral identification, it would seem very desirable to investigate microsamples of known optical character so that the findings may be applied to the solution of problems of nodule origin if possible. Information of this kind may also be of interest in the development of the most efficient commercial uses of manganese nodules.

Available data on major nodule minerals are presented in Table 6-I.

Scanning electron microscope

The SEM is unsurpassed for the study of *undisturbed* nodule material at any magnification beyond that of a medium-power binocular microscope. Whether the subject is the external habit or a fracture surface from within the nodule, preparation is simple and the operator can see and photograph objects less than one micron in size with remarkable depth of focus. Under suitable operating conditions, gold vapor-coating of the sample may be omitted (Margolis and Glasby, 1973), or a carbon coating may suffice if an electron microprobe is used in the SEM mode (C. Knowles, personal communication, 1973). Although the SEM generally shows little when a polished section is examined, study of an etched nodule section may be of value for certain problems (Margolis and Glasby, 1973).

TABLE 6-I

Properties of chief nodule minerals in polished sections

Mineral	Colour	Reflectivity[1]		Comments
Todorokite $(Mn,Ca)Mn_3O_7 \cdot 2H_2O$[2]	White to light gray (bireflectant)	(nm) 475 547 591 651	(%) 14.0 12.7 13.7 13.3	Randomly oriented microcrystalline intergrowths, appearing fibrous in places. Takes a fair polish.
Birnessite $(Na,Ca)Mn_7O_{14} \cdot 2.8H_2O$[3]				Anisotropic, colours bluish gray to bluish black. Talmage hardness B.
Opaque amorphous material (Impure Mn and Fe hydroxides)	White to gray	(nm) 475 547 591 651	(white) % 14.8 14.0 13.7 13.4 (gray) % 9.1 9.0 8.8 8.3	Massive, fine-grained, amorphous. Colour and reflectivity vary with composition. Sensibly isotropic. White takes a good polish. Talmage hardness B+. Gray polishes with some difficulty. Talmage hardness B.

[1] Measured by D. L. Banning, Washington State University, using Leitz MPV—1 Photometer with calibrated glass standard.

[2] After Larson (1962).

[3] After Brown et al. (1971).

X-ray diffraction

The mineralogical composition of manganese nodules is as yet poorly understood (see Chapter 7). At present, the most useful information is based on X-ray diffraction analysis of microsamples weighing less than a milligram (Sorem and Foster, 1972a) but much remains to be done. Serious problems of recognition and nomenclature (Sorem, 1972), which were discussed recently at a workshop seminar sponsored by the National Science Foundation (Battelle Seattle Research Centre, June 1973), will not be overcome easily. Work with very small microsamples, for which optical and chemical data are known, seems to hold the greatest promise for the development of routine mineral recognition criteria and for providing mineralogical data related to nodule origin. Crystal structure determinations based upon single crystal analysis are desirable, but the outlook for this work is not bright in view of the cryptocrystalline and amorphous nature of most nodule material. Electron diffraction analysis has been attempted (J. Zussman, personal communication, 1970), but as yet no useful results have been published.

Inter-laboratory standardization of X-ray diffraction methods and nomenclature has recently been recommended (Sorem, 1972), and round-robin analysis of standards was begun in 1975 among a number of laboratories.

Chemical analysis

The chemical composition of manganese nodules has been reported by many investigators. The results represent a wide spectrum of sample types and analytical techniques, however, and therefore all are not directly comparable. Both partial and complete analyses have been made, and both whole nodules and parts of nodules have been used as samples. The most common methods of bulk analysis are X-ray fluorescence and atomic absorption, although other techniques are also reported. Emission spectrography and neutron activation analysis seem to offer special advantages for small sample work.

Recently the importance of analyzing very small parts of nodules has been recognized, and several non-destructive probe methods of analysis have become popular. Not only do these methods permit the anlaysis of very small samples, but they are essentially non-destructive and leave the sample in condition for optical study. Both spot analyses and concentration variation curves along selected paths have been published. The finest spatial resolution reported thus far is obtained by use of the electron microprobe, with which areas as small as 1—3 microns across on polished sections may be analyzed quantitatively. Greater resolution (e.g., 100 Å) is claimed by several SEM manufacturers who are now marketing energy dispersive analyzing units, but quantitative work is difficult. Analyses of nodule materials by this method have not yet been published. As with the microprobe, both traverses and spot analyses should be possible eventually. At the other extreme of resolution is the X-ray macroprobe spectrograph, which can be used to determine major element composition of areas from 100 to 500 microns in size on polished sections and is especially well-suited to traversing distances of several centimetres across layered specimens relatively rapidly (Sorem and Foster, 1972b).

Semi-quantitative element distribution in polished sections can also be shown by characteristic X-ray scanning photographs obtainable with both microprobe and SEM instruments. This method effectively shows the spatial inter-relationships of numerous elements in an area of microscopic size.

NODULE MINERALOGY

General

The emphasis in this chapter is on mineral recognition in polished sections of manganese nodules. For this purpose both crystalline and X-ray amorphous materials are termed minerals, for both are abundant in nodules and generally are closely intergrown (Fig. 6-2). Although hydrous manganese and iron oxide minerals predominate in most nodules, minor amounts of silicates and other minerals are commonly present. All deserve careful study.

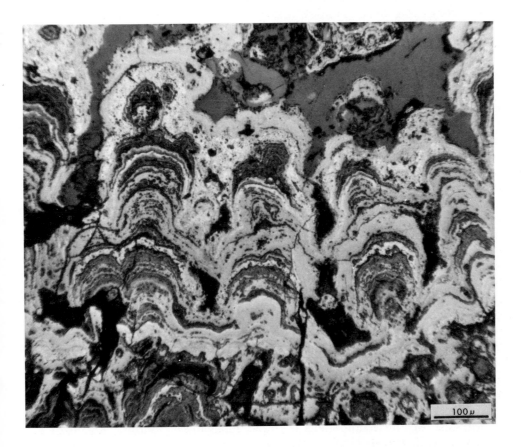

Fig. 6-2. Photomicrograph. Plain light. Typical colloform interlamination of iron-rich amorphous material (medium gray) and crystalline manganese minerals todorokite and birnessite (light gray). Dark gray and black irregular areas are fine-grained detritus, synthetic resin, and surface pits. Outer edge of nodule is just beyond field of view at top. Nodule NP 2—1B, East Pacific Ocean, depth 4,570 m.

Minerals identified

The most common crystalline oxide minerals in marine manganese nodules are apparently todorokite and birnessite. Almost without exception, these minerals are intimately intergrown, even in microsamples weighing less than a milligram. Proportions vary from place to place in a single nodule, however, and rarely monomineralic specimens are found (Banning, 1974). The optical properties of these minerals are poorly known, although some data have been presented for both the marine minerals and for specimens from terrestrial manganese deposits (Cameron, 1961; Larson, 1962; Sorem, 1967, 1972). At present the most useful optical properties are reflectivity, bireflectance, and anisotropism. Reliable chemical analyses are reported for terrestrial specimens (Straczek et al., 1960; Larson, 1962; Brown et al., 1971), and the general chemical formula for each mineral is well established, as noted in Table 6-1. Extensive cation substitution has been documented in todorokite.

Other crystalline oxide minerals rich in manganese which reportedly have been found in the growth layers of marine nodules include nsutite (Manheim, 1965), psilomelane, pyrolusite, woodruffite, and vernadite (Andrushchenko and Skornyakova, 1969), and ranciéite (Sorem, 1967). Details of the habit and occurrence of these minerals in nodules have not been described, however, and their significance cannot be evaluated at present. In any event, these minerals would be expected to be very fine-grained and difficult to recognize optically without prior identification by X-ray diffraction analysis of microsamples.

Non-crystalline or X-ray amorphous oxides are abundant in nodules and in fact commonly predominate. Poorly crystallized goethite or a related ferric hydroxide has been reported as a nodule constituent by several investigators (Arrhenius, 1963; Glasby, 1972c), but it has not been widely observed. Detailed X-ray diffraction study of iron-rich material in many eastern-Pacific nodules failed to reveal the presence of any crystalline iron compounds (Carr, 1970), even where the material was interlaminated with crystalline manganese minerals. These materials cannot be ignored in research on the origin of manganese nodules, however, and they deserve far more attention than has been reported in the literature. Thus far the only attempt to classify the amorphous oxides microscopically is that of Foster (1970), who used the general term "opaque amorphous material" and recognized both iron-rich and manganese-rich types.

The very fine-grained non-opaque minerals which form an essential but minor part of the oxide shells of most manganese nodules (Fig. 6-3) are poorly known, and in most published reports they are ignored or treated in only a cursory way. Arrhenius (1963) reported "crystallites" of opal, rutile, anatase, barite, and nontronite. Various authors have reported such minerals as montmorillonite, illite, and other less precise species, but identification criteria and mineral relationships are usually not described. According to D.

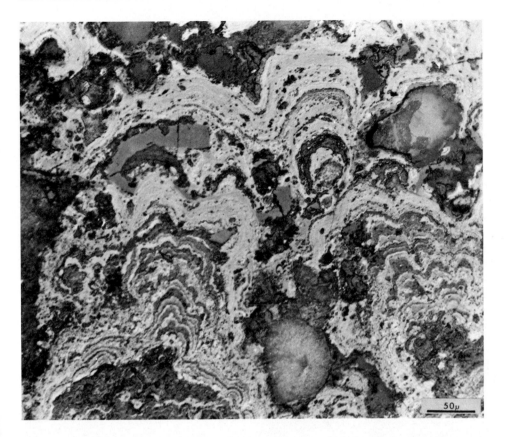

Figure 6-3. Photomicrograph. Plain light. Thinly laminated colloform structure, largely amorphous iron-rich oxides. Dark gray nonlayered areas are "clay" pods and coarser silicate minerals. Angular outlines indicate detrital origin. Nodule SP 2—1B, South-central Pacific Ocean, depth 5,180 m.

Pitzl and J. A. Kittrick (personal communication, 1973), both amorphous and finely crystalline minerals are present in the material usually characterized as "clay", but identification of the mineral species present is a difficult and tedious task. Pitzl has attacked this problem by applying mineral treatment techniques and X-ray diffraction analysis successively to standard API clay minerals, bulk concentrates of non-opaque nodule material, and finally to microsamples extracted from well-documented sites in a single nodule. He concludes that the minerals montmorillonite, illite and chlorite, as well as amorphous silica are present in the eastern-Pacific nodules he studied, and he believes that several other silicates (smectites?) will eventually be substantiated. Apparently there is not as yet any rapid method of identifying microsamples of these minerals with assurance.

Carbonate minerals are evidently rare in manganese nodules, probably because of the great depths at which most nodules form. Even in the relatively shallow deposits of the Blake Plateau, the only calcareous mineral recognized is manganocalcite, which forms thin veinlets which cut the oxide layers.

Clastic grains of many kinds of material have been reported in nodules, including minerals, rock fragments, fossils, and glass shards. Little detailed work has been reported on the detrital minerals, however, and there are as yet no good criteria for distinguishing between authigenic and detrital grains of clay-size or less except textural features revealed by SEM (Pitzl, 1974). Carr (1970) identified large angular grains of quartz, feldspar, and ilmenite by X-ray diffraction, and both he and Foster (1970) illustrated glass shards. Nodules from the Drake Passage show unusual textural features inherent in growth where influx of detritus is especially rapid (Fig. 6-4; Fewkes, 1972). Metallic spherulus of cosmic origin have been concentrated from nodules by Finkelman (1970, 1972), and Jedwab (1970) described a number of similar spherules in a *Challenger* nodule. In our experience, metallic spherules are rare in nodule sections, for we have seen only a few in the hundreds of sections we have studied (Fig. 6-5). Their occurrence is of more than passing

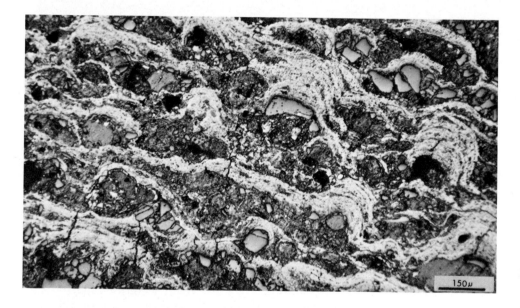

Fig. 6-4. Photomicrograph. Plain light. Abundance and angular shapes of clastic silicate grains (gray, smooth appearance) are typical of debris accumulated during nodule growth in region of high sedimentation rate and strong currents. Matrix layers are highly reflective where iron and manganese content is relatively great (approximately 16% Fe, 6% Mn). Nodule DP 11—2, Scotia Sea, depth ca. 1,980 m.

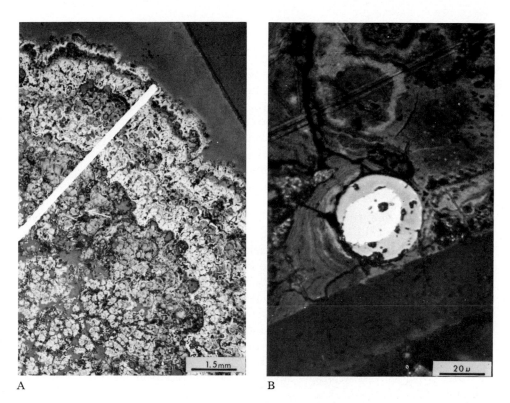

A   B

Fig. 6-5. A. Photomacrograph. Partial view of concentric shell structure, outside surface of nodule at top, showing location (arrow) of micro-spherule shown in B. White layers are rich in manganese; gray areas are rich in iron. White line shows probe traverse path. In places this nodule contains more than 1% nickel, yet only this one metallic spherule is visible. Spherules are not a likely source of nickel in manganese nodules. Nodule BR-1-1, Bermuda Rise, Atlantic Ocean, depth 5,390 m.
B. Photomicrograph of spherule indicate by arrow in A. Primary nature is shown by conformability of oxide layers (dark gray) around spherule. Microprobe analysis shows at least 4% nickel content for spherule (C. Knowles, personal communication, 1974).

interest, for the contribution of such particles to the nickel content of nodules has been speculated upon in the past. The paucity of metallic spherules and the unaltered character of the spherule illustrated, which is similar to those shown by Finkelman and Jedwab, suggest, however, that dissolved spherules are not a likely source for the nickel disseminated in the oxide portion of most marine nodules. Jedwab described spherules partially altered to hematite, but the absolute amount of nickel presumably liberated by such alteration would be very small unless many spherules were completely decomposed.

Mineral habit

Disregarding some of the included clastic debris, nodule materials are without exception extremely fine-grained. With the optical microscope it can be seen that botryoidal or colloform habit is predominant, both in material that is amorphous to X-rays and that which is crystalline. The opaque crystalline minerals are concentrated in relatively pure discrete layers in many places, but microscopic seams and patches of these minerals commonly appear also in dominantly amorphous masses in most nodules, and vice versa. In contrast, the non-opaque minerals rarely if ever form continuous layers but instead seem to fill interstices of opaque mineral aggregates.

In sections, even the crystalline minerals cannot generally be resolved microscopically as individual crystals, but their optical behaviour suggests that they generally form partially oriented intergrowths of platy or fibrous crystallites (Fig. 6-6). Crystallite size is commonly less than 0.2 μ to judge from the diffuse nature of X-ray patterns produced by single chips removed from polished sections. Rarely, tabular crystals a few microns across are revealed in polished sections, but crystals of this size are so fragile that they are likely to be destroyed when a nodule is sectioned, and their abundance is therefore difficult to estimate. Fortunately, the SEM does not require sectioning of specimens, and patient study of natural surfaces and fresh fractures with this instrument reveals well-formed crystals not otherwise visible.

Using micro-X-ray diffraction techniques to establish identification, it has been shown that birnessite is characterized by rounded clusters of thin slightly curved radiating lamellae forming a loose open structured array (Fig. 6-7A). In some areas individual clusters coalesce, resulting in large aggregates or continuous layers. Todorokite has been described by some authors as being in the form of fibrous aggregates (Levinson, 1960), as lath-like, flattened on (010) and elongate (001) with terminal edges inclined ca. 60° and 70° to (001) (Frondel et al., 1960a), or in columnar aggregates (Straczek et al., 1960). Crystallites with a morphology similar to todorokite have been observed in nodules (Fig. 6-7B), but because of their small size a positive identification has not been made. Phillipsite crystals are formed in nodules in association with oxides and non-oxides and can be recognized by their acicular or blocky habit and interpenetrant-twinning. (Fig. 6-9A).

Investigation of X-ray amorphous oxide material with the SEM is also of interest because not only is the lack of symmetrical crystals evident, but textures are revealed which strongly suggest deposition by flocculation. Undisturbed recent accumulations are commonly characterized by tiny floccules which occur on external nodule surfaces in the form of continuous layers and as small globular protrusions (Fig. 6-8A). The amount of admixed detritus is variable. Floccule protrusions (Fig. 6-8B) and loosely aggregated

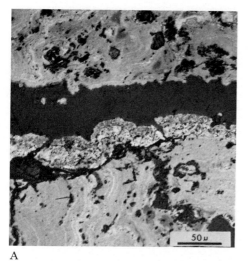
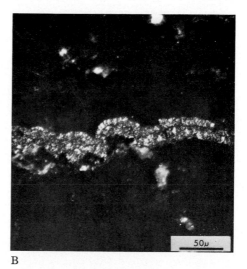

A   B

Fig. 6-6. Photomicrographs. A. Plain light. Encrustation of crystalline todorokite and birnessite (bireflectant, speckled white and gray) along fracture wall. Fracture cuts dense mass of nodule oxides, largely medium gray amorphous material. Very thin contorted white seams may be cryptocrystalline. Main fracture is filled with synthetic resin.
B. Crossed polarizers. Anisotropism of todorokite and birnessite is clearly visible. Wavy extinction suggests aggregates of preferentially oriented crystallites. Nodule NP 2—1B, East Pacific Ocean, depth 5,300 m.

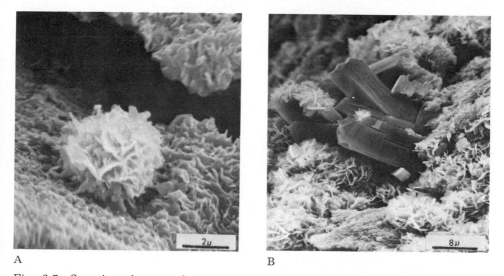

A   B

Fig. 6-7. Scanning electron photomicrographs of crystalline minerals along an open fracture in a manganese nodule. Nodule NP 2-4.
A. Birnessite in radiating clusters of thin lamellae.
B. Blocky crystals of phillipsite and todorokite (?) (lower right centre) intergrown with birnessite.

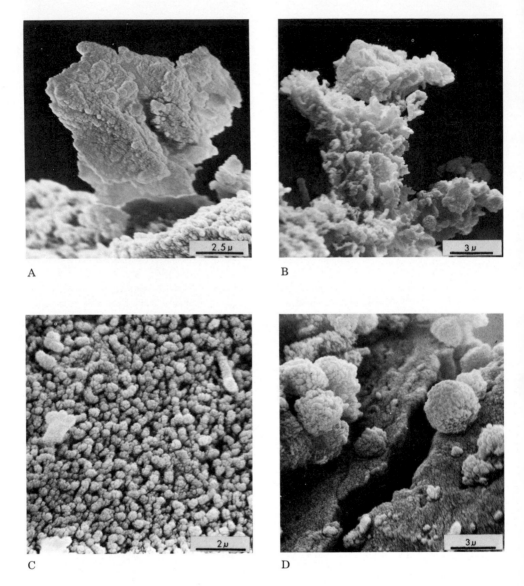

Fig. 6-8. Scanning electron photomicrographs of amorphous material found on external and internal nodule surfaces. This amorphous material is thought to consist primarily of iron and manganese hydroxides which have flocculated onto the nodule during nodule growth. The floccules may form a continuous layer covering everything on the outer surface of the nodule (A), occur as tiny protrusions within the protected environment of nodule cavities (B), form loosely aggregated clusters along open fracture surfaces (C), and agglomerate into larger botryoidal masses (D). Nodule NP 11—4 (A and B), Nodule NP 2—4 (C and D), East Pacific Ocean, depth ca. 4,570 m.

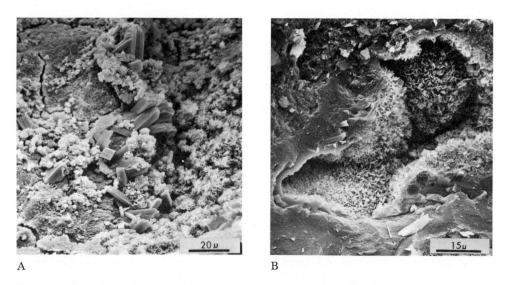

A   B

Fig. 6-9. Scanning electron photomicrographs of montmorillonite on internal nodule surfaces.
A. Intergrowth of birnessite, phillipsite, and todorokite (?) on a montmorillonite substrate. Nodule NP 2-4, East Pacific Ocean, depth ca. 4,570 m.
B. Secondary crystalline growth within a nodule cavity. The thin platelets with ragged edges are suggestive of a clay mineral. Nodule NP 12-1, East Pacific Ocean.

clusters (Fig. 6-8C) also occur within nodules along open cavities and fracture surfaces. The general appearance of the clusters suggests they resulted from the agglomeration of minute colloidal size masses, which in turn probably consist of even smaller particles tightly held together. The coalescing of the clusters may form larger botryoidal masses (Fig. 6-8D).

Clay minerals are also amenable to study by use of the SEM, and the habit of these minerals and their intricate relationship to the opaque minerals is clearly revealed in no other way. In places, birnessite, phillipsite, and todorokite (?) appear to have grown on a crystalline montmorillonite substrate (Fig. 6-9A), whereas elsewhere montmorillonite seems to have crystallized in cavities in the oxides (Fig. 6-9B); Fewkes, 1973).

Distribution of minerals

Considering the problems discussed in a preceding section concerning the identification and classification of nodule minerals, it is clear that published estimates of mineral abundance and distribution in nodules are difficult to evaluate. The most widely published estimates have to do with relative abundance of todorokite and birnessite (or the various "manganites" of Buser) and X-ray amorphous material. Among the earliest estimates are those of Barnes (1967a) and Cronan (1972c), who concluded, as have others more recently, that nodules from certain regions or depths contain predominantly one mineral or the other. The basic data for these estimates were obtained by diffractometer analysis of bulk powder samples, often by use of copper radiation (Glasby, 1972a). There are dangers in sampling, sample preparation, and contamination with this method, and it would be wise to use the estimates so derived with caution.

It is recommended that estimates of mineral abundance hereafter be made chiefly on the basis of microscopic study, combined with camera X-ray powder diffraction analysis of microsample chips. Any diffractometer work should involve carefully documented sampling, crushing under liquid nitrogen (Burns and Brown, 1972), and use of Fe radiation or some sort of monochromatization to avoid fluorescence. In addition, some means must be found to determine the proportion of amorphous material present.

Published illustrations of manganese nodule sections and diffractometer charts suggest that most manganese nodules contain a large proportion of amorphous material, that crystalline oxides are both randomly dispersed and in well-defined segregations, and that the distinction between the various possible oxide minerals is not an easy task.

COMMON INTERNAL FEATURES OF NODULES

Megascopic

Much detail can be seen in a good polished nodule section megascopically, especially if vertical illumination is used to emphasize variations in reflectivity and improve contrast (Fig. 6-1). Useful observations include thickness and continuity of major layers (shells), the number of major layers and the nature of their relationships, large textural features characteristic of certain layers, and the general distribution of oxide and non-oxide materials. In addition, other structures such as fractures and unconformities and the general nature of the nodule nucleus often can be clearly seen. Significant common megascopic internal features of manganese nodules are summarized below (after Sorem and Foster 1972c, pp. 192—194). (Figures referred to are in the original text and are not repeated here.)

Most nodule sections display prominent conformable layers, each ¼ to 1 mm thick, which represent the concentric shell structure so commonly visible on broken nodules. This gross layering, so-called to distinguish it from the much thinner and more delicate laminations always present in nodules, is visible chiefly because of reflectivity differences between adjacent layers or because of the presence of thin clay-rich partings between layers.* It is especially interesting that gross layers form a number of patterns which are not random and are found repeatedly in nodules from many localities.

---

*See Fig. 6-1.

In a single nodule, individual layers are generally extensive, but the pattern of the gross layering is rarely uniform throughout. The inner layers tend to conform to the shape of the core, which is commonly a fragment of an older nodule or a rock. Passing outward from the core, layers tend to follow the core shape less and less closely, but even in large nodules the exterior shape commonly reflects the shape of the core. Angular cores generate angular nodules, and equant or very small cores result in spheroidal ("cannonball") nodules (Fig. 1, 2, 3, 4).

Other common patterns of layering include crenulations, pinch-outs, facies changes and angular unconformities. Crenulated layers in the outer margins of a nodule show the internal structure of knobs or botryoidal forms common on most nodule exteriors. They are not restricted to the outer parts, however, and may be found at any level within the nodule. The other features mentioned actually result in the disappearance of a layer or group of layers. Layers may gradually thin and ultimately pinch out between adjacent layers, or they may change texture gradually as they are traced laterally, with or without a change in thickness (Fig. 3). The most striking termination of layers, however, is shown where the broken structure of an older nodule fragment is overlain at a sharp angle by successive layers which encrust the entire fragment (Fig. 5). These structures commonly pass into scarcely recognizable disconformities where the layers in the core fragment lie parallel to those of the encrusting material. The significance of these and similar features in working out the complex history of nodule growth has been summarized by Sorem and Foster (1968) and described in detail by Foster (1970).

Another striking feature of gross layering is the similarity in thickness from one layer to the next (Fig. 1, 2 and 3), and in many nodules there is in addition a similarity in the internal laminations of contiguous layers which suggests regular repetition or even cyclic deposition. A study of the fine details of the laminations has led to the recognition of texture zones, as described in the next section.

Microscopic

When one first views a polished manganese nodule section with the ore microscope, perhaps the most striking characteristic is the complex texture of many layers which megascopically appear to be structureless and homogeneous. As mentioned earlier, the predominant habit of nodule material is botryoidal, which is seen in polished cross-sections as colloform layering. Under the microscope, it is found that most gross layers have a complex colloform internal texture and are non-uniform mineralogically. Microscopically homogeneous layers are rarely thicker than 0.1 mm and most are as thin as 0.001 mm or less. Furthermore, a wide variety of layering patterns exists, ranging from broad arcs with a radius much greater than the thickness of an individual layer to intricate branching patterns where the radius of curvature is of the same order of magnitude as the thickness of the layers involved. Lateral changes in texture are common and correspond to the megascopic facies changes and unconformities previously described.

The layers vary greatly in denseness and continuity as well as in purity. Contacts between different materials are in places sharp, in others diffuse. Dense layers generally take a good polish, whereas impure or porous layers commonly are pitted. Either type of layer may be crystalline or X-ray amorphous, but in many if not most nodules amorphous material appears to predominate. It should perhaps be re-emphasized that the recognition of crystalline material optically may be difficult because of the extremely fine grain size in many aggregates and the lack of visible euhedral crystals.

In view of these complexities, it may appear that sorting out the mineralogy and structure of manganese nodules in an orderly fashion so that nodules may be characterized microscopically on a routine basis is a hopeless task. Fortunately, however, the fine details and the larger features upon which they are superimposed are not completely random in character, and certain patterns can be found repeatedly in many nodules. This accounts for the recurring use of terms like rind, core, cusp, and nucleus by different authors. At least two important difficulties attend the use of terms like these, however; first, no good definition of the terms is available, and, second, not all common nodule features are included. Probably the introduction of new terms should and can be kept to a minimum, and established petrographic and mineralogical terms will generally be adequate if used with care. To minimize confusion, however, it may be worthwhile soon to compile a glossary of nomenclature for nodule research.

Another approach to systematic usage is to try to categorize certain common major internal nodule features, so that a minimum of detailed description is needed to convey observations to others. The only attempt to do this that has come to our attention is Foster's zone classification, which is based largely on textures visible in sections at magnifications less than 100 X. Foster's research (1970) revealed that a clear relationship exists between many textural features, mineralogy, and chemical composition. The following summary of this classification was presented recently (Sorem and Foster, 1972c) (Corresponding figures in this chapter are noted in square brackets).

In detailed studies of nodules collected west of Baja California, five distinctive textural patterns called zones were recognized in the sequences of laminae. The zones, which differ in homogeneity, textural patterns and composition, are classified as massive, mottled, compact, columnar and laminated. The columnar and mottled zones are the most abundant; the massive, compact and laminated zones are less common. A detailed discussion of the origin and textural interpretations of the various zone types has been given by Foster (1970), from which the following descriptions are abstracted. Comparable zones have been observed in many nodules from a wide range of sampling localities.

Massive and Mottled Zones

The massive and mottled zones contain the greatest proportion of crystalline material, but they differ in textural pattern and content of clay and amorphous material (Fig. 6 and 7). These zones contain the greatest concentrations of Mn, Ni and Cu. The massive zone [Fig. 6-10A] is a dense unit composed predominantly of regular but diffuse laminae of intergrown microcrystalline todorokite and birnessite, with minor amounts of clastic debris. Where analyzed, the massive zone contains approximately 32% Mn, 4% Fe, 2% Ni and 0.8% Cu, and in places Mn may be as great as 60%, Ni as high as 7% and Cu as high as 2% (A. C. Dunham, personal communication, 1971). The mottled zone [Fig. 6-10B] differs in that it contains approximately 15% clay and amorphous material in which the laminae show a chaotic and discontinuous pattern. This zone type contains approximately 21% Mn, 12% Fe, 1% Ni and 0.5% Cu.

The other three zones are made up mostly of laminated opaque amorphous material and differ chiefly in clay content and prominence of intricate colloform layering. These zones contain the greatest concentrations of Fe, Ca, Ti and Si.

Compact Zone

The compact zone [Fig. 6-10C] is texturally similar to the massive zone, except that it is composed mostly of well-defined X-ray amorphous laminae (Fig. 7). However, lenses and pods of birnessite and todorokite intergrowths generally account for approximately 3% of the zone. The compact zone contains approximately 19% Mn, 17% Fe, 0.6% Ni and 0.2% Cu. This zone commonly contains the most highly reflective laminae found in the nodules.

Columnar and Laminated Zones

The columnar zone [Fig. 6-10D] consists of radially oriented columns of laminated X-ray amorphous material (Fig. 8). Clay fills the space between adjacent columns. The laminae composing the columns display a colloform texture and each column characteristically shows a delicate branching pattern. In places the radial columns are short, densely packed and relatively uniform laterally, giving a concentrically layered appearance. These units are called laminated zones (Fig. 7) [Fig. 6-10E]. The columnar and laminated zones are chemically similar, containing approximately 16% Mn, 16% Fe, 0.4% Ni and 0.25% Cu, but composition may vary greatly.

Foster (1970) also pointed out several other unique features which are common in internal nodule zones. Massive layers of todorokite and birnessite typically contain abundant replaced fossil remains, as well as angular fragments interpreted as volcanic glass shards, which are generally scarce elsewhere in the same nodule. The dense amorphous oxide layers (compact zones) contain a different kind of inclusion, however, which Foster termed "globules". He suggested (1970, p. 53) that these objects were possibly arenaceous fossil tests or perhaps "fossil bubbles" related to submarine volcanism (Fig. 6-11). Additional polished-section and SEM research in our department has since shown that many of these structures are tubular rather than spheroidal and strongly suggests an organic origin.

Another characteristic of massive and compact zones is the very irregular contact with older layers toward the nodule core, which contrasts strongly with the more regular broadly curved outer contact (with younger layers). Close examination shows that in places older laminations are actually truncated by the later material, and partial replacement of the older layers seems to be well documented (Fig. 6-12).

In his discussion of the zonal character of manganese nodules, Foster (1970) also mentioned the common occurrence of a group of well-developed broadly concentric layers which form an outer crustal zone a few millimetres thick on many nodules (Fig. 6-10F). This feature has been observed repeatedly since Foster's work and is present in many nodules from points as distant as several thousand kilometres in the Pacific Ocean and even in the Atlantic (Sorem, 1973).

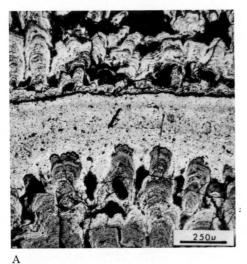

A

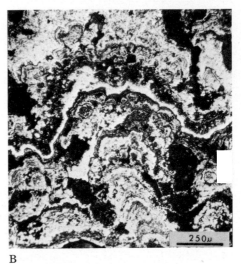

B

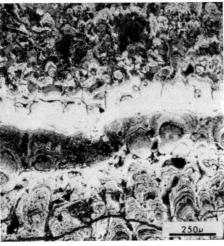

C

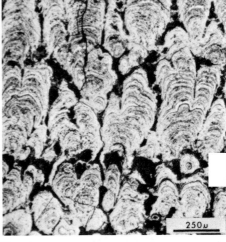

D

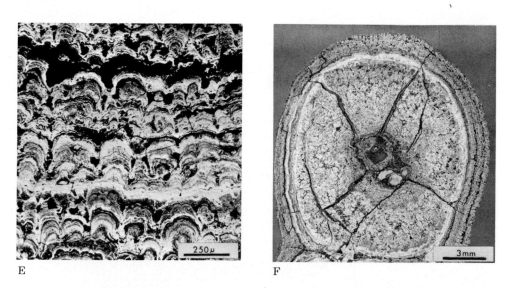

E   F

Fig. 6-10. Typical internal features of North Pacific Ocean manganese nodules. A, B, C, D, and E courtesy of A. R. Foster. Plain light.
A. Photomicrograph. Dense layer of crystalline todorokite and birnessite, with abundant very fine-grained mineral and fossil detritus, finely laminated cuspate oxides above and below. This is Foster's (1970) "massive zone". Irregular lower contact and even upper contact are typical. Crystalline layer was probably deposited as a primary coating on exterior of finely botryoidal nodule surface. Nodule DH 7—2A, depth 3,660 m.
B. Photomicrograph. Chaotic layering of crystalline todorokite and birnessite (white) and amorphous iron-rich oxides (gray). Some areas are pitted (black). This is called the "mottled zone" by Foster (1970). Nodule DH 7—1, depth 3,660 m.
C. Photomicrograph. Very dense white oxide zones roughly conformable with thinly laminated oxide matrix (gray). Both units are iron-rich and amorphous. Diffuse contacts and included pod-like features are characteristic of the highly reflective "compact zone" (Foster, 1970). Nodule DH 5—3, depth 3,840 m.
D. Photomicrograph. Intricate curved laminae of oxides of iron and manganese representing sections through botryoidal nodule structures. Thin white or light gray seams are partly todorokite and birnessite, darker gray seams are iron-rich and amorphous. Black areas are clay or pits filled with synthetic resin. This type of layering is called the "columnar zone" by Foster (1970). Nodule DH 7—2B, depth 3,660 m.
E. A variety of thinly-layered oxides similar to D but with well-developed larger-scale layering upon which the small colloform structures seem to be superimposed. This was recognized by Foster (1970) as the "laminated zone", in allusion to the thicker set of parallel layers. Nodule DH 7—2A, depth 3,660 m.
F. Photomacrograph. Part of a double-core nodule, emphasizing the parallel layers which form the outer shell, a few millimetres in thickness. Recent layers like these are so commonly recognizable as a discrete unit of structure that they are here called the "crustal zone". Fine structure of the zone varies from region to region. Here close examination reveals the texture is that of a "laminated zone". Nodule NP 15—1B.

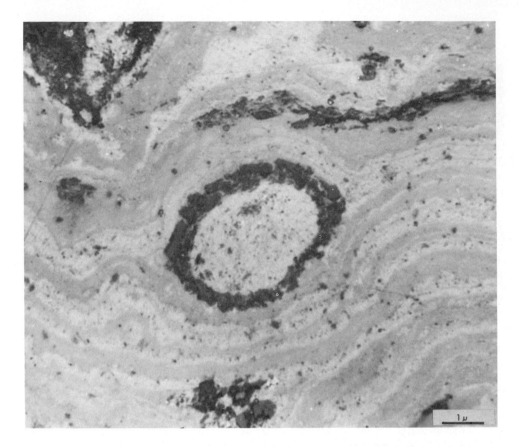

Fig. 6—11. Photomicrograph. Typical "fragment-rimmed pod" in oxide layer containing alternating laminae of amorphous iron-rich oxides (gray) and crystalline manganese oxides (light gray, pitted). Angular clastic grains (dark gray) are commonly associated preferentially with the manganese-rich minerals, whether they form pods or irregular masses. Pods are probably cross-sections through tubular organic structures. Fragments may be of volcanic origin. Nodule SP 2—2B, South Pacific Ocean, depth 4,570 m.

For discussion purposes, it is proposed to term this the "crustal zone". The layers represent the most recently deposited growth shells and deserve careful study.

In many nodules both oxide-rich and clay-rich shells a few microns thick alternate in this zone, suggesting that deposition was influenced by a crudely regular variation of environment with time. In some nodule sections, similar zones of great age may be observed virtually unchanged within nodules, but commonly the original fine structure appears to have been destroyed by diagenetic processes. In many nodules, however, ancient growth habit seems to have been quite different from that of the modern crustal zone.

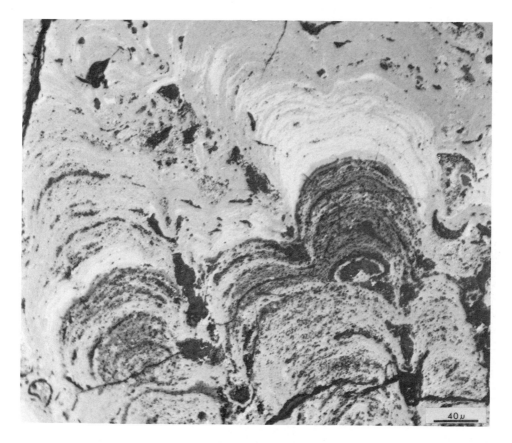

Fig. 6-12. Photomicrograph. Contact between dense oxide layer (well-polished, gray) and older colloform oxide laminations below. Vague colloform pattern in compact oxides above is interpreted as relict structure preserved after partial to complete replacement of the older laminated oxides. Additional evidence is needed to determine the time at which replacement occurred. Nodule NP 3—1A, North Pacific Ocean, depth 5,300 m.

In many nodules several or all of the above zone types may be found stacked one upon the other (Fig. 6-13), and Foster attempted to show a correlation in the sequence of occurrence of zones in nodules from the same dredge haul. He suggested that nodules from localities as distant as 300 km might show comparable similarities. Later work has shown that certain textural features are indeed comparable in nodules from Pacific Ocean localities separated by as much as 7,000 km (Sorem 1973), but correlation of zone succession in nodules on a broad scale has not been established. However, there is ample evidence that the general internal structural pattern of nodules is unique in several regions, and correlation of structure with

Fig. 6-13. Photomacrograph. Outer layers of large nodule showing a wide variety of structure. Close examination reveals most of the zones described by Foster (1970). Columnar, massive, and laminated zones are especially prominent. Relatively low reflectivity of columnar zone across centre of photograph, as compared to the thick zone below, is probably due to the presence of considerable amounts of dispersed clay. Even gray at top is synthetic resin in which nodule is imbedded. Photograph courtesy of A. R. Foster. Nodule DH 7-2A, North Pacific Ocean, depth 3,660 m.

geography or bottom topography may eventually be established. For example, very thin broad shells of clay-poor oxides seem to be typical of Blake Plateau nodules in the Atlantic Ocean, while crenulated oxide shells with appreciable clay are typical of Carlsberg Ridge nodules in the Indian Ocean.

The practicability or desirability of using Foster's terminology or a similar one is debatable, particularly when the internal features of nodules are examined at high magnification. Although "massive zone" logically describes a relatively thick continuous layer of todorokite and birnessite, it would not be a reasonable term for the thin seams and irregular patches of these minerals found in masses of colloform amorphous oxides. This sort of situation is not unusual in geology, however, and an analogy can be drawn with nomenclature problems with sedimentary rocks as an example. The terms sandstone, shale, and limestone are widely used to convey certain textural and mineralogical attributes, and they permit a sort of shorthand description of common rocks. Their wide usage, however, does not prevent the geologist from using more restrictive terms when describing features of special interest, like sandy lenses, shaly partings, and calcite veinlets. Clearly, there are definite advantages in the use of special terms if they are widely accepted. Whatever terms are developed for nodule materials, and there is already a good assortment in the literature, it is strongly urged that they be as truly descriptive as possible and not be based upon genetic interpretations.

In addition to the features already described, nodule growth also results in the development of open spaces, but these have so far received little attention in the literature. Recently, attention was called to the general nature and significance of pores and fractures which are associated with much oxide material (Sorem and Foster, 1967; Sorem, 1973). Pore spaces are commonly very irregular openings interstitial to oxide structures. They are extremely abundant and range from megascopic to sizes smaller than 100 Å. Fractures, which are both parallel to and transverse to the structure of the oxides, are less abundant and generally larger scale features than the pores. In many nodules, pores and fractures are wholly or partially filled with clay or oxides similar to those of the growth shells (Fig. 6-14). In Blake Plateau nodules, manganocalcite occupies some fractures. Fracture-filling minerals are properly called *veinlets* and clearly indicate late deposition within an existing nodule.

CHEMICAL COMPOSITION

General

Although the general chemical composition of marine manganese nodules has been known for many years, interest has recently been increasing in systematic sampling and analysis. Mero (1962, 1965a) published the first extensive compilation of analyses, which he used to delineate several ocean floor regions on the basis of major metal content of nodules. The areas rich in copper, nickel, cobalt, and manganese which Mero outlined in the Pacific Ocean have since been sampled extensively by both scientific groups and mining companies. Recent compilations of the analytical data now available (e.g., Frazer and Arrhenius, 1972; Horn et al., 1973a) provide additional details of the geographic variation of nodule composition and generally support Mero's original conclusions. A strong economic interest in the recent research is indicated by the fact that most of the analyses published since

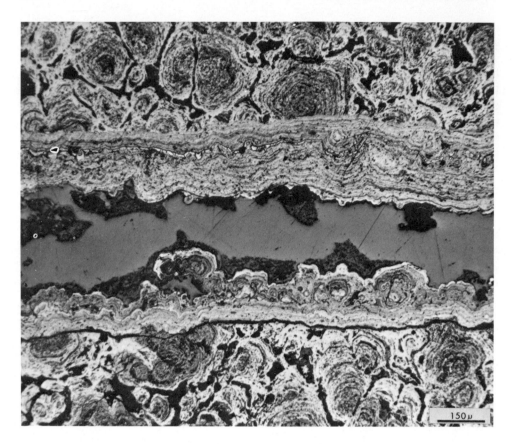

Fig. 6-14. Photomicrograph. A nodule fracture which has been partially filled by oxide material. The structural trend of the younger oxides deposited along the fracture is transverse to the older oxide structures. The abundance of similar textures thought to represent open-space filling in nodules suggests extensive internal migration of mineral depositing solutions. Nodule NP 3—1A, Central Pacific Ocean, depth 5,300 m.

Mero's work have given only the content of the major metals, whereas Mero presented essentially complete analyses. From the scientific point of view, it seems unfortunate that so few complete analyses are available.

Intra-nodule variations

Most nodule analyses published to date represent single whole nodules or a group of nodules from the same dredge haul. The results are useful for many purposes, both economic and scientific, but they camouflage the fact that most nodules are no more homogeneous chemically than they are texturally or mineralogically. Burns and Fuerstenau (1966) were perhaps the first to demonstrate that chemical variations on a microscopic scale exist in nodules, and their conclusions have been supported and amplified by later work (e.g., Cronan and Tooms, 1968; Friedrich et al., 1969; Foster, 1970; Raab, 1972; Sorem and Foster, 1972a; Dunham and Glasby, 1974; McKenzie, 1975). While others used non-destructive probe analysis methods and emphasized major element concentrations, Raab excavated small samples from selected layers of sawn nodules and analyzed them for as many as 12 elements by atomic absorption methods. Later, Piper (1972) and Rancitelli and Perkins (1973) also used small samples from parts of nodules for neutron activation analysis. All of the research cited showed that variations in composition within a single nodule are commonly great, far exceeding the variations in *average* composition of nodules from different bottom localities in the Pacific basin.

In spite of the fact that intra-nodule chemical variations are now well-documented, there is as yet no agreement on the general pattern of elemental distribution and associations, if any, and the cause of the variations. Certain relationships among concentrations of the major metals are found repeatedly, however, and these are now considered typical of nodules by many investigators. In summary, it can be said that iron and manganese are the most abundant metals, and concentration of each varies from place to place inversely with respect to the other, although in some materials they are present in about equal concentrations. In addition, there is a strong sympathetic variation of copper and nickel with manganese content and a definite antithetic relationship between iron and the concentration of copper and nickel, as described earlier by Arrhenius (1963). Burns and Fuerstenau recognized the same variations, and they also suggested a positive correlation between iron and concentrations of cobalt, titanium, and calcium. This correlation is still the subject of study and debate and must be considered uncertain. Most recently, Friedrich et al. (1973) have shown data which in general support the covariance of iron and cobalt within Pacific nodules, but exceptions are noted. Analysis of much smaller samples will probably be needed to establish the true pattern, if one exists.

Relation to mineralogy

It is one thing to establish the fact that certain chemical associations are characteristic of manganese nodules, and it is another thing to explain them. Although there are still gaps in our knowledge, several lines of evidence indicate that chemical variations within nodules are directly related to mineralogical variations. This would come as no surprise to petrologists, for this is the general case in most rocks. The chief reason that it has taken so long to establish the same idea in manganese nodule research lies in the underdeveloped state of nodule mineralogy and petrography.

Suggestions relating chemical composition to mineralogy in nodules have appeared sporadically in the literature for at least 15 years. The first convincing evidence was presented by Burns and Fuerstenau (1966, p. 901), who showed probe analyses relating high manganese, copper, and nickel content to "7-Å manganite", and suggested "isomorphic substitution" as a cause for this relationship. Barnes (1967a), also choosing to use Buser's terminology, concluded that nodules richer in "manganites" contain higher concentrations of nickel and copper than those containing only "δ MnO$_2$", but he also admitted to serious difficulty in estimating mineral proportions. Cronan (1972c), evidently using similar methods to establish mineral proportions in bulk samples, concluded that todorokite-rich nodules contained greater concentrations of copper and nickel than those rich in birnessite. He proposed that nickel and copper substitute for Mn^{2+} in todorokite and, further, that cobalt and lead substitute for either Mn^{4+} in birnessite or Fe^{3+} in FeOOH. He also noted a strong covariance of cobalt and lead. Since none of these investigators except Burns and Fuerstenau analyzed pure mineral samples of todorokite or birnessite but, rather, bulk samples which included unknown proportions of amorphous oxide material, the experimental results must be considered somewhat less than conclusive. The mineralogical identifications may also be open to questions, since evidently diffractometer charts were used and represent samples which were probably mixtures and may have suffered deleterious effects on mineral structure during grinding.

In any case, several investigators have recently attempted to improve the situation by extending the work of Burns and Fuerstenau by probe analysis of microscopic areas of known mineralogy (Carr, 1970; Foster, 1970; Sorem and Foster, 1972c). This work is especially valuable where analyzed material is also characterized optically and the relationships and proportions of the various minerals are known. To obtain the analytical data necessary for individual minerals, an electron probe microanalyzer with a 1–μ beam is essential, and in some instances greater resolution, such as that claimed for new SEM analyzers (100 Å), would be an advantage.

As an example of the capabilities of the electron microprobe analyzer, consider the small nodular feature shown in Fig. 6-15. This body was found in a fracture during a routine optical microscope examination of a polished section prepared from a Blake Plateau nodule. Obviously the mass is heterogeneous, and physical sampling for conventional analysis is impossible. An analysis of each component is desirable, for the relationships visible at high magnification indicate that alternating thin layers of crystalline material (todorokite and birnessite) and amorphous oxides developed by growth about a tiny nucleus. Analytical traverses with an ARL microprobe across the layering reveal that variations in composition are clearly related to the minerals present (Fig. 6-16). The crystalline material is rich in Mn, Cu, Ni, and K, as compared to the amorphous layers, which are rich in iron. The

Fig. 6-15. Photomicrograph. The complex nature of internal nodule features is well illustrated by this cross-section through a micronodule-like body found within a nodule fracture. An electron microprobe traverse was run perpendicular to the thin concentric layers of crystalline and amorphous material in order to determine element distribution (Fig. 6-16). Crystalline layers are light gray in colour with a mottled appearance whereas amorphous layers are more uniform light to dark gray. Nodule BP-1, Blake Plateau, Atlantic Ocean, depth 730 m.

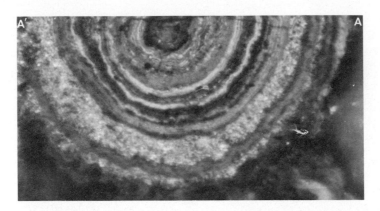

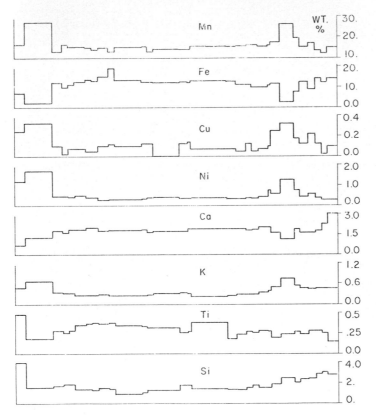

Fig. 6-16. Results of electron microprobe analysis of nodule layers. Variations in metal content along line A-A' of the nodular structure illustrated in Fig. 6-15. Metal variations can be correlated with optical properties by use of the photomicrograph at the top of the chart. Note that the crystalline layers are rich in Mn, Cu, Ni, and K as compared to the amorphous layers. This may suggest that Cu and Ni were incorporated into the crystal lattices by ionic substitution during nodule growth. Nodule BP—1.

value of the photomicrograph of the analyzed material is obvious. Without the optical data, it would be impossible to recognize the mineralogical units in the specimen. A back-scattered electron probe photograph cannot serve the same purpose, although it is a convenient way of recording the major structures analyzed.

In specimens where chemical variations on a larger scale are of interest, more rapid analyses for major metals can be made with the X-ray macroprobe (Fig. 6-17). Especially when the results are accompanied by a photograph of the section analyzed, much can be learned of the general distribution of major elements in the whole nodule structure. Note that even with the much coarser resolution of the macroprobe (spot size $250 \times 450\,\mu$) elemental and mineralogical relationships are clearly revealed. From this type of macro-analysis it has been learned that manganese-rich and iron-rich alternations are common not only in stacks of microscopic layers but also on a larger scale (Sorem and Foster, 1972a). It therefore appears that the so-called gross-layering in many nodules is not simply a reflection of textural features but represents chemical variations as well. Whether the large and small variations can be correlated from one nodule to the next is not yet known, but it is expected that nodules with similar textural zone sequences will show similar major chemical zonation.

NODULE CORES OR NUCLEI

It has often been reported that nodules commonly have a core or central zone which is in general different from the outer parts. This nucleus or core may be of igneous, sedimentary, or metamorphic rock, or it may be a fossil fragment or even a man-made object. Mero (1962, p. 750) concluded that the ". . . chemical nature of the nucleus does not seem to affect the deposition of manganese or the composition of the manganese and iron phases of the nodule." This opinion still seems to prevail, although there seems to be little doubt that nuclei of different kinds can affect the accumulation rate of oxide material. The extreme example of rapid growth is accretion about iron bolts, brass shell fragments, and sparkplugs, all of which have been reported in recent years.

It has long been assumed that nodule growth may begin on any relatively large or angular object on the ocean floor, and this is supported by the examination of many nodules from a wide variety of localities. In some areas certain types of core or nucleus seem to predominate over others. For example, nodules from the east-central Pacific Ocean commonly have a fragment of an older nodule as a nucleus, nodules from the Drake Passage have rock pebble nuclei, and nodules from the Carlsberg Ridge region of the Indian Ocean are often found to have a core of altered igneous rock. Local bottom topography, relief, geology, and currents, among other things, also

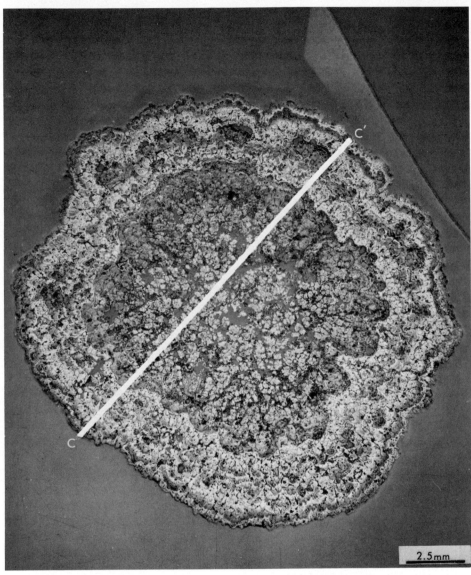

A

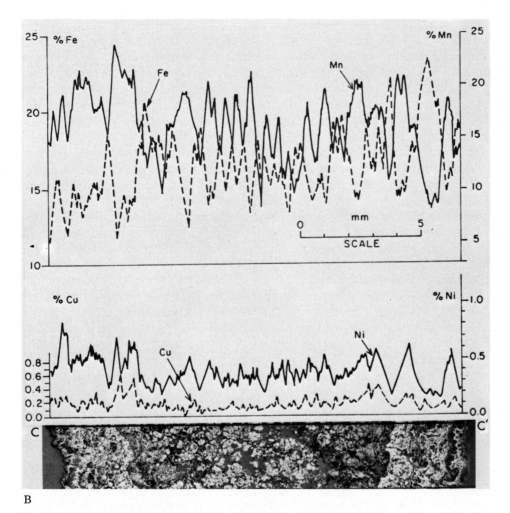

B

Fig. 6-17. Results of X-ray macroprobe analysis across whole nodule section.
A. Photomacrograph. Nodule shows well-developed layering in crustal zone but relatively massive core. Analysis of nodule material along line C-C' was obtained by traversing across stationary X-ray beam. A path about 450 μ wide was irradiated. Nodule BR-1-1, Bermuda Rise, Atlantic Ocean, depth 5,390 m.
B. Variations in metal content along line C-C'. Portion of nodule is shown at bottom to scale. Note traverse relationship of Fe and Mn, and generally sympathetic variation of Mn, Cu, and Ni. By projecting peaks vertically downward, it is found that manganese content is greatest where reflectivity is greatest. Highly reflective material also contains appreciable iron, but highest iron content is found in material of low reflectivity (medium gray). Nearly all oxide material in this nodule is amorphous.

seem to control the nature of the nucleus, however, and it is not surprising that a variety of types of nuclei are often found within a small area.

The relationship between the encrusting oxides and the nucleus in manganese nodules is of special interest, especially from the point of view of nodule genesis. In particular, close examination of the contact between the two may reveal evidence that basalt, palagonite, or other rocks supply some of the iron and manganese for crust formation. This might be expected in view of the work of Park (1946) and Krauskopf (1957) on spilitic rocks. Burns and Brown (1972) have proposed that the close study of nucleus-crust contacts may lead to a better understanding of the mechanics of crust formation, and Morgenstein (1973a) described a dating method ("hydration-rind") which involves determining the rate of crust accretion by a study of palagonite nuclei in nodules. As yet, however, there is no compelling published evidence that the relationship between nodule and nucleus has any more than very local importance. In polished sections, some nodules show veinlets of oxide material which cut both the oxide crust and the nucleus. Especially where the nucleus is rock, the only explanation for such a relationship seems to be migration of iron and manganese from the crust inward during diagenesis.

From the economic point of view, it is clear that rock nuclei decrease the value of nodules and must be considered waste material. It is also possible that nuclei consisting of old nodule fragments may, at least in some areas, have a different average metal content than the outer portions, and some sort of separation may be desirable before mining or extraction of metals. To date, however, there are apparently few data available on this problem.

DISCUSSION

Significant nodule features

We have tried to emphasize the conviction that microscopic features of manganese nodules are complex and deserve detailed study. All features of nodules are significant, however, if they provide an insight into problems of nodule genesis or economic use of nodules. The chief danger that exists if we do not study the fine structure is that we may try to interpret associations found in bulk samples or large units as if they were products of a single reaction or environment, whereas a sample a few grams in weight may represent a million years of the earth's history. If we assume an average rate of accretion of 10 mm/million years, for example, even a layer or shell 1 μ thick may record seafloor reactions over a period of 100 years. Unfortunately, chemical analyses are rarely reported for pure samples of shells even ten times that thickness.

If the fine details of nodule structure are to be used to understand nodule growth, however, it is clear that we must distinguish primary features from those formed later during diagenesis. It appears fairly certain that the gross layering, and much of the textural detail visible at low magnification in the layers, reflects growth conditions in the past, for many of these textures found in the older interior shells are very similar to those present in the crustal zone. In some nodules old shells have unique textures, however, and their origin is problematical. Some changes clearly do occur following accretion, as shown by cracks, veinlets, and replacement features. The generally high porosity of nodules would indeed be expected to promote secondary changes, but the true extent of diagenetic alteration is difficult to estimate. It seems safe to assume that most nodules have experienced changes to some extent in texture, chemical composition, and mineralogy, however, but not in overall structure.

Dynamic hypothesis of nodule origin

A satisfactory hypothesis of nodule origin must account for all significant nodule characteristics, large or small, primary or secondary. We must look at the physical, chemical, and mineralogical features of nodules as the end products of deep-sea processes and then try to visualize appropriate processes operating on a geologic time scale. The following suggestions attempt to outline some major concepts to consider in the development of such a genetic hypothesis, with the expectation that revisions will be made as new data on nodules and the marine environment become available.

The *source* of the chief metals is a major problem, since Fe, Mn, Cu, and Ni occur in very low concentrations in sea water. Various authors have appealed to sea water as a source, however, as well as substrate pore waters and igneous rocks and emanations on the sea floor, using chiefly chemical data for evidence. Agreement has not been reached, and one wonders, therefore, if there is any evidence in nodules that has so far been overlooked which might help clarify the situation. We suggest that the nature and distribution of very fine-grained detritus in nodules should be given closer attention; the genetic significance of this material has not been fully explored as yet.

Minute glass shards and mineral fragments have been reported dispersed in nodules in the past, but it is the unusual concentrations of these particles which may be of special interest. It seems reasonable to believe that these particles rain down slowly but continuously on the sea floor as a result of the settling of windborne dust, but how does this explain the origin of unusually rich concentrations of detritus in thin shells at many levels in a nodule? Since the actual relationship is simply an increase in the ratio of detritus to oxides, two possibilities exist; either the rate of infall of detritus increased periodically, or the rate of oxide accretion relative to the rate of detritus infall decreased from time to time. If the cause is a sporadic increase in the supply of detritus and glass shards and igneous minerals are abundant, then a logical source may be local volcanism (Carr, 1970, pp. 86—87). This explanation may also account for the ubiquitous association of crystalline manganese minerals with the plentiful detritus; igneous sources may at the same time supply solutions rich in manganese, copper, nickel. On the other hand, if the detritus-rich layers indicate periodic abatement of normal oxide accretion, it remains to be explained how at those times the iron oxide deposition is reduced to essentially zero while manganese deposition continues. A solution may be found by a careful study of the iron-rich material which immediately overlies such layers, but no data are presently available to resolve the problem. At present, it appears likely that the iron-rich material is deposited, probably from "normal" ocean bottom water, more or less uniformly with time.

The suggestion has been made in the past that solid state diffusion rather than primary deposition may account for the common layered colloform structures in nodules. The wide occurrence of very sharp textural and chemical boundaries between adjacent layers, among other things, seems to limit this possibility greatly.

In summary, we believe that not only chemical, mineralogical, and structural data on oxide shells are still needed, but that much more must be learned of the detrital particles as well. To dismiss them as "the insoluble fraction" is probably throwing good evidence away. We may speculate, in fact, that our data thus far indicate that many pulsations, both physical and chemical, occur during nodule growth. The scale ranges from that of the gross layers to that of micron-thick layers in many places. The key to understanding the cause of these pulsations may lie in examining the sequence of layers in many nodules from the same locality, as proposed by Foster (1970). If chemical as well as physical correlations can be found, variations in the general environment may well be indicated. Absence of a correlation would suggest, on the other hand, that other factors, perhaps very local in nature, cause the variations in question in nodule layers.

With this tentative picture of long-term deposition, we must consider the problem of the *mechanism* of deposition. Both inorganic and organic processes have been proposed in the past, but there are few real data. Many investigators who have touched on the problem have suggested colloidal deposition of one kind or another. We wish to emphasize that there is evidence in nodules that colloidal deposition and direct crystallization both probably occur. Iron-rich shells are essentially non-crystalline, even in the inner recesses of nodules, and the sub-microscopic textures revealed by the SEM suggest deposition by flocculation, presumably of colloidal particles. Botryoidal aggregates are common. The manganese-rich shells are crystalline, however, even on the modern exterior of some nodules, and it appears that todorokite and birnessite form largely by ionic crystallization from solution.

In protected cavities, relatively large euhedral crystals are formed, but in general crystals are sub-microscopic. It is possible that associated glass shards or minor elements catalyzed crystallization, but as yet this is pure speculation.

Rate of deposition cannot be judged quantitatively from nodule features, but as better dating methods are applied to small samples from well-documented nodule layers (so that periods of non-deposition can be taken into account), a relationship between rate of deposition and textures may be established. To judge from the textures in nodules and observations of the outer surface, it may be speculated that the familiar delicate botryoidal surface of many nodules represents very slow deposition. Where this is seen as columnar structure in section, the interlamination of oxides and clay seams suggests that there may have been some sort of cyclic pattern in the primary deposition. One might also propose, however, that the clay seams represent periodic exclusion of impurities by gel-like oxides soon after accretion. It seems likely that the dense and less irregular shells rich in manganese, in contrast, may represent short periods of relatively rapid deposition.

Space is too limited to try to account for other common events in nodule history, such as nodule diagenesis, movement, growth in the substrate, and causes of regional geographic variations, but all evidence points the same general way. We must consider nodule growth as a dynamic process, constantly changing, even on a small scale, and never really coming to a halt. Even after many shells have accreted, post-accretion adjustment and internal circulation and deposition of oxides continues. Other changes probably occur upon aging. For example, manganese-rich laminations within iron-rich shells, here interpreted as primary, may actually be in part due to filling of shrinkage cracks.

In spite of the many uncertainties which remain, we believe that it must be concluded that nodules several centimetres in size record the evidence of variations in sea-bottom environment over periods of millions of years. Furthermore, each nodule has probably been responsive to minor influences above, below, and around it, just as it has to major environmental changes.

In summary, nodules reveal a very complex growth history over a long period of time, and we call upon interested investigators to view their data in this context. Ancient environments are probably the key to understanding the general problem of nodule origin, and we must learn about them largely by the methods of historical geology and perhaps, palaeogeochemistry. Modern environments are important but may be of real significance in understanding only the crustal zone of most nodules.

CHAPTER 7

MINERALOGY

R. G. BURNS and V. M. BURNS

INTRODUCTION

The mineralogy of ferromanganese nodules determines most of the physical and chemical properties of this potential ore deposit. The constituent minerals not only control the authigenesis, growth, and structure of the nodules, but they also influence the uptake of certain strategic metals into the concretions. Furthermore, the metallurgical processes used to extract the metals from manganese nodules mined from different localities on the sea floor are likely to be governed by the minerals present. Therefore, it is essential to be able to characterize the minerals constituting manganese nodules from each marine deposit.

Manganese nodules are not monomineralic. Instead they consist of a complex mixture of materials, including crystallites of several minerals of detrital and authigenic origins, organic and colloidal matter, and igneous and metamorphic rocks of varying degrees of degradation. Some of the constituents of manganese nodules appear to be essential for the nucleation and growth of the component minerals. In most cases the concretions have a nucleus which may be pumice, altered basaltic fragments or glass, clays or tuffaceous material, or hard-parts of organisms (e.g., sharks' teeth, whale bones, Globigerina or radiolarian tests, coral, etc.). The phases in manganese nodules are fine-grained and intimately intergrown, giving rise to the complex textures described in the previous chapter. Such heterogeneities and complicated internal structures make it very difficult to characterize the very small mineral crystallites in manganese nodules. It is extremely difficult or impossible to extract a homogeneous, single-phase mineral sample from a manganese nodule for mineralogical studies. This has led to ambiguities over mineral identifications in manganese nodules. Another problem has been to distinguish between authigenic and detrital minerals, particularly the clay, silica, and iron oxyhydroxide phases, which either formed in situ or were transported as suspensions in seawater from terrigenous sources.

Most interest centers on the ferromanganese oxide phases in the nodules because inter-element correlations have demonstrated that Co, Ni, and Cu are associated with Fe and Mn (Goldberg, 1954; Mero, 1962; Burns and Fuerstenau, 1966; Cronan and Tooms, 1968; Sano and Matsubara, 1970; Ostwald and Frazer, 1973; Bezrukov and Andrushchenko, 1974). Therefore, this chapter will concentrate principally on the oxide minerals of manganese and iron. Only a brief mention will be made of other minor authigenic and detrital accessory phases.

Confusion exists in the literature on manganese nodules over the nomenclature of the constituent manganese and iron oxide phases. This may be attributed directly to the minute grain sizes of some of the phases, which not only makes them apparently amorphous to X-rays, but also renders X-ray crystal structure determinations of them impossible. Attempts have been made to correlate the nodule phases with terrestrial minerals on the one hand, and with products of synthesis on the other. As a result, at least three classification schemes exist in the literature on manganese nodules, leading to ambiguities between the variously named species. An attempt is made in this chapter to clarify the terminology of the manganese and iron oxide phases in manganese nodules.

Certain minimum criteria must be met before a phase can be classed as a mineral. Before enunciating them, the concept of a mineral should be defined. One definition of a mineral is "a solid possessing a characteristic chemical composition or a limited range of compositions and a systematic three dimensional atomic order" (Ernst, 1969). Some authors (Mason and Berry, 1968; Hurlbut, 1971) also specify the origin as "naturally occurring" and "inorganically produced". The ordered atomic arrangement of the crystal structure leads to a crystalline solid with smooth planar faces and regular geometric habits. Depending on the dimensions of the crystals, a mineral may be described as crystalline, microcrystalline, or crypto-crystalline. If the solid lacks an ordered internal structure of atoms, it is amorphous and is classed as a mineraloid. A problem encountered in manganese nodule mineralogy is to delineate the boundary between a cryptocrystalline mineral and an amorphous mineraloid which depends upon the extent of long-range ordering and the population of defects and vacancies in the crystal structure. If particle sizes are smaller than about 100 Å, coherent scattering of X-rays from the lattice planes is improbable so that the solid appears to be amorphous. However, crystallites smaller than 100 Å may possibly be recognized by electron microscopy. Information on dimensions and shapes of cryptocrystalline minerals is therefore particularly important for elucidating the mineralogy of manganese nodules. Such information, which may be obtained from electron microscopy, Langmuir adsorption isotherms, magnetic and Mössbauer measurements, are reviewed in this chapter. Although crystalline phases as well as colloidal matter and gels may all be present in manganese nodules, and some of the phases may be organically produced (Graham and Cooper, 1959; Ehrlich, 1963; Sorokin, 1972; Monty, 1973; Greenslate, 1974a, b), the coverage in this chapter is confined to the minerals, regardless of origin, whose crystalline particles have dimensions exceeding about 30—40Å (representing ordered stacking of several unit cells).

Additional criteria, some or all of which may be necessary to characterize a mineral completely, include: (1) lattice or d-spacing data derived from X-ray and electron diffraction measurements; (2) chemical composition and formula; (3) unit cell parameters; and (4) crystal structure data if available. Correlations with products of synthesis are also useful. Sometimes the identification of functional groups such as H_2O molecules, OH^- ions, and $CO_3{}^{2-}$ groups in minerals, as well as the oxidation states of the metals, are essential for the complete chemical characterization of a mineral phase. Such data may be obtained from various spectral techniques, including differential thermal analysis (DTA), infrared, Mössbauer and electron paramagnetic resonance (EPR) spectroscopy. Selected information derived from some of these techniques and pertaining to the ferromanganese oxide phases in manganese nodules is reviewed in this chapter.

Following a survey of the mineralogical and X-ray data for the principal oxide phases of manganese and iron and prominent accessory minerals, information on crystallinity, X-ray powder diffraction data, Mössbauer and other spectral measurements of manganese nodules is reviewed. Structural correlations between the Mn and Fe oxide phases are discussed, together with aspects of the crystal chemistry of Mn, Fe, Co, Ni, and Cu. These data form the basis of a hypothesis for the nucleation and growth of manganese nodules. The chapter concludes with recommendations for future mineralogical research on manganese nodules.

OXIDES OF MANGANESE

Glossary and X-ray data

Manganese forms a large number of compounds with oxygen. These range from refractory anhydrous phases to hydrated minerals stable at low temperatures in aqueous environments. The emphasis of this review is on relationships between the low temperature oxides of manganese which are relevant to the mineralogy of manganese nodules formed in situ in water. Although oxides such as jacobsite ($MnFe_2O_4$), a-vredenburgite [$(Mn,Fe)_3O_4$], hausmannite (Mn_3O_4), bixbyite [a- $(Mn,Fe)_2O_3$], partridge-ite (a-Mn_2O_3), and braunite ($Mn^{2+}Mn^{3+}{}_6O_8$ [SiO_4]) are formed when ferromanganese nodules are heated to high temperatures (Smith et al., 1968; Okada et al., 1972a; Hrynkiewicz et al., 1972b; Carpenter et al., 1972), they are not discussed further in this chapter.

Many naturally occurring and synthetic oxide phases of manganese relevant to manganese nodule mineralogy are listed in Table 7-I, together with available crystallographic data and current information on chemical composition and crystal structures. The mineral data are based on compilations by Strunz (1970) and Fleischer (1967, 1971). The classification scheme adopted in Table 7-I follows an approximate order of increasing

TABLE 7-I

Oxides of manganese

Group	Mineral or compound	Formula	Crystal class (space group)	Cell parameters
β-MnO$_2$	pyrolusite (polianite)	MnO_2	tetragonal (P4$_2$/mn2)	a_0=4.39; c_0=2.87; z=2
—	ramsdellite	MnO_2	orthorhombic (Pbnm)	a_0=4.53; b_0=9.27; c_0=2.87; z=4
γ-MnO$_2$	nsutite (yokosukaite)	$(Mn^{2+},Mn^{3+},Mn^{4+})(O,OH)_2$	hexagonal	a_0=9.65; c_0=4.43; z=12
Hollandite	synthetic α-MnO$_2$	$MnO_2 \cdot xH_2O$	tetragonal	a_0=9.88; c_0=2.843; z=8
Hollandite	hollandite	$(Ba,K)_{1-2}Mn_8O_{16} \cdot xH_2O$	tetragonal (I4/m) or monoclinic (P2$_1$/n)	a_0=9.96; c_0=2.86; z=1 a_0=10.03; b_0=5.76; c_0=9.90; β=90°42'; z=2
Hollandite	cryptomelane	$K_{1-2}Mn_8O_{16} \cdot xH_2O$	tetragonal (I4/m) or monoclinic (I2/m)	a_0=9.84; c_0=2.86; z=1 a_0=9.79; b_0=2.88; c_0=9.94; β=90°37'; z=1
—	(ishiganeite)	$4MnO_2 \cdot RO \cdot H_2O$		
Hollandite	manjiroite	$(Na,K)_{1-2}Mn_8O_{16} \cdot xH_2O$	tetragonal (I4/m)	a_0=9.92; c_0=2.86; z=1
Hollandite	coronadite	$Pb_{1-2}Mn_8O_{16} \cdot xH_2O$	tetragonal (I4/m)	a_0=9.89; c_0=2.86; z=1
Psilomelane	psilomelane or romanèchite	$(Ba,K,Mn,Co)_2Mn_5O_{10} \cdot xH_2O$	monoclinic (A2/m) or orthorhombic (P2$_1$2$_1$2)	a_0=9.56; b_0=2.88; c_0=13.85; β=92°30'; z=2 a_0=8.254; b_0=13.40; c_0=2.864; z=1
Birnessite	synthetic Na manganese(II,III) manganate(IV)	$Na_4Mn_{14}O_{27} \cdot 9H_2O$	orthorhombic	a_0=8.54; b_0=15.39; c_0=14.26; z=60
	synthetic manganese(III) manganate(IV)	$Mn_7O_{13} \cdot 5H_2O$	hexagonal	a_0=2.84; c_0=7.27; z=2
	("7-Å manganite" or manganous manganite)	$(4MnO_2 \cdot Mn(OH)_2 \cdot 2H_2O)$	(hexagonal)	(a_0=5.82; c_0=14.62)
	synthetic δ-MnO$_2$		hexagonal	
	birnessite	$(Ca,Na)(Mn^{2+},Mn^{4+})_7O_{14} \cdot 3H_2O$		

Structure-type (isostructural compounds)	References to compound or mineral	Manganese nodule occurrences
rutile	1,2	Bezrukov and Andrushchenko (1972)
own, diaspore (goethite, groutite)	1,3,4	Manheim (1965); Hering (1971)
pyrolusite + ramsdellite structural intergrowths	5,6,7,8,9	Manheim (1965); Gattow and Glemser (1961b) and Schweisfurth (1971), but see Giovanoli et al. (1973a)
hollandite (cryptomelane, akaganéite)	10,11	
hollandite (cryptomelane, manjiroite, akaganéite, coronadite, etc.)	1,12,13,14,15	
hollandite (manjiroite, etc.)	1,11,16,17,18	Nohara (1972); Gattow and Glemser (1961b), but see Giovanoli et al. (1973a)
mixture of cryptomelane and birnessite	19	
hollandite (cryptomelane, etc.)	20	
hollandite (cryptomelane, etc.)	1,21	
psilomelane	1,17,22,23	Murata and Erd (1964); Andrushchenko and Skornyakova (1965); Hering (1971); Bezrukov and Andrushchenko (1972)
unknown, but may be related to chalcophanite	24,25	Giovanoli et al. (1973a); Von Heimendahl et al. (1976)
unknown, but may be related to chalcophanite	24,25	
	26,27	Buser (1959); Burns and Fuerstenau (1966); Barnes (1967); Price and Calvert (1970); Brown (1972); Glasby (1972a)
fine-grained, disordered birnessite	11,27,28,29,30	Buser and Grütter (1956); Burns and Fuerstenau (1966); Barnes (1967a); Price and Calvert (1970); Brown (1972); Ostwald and Frazer (1973)
unknown	13,31,32,33,34,35	Manheim (1965); Cronan and Tooms(1969); Sorem and Foster (1972a,b); Scott et al. (1972a); Crerar and Barnes (1974); Hering (1971); Sorem (1967)

TABLE 7-I (continued)

Group	Mineral or compound	Formula	Crystal class (space group)	Cell parameters
Chalcophanite	chalcophanite	$ZnMn_3O_7 \cdot 3H_2O$	triclinic (P$\bar{1}$)	$a_0=7.54; b_0=7.54; c_0=8.22; \alpha=90°; \beta=117°12'; \gamma=120°; z=2$
Chalcophanite	aurorite	$(Ag,Ba,Ca)Mn_3O_7 \cdot 3H_2O$	triclinic (?)	
Ranciéite	ranciéite	$(Ca,Mn)Mn_4O_9 \cdot 3H_2O$	hexagonal (?)	
Ranciéite	takanelite	$Mn^{2+}Mn_4O_9 \cdot 1 \cdot 3H_2O$	hexagonal (?)	
Buserite	buserite or synthetic sodium (manganese(II,III) manganite(IV) hydrate)		orthorhombic	
	(10-Å manganite)	$(Na,Mn)Mn_3O_7 \cdot nH_2O$ or $3MnO_2 \cdot Mn(OH)_2 \cdot nH_2O)$	(hexagonal)	$(a_0=8.41; c_0=10.01; z=2)$
Todorokite	todorokite (delatorreite)	$(Na,Ca,K,Ba,Mn^{2+})_2 Mn_5O_{12} \cdot 3H_2O$ or $R^{2+}Mn_3O_7 \cdot 1-2H_2O$	monoclinic	$a_0=9.75; b_0=2.849; c_0=9.59; \beta=90°$
Todorokite	woodruffite	$(Zn,Mn)_2 Mn_5O_{12} \cdot 4H_2O$	tetragonal	$a_0=8.42; c_0=9.28; z=2$
Lithiophorite	lithiophorite	$(Al,Li)(OH)_2 MnO_2$ or $[Mn_5{}^{4+}Mn^{2+}O_{12}]^{2-} [Al_4 Li(OH)_2]^{2+}$	monoclinic (C2/m)	$a_0=5.06; b_0=2.91; c_0=9.55; \beta=100°30'$ z=4 or $a_0=5.06; b_0=8.70; c_0=9.61; \beta=100°7'; z=12$
—	quenselite	$Pb(OH)MnO_2$	monoclinic (P2$_1$/a)	$a_0=5.61; b_0=5.68; c_0=9.13; \beta=93°29'; z=8$
—	synthetic	$CdMn_3O_8$ and Mn_5O_8 $(=Mn_2{}^{2+}Mn_3{}^{4+}O_8)$	monoclinic (C2/m) or C2)	$a_0=10.34; b_0=5.72; c_0=4.85; \beta=109°25'; z=2$
Oxide Hydroxides:				
α-MnOOH	groutite	$MnOOH$	orthorhombic (Pbnm)	$a_0=4.56; b_0=10.70; c_0=2.85; z=4$
β-MnOOH	feitknechtite	$MnOOH$	hexagonal (P$\bar{3}$m1)	$a_0=3.32; c_0=4.71; z=1$
γ-MnOOH	manganite	$MnOOH$	monoclinic (B2$_1$/d)	$a_0=8.88; b_0=5.25; c_0=5.71; \beta=90°$
—	hydrohausmannite			
—	hydrohetaerolite	$Zn_2 Mn_4O_{18} \cdot H_2O$	tetragonal (I4$_1$/amd)	$a_0=5.72; c_0=9.06; z=4$

Structure-type (isostructural compounds)	References to compound or mineral	Manganese nodule occurrences
chalcophanite	36,37	
chalcophanite	38	
unknown	1,34,39,40	Sorem (1967)
unknown	41	
unknown, but may be related to birnessite	24,25,42, 79,80	Giovanoli et al. (1971; 1973a,b; 1975); Giovanoli and Burki (1975)
	27,78	Buser (1959); Barnes (1967); Brown (1972); Glasby (1972a); Ostwald and Frazer (1973)
unknown	11,19,38,43,44,45, 46,47,48,49,50,51, 52,53,54,55	Hewett et al. (1963); Manheim (1965); Sorem (1967); Grill et al. (1968b); Price and Calvert (1970); Calvert and Price (1970); Sorem and Foster (1972a,b); Andrushchenko and Skornyakova (1969); Hering (1971)
unknown, may be related to todorokite	56,57	Andrushchenko and Skornyakova (1969)
lithiophorite	1,58,59	
	42	
quenselite	60,61	
Mn_5O_8	62,63,64	
diaspore (goethite, ramsdellite)	64,65	Goodier (1972); Nohara (1972)
pyrochroite	30,66,67	
manganite (distorted marcasite; ϵ-FeOOH)	1,11,26,30,68,69	Hering (1971) — however, probably an error over nomenclature (i.e. 10Å or 7Å manganite implied)
is a mixture of feitknechtite and hausmannite	26,30,56,70	
	71	

TABLE 7-I (continued)

Group	Mineral or compound	Formula	Crystal class (space group)	Cell parameters
—	pyrochroite	$Mn(OH)_2$	hexagonal (P$\bar{3}$m1)	a_0=3.322; c_0=4.734; z=1
—	backströmite	$Mn(OH)_2$		
Spinel	jacobsite	$MnFe_2O_4$	cubic (Fd3m)	a_0=8.474
Spinel	hausmannite	Mn_3O_4	tetragonal (I4$_1$/amd)	a_0=5.76; $c_0\doteq$9.44; z=4
Spinel	α-vredenbergite	$(Mn,Fe)_3O_4$	tetragonal (I4$_1$/amd)	a_0=5.78; c_0=9.35; z=4
Spinel	hetaerolite	$ZnMn_2O_4$	tetragonal (I4$_1$/amd)	a_0=5.75; c_0=9.17; z=4
—	crednerite	$CuMnO_2$	monoclinic (C2/m, C2, or Cm)	a_0=5.58; b_0=2.87; z=4; β=104°
	partridgeite	α-Mn_2O_3	cubic Ia3	a_0=9.43; z=16
	bixbyite	α-$(Mn,Fe)_2O_3$	cubic Ia3	a_0=9.39; z=16
	braunite	$Mn^{2+}Mn_6{}^{3+}O_8\cdot[SiO_4]$	tetragonal (I4$_1$/acd)	a_0=9.44; c_0=18.76; z=8

wad = soft, black, amorphous, naturally-occurring manganese oxide

asbolane = cobalt-bearing wad

lampadite = copper-bearing wad

vernadite = colloidal hydrated manganese oxide occurring as a weathering product of manganese ores

nickelmelane

cobaltmelane

elizavetinskite

buryktalskite

discredited names for manganese oxides formed by weathering of ultrabasic rocks; contain appreciable amounts of Co and Ni, but considered to be mixtures of pyrolusite, cryptomelane, lithiophorite, and goethite.

References
1. M. Fleischer and W. E. Richmond (1943). Econ. Geol., 38: 269—286.
2. H. Strunz (1943). Naturwiss., 31: 89—92.
3. M. Fleischer, W. E. Richmond, and H. T. Evans (1962). Am. Miner., 47: 47—58.
4. A. M. Byström (1949). Acta Chem. Scand., 3: 163—173.
5. P. M. de Wolff (1959). Acta Cryst., 12: 341—345.
6. R. K. Sorem and E. N. Cameron (1960). Econ. Geol., 55: 278—310.
7. W. K. Zwicker, W. O. J. Groeneveld Meijer and H. W. Jaffe (1962). Am. Miner., 47: 246—266.
8. G. M. Faulring (1965). Am. Miner. 50: 170—179.
9. R. Giovanoli, R. Mauer and W. Feitknecht (1967). Helv. Chim. Acta, 50: 1072—1080.
10. G. Butler and H. R. Thirsk (1952). Acta Cryst., 3: 288—289.
11. R. M. McKenzie (1970). Miner. Mag., 38: 493—502.
12. A. Byström and A. M. Byström (1950). Acta Cryst., 3: 146—154; ibid., 4: 469 (1951).
13. C. Frondel, U. B. Marvin and J. Ito (1960b). Am. Miner., 45: 871—875.
14. B. Mukherjee (1964). Acta Cryst. 17: 1325.
15. M. Fleischer (1964). Adv. Frontiers Geol. Geophys., 221—232.
16. A. McL. Mathieson and A. D. Wadsley (1950). Am. Miner., 35: 99—101.
17. B. Mukherjee (1959). Miner. Mag., 32: 166—171.
18. G. M. Faulring, W. K. Zwicker and W. D. Forgeng (1960). Am. Miner., 45: 946—959.

Structure-type (isostructural compounds)	References to compound or mineral	Manganese nodule occurrences
pyrochroite (feitknechtite)	30.72	
backströmite; believed to be identical to hydrohausmannite which is a mixture	30,56,73	
spinel	34,74	
distorted spinel	1,11,26,30	
distorted spinel	74	
distorted spinel	74	
	34,75	
partridgeite	11,74	
partridgeite	1,74	
braunite	1,34,74	
	23,78	Murray and Renard (1891); Mero (1965a)
	74	
	42,74	
	74,76	Andrushchenko and Skornyakova (1965)
	77	

19. Y. Hariya (1963). Am. Miner., 48: 952—954.
20. M. Nambu and K. Tanida (1967). Jap. J. Assoc. Miner. Petrol. Econ. Geol., 58: 39—54.
21. D. F. Hewett (1971). Econ. Geol., 66: 164—177.
22. A. D. Wadsley (1953). Acta Cryst., 6: 433—438.
23. M. Fleischer (1960). Am. Miner., 45: 176—187.
24. R. Giovanoli, E. Stähli and W. Feitknecht (1970b). Helv. Chim. Acta, 53: 209—220; ibid., 53: 453—464.
25. R. Giovanoli, W. Feitknecht and F. Fischer (1971). Helv. Chim. Acta, 54: 1112—1124.
26. W. Feitknecht and W. Marti (1945). Helv. Chim. Acta, 28: 129—147; 148—156.
27. W. Buser, P. Graf and W. Feitknecht (1954). Helv. Chim. Acta, 37: 2322—2333.
28. H. F. McMurdie (1944). Trans. Electrochem. Soc., 86: 313—326.
29. H. F. McMurdie and E. Golovato (1948). J. Res. Nat. Bur. Stand., 41: 589.
30. O. P. Bricker (1965). Am. Miner., 50: 1296—1354.
31. L. H. P. Jones and A. A. Milne (1956). Miner. Mag., 31: 283—288.
32. Y. Hariya (1961). Jap. J. Assoc. Miner. Petrol. Econ. Geol., 45: 219—230.
33. A. A. Levinson (1962). Am. Miner., 47: 790—791.
34. R. K. Sorem and D. W. Gunn (1967). Econ. Geol., 62: 22—56.
35. F. H. Brown, A. Pabst and D. L. Sawyer (1971). Am. Miner., 56: 1057—1064.
36. A. D. Wadsley (1955). Acta Cryst., 8: 165—172.

TABLE 7-I (continued)

37. D. R. Dasgupta (1974). Z. Krist., 139: 116—128.
38. A. S. Radtke, C. M. Taylor and D. R. Hewett (1967). Econ. Geol., 62: 186—206.
39. W. E. Richmond, M. Fleischer and M. E. Morse (1969). Bull. Soc. Franç. Min. Crist., 92: 191—195.
40. E.-A. Perseil (1967). C. R. Acad. Sci., 264D: 1241—1244.
41. M. Nambu and K. Tanida (1971). J. Jap. Assoc. Miner. Petrol. Econ. Geol., 65: 1—15.
42. R. Giovanoli, H. Bühler and K. Sokolowska (1973). J. Microscopie, 18: 271—284.
43. T. Yoshimura (1934). J. Fac. Sci. Hokkaido Univ., Sapporo, Ser. 4, 2: 289—297.
44. F. S. Simons and J. A. Straczek (1958). U.S. Geol. Surv., Bull., 1057.
45. P. Ljunggren (1960). Am. Miner., 45: 235—238.
46. A. A. Levinson (1960). Am. Miner., 45: 802—807.
47. C. Frondel, U. B. Marvin and J. Ito (1960a). Am. Miner., 45: 1167—1173.
48. J. A. Straczek, A. Horen, M. Ross and C. M. Warshaw (1960). Am. Miner., 45: 1174—1184.
49. L. T. Larson (1962). Am. Miner., 47: 59—66.
50. G. M. Faulring (1962). Adv. X-Ray Analysis, 5: 117—126.
51. F. J. Eckhardt and H. W. Walther (1963). Geol. Jahrb., 70: 867—882.
52. M. Nambu and K. Okada (1963). Bull. Res. Inst. Min. Dress. Metall., Tohoku Univ., 19: 1—12.
53. M. Nambu, K. Okada and K. Tanida (1964). Jap. J. Assoc. Miner. Petrol. Econ. Geol., 51: 30—38.
54. K. Okada and K. Tanida (1965). Bull. Res. Inst. Min. Dress. Metall., Tohoku Univ., 21: 194.
55. E.-A. Perseil (1966). C. R. Acad. Sci., 262D: 949—951.
56. C. Frondel (1953). Am. Miner., 38: 761—769.
57. C. Naganna and C. Bouska (1963). Miner. Mag., 33: 506—507.
58. A. D. Wadsley (1952). Acta Cryst., 5: 676—680.
59. M. Fleischer and G. T. Faust (1963). Schweiz. Miner. Petrogr. Mitt., 43: 197—215.
60. A. Byström (1945). Ark. Kem. Min. Geol., 19A, No. 5.
61. E. Welin (1968). Ark. Min. Geol., 4: 499—541.
62. H. R. Oswald, W. Feitknecht and M. J. Wampetich (1965). Nature, 207: 72.
63. H. R. Oswald and M. J. Wampetich (1967). Helv. Chim. Acta, 50: 2023—2034.
64. R. Giovanoli and U. Leuenberger (1969). Helv. Chim. Acta, 52: 2333—2347.
65. L. S. Dent Glasser and L. Ingram (1968). Acta Cryst., B24: 1233—1236.
66. M. Nambu, K. Tanida, T. Kitamura and T. Komura (1969). Ganseki Kobutsu Kosho Gakkaishi, 59: 91—107.
67. R. Meldau, H. Newesely, and H. Strunz (1973). Naturwiss., 60: 387.
68. M. J. Buerger (1936). Z. Krist., 95: 163—174.
69. H. Dachs (1963). Z. Krist., 118: 303—326.
70. W. Feitknecht, P. Brunner and H. R. Oswald (1962). Z. Anorg. Allgem. Chem., 316: 154—160.
71. J. McAndrew (1956). Am. Miner., 41: 276—287.
72. A. N. Christensen (1965). Acta Chem. Scand., 19: 1765—1766.
73. G. Aminoff (1919). Geol. For. Forh., 41: 473—491.
74. H. Strunz (1970). Mineralogische Tabellen (Akad. Vestegsgesellschaft, Leipzig), 5th ed.
75. J. McAndrew (1956). Am. Miner., 41: 268—275.
76. A. G. Betekhin (1944). Bull. Akad. Nauk. SSSR., Ser. Geol., 3—46.
77. M. Fleischer (1961). Am. Miner., 46: 766—767.
78. A. D. Wadsley (1950a,b). J. Am. Chem. Soc., 72: 1782—1784; Am. Miner., 35: 485—499.
79. R. Giovanoli and P. Burki (1975). Chimia, 29: 266—269.
80. R. Giovanoli, P. Burki, M. Giuffredi and W. Stumm (1975). Chimia, 29: 517—520.

complexity of crystal structure and decreasing oxidation state of the manganese. Structural correlations between the phases and oxidation states of metal ions in the manganese oxide compounds are discussed later. The references footnoting Table 7-I provide additional information on occurrences and the current status of knowledge about the crystal chemistry and structure of each mineral or synthetic compound.

X-ray powder diffraction data for the manganese oxide phases are summarized in Table 7-II, in which the Fink Indexing System is employed. Thus, up to eight of the strongest lines, observed in powder photographs or diffractometer patterns, are listed in order of *decreasing d-spacing*. The Fink System is favoured over the Hanawalt Indexing Method, which lists d-spacings in order of *decreasing intensity*, because it overcomes discrepancies induced by preferred orientation of constituent grains which may give rise to anomalous increases in intensity of certain d-spacings. This is particularly important for todorokite (Faulring, 1962; see Table 7-VI.) The Fink System is also preferable for mineral identification from electron diffraction data. The d-spacing data are taken from *Selected Powder Diffraction Data for Minerals* (Joint Committee on Powder Diffraction Stands, 1974), which reproduces in one volume information appearing on the powder diffraction file cards for minerals. A noteworthy feature of the X-ray diffraction data for manganese (IV) oxide phases is the frequent occurrence of lines at approximately 2.40 Å and 1.42 Å. As discussed later, these lines represent diffraction of X-rays from atomic planes of hexagonally close-packed oxygens containing Mn^{4+} ions in octahedral coordination with oxygen, in which the $[MnO_6]$ octahedra share edges. The 2.40 Å and 1.42 Å spacings thus correspond to the $(10\bar{1}0)$ and $(11\bar{2}0)$ planes of the hexagonal close-packed system. The Mn—Mn interatomic distance across the shared $[MnO_6]$ octahedra is 2.84—2.88 Å, and this dimension (or multiples of it) commonly appears in the cell parameter data summarized in Table 7-I.

Nomenclature

Several of the minerals and synthetic compounds listed in Tables 7-I and 7-II have generated confusion over nomenclature and identity. In some older publications, including Strunz (1970), for example, ramsdellite is formulated as γ-MnO_2 instead of unprefixed MnO_2. The γ-MnO_2 designation is now applied to the nsutite group (Giovanoli, 1969). Naturally occurring nsutites, however, give a variety of X-ray powder patterns (for example, Sorem and Cameron, 1960; Zwicker et al., 1962; Faulring, 1965), often containing some of the lines found also in X-ray patterns of pyrolusite and ramsdellite. It has been demonstrated (Byström, 1949; De Wolff, 1959; Giovanoli et al., 1967) that the nsutite group or γ-MnO_2 phases in general consist of structural intergrowths of ramsdellite and pyrolusite units (see Fig. 7-3). These microdomains are disordered and do not form a super-lattice, with the result that nsutites may give rise to an infinite variety of X-ray diffraction patterns.

Perhaps the major source of confusion over the terminology of manganese oxide compounds occurring in ferromanganese nodules concerns the phases variously called "10-Å manganite" — "todorokite" — "buserite", and "7-Å manganite" — "δ-MnO_2" — "manganous manganite" — "birnessite". Buser (1959; see also Buser and Grütter, 1956; Grütter and Buser, 1957) originally coined the terms "10-Å manganite and "7-Å manganite", as well as "δ-MnO_2", for specific phases in manganese nodules, the d-spacings of which matched the prominent lines observed in X-ray powder patterns of correlative synthetic compounds. This terminology was considered unsatisfactory (Arrhenius, 1963; Burns and Fuerstenau, 1966) because it led to

TABLE 7-II

X-ray diffraction data for oxide compounds of manganese (Fink Classification System)

Mineral or compound	The eight most intense d-spacings (intensity ⩾ 10%)								JCPDS card or reference
Pyrolusite (β-MnO$_2$)	3.14 (100)	2.41 (50)	2.13 (25)	1.98 (15)	1.63 (50)	1.56 (25)	1.43 (15)	1.31 (20)	12—716
Ramsdellite (MnO$_2$)	4.07 (100)	2.55 (100)	2.44 (70)	2.19 (70)	1.907 (70)	1.660 (80)	1.621 (80)	1.473 (80)	7—222* (11—55)
Nsutite (γ-MnO$_2$)	4.00 (95vb)	2.42 (65)	2.33 (70b)	2.13 (45)	1.635 (100)	1.603 (45b)	1.367 (40bb)		17—510* (14—614,14—615)
Hollandite (Ba$_{1-2}$)Mn$_8$O$_{16}$)	6.98 (50)	4.93 (30)	3.47 (80)	3.13 (100)	2.40 (90)	2.15 (80)	1.83 (60)	1.55 (70)	13—115 (12—514
Cryptomelane (K$_{1-2}$Mn$_8$O$_{16}$)	6.90 (90)	4.90 (80)	3.10 (80)	2.39 (100)	2.15 (60)	1.83 (60)	1.54 (60)	1.35 (50)	20—908 (4—778)
Manjiroite (Na$_{1-2}$Mn$_8$O$_{16}$)	7.02 (95)	4.94 (75)	3.49 (30)	3.14 (90)	2.406 (100)	2.160 (65)	1.839 (45)	1.548 (45)	21—1153
Coronadite (Pb$_{1-2}$Mn$_8$O$_{16}$)	3.466 (60)	3.104 (100)	2.400 (40)	2.205 (40)	2.155 (20)	1.836 (20)	1.642 (20)	1.542 (50)	7—361
Psilomelane (romanèchite)	9.68 (30)	6.96 (55)	3.481 (60)	2.882 (40)	2.408 (100)	2.366 (50)	2.188 (85)	1.824 (40)	14—627 (18—174)
Vernadite (MnO(OH)$_2$)	6.81 (30)	3.11 (60)	2.45 (20)	2.39 (100)	2.15 (60)	1.649 (30)	1.537 (40)	1.422 (40)	15—604
Birnessite (Ca,Na)Mn$_7$O$_{14}$·3H$_2$O	7.27 (s)	3.60 (w)	2.44 (m)				1.412 (m)		(1)
Syn. birnessite Mn$_7$O$_{13}$·5H$_2$O	7.21 (100)	3.61 (80)	2.46 (100)	2.33 (100)	2.04 (80)	1.723 (80)	1.422 (60)	1.391 (40)	23—1239
Syn. Na birnessite Na$_4$Mn$_{14}$O$_{27}$·9H$_2$O	7.09 (100)	3.56 (80)	2.51 (70)	2.42 (60)	2.21 (40)	2.15 (40)	1.47 (60)	1.43 (50)	23—1046
Chalcophanite (ZnMn$_3$O$_7$·3H$_2$O)	6.96 (100)	4.08 (50)	3.50 (60)	2.57 (40)	2.24 (50)	1.900 (30)	1.597 (40)	1.431 (30)	15—807
Aurorite (Ag$_2$Mn$_3$O$_7$·3H$_2$O)	6.94 (100)	4.06 (50)	3.46 (70)	2.54 (50)	2.45 (40)	2.23 (50)	1.560 (50)	1.429 (50)	19—88
Ranciéite (Ca,Mn)Mn$_4$O$_9$·3H$_2$O	7.49 (100)	3.74 (14)	2.463 (10)	2.342 (6)	1.425 (46)				22—718
Takanelite (Mn,Mn$_4$O$_9$)·xH$_2$O	7.57 (100)	4.43 (18b)	3.765 (25)	2.462 (15)	2.349 (20)	2.065 (10)	1.420 (17)		(2)
Syn. (Na,Mn)Mn$_3$O$_7$·nH$_2$O	10.00 (s)	7.28 (w)	5.06 (m)	3.47 (w-b)	3.372 (w)	2.50 (w-b)	2.429 (w)	1.472 (mw)	(3)
Todorokite (Na,Ca,Mn)$_2$Mn$_5$O$_{12}$·3H$_2$O	9.68 (100)	4.80 (80)	2.46 (20)	2.39 (40)	2.22 (20)	1.98 (20)	1.42 (30)	1.331 (50)	13—164* (18—1411,19—83, 21—553)
Woodruffite (Zn,Mn)$_2$Mn$_5$O$_{12}$·4H$_2$O	9.34 (50)	4.66 (100)	2.66 (60)	2.48 (60)	1.86 (40)	1.69 (40)	1.48 (50)		16—338
Lithiophorite (Li,Al)MnO$_2$(OH)$_2$	9.45 (50)	4.71 (100)	2.37 (70)	1.88 (70)	1.57 (50)	1.45 (50)	1.39 (50)	1.23 (50)	16—364

TABLE 7-II (continued)

Mineral or compound	The eight most intense d-spacings (intensity $\geqslant 10\%$)								JCPDS card or reference
Quenselite $PbMnO_2(OH)$	3.68 (70)	3.60 (70)	3.04 (100)	2.95 (60)	2.72 (80)	2.44 (40)	2.08 (40)	1.994 (40)	23—351
Groutite (α-MnOOH)	4.20 (100)	2.81 (70)	2.67 (70)	2.38 (40)	2.30 (60)	1.737 (40)	1.695 (50)	1.608 (40)	12—733
Feitknechtite (β-MnOOH)	4.62 (100)	2.64 (50)	2.36 (20b)	1.96 (10b)					18—804
Manganite (γ-MnOOH)	3.40 (100)	2.64 (60)	2.28 (50)	1.708 (40)	1.672 (30)	1.636 (40)	1.437 (30)	1.029 (30)	8—99 (18—805)
Pyrochroite ($Mn(OH)_2$)	4.72 (90)	2.87 (40)	2.45 (100)	1.826 (60)	1.658 (40)	1.565 (30)	1.382 (20)	1.374 (20)	18—787
Hausmannite (Mn_3O_4)	3.09 (50)	2.77 (90)	2.49 (100)	2.36 (40)	2.04 (40)	1.795 (50)	1.579 (50)	1.544 (80)	16—154

References:
(1) L. H. P. Jones and A. A. Milne (1956). Miner. Mag., 31: 283—288.
(2) M. Nambu and K. Tanida (1971). Jap. J. Assoc. Miner. Petrol. Econ. Geol., 65: 1—15.
(3) A. D. Wadsley (1950a, b). J. Am. Chem. Soc., 72: 1782—1784; Am. Miner., 35: 485—499.

confusion with the mineral manganite, γ-MnOOH, a manganese (III) oxide hydroxide which does not give the prominent basal plane reflections around 7 Å and 10 Å. The confusion is enhanced by the observation (Giovanoli et al., 1970b) that manganite (γ-MnOOH) is formed as a decomposition product of synthetic "7-Å manganite" (birnessite). Attempts to propose alternative nomenclatures have led to further confusion over the use of the names birnessite and todorokite or buserite.

The birnessite problem
    Originally McMurdie (1944) synthesized δ-MnO$_2$ with an X-ray powder pattern consisting of only two lines at 2.39 Å and 1.40 Å, consistent with diffraction from the $(10\bar{1}0)$ and $(11\bar{2}0)$ planes of layers of close-packed oxygens containing edge-shared [MnO$_6$] octahedra. Later, McMurdie and Golovato (1948) reported the synthesis of another synthetic δ-MnO$_2$ with two additional lines at 7.0 Å and 3.64 Å, and suggested that a natural δ-MnO$_2$ phase having lines at 7 Å and 2.40 Å occurred in a manganese ore from Canada. Feitknecht and Marti (1945) had synthesized a series of compounds with atomic ratios of oxygen: manganese ranging from 1.74 to 1.96 which contained extra X-ray diffraction lines compared to McMurdie's (1944) original δ-MnO$_2$. Feitknecht and Marti called the compounds "manganous manganites" with formulae approximating $4MnO_2 \cdot Mn(OH)_2 \cdot 2H_2O$. They were postulated to be double-layer compounds consisting of sheets of $Mn(OH)_2 \cdot 2H_2O$ interspersed with layers of $4MnO_2$, approximately 7 Å apart, which produced the basal reflections around 7 Å and 3.5 Å. The most oxidized forms of "manganous manganite" were

considered to be disordered from this idealized "sandwich" structure because they produced X-ray diffractions around 2.40 Å and 1.40 Å only. Subsequently, Cole et al. (1947) and Copeland et al. (1947) independently synthesized these manganese oxide compounds and suggested that McMurdie's (1944) original "δ-MnO_2" was similar to the "manganous manganite". Buser et al. (1954) re-investigated "δ-MnO_2" and "manganous manganite". These authors observed only two diffraction lines (corresponding to McMurdie's (1944) "δ-MnO_2") for more highly oxidized samples having atomic proportions of oxygen above $MnO_{1.90}$. Below this composition, patterns with additional lines similar to those described by Feitknecht and Marti (1945) for "manganous manganite" were observed. Buser et al. (1954) again suggested that "δ-MnO_2" and "manganous manganite" were the same phase. They also attributed the different X-ray diffraction patterns to disordering of the structure when Mn^{2+} ions in the intermediate layer of "manganous manganite" were oxidized, thereby removing the $Mn(OH)_2$ sheets and the periodicity of the basal planes (but see pp. 237—242). Buser et al. (1954) recommended the retention of the names "δ-MnO_2" and "manganous manganite" to indicate oxygen: manganese ratios above and below 1.90, respectively. However, discrepancies were observed by Glemser and Meisiek (1957) and Glemser et al. (1961). They synthesized compounds with formulae up to $MnO_{1.99}$ that gave X-ray patterns containing the basal reflections around 7.2 Å and 3.6 Å. This led Bricker (1965) to suggest that particle size rather than the degree of oxidation influences the X-ray powder patterns, the basal reflections around 7 Å and 3.5 Å being absent for very fine crystallites.

In 1956, Jones and Milne reported a new manganese oxide mineral, birnessite, in a fluvio-glacial deposit near Birness, Scotland, having a formula approximating to $(Na, Ca)Mn_7O_{14} \cdot 2 \cdot 8H_2O$ and an X-ray pattern similar to that reported for "manganous manganite" (Feitknecht and Marti, 1945; Cole et al., 1947; McMurdie and Golovato, 1948; Buser et al., 1954). The X-ray data for the natural birnessite and synthetic analogues are summarized in Table 7-III. Subsequently, birnessite has been identified by its X-ray diffraction in other terrestrial deposits (Frondel et al., 1960b; Hariya, 1961; Levinson, 1962; Sorem and Gunn, 1967; Brown et al., 1971). Birnessite is now generally considered to be a low-temperature manganese oxide formed by the supergene weathering and oxidation of rocks containing manganese (Bricker, 1965). Because of the similarity of X-ray powder patterns between natural birnessites and the synthetic "manganous manganite" and "δ-MnO_2" phases, the latter are now commonly regarded as synthetic analogues of birnessite (Bricker, 1965; Giovanoli, 1969).

Recently, the synthetic birnessite group has been investigated further using high resolution X-ray and electron diffraction techniques. Giovanoli et al. (1970a, b) showed that the certain synthetic "manganous manganite" phases of earlier workers contain sodium, as do natural birnessites. They found the analytical formula to be $Na_4Mn_{14}O_{27} \cdot 9H_2O$ and proposed the chemical name sodium manganese (II, III) manganate (IV) for a phase with an orthorhombic superlattice, in accord with nomenclature rules of the IUPAC*. The sodium-rich phase readily undergoes a topotactic transition to a hexagonal phase called manganese (III) manganate (IV) and formulated as $Mn_7O_{13} \cdot 5H_2O$ or $(Mn_5^{4+}Mn_2^{3+}O_{12})(OH)_2 \cdot 4H_2O$. Measurements of X-ray powder patterns with a focussing Guinier-de Wolff camera have yielded a large number of d-spacings for these synthetic birnessites, and they are also listed in Table 7-III. It is suggested on JCPDS file cards 23-1046 and 23-1239 that many of the closely spaced lines measured for these synthetic birnessites overlap and coincide in finely divided specimens, accounting for the extremely broad and diffuse characters of the reflections observed for natural birnessites. Giovanoli et al. (1973a) have suggested that the synthetic birnessite group, by analogy with the nsutite group, is inherently non-stoichiometric and consists of "an infinite number of varieties of one and the same crystal lattice".

The connection between birnessite and the synthetic "7-Å manganite", "manganous manganite" $(4MnO_2 \cdot Mn(OH)_2 \cdot 2H_2O)$, and sodium manganese (II, III) manganate (IV) $(Na_4Mn_{14}O_{27} \cdot 9H_2O)$ or manganese (III) manganate (IV) $(Mn_7O_{13} \cdot 5H_2O)$ phases appears to be well substantiated. It is therefore recommended that the terminology "birnessite" be adopted instead of "7-Å manganite" for the phase in manganese nodules that gives diagnostic X-ray powder diffractions around 7.0—7.2 Å and 3.5—3.6 Å.

There is also good evidence that the poorly crystalline synthetic "δ-MnO$_2$", giving X-ray lines around 2.40 Å and 1.42 Å only, may be a disordered variety of birnessite. Some manganese nodules give X-ray powder diffraction patterns containing only two broad diffuse lines at about 2.40 Å and 1.42 Å, with no suggestion of additional lines around 7.0—7.2 Å and 3.5—3.6 Å which are diagnostic for terrestrial birnessites. However, Brown (1971) has reported that this poorly crystalline phase in manganese nodules may be recrystallized under water to a phase giving additional lines around 9.6 Å and 4.8 Å, which clearly differ from the d-spacings of the birnessite group (Table 7-III). R. K. Sorem (personal communication, 1974 has suggested that these results may be explained by crystallization of incipient todorokite present in the δ-MnO$_2$ samples. Similar pressure-induced hydration studies of birnessites (Burns et al., 1974) produced no phase changes. Woo (1973) proposed the name "protobirnessite" for the natural δ-MnO$_2$ phase. If Brown's (1971) pressure- and time-induced hydration measurements are correct, the name "protobirnessite" is unsatisfactory because δ-MnO$_2$ rehydrates to todorokite and not birnessite. Furthermore, it has been suggested that the poorly crystalline manganese oxide phase, and not the more ordered birnessite, plays an essential role in the nucleation and authigenesis of manganese nodules through its ability to form epitaxial intergrowths with an isostructural iron (III) oxide hydroxide phase (Burns

---

*International Union of Pure and Applied Chemistry.

TABLE 7-III

Correlations of X-ray powder diffraction data between compounds of the birnessite group

Synthetic δ-MnO₂		Synthetic manganous manganite		Synthetic δ-MnO₂ or manganous manganite		Synthetic δ-MnO₂		Synthetic manganous manganite ("7-Å manganite")		Birnessite (Scotland)		Birnessite (Massachusetts)	
(1)		(2)		(3)		(4)		(5)		(6)		(7)	
		6.90	(ms)	7.13	(ms)	7.0	(ms)	7.4	(m)	7.27	(s)	7.31	(100)
		3.49	(w)	3.53	(w)	3.64	(vw)	3.71	(w)	3.60	(w)	3.60	(w)
2.39	(b)	2.42	(m)	2.41	(m)	2.41	(mw)	2.49	(m)	2.44	(m)	2.44	(70)
				2.14	(vw,b)								
		1.69	(m)										
		1.50	(w)										
1.40	(b)			1.418	(vw)			1.44	(m)	1.412	(m)	1.418	(80)

*d-spacings with intensities 10% and lower are omitted.

References:

1. H. F. McMurdie (1944). Trans. Electrochem. Soc., 86: 313.
2. W. Feitknecht and W. Marti (1945). Helv. Chim. Acta, 28: 129.
3. W. F. Cole, A. D. Wadsley and A. Walkley (1947). Trans. Electrochem. Soc., 92: 133—154.
4. H. F. McMurdie and E. Golovato (1948). J. Res. Nat. Bur. Stand., 41: 589.
5. W. Buser, P. Graf and W. Feitknecht (1954). Helv. Chim. Acta, 37: 2322.
6. L. H. P. Jones and A. A. Milne (1956). Miner. Mag., 31: 283—288.

and Burns, 1975). This hypothesis is discussed further later in the chapter. Therefore, it is recommended that the name "δ-MnO₂" be retained for the phase in manganese nodules giving rise to X-ray diffractions around 2.40 Å and 1.42 Å. Thus, "birnessite" and "δ-MnO₂" are regarded as two distinct phases in this chapter. The coexistence of δ-MnO₂ and birnessite or todorokite in manganese nodules is indicated when the 2.40-Å and 1.42-Å lines are more intense than the reflections around 7 Å or 9.6 Å, provided that effects due to preferred orientation are eliminated.

The todorokite—buserite problem

Greater confusion exists over the compound variously called "10-Å manganite" ($3MnO_2 \cdot Mn(OH)_2 \cdot xH_2O$), "todorokite", "buserite", and sodium manganese (II, III) manganate (IV) hydrate. Wadsley (1950a, b)

Birnessite (New Jersey)		Birnessite (Japan)		Birnessite (Mexico)		Birnessite (Washington)		Birnessite (California)		Synthetic* $Na_4Mn_{14}O_{27}\cdot9H_2O$		Synthetic* $Mn_7O_{13}\cdot9H_2O$	
(7)		(8)		(9)		(10)		(11)		(12)		(13)	
7.36	(100)	7.37	(100)	7.2	(100)	7.2	(100)	7.24	(s)	7.09	(100)	7.21	(100)
		4.69	(10)										
3.67	(90)	3.69	(90)	3.60	(50)	3.6	(50)	3.55	(mw)	3.56	(80)	3.61	(80)
		3.32	(50)										
		3.27	(40)										
2.46	(80)	2.45	(50)	2.40	(60)	2.45	(10)	2.46	(m)	2.51	(70)	2.46	(100)
		2.37	(70)							2.42	(60)		
										2.21	(40)	2.33	(100)
										2.15	(40)		
		2.09	(50)							2.14	(40)	2.04	(80)
										1.86	(40)		
										1.82	(40)		
										1.81	(40)		
										1.77	(20)	1.723	(80)
										1.66	(20)		
										1.63	(20)		
										1.47	(60)	1.454	(20)
										1.43	(50)	1.422	(60)
		1.41	(80)	1.42	(10)	1.42	(10)	1.424	(m)	1.41	(40)	1.391	(20)
										1.37	(20)	1.321	(20)

7. C. Frondel, U. B. Marvin and J. Ito (1960b). Am. Miner., 45: 871—875.
8. Y. Hariya (1961). Jap. J. Assoc. Miner. Petrol. Econ. Geol., 45: 219—230.
9. A. A. Levinson (1962). Am. Miner., 47: 790—791.
10. R. K. Sorem and D. W. Gunn (1967). Econ. Geol., 62: 22—56.
11. F. H. Brown, A. Pabst, and D. L. Sawyer (1971). Am. Miner., 56: 1057—1064.
12. JCPDS Powder File Card No. 23—1046.
13. JCPDS Powder File Card No. 23—1239.

claimed to have synthesized for the first time an oxide of approximate formula $(Na, Mn)Mn_3O_7\cdot xH_2O$ consisting of thin sheets which gave an X-ray pattern indexed for a hexagonal phase with prominent basal plane reflections at 10.0 Å and 5.06 Å R. Giovanoli (personal communication, 1974) suggests that Wadsley reproduced products synthesized originally by Marti (1944). Wadsley (1950a, b) demonstrated that this hydrous manganese oxide had cation exchange properties, whereby the sodium is irreversibly exchanged by other smaller cations such as Al^{3+}, Zn^{2+}, and Mn^{2+} to give X-ray patterns resembling that of the untreated oxide but with smaller basal plane spacings. However, the strong line around 9.5—10.0 Å disappeared from the patterns for the K^+, Pb^{2+}, Ba^{2+}, Ca^{2+}, and Cu^{2+} derivatives, which instead resembled "manganous manganite" (synthetic birnessite). Wadsley (1950a, b) also noted that the original oxide, $(Na, Mn)Mn_3O_7\cdot xH_2O$, itself

TABLE 7-IV

Correlations of X-ray powder diffraction data for todorokite and related compounds

Synthetic (Na,Mn) Mn_3O_7 (1)	Todorokite (Sweden) (2)	Todorokite (Cuba) (3)	Todorokite (Cuba) (4)	Todorokite (Portugal) (4)	Todorokite (Japan) (4)	Todorokite (Austria) (4)	Todorokite (France) (4)
10.0 (s)	9.67 (vs)	9.6 (10)	9.60 (10)	9.56 (10)	9.68 (10)	9.60 (10)	9.58 (10)
7.28 (w)		7.1 (1/2)	7.13 (1/4d)		7.15 (1/4d)	7.02 (3)	6.98 (2)
5.06 (m)	4.78 (m)	4.77 (8)	4.80 (8)	4.80 (8)	4.80 (8)	4.79 (8)	4.79 (8)
4.76 (vw-d)	4.47 (vw)	4.45 (2b)	4.45 (1d)		4.45 (1/2d)		
3.47 (w)		4.2	3.40 (1/2d)			3.48 (1)	3.48 (3)
3.372 (w)	3.22 (vw)	3.19 (2)	3.20 (1)	3.19 (4)	3.22 (1-1/2)	3.21 (1/2)	3.22 (1)
			3.10 (1)	3.11 (1/2)		3.10 (3)	
							2.98 (1/2d)
2.50 (w-d)							2.88 (1/2d)
2.429 (w)	2.43 (vw)	2.46 (3)	2.46 (3)	2.46 (2)	2.46 (2)	2.46 (2)	
2.38 (w-d)	2.39 (vw)	2.40 (5)	2.40 (5)	2.40 (4)	2.39 (4)	2.40 (5)	2.41 (4)
		2.35 (3b)	2.34 (4)	2.34 (3)	2.34 (1-1/2)	2.33 (2)	2.36 (2)
2.14 (vw-d)	2.21 (vw)	2.22 (4)	2.23 (2)	2.23 (3)	2.22 (2)	2.23 (2)	2.19 (3)
		2.16 (1)	2.13 (1/2d)		2.15 (1/2d)	2.15 (2)	2.15 (1/4)
		2.04 (1/2)	1.98 (2)	1.98 (3)	1.98 (1/2)	2.00 (1)	
	1.97 (vw)	1.98 (2)	1.92 (1/2)	1.93 (2)			
							1.83 (1/2)
1.85 (vw-d)			1.78 (1)	1.78 (1)	1.75 (1)	1.73 (1/4d)	
	1.74 (vvw)	1.74 (2)	1.74 (1)	1.74 (1)			
	1.68 (vvw)	1.68 (1/2b)	1.68 (1/2)				
		1.64	1.53 (1)	1.54 (1)	1.54 (1/2)	1.55 (1)	1.56 (1/2)
1.472 (mw)		1.53 (2)	1.49 (3)				
1.42 (w)	1.42 (vw)	1.42 (5)	1.42 (2)	1.42 (3)	1.42 (3)	1.43 (4)	1.43 (1)
		1.38 (2)	1.39 (1)	1.39 (2)	1.39 (1)	1.40 (1/2)	1.40 (1/4)

*X-rays diffracted parallel to the fiber axis.
**X-rays diffracted perpendicular to the fiber axis.

References:

1. A. D. Wadsley (1950). J. Am. Chem. Soc., 72: 1781—1784.
2. P. Ljunggren (1960). Am. Miner., 45: 235—238.
3. A. A. Levinson (1960). Am. Miner., 45: 802—807.
4. C. Frondel, U. B. Martin and J. Ito (1960a). Am. Miner., 45: 1167—1173.
5. J. A. Straczek, A. Horen, M. Ross and C. M. Warshaw (1960). Am. Miner., 45: 1174—1184.

readily dehydrates or oxidizes to "manganous manganite". This instability was utilized in the preparation of synthetic birnessites (Buser et al., 1954; Giovanoli et al., 1970a, b). Thus, Buser et al. (1954) reproduced Wadsley's (1950a) synthetic compound, formulating it as $3MnO_2 \cdot (Na, Mn)(OH)_2 \cdot xH_2O$, and also prepared the zinc analogue. This phase constituted the mineral identified as "10-Å manganite" in manganese nodules by Buser (1959).

Levinson (1960) drew attention to the close similarities of X-ray diffraction data between Wadsley's (1950a) synthetic hydrated manganese oxide, and the mineral todorokite originally described in a terrestrial deposit from Japan (Yoshimura, 1934). Subsequently, todorokite was identified by its X-ray powder diffraction pattern in a variety of terrestrial sources. At the type locality in Japan, for example, todorokite is formed as an alteration product of inesite $(Ca_2Mn_7Si_{10}O_{28}(OH)_2 \cdot 5H_2O)$. At Charco Redondo in Cuba, todorokite appears to be a fissure or breccia filling of hypogene origin and has commonly crystallized around kernels of volcanic tuff or limestone

Todorokite (Cuba) (5)		Todorokite (Japan) (6)		Todorokite Zn-bearing (Montana) (7)		Todorokite (France) (8)		Todorokite Ag-bearing (Nevada) (9)		Todorokite (Cuba) (10)			Synthetic Cu-"buserite" (11)		Synthetic Ni-"buserite" (11)	
9.6	(s)	9.66	(10)	9.60	(10)	9.65	(10)	9.43	(10)	9.65	vs*	vw**	9.6	vs	9.7	vs
				7.0	(1d)			6.75	(1)	7.02	w	—				
4.77	(s)	4.81	(8)	4.8	(6)	4.80	(4)	4.76	(4)	4.82	m	—	4.78	vs	4.84	s
				4.45	(1d)	4.15	(1)	4.48	(<1)	4.48	m	--				
3.19	(w)	3.23	(7)	3.20	(1)			3.20	(1)	3.20	vw	—	3.18	w	3.23	m
3.11	(b)							3.11	(5)	3.11	w	—	3.06	w		
2.95	(b)							2.66	(b)	2.66	vw	—	2.75	w	2.751	w
2.7	(b)					2.85	(4)	2.49	(1)							
2.448	(m)			2.46	(2)	2.43	(4)	2.45	(2)	2.46	w	mw	2.46	m	2.567	w
2.398	(s)	2.40	(3)	2.405	(4)			2.40	(3)	2.42	vw	ms	2.40	m	2.49	m
2.34	(m)			2.34	(1)	2.36	(4)	2.35	(<1)	2.35	w	w			2.41	w
2.21	(m)			2.22	(2)	2.20	(1)	2.23	(3)	2.20	—	vw	2.20		2.29	w
2.16	(w-b)	2.173	(4)					2.16	(1)	2.13	vw	—			2.21	m
2.11	(w-b)															
1.98	(m)			1.99	(2)			1.995	(<1)	1.98	vw	—	1.97	w	1.97	w
1.92	(w)	1.923	(3)					1.914	(<1)	1.91	—	—				
1.74	(mw)			1.75	(1/2)			1.756	(<1)	1.77	vw	—	1.73	w	1.73	w
1.69	(f-b)	1.662	(3)					1.691	(2)	1.67	vw	—				
		1.636	(4)					1.630	(1)	1.64	vw	—				
1.53	(w)	1.599	(5)	1.54	(1)			1.593	(<1)	1.54	vw	—			1.537	w
								1.471	(1)						1.469	m
1.423	(m)			1.42	(3)			1.456	(1)	1.42	—	vs	1.416	m	1.421	m
								1.428	(2)	1.400	—	--				
								1.362	(1)							
								1.348	(<1)							

6. Y. Hariya (1961). Jap. J. Assoc. Miner. Petrol. Econ. Geol., 45: 219—230.
7. L. T. Larson (1962). Am. Miner., 47: 59—66.
8. E.-A. Perseil (1966). C. R. Acad. Sci., 262D: 949—951.
9. A. S. Radtke, C. M. Taylor and D. F. Hewett (1967). Econ. Geol., 62: 186—206.
10. G. M. Faulring (1962). Adv. X-Ray Analysis, 5: 117—126.
11. W. Sung, R. G. Burns and V. M. Burns (unpubl. data).

matrix (Levinson, 1960; Straczek et al., 1960). In other localities, todorokite is clearly of secondary origin (Frondel et al., 1960a; Larson, 1962). X-ray data for todorokite are compared in Table 7-IV. It is apparent from the d-spacings in Table 7-IV that considerable variation exists between relative intensities of comparable lines for different todorokite samples. Faulring (1962) attributed these large variations in intensities, as well as the diffuseness of certain reflections, to preferred orientation of the fine fibrous crystallites of todorokite from Charco Redondo. Thus, the most intense lines at 9.65 Å and 4.82 Å observed when X-rays are diffracted parallel to the fiber axis were weak or undetected for X-rays diffracted perpendicular to the crystallite axis. The most intense lines in the latter orientation occur at 2.42 Å and 1.42 Å (Table 7-IV). Faulring (1962) also noted that manganite (γ-MnOOH) was topotactically intergrown with the Cuban todorokite, and suggested that the manganite either formed simultaneously with the todorokite or resulted from the decomposition of todorokite. The latter observation constituted one reason for Giovanoli et al. (1973a) to reject todorokite as a valid mineral and to propose the name "buserite" for the primary mineral, after W. Buser, the Swiss chemist. Other reasons were based on studies of synthetic analogues.

Giovanoli et al. (1970a, 1971, 1973a) re-synthesized Marti's (1944) or Wadsley's (1950a, b) phase, and proposed the names sodium manganese (II, III) manganate (IV) hydrate or "buserite". The cation exchange properties of this phase were confirmed and Co^{2+}, Cu^{2+}, and Ni^{2+} varieties were prepared. These derivatives, as well as the original Na derivative, are unstable and, as demonstrated by the earlier workers, readily oxidize and lose water to form birnessite phases. Electron microscopy and diffraction measurements of the synthetic birnessites showed that under reducing conditions at room temperature, the birnessite platelets decomposed to perfectly orientated pseudomorphs of topotactically produced and extremely thin γ-MnOOH crystallites, which were not detected by X-ray diffraction (Giovanoli et al., 1971). This result, together with the observation that synthetic birnessite is a dehydration product of "buserite", led Giovanoli et al. (1973a) to conclude that todorokite is a complex mixture of several compounds produced by dehydration and reduction of "buserite". They have obtained electron micrographs of minerals identified as todorokite on the basis of their X-ray patterns which they interpret as showing the remains of unaltered buserite coexisting with needles of manganite (γ-MnOOH). The evidence was submitted to the IMA Commission on New Minerals in 1970 and accepted by a majority. Some mineralogists on the Commission believe that much more detailed evidence is necessary before todorokite is discredited in favour of buserite (M. H. Hey, personal communication, 1974; M. Fleischer, personal communication, 1974).

Fleischer has pointed out that the suggestion by Giovanoli that todorokite consists of a mixture of buserite partly dehydrated to birnessite and partly reduced to manganite has puzzling ramifications. It is not in accord with the widespread occurrence of todorokite, sometimes in tonnage amounts, nor with the reproducible X-ray powder patterns (Table 7-IV), nor with the absence of the strongest lines for either manganite or birnessite in all todorokites. Levinson (1960) has suggested that the weak line around 7 Å observed in some todorokites may be due to impurities such as birnessite or kaolinite, but Faulring (1962), Brown (1972), and Burns and Brown (1972) attributed it to a prismatic reflection of todorokite. Further, P. B. Moore (personal communication, 1974) has obtained small single crystals of todorokite from a locality in Arkansas which, when fresh, are light brown, translucent laths. The mineral dehydrates spontaneously when exposed to the air and turns black.

Giovanoli (personal communication, 1974) contends that X-ray evidence alone can, in the case of poorly crystallized manganese oxide ores, lead to errors when one constituent of a mixture fails to diffract X-rays. Giovanoli et al. (1973a) do concede that in manganese nodules the finely divided crystallites and intimate association of hydrated manganese oxide and hydrated iron oxide hydroxide phases has made it difficult to investigate directly the "buserite" phase in manganese nodules. Burns and Burns (1975), however, suggest that such intimate associations of iron and manganese oxides hold the key to the authigenesis of ferromanganese nodules.

It is obvious that further work is required before the todorokite— "buserite" problem is satisfactorily resolved. In the meantime, the name "todorokite" is recommended for the hydrated manganese oxide phase giving the reasonably reproducible d-spacings listed in Table 7-IV. Todoro- kite, which is finding increasing acceptance in the literature on manganese nodules, is therefore adopted in this chapter over the original "10-Å manganite" nomenclature of Buser (1959). The terminology may have to be revised, however, if todorokite proves to be a discredited mineral.

Other nomenclature problems

Confusion also exists over usage of the terms psilomelane, wad, and vernadite. Psilomelane is often reported as a constituent of manganese nodules in the Russian literature (Andrushchenko and Skornyakova, 1965, 1969, Bezrukov and Andrushchenko, 1972). Although psilomelane was the name originally given to a specific mineral, the term psilomelane has also been used to refer to any black, hard, botryoidal, unidentified manganese oxide mineral, just as the designation wad refers to soft, unidentified samples (Wadsley, 1950b; Fleischer, 1960). Because psilomelane had ceased to be a well-defined mineral, the IMA Commission on New Minerals has recom- mended that the name romanèchite be adopted for the specific hydrated barium manganese oxide mineral formerly called psilomelane. Thus, it is necessary to be aware of the dual role of the name psilomelane in the mineralogical literature. It may refer either to a specific mineral with a well-defined composition and crystal structure, or to an unidentified manganese oxide mineral.

The term vernadite is used frequently in the Russian literature on manganese nodules (Andrushchenko and Skornyakova, 1965; 1969; Bezrukov and Andrushchenko, 1972). It is described as a colloidal hydrate of manganese dioxide, and yet a well-defined X-ray pattern resembling that of pyrolusite is reported in the JCPDS powder diffraction file (Table 7-II). Strunz (1970) suggests that vernadite may be a doubtful mineral.

Occurrences in manganese nodules

Many of the manganese oxide minerals listed in Table 7-I have been positively identified or tentatively suggested to occur in a variety of manganese nodules. Examples of cited occurrences are included in Table 7-I. For reasons discussed later, the basis of some of these phase identifications may be suspect.

Originally, Murray and Renard (1891), on the basis of chemical analyses, concluded that the nodules consisted of wad or bog manganese ore, which they noted were impure varieties of manganese ore related to psilomelane. Following the classic studies of Buser and Grütter (1956) and Grütter and Buser (1957), Buser (1959) summarized results for twelve manganese nodules and named the three principal manganese oxide minerals occurring in manganese nodules "10 Å manganite", "7-Å manganite", and "δ-MnO_2", in recognition of prominent features found also in X-ray powder diffraction patterns of synthetic analogues. Manheim (1965) noted close similarities of X-ray powder diffraction data between five marine manganese nodules and the terrestrial minerals todorokite and birnessite. His data are reproduced in Table 7-V. He suggested that todorokite was a dominant manganese-bearing mineral in marine and non-marine concretions. Hewett et al. (1963) had also tentatively identified todorokite in ten Pacific Ocean nodules, but cited no X-ray data. Hewett et al. (1963) and Manheim (1965) therefore favoured the mineral names todorokite and birnessite over the correlative synthetic phases "10-Å manganite" and "7-Å manganite". Subsequent investigators have been more or less equally divided over these two nomenclature schemes (for examples see Table 7-I). Manheim's (1965) X-ray data also suggested that ramsdellite, or its poorly crystalline equivalent nsutite (γ-MnO_2) was present in some nodules. A similar conclusion was drawn by Gattow and Glemser (1961b) and by Schweisfurth (1971), although Giovanoli et al. (1973a) refuted these particular occurrences of γ-MnO_2. None of Manheim's (1965) X-ray data resembled the two-line patterns obtained by Buser and Grütter (1956) for nodules containing δ-MnO_2.

. Barnes (1967a), adopting the terminology of Buser (1959), showed that δ-MnO_2 was a common constituent of some 67 ferromanganese nodules studied by him. Many of the nodules, particularly those from shallow depths, contained only δ-MnO_2 while others contained δ-MnO_2 coexisting with "10-Å manganite" and/or "7 Å manganite". The "10 Å manganite" and "7 Å maganite" phase appeared to occur most frequently in nodules from greatest depths, in accord with Brown's (1971) observation that pressure-induced hydration of δ-MnO_2 leads to the formation of "10-Å manganite". However, thermodynamic calculations (Glasby, 1972d) do not support the phase conversion of δ-MnO_2 to todorokite at depths in the ocean. The "δ-MnO_2" phase has been accepted as a separate entity by several investigators (for example, Price and Calvert, 1970; Brown, 1972; Ostwald and Frazer, 1973; Woo, 1973).

Other investigators have regarded the "δ-MnO_2" phase of Buser and Grütter (1956) as a variety of birnessite (for example, Sorem and Foster, 1972a, c; Scott et al., 1973; Crerar and Barnes, 1974). Thus, Cronan and Tooms (1969), who studied more than 60 nodules, provided selected X-ray powder diffraction data for four typical nodules which are summarized in Table 7-VI. Cronan and Tooms (1969) adopted the todorokite—birnessite

TABLE 7-V

X-ray powder diffraction data reported by Manheim (1965)

1 d(Å)	I	2 d(Å)	I	3 d(Å)	I	7 d(Å)	I	8 d(Å)	(I)	T d(Å)	(I)	B d(Å)	(I)	R d(Å)	(I)	G d(Å)	(I)
9.5	100b	7.10	11b	9.7	10b	9.6	30b	9.7	100	9.6	10						
						7.0	10	7.2	60b	7.17	1b	7.27	s				
4.8	90vb			4.87	15b	4.8	30b	4.85	24b	4.76	4			4.08	10	4.02	s
4.48	20vb							4.56	6b	4.42	1						
		3.9	100vb														
								3.58	6b			3.60	w				
		3.13	5b											3.10	9		
		2.56	4b											2.53	8	2.53	vvw
2.47	40vb	2.42	59b	2.46	100b	2.45	100vb	2.45	96b	2.45	1	2.44	m	2.43	4	2.41	m
2.39	10b	2.34	7b	2.39	10b			band		2.39	2			2.32	4		
				2.34	20b			band		2.34	1						
				2.24	10b			2.23	3b	2.22	1/2						
		2.13	35b											2.13	5	2.10	ms
		2.05	?					2.06	5b					2.04	2		
								2.00	7								
		1.88	2b											1.88	5		
		1.82	3b											1.82	2		
		1.63	60b											1.64	6	1.62	vs
														1.60	7		
		1.54	4b											1.53	3		
		1.49 ?	6b														
		1.46	3b											1.46	5		
1.42	30b			1.41	40b	1.41	20vb	1.42	21b	1.42	3	1.412	m				
1.39	20b	1.385	25b	1.385	10b					1.39	2					1.38	w

1. Blake Plateau "potato ore".
2. Pacific Ocean nodule, *Challenger* Sta. 286.
3. Pacific Ocean nodule, *Challenger* Sta. 176.
7. Gulf of Maine, 100—200 m (near shore).
8. Baltic Sea, near Bornholm.
T. well-crystallized todorokite from Charco Redondo, Cuba.
B. birnessite, reported by Jones and Milne (1956).
R. ramsdellite, reported by Fleischer and Richmond (1943).
G. γ-MnO$_2$ (poorly crystalline ramsdellite).

TABLE 7-VI

Powder patterns of selected manganese nodules (Cronan and Tooms, 1969)

1		2		3		4	
d(Å)	I	d(Å)	I	d(Å)	I	d(Å)	I
9.80	100			9.70	100		
		7.18	100	7.20	80		
4.81	75			4.80	50		
4.45	50			4.45	20		
		3.57	30	3.57	20		
2.45	55	2.44	20	2.44	30	2.43	broad
2.40	50			2.39	25		
2.36	30			2.35	15		
2.23	40			2.25	20		
2.13	25						
1.97	25						
1.91	10						
1.77	15						
1.53	35						
1.42	35	1.41	15	1.42	30	1.41	broad
1.40	10			1.39	10		

1. Station *Challenger* 160; 42°42'S 134°10'E; depth 4,760 m. Contains todorokite.
2. Station MV—65—1—41; 24°34'N 113°28'W; depth 3,510 m. Contains birnessite.
3. Station Mag. Bay A35; 24°23'N 113°18'W; depth 3,550 m. Contains todorokite and birnessite.
4. Station M.P. 25 Fl; 19°05'N 169°45'W; depth 1,777 m. Contains "two line form of birnessite" (i.e., δ-MnO_2).

terminology and, following the suggestion of Bricker (1965), subdivided the birnessites in manganese nodules into two-line (smaller particle sizes) and four-line (more ordered varieties) forms. However, on the basis of hydration studies described earlier (Brown, 1971; Burns et al., 1974), the four-line phase, birnessite, should now be distinguished from δ-MnO_2 which has two diffraction lines, and undergoes pressure-induced hydration to todorokite.

Although the todorokite—birnessite—δ-MnO_2 terminology is currently recommended for the three principal manganese oxide minerals in manganese nodules, it should be noted that Giovanoli et al. (1970b, 1971, 1973a, b) regard the "10-Å manganite" phase of Buser (1959) to be equivalent to the new synthetic phase, buserite. They maintain that buserite is a primary phase in manganese nodules and that todorokite is a mixture of buserite, birnessite, and manganite (γ-MnOOH). More conclusive evidence is required before todorokite is discredited (M. H. Hey, personal communication, 1974; M. Fleischer, personal communication, 1974), and before buserite can be taken to be one of the principal minerals of manganese nodules.

Certain of the phase identifications in the heterogeneous manganese nodules have been based on subtle correlations of d-spacing intensities. This is particularly well illustrated by the evidence for the presence of pyrolusite, psilomelane, woodruffite, and vernadite (amorphous) in manganese nodules (Andrushchenko and Skornyakova, 1965, 1969; Bezrukov and Andrushchenko, 1972). Some of the data are reproduced in Table 7-VII. Other important mineral assemblages have been described without essential corroborative evidence. For example, Stevenson and Stevenson (1970) reported the occurrence of birnessite on the outer surfaces of manganese nodules that had dried out during storage in a museum. However, it is not clear from this report whether dehydration had produced the well-crystallized phase (birnessite) or a disordered form (δ-MnO_2). The significance of this result is that it suggests that mineralogical changes may occur in manganese nodules removed from seawater and exposed to the atmosphere after long periods of time such as dehydration during X-ray diffraction and electron microscopy measurements in vacuo.

OXIDES OF IRON

Glossary and X-ray data

Several oxide, oxide hydroxide or oxyhydroxide, and hydrated oxide phases of iron are known, many of them occurring as minerals in terrestrial deposits. Most of the phases relevant to manganese nodule mineralogy are listed in Table 7-VIII. This table includes crystallographic data and references to current information on chemical composition and crystal structures. Once again, the information contained in Table 7-VIII is based on compilations by Fleischer (1967, 1971) and Strunz (1970).

X-ray powder diffraction data based on the Fink Indexing System are summarized in Table 7-IX. By analogy with manganese oxides, discussed earlier, the structures of many of the iron oxides consist of close-packed oxygens containing Fe^{3+} ions in octahedral coordination with oxygen, in which [FeO_6] octahedra share edges. Certain iron (III) oxide hydroxides are isostructural with manganese (IV) oxides listed in Table 7-I. The former contain Fe^{3+} ions octahedrally coordinated O^{2-} and OH^- ions. The larger ionic radius of Fe^{3+} compared to Mn^{4+} results in larger spacings for the ($10\bar{1}0$) and ($11\bar{2}0$) planes of the hexagonal close-packed system (approximately 2.50—2.56 Å and 1.48—1.54 Å, respectively), and for Fe—Fe interatomic distances across edge-shared [$Fe(O,OH)_6$] octahedra (2.95—3.05 Å). These dimensions again figure prominently in X-ray powder diffraction and cell parameter data, respectively, of iron oxide phases.

TABLE 7-VII

X-ray diffraction data for manganese nodules from Russian literature, together with data for the phases identified

Mn nodule (Station 5414)[1]		Pyrolusite (cited ref.)[2]		Mn nodule (Station 5202)[3]		Psilomelane (cited ref.)[2]		Mn nodule (Station 5398)[1]		Woodruffite (cited ref.)[4]		Mn nodule (Station 5414)[1]		Todorokite (cited ref.)[5]	
d(A)	I	d(A)	I	d(A)	I	d(A)	I	d(A)	I	d(A)	I	d(A)	I	d(A)	I
								9.62	70	9.51	50	9.60	100	9.65	100
								6.86	40	6.99	40				
4.87	20							4.98	80	4.77	100	4.88	90	4.81	80
										4.40	30			4.46	30
41.6	20														
3.69	20														
								3.56	70	3.48	20				
				3.35	10	3.35	40			3.33	10	3.32	40		
3.18	100	3.11	100	3.22	30	3.12	40	3.18	50	3.13	20			3.20	40
3.01	70									3.03	10				
2.73	10					2.70	80			2.63	20				
2.55	90							2.56	10	2.56	10				
2.45	90	2.40	50	2.45	100	2.43	100	2.395	40	2.466	40	2.43	40	2.45	30
										2.404	50	2.39	40	2.40	40
2.23	20	2.20	5	2.22	10			2.27	40	2.225	50			2.16	40
2.13	20	2.11	40			2.15	40	2.19	10	2.131	20	2.10	80	2.15	10
								1.95	10	1.984	30	1.91	40	1.98	10
										1.895	20				
1.72	10			1.72	10	1.72	40	1.714	40	1.747	20				
						1.63	60	1.657	40	1.660	20				
1.677	10	1.623	70												
1.637	10														
1.419	90	1.438	20	1.42	100	1.42	20	1.440	40	1.423	50	1.42	30	1.419	40
						1.35	40					1.329	40	1.331	50
								1.307	70			1.264	40		
								1.25	40						
								1.207	20			1.121	50		

References:
1. Andrushchenko and Skornyakova (1969).
2. Mikheyev (1957).
3. Bezrukov and Andrushchenko (1972).
4. Berry and Thompson (1962).
5. Frondel et al. (1960a).

Nomenclature

The list of iron oxide phases in Table 7-VIII contains five polymorphs of "FeOOH", which according to the International Union of Pure and Applied Chemistry (IUPAC) convention are more correctly termed iron (III) oxide hydroxides, but are commonly called ferric oxyhydroxides. The most common of these are the minerals goethite (a-FeOOH) and lepidocrocite (γ-FeOOH). Note that among the oxyhydroxides of Mn(III) and Fe(III) listed in Table 7-I and 7-VIII, only goethite and groutite, and perhaps ϵ-FeOOH and manganite (γ-MnOOH), are isostructural. Feitknechtite (β-MnOOH) is isostructural with $Fe(OH)_2$ (amakinite) and $Mn(OH)_2$ (pyrochroite).

A sixth oxide hydroxide phase, variously called "amorphous $Fe(OH)_3$", "iron (III) oxide hydrate gel", "colloidal ferric species", "hydrated ferric oxide polymer" may be identical to a natural ferric gel (Coey and Readman, 1973) and the proposed new mineral ferrihydrite (Chukhrov et al., 1973). The synthetic hydrated ferric oxyhydroxide polymer has been extensively studied (for example, Towe and Bradley, 1967; Van der Giessen, 1966; Okamoto et al., 1972; Giovanoli, 1972; Feitknecht et al., 1973) and is generally considered to contain crystallites with dimensions between 30 Å and 100 Å which recrystallize to different iron (III) oxyhydroxides upon aging in various solutions (Giovanoli, 1972; Feitknecht et al., 1973). The fine-grained and disordered nature of these synthetic particles reduce the ability of the crystallites to scatter X-rays and electrons coherently. They therefore appear to be amorphous to X-rays. Other techniques, such as Mössbauer spectroscopy and electron microscopy, have been used to characterize the very fine crystallites. Similar studies of manganese nodules indicate that the principal iron-bearing phases in manganese nodules are also very fine grained. In this chapter, the apparently X-ray amorphous hydrous iron oxide phase occurring in manganese nodules will be called hydrated ferric oxyhydroxide polymer and formulated as $FeOOH \cdot xH_2O$.

Occurrences in manganese nodules

Of the phases listed in Table 7-VIII, those iron-bearing minerals which have been reported in manganese nodules include goethite, lepidocrocite, maghemite, hematite, and akaganéite. In most descriptions of manganese nodules, however, the iron-bearing phases are usually described as being cryptocrystalline or amorphous hydrated iron oxides (Goodell et al., 1971; Glasby, 1972a, c; Crerar and Barnes, 1974). This is a consequence of the poor or negligible X-ray or electron diffraction patterns obtained for the fine-grained, or disordered particles. Evidence from Mössbauer spectroscopy described later, together with recent transmission electron microscope data (Von Heimendahl et al., 1976), indicates that a hydrated ferric oxyhydroxide polymer predominates in many nodules.

TABLE 7-VIII

Oxides of iron

Compound	Mineral	Formula	Space group/ crystal class	Cell parameters (Å)
α-FeOOH	goethite	FeOOH	Pbnm orthorhombic	$a_0 =4.65$; $b_0 =10.02$; $c_0 =3.04$; $z=4$
β-FeOOH	akaganéite	$(OH,Cl,H_2O)_{1-2}$ $Fe_8(O,OH)_{16}$	I4/m tetragonal	$a_0 =10.48$; $c_0 =3.028$; $z=1$
γ-FeOOH	lepidocrocite	FeOOH	Amam orthorhombic	$a_0 =3.88$; $b_0 =12.54$; $c_0 =3.07$; $z=4$
δ-FeOOH	synthetic	FeOOH	hexagonal	$a_0 =2.95$; $c_0 =4.53$
ϵ-FeOOH	synthetic	FeOOH	$P2_{1/c}$ monoclinic	$a_0 =8.731$; $b_0 =5.164$; $c_0 =5.680$; $\beta=90°$; $z=8$
HP-FeOOH	synthetic	FeOOH	$P2_1 nm$ orthorhombic	$a_0 =4.937$; $b_0 =4.432$; $c_0 =2.994$; $z=2$
hydrated ferric oxyhydroxide polymer	synthetic	$FeO_{(3-x)/2}(OH)_x$	hexagonal	$a_0 =5.88$; $c_0 =9.4$
		$Fe_5 HO_8 \cdot 4H_2O$ $\rightarrow$	hexagonal	$a_0 =5.08$; $c_0 =9.4$
		$Fe_2O_3 \cdot 1\cdot 2H_2O$ $\rightarrow$	cubic	$a_0 =8.37$
Natural ferric oxide hydrate gel	ferrihydrite	$5Fe_2O_3 \cdot 9H_2O$		
Limonite or hydrogoethite				
α-Fe_2O_3	hematite	Fe_2O_3	R$\bar{3}$c hexagonal	$a_0 =5.04$; $c_0 =13.77$; $z=6$
γ-Fe_2O_3	maghemite	Fe_2O_3	$P2_1 3$ (cubic) or $P4_1 2_1 2$ (tetragonal)	$a_0 =8.32$; $z=8$ $a_0 =8.338$; $c_0 =25.014$
Spinel	magnetite		Fd3m cubic	$a_0 =8.391$; $z=8$
Green rust	synthetic	$xFe(OH)_2 \cdot yFeOCl$ zH_2O	hexagonal	$a_0 =3.22$; $c_0 =24.0$
$Fe(OH)_2$	amakinite	$(Fe,Mg,Mn)(OH)_2$	P$\bar{3}$m1	$a_0 =3.465$; $c_0 =4.85$; $z=2$

References:

1. H. Strunz (1970). *Mineralogische Tabellen*. Akad. Verlagsgesellschaft, Leipzig, 5th ed., p. 216.
2. A. L. MacKay(1962). Miner. Mag., 33: 270—280.
3. K. J. Gallagher (1970). Nature, 226: 1225—1228.
4. P. Keller (1970). Neues Jahrb. Miner. Abh., 113: 29—49.
5. H. Strunz (1970). idem, p. 218.
6. S. Okamoto (1968). J. Am. Ceram. Soc., 51: 594—599.
7. M. H. Francombe and H. P. Rooksby (1959). Clay Miner. Bull., 4: 1—14.
8. J. D. Bernal, D. R. Dasgupta and A. L. MacKay (1959). Clay Miner. Bull., 4: 15—30.
9. N. A. Bendeliani, M. I. Baneyeva and D. S. Poryvkin (1972). Geochem. Int., 9: 589—590.
10. J. Chenavas, J. C. Joubert and J. J. Capponi (1973). J. Solid State Chem., 6: 1—15.
11. M. Pernet, J. Chenavas and J. C. Joubert (1973). Solid State Comm., 13: 1147—1154.

Structure-type	Reference	Manganese nodule occurrence
diaspore, or ramsdellite (also groutite)	1	Buser and Grütter (1956); Gager (1968); Aumento et al. (1968); Johnson and Glasby (1969); Goodell et al. (1971); Carpenter et al. (1972)
hollandite	2,3,4	Goncharov et al. (1973)
boehmite	5	Gager (1968); Johnson and Glasby (1969); Goodell et al. (1971), but see Carpenter et al. (1972)
disordered CdI$_2$	6,7,8	
manganite	9	
InOOH	10,11	
	12	Herzenberg and Riley (1969); Hrynkiewicz et al. (1970, 1972a,b)
	13	Carpenter and Wakeham (1973); Von Heimendahl et al. (1976)
	14	
	15,16	
mixture of hydrous iron oxides	17	Andrushchenko and Skornyakova (1965) — probably goethite, however, since X-ray data correlate well
corundum	18	
defect spinel	19	Goodell et al. (1971); Von Heimendahl et al. (1973); Carpenter et al. (1972)
spinel	20	Carpenter et al. (1972)
—	8	
brucite (also pyrochroite)	21	

12. S. Okamoto, H. Sekizawa and S. I. Okamoto (1972). In: *Reactivity of Solids*. Chapman and Hall, pp. 341—350.
13. K. M. Towe and W. F. Bradley (1967). J. Colloid Interface Sci., 24: 384—392.
14. A. A. van der Giessen (1966). J. Inorg. Nucl. Chem., 28: 2155—2159.
15. F. V. Chukhrov, B. B. Zvyagin, A. I. Gorshkov, L. P. Yermilova and V. V. Balashova (1973). Dokl. Akad. Sci. SSSR, 4: 23—33.
16. J. M. D. Coey and P. W. Readman (1973). Earth Planet. Sci. Lett., 21: 45—51.
17. M. Fleischer (1971). *Glossary of Mineral Species.* Mineral Record Inc., Maryland.
18. R. L. Blake, R. E. Hessevick, T. Zoltai and L. W. Finger (1966). Am. Miner., 51: 123—129.
19. H. Strunz (1970). idem, p. 186.
20. H. Strunz (1970), idem, p. 177.
21. I. T. Kozlov and P. P. Levshov (1962). Zap. Vses. Obshch., 91: 72.

TABLE 7-IX

X-ray diffraction data for oxide compounds of iron (Fink Classification System)

Mineral or compound	The eight most intense d-spacings (intensity ⩾ 10%)								JCPDS card or reference
Goethite (α-FeOOH)	4.18 (100)	2.69 (30)	2.490 (16)	2.452 (25)	2.192 (20)	1.721 (20)	1.564 (16)		17—536
Akaganéite (β-FeOOH)	7.40 (100)	5.25 (40)	3.311 (100)	2.616 (40)	2.543 (80)	1.944 (60)	1.635 (100)	1.438 (80)	13—157
Lepidocrocite (γ-FeOOH)	6.26 (100)	3.29 (90)	2.47 (80)	1.937 (70)	1.732 (40)	1.524 (40)	1.367 (30)	1.075 (40)	8—98
δ-FeOOH	4.61 (20)	2.545 (100)	2.255 (100b)	1.685 (100vb)	1.471 (100)	1.271 (20)	1.223 (20)		13—87
ϵ-FeOOH	3.314 (100)	2.567 (50)	2.489 (80)	2.223 (50)	1.755 (70)	1.682 (80)	1.658 (50)	1.447 (50)	(1)
Hydrated ferric oxy-hydroxide polymer	2.52 (sss)		2.25 (s)	1.97 (s)	1.72 (s)		1.48 (sss)		(2), (3)
Idem	2.54 (s)	2.47 (m)	2.24 (ms)	1.98 (m-b)	1.725 (w)	1.515 (m)	1.47 (s)		(4)
Idem	2.54					1.52			(5)
Ferrihydrate	2.50 (m)	2.21 (m-w)	1.96 (w)	1.72 (vw)	1.51 (m-w)	1.48 (m)			(6)
Hydrogoethite	4.16 (90)	2.67 (70)	2.44 (100)	1.716 (80)	1.564 (40)	1.510 (50)	1.454 (40)	1.318 (40)	(7)
Hematite (α-Fe$_2$O$_3$)	3.66 (25)	2.69 (100)	2.51 (50)	2.201 (30)	1.838 (40)	1.690 (60)	1.484 (35)	1.452 (35)	13—534
Maghemite (γ-Fe$_2$O$_3$)	5.95 (60)	3.75 (100)	3.42 (65)	2.950 (100)s	2.521 (100)s	2.089 (100)s	1.702 (100)s	1.608 (100)s	15—615
Magnetite (Fe$_3$O$_4$)	4.85 (8)	2.967 (30)	2.532 (100)	2.424 (8)	2.096 (20)	1.715 (10)	1.616 (30)	1.485 (40)	19—629 11—614
Green rust (xFe(OH)$_2$·yFeOCl·zH$_2$O)	8.02 (100)	4.01 (80)	2.701 (60)	2.408 (60)	2.037 (30)	1.598 (40)	1.567 (40)		13—88
Amakinite (Fe,Mg,Mn)(OH)$_2$	5.49 (70)	2.80 (80)	2.30 (100)	1.728 (90)	1.551 (70)	1.530 (80)	1.386 (70)	1.265 (70)	15—125

References:
1. N. A. Bendeliani, M.I. Baneyeva and D. S. Poryvkin (1972). Geochem. Int., 9: 589—590.
2. A. A. van der Giessen (1966). J. Inorg. Nucl. Chem., 28: 2155—2159.
3. S. Okamoto, H. Sekizawa and S. I. Okamoto (1972). In: *Reactivity of Solids.* Chapman and Hall, pp. 341—350.
4. K. M. Towe and W. F. Bradley (1967). J. Colloid Interface Sci., 24: 384—392.
5. W. Feitknecht, R. Giovanoli, W. Michaelis and M. Müller (1973). Helv. Chim. Acta, 56: 2847—2856.
6. F. V. Chukhrov, B. B. Zvyagin, A. I. Gorshkov, L. P. Yermilova and V. V. Balashova (1973). Proc. Acad. Sci., USSR. 4: 23—33.
7. P. F. Andrushchenko and N. S. Skornyakova (1965). In: Manganese Deposits of the Soviet Union. Israel Progr. Sci. Transl., Jerusalem, 1970, pp. 101—124.

The early work of Buser and Grütter (1956; see also Grütter and Buser, 1957; Buser, 1959) established the presence of goethite in manganese nodules by X-ray and electron diffraction. The patterns for goethite were accentuated by leaching the nodules with acidified hydroxylamine. This process not only dissolves most of the manganese oxide phases but may also recrystallize the amorphous $FeOOH \cdot nH_2O$ to goethite (W. Buser, personal communication, cited by Goldberg, 1961). From syntheses of "iron (III) manganites", Buser and Grütter (1956) suggested that some of the iron in manganese nodules occurred in the disordered hydroxyl sheet between the basal layers of the "manganite" structures. It was suggested (Goldberg and Arrhenius, 1958; Goldberg, 1959) that goethite appears in manganese nodules after the disordered layer of "manganite" is saturated by iron, its degree of crystallinity increasing with age. Analyses of terrestrial todorokites and birnessites, however, show low abundances of iron (Frondel et al., 1960a, b; Nambu et al., 1964; Brown et al., 1971). Other evidence for goethite in manganese nodules is based on X-ray diffraction (Manheim, 1965; Bonatti and Joensuu, 1966; Aumento et al., 1968; Goodell et al., 1971), electron diffraction (Buser and Grütter, 1956), and magnetic and Mössbauer spectroscopy measurements (Gager, 1968; Johnson and Glasby, 1969; Georgescu and Nistor, 1970).

Russian workers (Andrushchenko and Skornyakova, 1965; Bezrukov and Andrushchenko, 1972) suggest that the dominant iron-bearing phases in manganese nodules are hydrogoethite and X-ray amorphous material. The X-ray diffraction data for hydrogoethite are identical to those for goethite, however. "Ferrous manganite" is also frequently listed in the Russian literature. However, as discussed later, Mössbauer spectroscopy has established that there are negligible amounts of Fe^{2+} ions in manganese nodules.

ACCESSORY MINERALS

In addition to the hydrated oxides and oxyhydroxides of manganese and iron described in the previous sections, manganese nodules always contain appreciable amounts of other materials (Goldberg and Arrhenius, 1958; Arrhenius, 1963; Mero, 1965a), some of which appear to be essential for the nucleation and growth of the ferromanganese oxides. Most sectioned nodules contain a core consisting of fresh or altered volcanic rocks, pumice or glass, sharks' teeth, fish bones, siliceous and calcareous tests, etc. However, several detrital and authigenic accessory minerals have been identified in the ferromanganese oxide concretions surrounding these cores. A compilation of accessory minerals reported in manganese nodules is given in Table 7-X, together with their diagnostic d-spacings in X-ray powder diffraction patterns. Many of the accessory minerals have been identified in the residuals of acid-leached nodules, which appear to concentrate, recrystallize, or

TABLE 7-X

X-ray diffraction data for accessory minerals in manganese nodules

Mineral	d-spacings of the four most intense lines				JCPDS powder file card no.	Manganese nodule occurrence references
Quartz	4.26 (35)	3.343 (100)	1.817 (17)	1.541 (15)	5—490	3,5,6,7,9, 12,18,19, 20,21,22
Feldspar:						3,5,7,9,12, 20,21,22
K-feldspar (orthoclase)	4.22 (70)	3.77 (80)	3.31 (100)	3.24 (65)	19—931	5,6,21
Sanidine	3.27 (75)	3.26 (100)	3.25 (75)	3.22 (90)	19—1227	17?
Plagioclase						17?,21
Labradorite	3.759 (70)	3.210 (70)	3.203 (70)	3.181 (100)	18—1202	5
Mica:						3,5,7,20
Biotite	10.1 (100)	3.37 (100)	2.66 (80)	2.45 (80)	2—45	16
Olivine	2.791 (100)	2.533 (60)	2.475 (60)	1.761 (50)	7—159	5,17
Stilpnomelane	12.3 (100)	4.16 (100)	2.69 (70)	2.55 (100)	18—634	20
Pyroxene (augite)	2.991 (100)	2.893 (30)	2.528 (40)	2.518 (30)	11—654	3,5,12,16, 17,20
Amphibole:						16,17,20,21
Hornblende	8.51 (55)	3.29 (25)	3.14 (100)	2.720 (35)	21—149	3,16
Prehnite	3.48 (90)	3.08 (100)	2.55 (100)	1.77 (70)	7—333	17?
(Opal)					amorphous	3,16
Clay minerals:						1,12,19,20
Kaolinite	7.17 (100)	4.366 (60)	3.579 (80)	2.495 (45)	14—164	20,21
Pyrophyllite	9.14 (40)	4.57 (50)	3.04 (100)	2.40 (40)	2—714	17?
Talc	9.35 (100)	4.59 (45)	3.12 (40)	2.479 (30)	19—770	17?
Chlorite	30.0 (60)	15.0 (90)	4.97 (75)	4.53 (100)	12—231	1,5,7
Montmorillonite	30.0 (60)	15.0 (90)	4.97 (75)	4.53 (100)	12—231	1,2,4,5,8,11, 10,17,18
Nontronite	13.1 (100)	13.5 (100)	4.51 (60)	3.49 (50)	13—508	3,5,16,20
Illite	25.8 (100)	12.4 (80)	4.47 (80)	3.33 (50)	7—333	17,21,22
Zeolites:						2,5,12
Mordenite	9.10 (90)	6.61 (90)	3.48 (100)	3.22 (100)	6—239	2,17?

Mineral	d-spacings of the four most intense lines				JCPDS powder file card no.	Manganese nodule occurrence references
Phillipsite	7.19 (100)	4.13 (40)	3.19 (85)	3.14 (35)	20—923	1,2,5,10, 18,19,20,21
Erionite	11.4 (100)	6.61 (75)	4.32 (65)	3.75 (65)	22—854	17?
Epistilbite	8.89 (90)	3.87 (70)	3.45 (100)	3.21 (90)	19—213	17?
Analcite	5.60 (60)	3.43 (100)	2.927 (50)	2.226 (40)	19—1180	12?
Clinoptilolite	8.92 (100)	3.96 (55)	3.90 (55)	2.974 (80)	22—1236	17
Apatite	3.08 (25)	2.811 (80)	2.717 (100)	2.261 (35)	21—145	1,3,15,16
Calcite	3.035 (100)	2.285 (18)	2.095 (18)	1.913 (17)	5—586	13,15
Mangano-calcite	2.95 (100)	2.24 (50)	1.85 (80)	1.81 (70)	2—714	15
Aragonite	3.396 (100)	3.273 (52)	2.700 (46)	1.977 (65)	5—453	14
Rutile	3.25 (100)	2.487 (50)	2.188 (25)	1.6874 (60)	21—1276	3,12,16
Anatase	3.52 (100)	2.378 (20)	1.892 (35)	1.6999 (20)	21—1272	3,12,16
Barite	3.442 (100)	3.101 (97)	2.120 (80)	2.104 (76)	5—448	3,12,16
Spinels:						3,16
Magnetite	2.967 (30)	2.532 (100)	1.616 (30)	1.485 (40)	19—629	17
Ilmenite	2.74 (100)	2.54 (85)	1.86 (85)	1.72 (100)	3—781	17?

References:
1. P. F. Andrushchenko and N. S. Skornyakova (1965). In: Manganese Deposits of the Soviet Union. Israel Progr. Sci. Transl., Jerusalem, 1970, pp. 101—124.
2. P. F. Andrushchenko and N. S. Skornyakova (1969). Oceanology, 9: 229—242.
3. G. Arrhenius (1963). The Sea, 3. The Earth Beneath the Sea, pp. 655—727.
4. F. Aumento, D. E. Lawrence and A. G. Plant (1968). Geol. Surv. Can., Pap., 68—32: 30 pp.
5. E. Bonatti and Y. R. Nayudu (1965). Am. J. Sci., 263: 17—39.
6. W. Buser and A. Grütter (1956). Schw. Min. Petrogr. Mitt., 36: 49—62.
7. S. E. Calvert and N. B. Price (1970). Contr. Miner. Petrol., 29: 215—233.
8. G. P. Glasby (1972a). Mar. Geol., 13: 57—72.
9. H. G. Goodell, M. A. Meylan and B. Grant (1971). Antarc. Res. Ser., 15: 27—92.
10. J. B. Grant (1967). M.S. Thesis, Florida State University, 100 pp.
11. M. von Heimendahl, G. L. Hubred, D. W. Fuerstenau and G. Thomas (1976). Deep-Sea Res., 23: 69—79.
12. V. N. Hering (1971). Stahl Eisen, 91: 452—459.
13. D. F. Hewett, M. Fleischer and N. Conklin (1963). Econ. Geol., 58: 1—51.
14. P. F. McFarlin (1967). J. Sedim. Petrol., 37: 68—72.
15. F. T. Manheim (1965). Occ. Publ. Univ. Rhode Island, 3: 217—76.
16. J. L. Mero (1965a). The Mineralogical Resources of the Sea. 312 pp.
17. M. A. Meylan (1968). M.S. Thesis, Florida State University, 177 pp.
18. A. Okada and M. Shima (1970). J. Oceanogr. Soc. Jap., 26: 151—158.
19. A. Okada, T. Minakuchi and M. Shima (1972a). J. Oceanogr. Soc. Jap., 28: 39—47.
20. J. Ostwald and F. W. Frazer (1973). Miner. Deposita, 8: 303—311.
21. R. E. Smith, J. D. Gassaway and H. N. Giles (1968). Science, 161: 780—781.
22. J. S. Stevenson and L. S. Stevenson (1970). Can. Mineral., 10: 599—615.

flocculate some of the accessory minerals with goethite while leaching out the manganese oxides. This suggests that some of the accessory minerals are poorly crystalline and apparently amorphous to X-rays in the unleached nodules.

By far the most common accessory mineral in manganese nodules is quartz of detrital and authigenic origins (Harder and Menschel, 1967), which may be associated with opaline silica. Other commonly reported detrital minerals include feldspars, pyroxene, and amphibole, together with accessory minerals from volcanic rocks such as rutile, barite, sphene, anatase. Submarine alteration of the primary minerals in basalts or of volcanic glass leads to several clay and zeolite minerals, the most common of which are illite, montmorillonite, nontronite, and phillipsite (Goodell et al., 1971). Electron micrographs (Fewkes, 1973; Margolis and Glasby, 1973; Woo, 1973), showing euhedral crystals indicate that some of the clay and zeolite minerals may have authigenic origins and grew contemporaneously with the ferromanganese oxides. Minerals of biological origin include calcite, aragonite, and apatite, in addition to some quartz or opal. The carbonates and phosphates are more abundant in manganese nodules from shallow depths. Cosmic spherules constitute another non-authigenic inclusion in manganese nodules (Jedwab, 1970). Minerals identified in the spherules, which together constitute less than 10^{-6} of the total mass of seafloor nodules, include hematite, trevorite ($NiFe_2O_4$), metallic iron, and goethite.

THE CRYSTALLINITY OF MANGANESE NODULES

Information from a variety of sources testifies to the presence of very small crystallites in manganese nodules. This is demonstrated by the diffuseness of lines in X-ray powder diffraction patterns, effects on the parameters in Mössbauer spectra, estimates of particle and crystallite sizes from gas adsorption isotherms, and direct observations of constituent minerals in the nodules by electron microscopy.

X-ray diffraction measurements

Manganese nodules are renowned for the poor quality X-ray powder diffractograms or photographs they produce, in which there is a general paucity of sharp peaks and lines. The X-ray data for manganese nodules show that those lines attributed to manganese oxide minerals (and rarely ferric oxyhydroxide phases) are almost invariably broader than lines derived from impurities of detrital silicate minerals. Frequently, the X-ray diffraction peaks are so diffuse and weak as to yield featureless patterns. The diffuseness, low intensity, and apparent absence of X-ray diffraction lines is a direct consequence of the poor crystallinity and very small particle sizes of the iron and manganese oxide minerals in the nodules.

X-rays incident at an angle θ to a stack of parallel planes of atoms in a crystal structure are diffracted when the Bragg equation is obeyed, namely $\theta = \sin^{-1} (n\lambda/2d)$, where d is the spacing between parallel planes, λ is the wavelength of incident X-rays, and n is an integer. Each plane of atoms may be pictured as reflecting a fraction of the incident beam. In an ideal crystal, X-rays reflected from successive planes combine to form a coherent diffracted beam, the intensity of which depends on the depth of penetration of the X-rays and the periodicity of the atomic planes. If there is only a small number of parallel planes of atoms in a crystal, the intensity of the diffracted X-ray beam will be correspondingly weak. Thus, in crystallites with dimensions smaller than about 100 Å, there is inadequate periodicity of atomic planes for diffracted X-rays of sufficient intensity to be detected, particularly if they are diffuse.

The diffuseness of diffracted X-rays results from defects, vacancies, chemical heterogeneities, and disordering of atoms in the crystal structures. These effects all combine to cause small variations of d-spacings in a stack of atomic planes, so that the range of θ-angles is extended over which the Bragg equation is obeyed. As a result, a broad diffraction peak or line is observed. Small crystallites are particularly susceptible to imperfections of crystal structure. Thus, the small number of parallel perfect atomic planes and the variability of interplanar spacings in minute crystals are responsible for the broad and weak X-ray diffraction lines. Conversely, their presence in powder diffraction patterns of manganese nodules is indicative of the small crystallite sizes of iron and manganese oxide minerals.

Adsorption isotherm data

Since surface areas are inversely proportional to particle sizes, information on the dimensions of crystals and matter constituting manganese nodules may be obtained from measurements of specific surface areas. One of the most widely used methods to measure surface areas of solids is the gas adsorption technique of Brunauer, Emmett, and Teller (1938). The B.E.T. method is based on the adsorption of a monatomic layer of argon or nitrogen gas on the solid surface at several different gas pressures. The particle size of a known mass of dry, powdered solid may be calculated from experimentally determined specific surface area (e.g., argon 13.2 Å^2; nitrogen 16.3 Å^2). Surface areas of solids may also be determined by the t-plot method (De Boer et al., 1966), which is a variation of the N_2 adsorption method, and the glycol retention method (Bower and Goertzen, 1959; Loganathan and Burau, 1973).

In a B.E.T. adsorption isotherm study of synthetic manganese oxides, Buser and Graf (1955a) found that δ-MnO_2 (giving two-line X-ray diffraction patterns) had specific surface areas of about 300 m^2 g^{-1}, which is almost an order of magnitude larger than the 30—50 m^2 g^{-1} values measured for "manganous manganite" (birnessite). Similar results were obtained by Healy et al. (1966). Buser and Graf (1955a) calculated that the surface area for a mono-atomic layer amounts to 720 m^2 g^{-1}, and concluded that the thickness of the two-dimensional δ-MnO$_2$ crystallites is, on the average, no more than two to three atomic layers. In studies of iron (III) oxide-hydrate gel, Van der Giessen (1966) measured surface areas of 265—270 m^2 g^{-1} for specimens aged for 1 month in solution, which corresponded to particles having diameters of about 40 Å. Aging for a year yielded coarser-grained crystallites, approximately 90 Å in diameter, having surface areas of 215—220 m^2 g^{-1}. These data form the basis for interpreting B.E.T. adsorption isotherm measurements of manganese nodules.

In their classic paper, Buser and Grütter (1956) reported surface area data for four manganese nodules from the Pacific, namely 54, 126, 190, and 191 m^2 g^{-1}. Highest values were obtained for those nodules containing δ-MnO$_2$, while smaller surface areas were reported in "10-Å manganite" (todorokite)-bearing nodules. Similar results were reported by Weisz (1968), Johnson and Glasby (1969), Brooke and Prosser (1969), and Glasby (1970). One concludes from the surface area data of corroborative oxide phases of iron and manganese that particle sizes of some of the constituent minerals in many manganese nodules are below diameters of 100 Å.

Mössbauer measurements

Mössbauer spectroscopy, which is applicable only to iron-bearing phases, · forms a third method for estimating mean particle sizes of certain minerals in manganese nodules. The technique is based on the temperature dependence of magnetic hyperfine splitting of antiferromagnetic or ferromagnetic minerals, such as hematite or goethite, which become superparamagnetic when present as very small crystallites (see p. 228). Calculations based on goethite by Johnson and Glasby (1969) yielded mean spherical diameters of about 90 Å and 200 Å for two different nodules from the Gulf of Aden and the Indian Ocean, respectively. Herzenberg and Riley (1969) claim to have established independently the presence of Fe^{3+} ions in fine-grained particles of dimensions smaller than a few hundred angstroms. Carpenter and Wakeham (1973) also estimated that most of the goethite present in many of the nodules they examined must have mean spherical diameters smaller than about 260 Å.

Electron microscopy

The three techniques just described give indirect information on particle sizes of materials in manganese nodules. Observations with electron microscopes not only enable constituent materials to be examined directly, but also provide some information on associations of the intimately intergrown crystallites. Two different techniques, scanning electron micro-scopy (S.E.M.) and transmission electron microscopy (T.E.M.), have been applied to manganese nodules. The SEM reveals surface topography and morphology of associated minerals, whereas TEM enables crystal symmetry and habits of very small dispersed particles to be examined. In addition, selected area diffraction patterns may be obtained from the crystallites, yielding information on lattice symmetry and dimensions. These electron diffraction patterns consist of a series of spots for single crystals with diameters exceeding about 1,000 Å, or rings and striations from clusters of smaller crystallites.

Electron microscopy has complemented morphological studies of minerals in hand specimens. For example, terrestrial todorokites typically consist of aggregates of fine fibrous crystallites elongated along the one axis (Frondel et al., 1960a; Faulring, 1962). Electron microscopy, however, reveals the todorokite crystals to be thin plates flattened perpendicular to the c-axis (Straczek et al., 1960; Hariya, 1961). Many of the plates are broken into narrow laths or blades less than 1 μ in diameter which are elongated parallel to the b-axis. SEM studies of todorokites in geodes (Finkelman et al., 1972, 1974) reveal that the dominant habits are fibrous mats and stubby needles. Birnessite, on the other hand, is generally soft and very fine-grained in hand specimens (Jones and Milne, 1956; Frondel et al., 1960b). SEM studies, however, reveal birnessite clusters with a cellular structure formed by a more or less haphazard array of thin, slightly bent, lamellae (Brown et al., 1971; Sorem and Fewkes, chapter 6, this volume, Fig. 6-7). The lamellae are about 2,000 Å thick and spaces between them have dimensions of about 10,000 Å. Synthetic birnessites have also been shown to consist of platelets (Giovanoli et al., 1970a, 1971). Electron micrographs of the ferric oxyhydroxide minerals (for example, Gallagher, 1970; Langmuir and Whittemore, 1971; McKenzie et al., 1971; Finkelman et al., 1972) show that: goethite most commonly occurs as acicular and narrow prismatic crystals elongate along the c-axis; lepidrocrocite is typically present in bladed prismatic or micaceous forms flattened perpendicular to the b-axis; and akaganéite consists of porous bundles of needles elongated parallel to the c-axis. The hydrated ferric oxyhydroxide polymer, however, contains aggregates of minute particles 30—50 Å in diameter (Van der Giessen, 1966; Towe and Bradley, 1967; Feitknecht et al., 1973).

Early observations of manganese nodules under the electron microscope revealed a number of different morphologies and habits of the dispersed particles (Buser and Grütter, 1956; Aumento et al., 1968; Okada and Shima, 1969). The most common were agglomerations of grains consisting of featureless and ill-defined particles which did not give crystal diffraction patterns. Other particles showed mesh structures of very thin fibers of variable lengths and straightness, but not exceeding thicknesses of 80 Å, which were too small to give individual diffraction patterns. By leaching out manganiferous material, Buser and Grütter (1956) showed that X-ray diffraction lines of some of these needle-like clusters corresponded with the goethite pattern. Some of the acicular crystals may have been todorokite, however. Aumento et al., (1968) reported that a number of extremely thin, overlapping anhedral lamellae were randomly dispersed throughout the agglomerate and fibrous particles. Three different phases could be identified from the electron diffraction patterns which showed hexagonal or pseudo-hexagonal symmetry. One of these was positively identified as δ-MnO_2. Aumento et al. (1968) also reported the presence of particles with lath-like or tubular morphologies occurring in intimate association with the flakes, fibres, and agglomerates. The laths gave "rotation" electron diffraction patterns characteristic of crystals with tubular morphology such as the serpentine mineral, chrysotile. This led Aumento et al. (1968) to postulate that chrysotile was either authigenic and forming contemporaneously with the ferromanganese deposits or a detrital mineral originating from serpentinized submarine basalts.

Recently, Von Heimendahl et al. (1976) studied two Pacific manganese nodules from seamount and deep-sea environments by transmission electron microscopy. Thin, pseudo-hexagonal flakes of montmorillonite were found to be ubiquitous in the nodules, being most abundant in the deep-sea specimen. The predominant manganese phase in the two nodules had "rafted fine structure" and gave electron diffraction patterns comparable with the synthetic sodium birnessite or Na manganese (II, III) manganate (IV) (Giovanoli et al., 1970a). Other crystallites could be correlated with another synthetic birnessite, manganese (III) manganate (IV) (Giovanoli et al., 1970b, 1971). Todorokite was identified once, as was the discredited mineral hydrohausmannite (= hausmannite + feitknechtite), in selected area diffraction patterns. Although maghemite was found once, the predominant iron-bearing mineral was identified as "ferric hydroxide" (hydrated ferric oxyhydroxide polymer) from comparisons of electron micrographs and ring diffraction patterns with those of a synthetic material. Most of the $FeOOH \cdot xH_2O$ consisted of particles 30—60 Å in diameter, but crystals 100—450 Å were also found. The much larger manganese oxide minerals were embedded in a general mass of $FeOOH \cdot xH_2O$ and only occasionally occurred as isolated particles. SEM and electron microprobe measurements also showed that the iron oxide phase is present everywhere and that many submicron manganese oxide crystals were embedded in the particles of $FeOOH \cdot xH_2O$.

Other SEM measurements of manganese nodules have provided information on their textures and growth histories. Margolis and Glasby (1973) in a study of several nodules from the Pacific and Indian oceans, showed that the microlaminations range in size from less than 0.25 μm to greater than 10 μm. These microlaminations are believed to be primary structures as they are laterally continuous around a nodule and remain uniform in thickness in both straight and cuspate regions. Several authors have photographed euhedral crystals and epitaxial growths of minerals in the nodules. For example, Margolis and Glasby (1973) illustrate phillipsite crystals showing penetration twinning, while Sorem and Fewkes (Chapter 6, this volume) and Woo (1973) have obtained SEM photographs of platey, butterfly-shaped crystal agglomerates showing hexagonal outlines believed to be birnessite.

MÖSSBAUER, MAGNETIC, AND OTHER SPECTRAL DATA

Since transition metals are the major chemical constituents of ferromanganese concretions, it is essential to be able to establish the identity and role of the various oxidation states of these elements in the mineral phases of manganese nodules. Sometimes X-ray diffraction analysis of the constituent minerals imply the existence of oxidation states necessary for charge balance. Examples are essential Mn^{2+}, Mn^{4+}, and perhaps Mn^{3+}, ions in birnessites and todorokites, or Fe^{3+} ions in the iron oxyhydroxide phases. Similarly, some Mn^{2+} ions must be present in minerals of the hollandite group in order to compensate for Ba^{2+}, K^+, Na^+, Pb^{2+}, etc., ions in the structure.

Most information on cation valencies of the major transition elements stems from various magnetic and spectral measurements. In this section, Mössbauer and magnetic data for iron, and Electron Spin Resonance (ESR) and Photoelectron (ESCA) results for manganese, are summarized. The oxidation states of less abundant cobalt, nickel, and copper ions in manganese nodules are discussed on pp. 244 and 245.

Mössbauer measurements

The presence of about 2.19% of the ^{57}Fe isotope in natural iron renders iron-bearing compounds and minerals amenable to Mössbauer spectroscopy. This technique involves the measurement of recoil-free absorption of 14.41 keV gamma rays emitted by a vibrating radioactive source by nuclei of ^{57}Fe isotopes chemically bound in solids. Mössbauer spectroscopy has been applied to numerous geochemical problems to elucidate oxidation states, coordination numbers, magnetic properties, and electronic configurations of iron atoms in minerals (Bancroft, 1973). In ideal cases, the technique may be used to identify iron-bearing minerals, as well as to obtain accurate ferrous/ferric ratios and cation distributions in mineral structures.

Although confined to the solid state, Mössbauer spectroscopy is applicable to the minute crystallites found in manganese nodules because each ^{57}Fe nucleus absorbs gamma rays independently and is not limited by particle size in the same way as is the observation of Bragg's law reflections in diffraction measurements. However, as discussed later, particle sizes do affect certain

TABLE 7-XI

Mössbauer data for ferromanganese nodules

Reference	Specimens	Temperature	Isomer shift[1] (mm/sec)	Quadrupole splitting (mm/sec)	Half width (mm/sec)
Gager (1968)	1 Blake Plat. 2 Pacific	RT	0.35—0.37	0.58—0.65	
Johnson and Glasby (1969)	1 Gulf. Aden 1 Indian Oc.	77°K	0.43	0.78	1.0
Herzenberg and Riley (1969); Herzenberg (1969)	suite, un-specified				
Georgescu and Nistor (1970)	4 Black Sea 1 Pacific	RT	0.32—0.35	0.71—0.93	large
Hrynkiewicz et al. (1970)	4 Pacific	RT	0.38±0.01	0.64—0.67	
Hrynkiewicz et al. (1972a, b)	11 Pacific	RT	0.396±0.01	0.62—0.75	0.49—0.57
Okada et al. (1972b, 1973)	4 Pacific	RT 77°K 4.2°K	0.37—0.40 0.51 0.49	0.65—0.73 0.73 0	
Georgescu et al. (1973)	2 Black Sea	RT 77°K	0.34—0.35 0.43	0.76—0.77 0.77—0.81	0.52—0.55 0.70—0.83
Goncharov et al. (1973)	1 Pacific	RT	0.36±0.02 0.33±0.02	0.71±0.04 0.72±0.04	0.48±0.04 0.52±0.04
Carpenter and Wakeham (1973)	40 world-wide fresh and marine	RT	0.37±0.05 0.46±0.03	0.67—0.86 incr. 0.01—0.07[3]	0.44—0.68

[1] Relative to metallic iron standard.
[2] At 4.2°K.
[3] Q.S. only 0.01—0.07 higher than RT values.

Mössbauer parameters, notably quadrupole splittings and magnetic hyperfine structure.

A variety of nodules have been studied by Mössbauer spectroscopy. The Mössbauer data obtained by several investigators are summarized in Table 7-XI. Corroborative Mössbauer parameters for well-characterized iron oxide and oxyhydroxide phases are listed in Table 7-XII. The values of the isomer shift parameter cited in the literature depend on the calibration standard, which may be metallic iron, stainless steel, sodium nitroprusside, or palladium, to mention just a few. The isomer shift data in Tables 7-XI and

Magnetic field (kG)	Suggested Fe-bearing phases in the nodules
	Fe^{3+} ions in very fine particles (<70Å diameter) (e.g., lepidocrocite or goethite in a finely divided state).
500*2	Fe^{3+} ions octahedrally coordinated in two or more sites or phases (e.g., α-FeOOH and γ-FeOOH, or other combinations of mixed ferric oxide compounds).
	Fine-grained particles related to ferric oxide hydrate gels and hydrated ferric oxide polymers which are precursors to goethite; up to 0.25 wt.% Fe^{2+} observed.
	Fe^{3+} ions principally in goethite with small amounts of ferromanganese oxides.
457±15*2	Similar to $Fe(OH)_3$ gels. Excluded γ-FeOOH and α-, β-, and δ-FeOOH on the basis of Q.S. and magnetic data, respectively (but ignored superparamagnetism).
	Similar to $Fe(OH)_3$ gels. Variable Q.S. attributed to distortions due to sorbed cations or H_2O. Outer regions of nodules have higher Q.S. attributed to smaller particle sizes at exterior than interior. Parameters at higher T support $Fe(OH)_3$ gel phase.
	Colloidal ferric species.
440	
	Trivalent iron.
	At least two structural states of Fe^{3+}. Inner layers of concretion contained β-FeOOH (akaganéite).
	Most iron present as Fe^{3+} in paramagnetic or superparamagnetic phases (e.g. goethite crystallites $\leqslant$ 260 Å for spherical particles).

7-XII are expressed relative to metallic iron. To compare these isomer shift values with data based on other standards, one adds (or subtracts) the following amounts (in mm/sec) to the values based on iron foil: stainless steel, + 0.0901; sodium nitroprusside, + 0.257; Pd, - 0.185; Pt, - 0.347; Cu, - 0.226; Cr, + 0.198.

Considering the diversity of localities and paragenesis of the ferromanganese nodules studied, there is a remarkable similarity of Mössbauer spectra measured at either liquid nitrogen or room temperatures. The spectra typically consist of two broad peaks and the relatively narrow range of

TABLE 7-XII

Mössbauer parameters for oxide phases containing iron

Mineral or phase	Isomer shift (mm/sec)[1]	Quadrupole splitting[2] (mm/sec)	T_N or T_C (°K)	H_{eff}(at 0°K) (kOe)	References
Goethite α-FeOOH	0.35	0.55—0.60[3]	383—403	504—510	1,2,3,4
Akaganéite β-FeOOH	0.35	0.50	290—300	487	5
Lepidocrocite γ-FeOOH	0.30	0.55	73	460±5[4]	6,7,8
δ-FeOOH	0.50[5]	0	455 and 420[6]	525 and 505[7]	5,9
HP-FeOOH	0.30	0	570	525	10
Hydrated ferric oxyhydroxide polymer	0.34	0.60—0.65	450	507	11,12,13
Natural ferric gel	0.35	0.72		484[4]	14
Hematite crystals α-Fe$_2$O$_3$	0.34	0.20 or 0.43[8]	958±5	517±5	15
<100 Å crystallites	0.30	0.68			16,17,18,
>180 Å crystallites	0.38	0.44			19
γ-Fe$_2$O$_3$ oct Fe^{3+}	0.40	0		499±5	
tet Fe^{3+}	0.26	0		488±5	
Illite	0.32±0.03	0.54			20
Montmorillonite	0.34±0.03	0.57			20

[1] Room temperature, relative to velocity zero for metallic iron.
[2] Room temperature.
[3] Above T_N or for crystallite size <130 Å diameter.
[4] At 4.2°K.
[5] At 77°K.
[6] Estimated; extrapolated to above the decomposition temperature.
[7] For Fe^{3+} in two sites.
[8] Above and below T_C, respectively.

TABLE 7-XII (continued)

References:
1. J. B. Forsyth, I. G. Hedley and C. E. Johnson (1968). J. Phys., 2C: 178—188.
2. T. Shinjo (1966). J. Phys. Soc. Jap., 21: 917—922.
3. F. van der Woude and A. J. Dekker (1966). Phys. Stat. Sol., 13: 181—193.
4. I. G. Hedley (1971). Z. Geophys., 37: 409—420.
5. I. Dezsi, L. Keszthelyi, D. Kulgawczuk, B. Moinor and N. A. Eissa (1967). Phys. Stat. Sol., 22: 617—629.
6. C. E. Johnson (1969). J. Phys., 2C: 1996—2002.
7. J. H. Terrell and J. J. Spijkerman (1968). Appl. Phys. Lett., 13: 11—13.
8. M. J. Rossiter and A. E. M. Hodgson (1965). J. Inorg. Nucl. Chem., 27: 63—71.
9. S. Okamoto (1968). J. Am. Ceram. Soc., 51: 594—599.
10. M. Pernet, J. Chenavas and J. C. Joubert (1973). Solid State Comm., 13: 1147—1154.
11. S. Okamoto, H. Sekizawa and S. I. Okamoto (1972). Reactivity of Solids. Chapman and Hall, pp. 341—350.
12. A. Z. Hrynkiewicz, A. J. Pustówka, B. D. Sawicka and J. A. Sawicki (1972). Phys. Stat. Sol., 10: 281—287.
13. Z. Mathelone, M. Ron and A. Biran (1970). Sol. State Comm., 8: 333—336.
14. J. M. D. Coey and P. W. Readman (1973). Earth Planet. Sci. Lett., 21: 45—51.
15. G. Shirane, D. E. Cox, W. J. Takei and S. L. Ruby (1962). J. Phys. Soc. Jap., 17: 1598.
16. W. Kundig, H. Bommel, G. Constabaris and R. H. Lindquist (1966). Phys. Rev., 142: 327—333.
17. G. Constabaris, R. H. Lindquist and W. Kundig (1965). Appl. Phys. Lett., 7: 59—60.
18. T. Nakamaru and S. Shimizu (1964). Bull. Inst. Chem. Phys., Kyoto Univ., 42: 299.
19. Y. F. Krupyanskii and I. P. Suzdalev (1973). Zh. Eksp. Teor. Fiz., 65: 1715—1725.
20. N. Malathi, S. P. Puri and I. P. Saraswat (1969). J. Phys. Soc. Jap., 26: 680—683.

isomer shifts (Table 7-XI) indicate that Fe^{3+} ions, octahedrally coordinated to oxygens, predominate in manganese nodules. Negligible amounts (< 0.25 wt.% FeO) of Fe^{2+} ions have been detected in manganese nodules by Mössbauer spectroscopy (Herzenberg, 1969; Brown, 1971; Burns and Brown, 1972; Carpenter and Wakeham, 1973). The large widths of the Mössbauer peaks and the range of quadrupole splittings observed for manganese nodules may be explained either by mixtures of different minerals or by variable particle sizes. On the one hand, Fe^{3+} ions may occur in a diversity of octahedral sites either in different crystal structures (i.e. different minerals) or in different ligand environments in one phase as a result of variations of the type (e.g., O^{2-}, OH^-, H_2O) and symmetry of oxygen ligands surrounding the Fe^{3+} ions. On the other hand, data for hematite show that quadrupole splittings increase linearly with the reciprocal of particle size (Kundig et al., 1966, Krupyanskii and Suzdalev, 1973), which suggests that a range of crystallinity of iron-bearing minerals may affect the Mössbauer spectra of manganese nodules in a similar manner.

The Mössbauer spectra of manganese nodules show magnetic hyperfine (six-peak) structure at liquid helium temperatures, suggesting that the Neél temperatures of the iron phases in nodules lie somewhere between 4.2°K and 77°K. This evidence appeared to eliminate the occurrence of goethite, akaganéite, δ-FeOOH, magnetite, maghemite, and hematite in the nodules (Hrynkiewicz et al., 1970; Okada et al., 1972b) since these phases have Neél or Curie temperatures above 77°K. However, it is now recognized that fine-grained magnetic materials display superparamagnetism, so that Mössbauer spectra of crystalline materials measured below the magnetic ordering temperatures may consist of simple quadrupole doublets found also for paramagnetic Fe (III) compounds. Indeed, studies of goethite and hematite have demonstrated that particles with diameters smaller than 100—200 Å are superparamagnetic at 77°K, well below their Neél temperatures (Kundig et al., 1966; Shinjo, 1966; Suzdalev, 1969; Krupyanskii and Suzdalev, 1973). Conversely, the general lack of magnetic hyperfine splitting in manganese nodules above 77°K has formed the basis for estimating particle sizes of the iron-bearing phases in the nodules. For example, Johnson and Glasby (1969) assumed that fine-grained, superparamagnetic goethite particles were present in manganese nodules, and calculated their mean diameters to be of the order 90 Å and 200 Å in two different nodules (see also p. 220).

The corroborative data summarized in Table 7-XII show that Fe^{3+} ions in a variety of oxygen environments have Mössbauer parameters falling in the range observed for manganese nodules. These not only include superparamagnetic or cryptocrystalline goethite, but also the clay minerals illite and montmorillonite (nontronite) and the various hydrated ferric oxyhydroxide polymers or gels. Although no Mössbauer data exist for iron in todorokite or "Fe (III)-manganite", they are expected to be similar to those for Fe^{3+} ions in other oxide compounds listed in Table 7-XII. It is therefore impossible to identify the iron-bearing minerals in manganese nodules unambiguously by Mössbauer spectroscopy. A widely held view based on recent Mössbauer measurements is that the dominant iron-bearing phase in manganese nodules is hydrated ferric oxyhydroxide polymer (Herzenberg, 1969; Herzenberg and Riley, 1969; Hrynkiewicz et al., 1970, 1972a, b; Okada et al., 1972b; Carpenter and Wakeham, 1973).

Other notable conclusions drawn from the Mössbauer spectra of manganese nodules include the following. First, freshwater nodules have slightly greater quadrupole splittings and line widths than marine nodules. Carpenter and Wakeham (1973) suggest that this is due to different conditions of formation. Alternatively, the data could indicate a greater range of mineral constituents or particle sizes in freshwater nodules. Second, higher quadrupole splittings were found for outer portions than for inner portions of nodules (Hrynkiewicz et al., 1972a), which again was related to differences in average size of the particles (Hrynkiewicz et al., 1972a). This result suggests that either older (central) parts of manganese nodules formed under conditions different from those of younger (outer) parts, or that recrystallization of particles to coarser-grained crystallites occurs inside manganese nodules in the course of time.

Magnetic measurements

Magnetic measurements, excluding results derived from Mössbauer spectroscopy, have also been used to elucidate the crystalline forms of iron in manganese nodules. Techniques used include studies of natural remanent and saturation magnetization, the temperature dependence of the strong-field magnetization, and magnetization versus applied magnetic field hysteresis loops. The interpretations of the results are based on the fact that above $70°K$ all crystalline iron oxides and oxyhydroxides display remanent magnetism and are antiferromagnetic (e.g., hematite, goethite), ferrimagnetic (e.g., magnetite, maghemite, δ-FeOOH), or ferromagnetic (e.g, akaganéite parallel to c_o). Very small crystallites may be superparamagnetic and display negligible magnetic ordering. All pure manganese oxides, on the other hand, are paramagnetic.

Early magnetic measurements of a dredged ferromanganese crust (Ozima, 1967) indicated that paramagnetic materials predominated in manganese nodules. Crecelius et al. (1973) studied natural remanent magnetization of some 24 marine and freshwater nodules, and found small but reasonably stable remanence in most of the nodules, suggesting the presence of small amounts of ferrimagnetic material. This was confirmed by saturation magnetization measurements (Carpenter et al., 1972), which indicated upper limits of 0.2—0.6 wt.% magnetite and/or maghemite in over 50 manganese nodules. Carpenter et al. (1972) also measured a series of hysteresis loops at $20°C$ and confirmed that the nodules contained predominantly paramagnetic phases, but small amounts of ferrimagnetic material were deduced to be present in all samples. Nodules from Antarctica contained the largest amounts of the ferrimagnetic phases, in agreement with earlier evidence from X-ray diffraction measurements (Goodell et al., 1971). The presence of substantial amounts of paramagnetic phases in manganese nodules lowers the upper detectability limits for maghemite and magnetite by the magnetic measurements, which are nevertheless more sensitive than X-ray diffraction or Mössbauer spectroscopy techniques. Carpenter et al. (1972) concluded that the decrease of magnetization on heating is consistent with the presence of goethite and small amounts of maghemite and/or magnetite in the nodules, but ruled out the presence of lepidocrocite in significant amounts, contrary to the speculations of Goodell et al. (1971).

Electron spin resonance

Electron spin resonance spectroscopy provides information on species having electrons with unpaired spins, and is thus applicable to cations such as Mn^{4+}, Mn^{3+}, Mn^{2+}, Fe^{3+}, Fe^{2+}, Co^{2+}, Ni^{2+}, and Cu^{2+}. Optimum conditions for maximum detail and resolution in ESR spectra necessitate the paramagnetic cations being present in very dilute concentrations in a diamagnetic host phase. The spectra then consist of multiple absorption bands and fine structure splittings. Manganese nodules, however, with high Mn and Fe contents, have such large concentration of paramagnetic ions as to generate ESR spectra with broad resonance lines that obscure both the fine and hyperfine structure and contributions from minor elements. Nevertheless, Wakeham and Carpenter (1974), in an ESR study of 80 ferromanganese nodules from a variety of marine and freshwater localities, were able to identify and estimate relative amounts of Mn^{2+} in many of the nodules. In most cases, the Mn^{2+} ion estimates by ESR agreed with observations based on the presence or absence of "10-Å manganite" (todorokite) and "7 Å manganite" (birnessite). Major contributions to the spectra were also attributed to Fe^{3+} ions, but concentrations of other cations such as Fe^{2+}, Co^{2+}, Ni^{2+}, and Cu^{2+} were too low to be detected. No suggestion was offered for Mn^{3+} ions being present in manganese nodules.

Photoelectron (ESCA) spectroscopy

Binding energies of electrons are affected by the oxidation states of transition metal ions and generally increase with rising cation valence. Photoelectron spectroscopy or ESCA, which measures the binding energies of core and valence electrons, is a potential method for analyzing transition metals in mixed oxidation states. For example, in a preliminary study of the ESCA spectra of manganese nodules and corroborative chemical compounds, Andermann (1973) demonstrated that co-existing Mn^{2+} and Mn^{4+} ions could be resolved in nodules containing "10-Å manganite" (todorokite). The capability of using ESCA for semiquantitative estimates of mixed oxidation states of Mn and other transition elements is currently limited by problems of specimen oxidation during sample preparation and electron bombardment in the ESCA spectrometer, and by computer fitting of the spectra.

GROWTH OF MANGANESE NODULES

The authigenesis of manganese nodules and the uptake of certain metals into the ferromanganese oxide phases is fundamentally controlled by the crystal structures of the host minerals and the crystal chemistries of the substituent cations. In this section, structural correlations are made between some of the key manganese oxide and iron oxyhydroxide phases listed in Tables 7-I and 7-VIII, particularly those minerals whose crystal structures have been determined. Following this, possible crystal structures of birnessite, todorokite, and the hydrated ferric oxyhydroxide polymer are discussed. Attention is then focussed on features of the crystal chemistry of those transition metal ions which are enriched in manganese nodules. These correlations of crystal structure and crystal chemistry form the basis of a working hypothesis for the nucleation and growth of manganese nodules, which is discussed towards the end of the section.

Structural correlations between oxide and oxyhydroxide phases of Mn and Fe

Available crystal structural data for the various manganese oxides and iron oxyhydroxides indicate that there is a hierarchy of structure-types, somewhat resembling the classification of silicates. Thus, by analogy with the independent, chain, ring, framework, and layer silicates resulting from different linkages of $[SiO_4]$ tetrahedra, there are different ways of connecting $[MnO_6]$ or $[Fe(O,OH)_6]$ octahedra in the ferromanganese oxides. In silicate mineralogy, corner sharing forms the basis of the linkages of the $[SiO_4]$ tetrahedra, whereas the $[MnO_6]$ octahedra are edge-shared in manganese oxide mineralogy.

Chain structures

The basis for describing the crystal structures of manganese (IV) oxides and iron (III) oxyhydroxides is the pyrolusite (β-MnO_2) structure. It has the rutile (TiO_2) structure, in which every metal atom is surrounded by six oxygen atoms located at the vertices of a distorted octahedron with Mn at the centre. The $[MnO_6]$ octahedra share edges to form single chains of octahedra extending along the c crystallographic axis. All octahedra are equivalent, and the average Mn—O distance is 1.88 Å. The unit cell c_o dimension (2.87 Å) represents the Mn—Mn internuclear distance across the shared octahedral edge, and as may be noted from Table 7-I, is a common cell parameter found in manganese (IV) oxide mineralogy. The chains of $[MnO_6]$ are cross-linked with neighbouring chains through corner sharing of oxygen atoms of adjacent octahedra to give tetragonal symmetry to pyrolusite. The crystal structure of pyrolusite is shown in Fig. 7-1. The single chains of linked $[MnO_6]$ octahedra in pyrolusite thus bear resemblances to the $[SiO_4]$ chains in pyroxenes in silicate mineralogy.

Ramsdellite (MnO_2) is built up of alternating double chains of linked $[MnO_6]$ octahedra (Fig. 7-2), and therefore resembles the linkages in amphiboles in silicate mineralogy. The octahedra are linked together by sharing opposite edges, thus producing continuous pyrolusite chains along the c-axis. Two such chains are cross-linked by edge sharing, one chain being displaced $c_o/2$ with respect to the other, so that an octahedron from one chain shares an edge with each of two octahedra from the other chain (Byström, 1949). The double chains of linked octahedra are further cross-linked to adjacent double chains through corner sharing of oxygen atoms, to give orthorhombic symmetry to ramsdellite. These features are illustrated in Fig. 7-2. All octahedra have identical configurations with an average Mn—O distance of 1.89 Å.

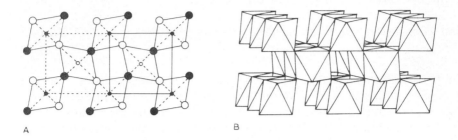

Fig. 7-1. The pyrolusite structure. A. Projection onto (0$\bar{0}$1); black circles, atoms at zero level; white circles, atoms at level $c/2$ (after Byström, 1949). Small circles Mn; big circles oxygen. B. The single chain of edge-shared [MnO$_6$] octahedra parallel to c (after Clark, 1972).

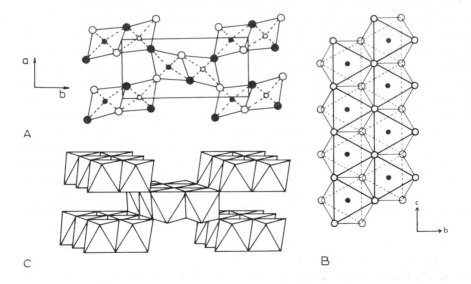

Fig. 7-2. The ramsdellite structure. A. Projection onto (001); white circles, atoms at level $1/4c$; black circles, atoms at level $3c/4$ (after Byström, 1949). B. A double chain of [MnO$_6$] octahedra running along c (after Byström, 1949). C. The double chains of edge-shared [MnO$_6$] octahedra parallel to c (after Clark, 1972). Small circles Mn; big circles oxygen.

The nsutite or γ-MnO_2 group consists of irregular structural intergrowths between pyrolusite and ramsdellite units (De Wolff, 1959; Giovanoli et al., 1967; Giovanoli, 1969). The alternating c-axis chain segments of the basic single- and double-chain units shown in Fig. 7-3 is random, so that no regular periodicity or superstructure is apparent. The lattice disorder, together with the small crystallite sizes of natural and synthetic phases, thus gives rise to an infinite number of X-ray powder diffraction patterns, as well as the frequently observed asymmetric and selective line broadening for nsutites.

Goethite (a-FeOOH) is isostructural with ramsdellite and consists of double chains of linked $[Fe(O,OH)_6]$ octahedra in which hydrogen-bonding also plays a significant role (Giese et al., 1971). The recently synthesized high-pressure form of FeOOH has a deformed rutile structure (Pernet et al., 1973). The lepidocrocite (γ-FeOOH) structure does not resemble any known manganese oxide or hydroxide phase.

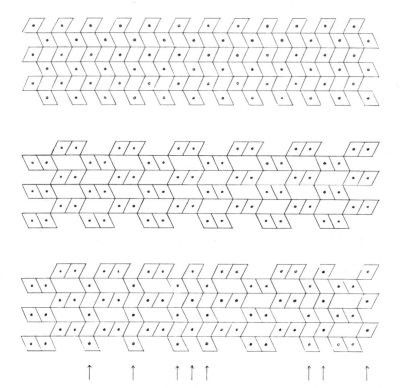

Fig. 7-3. Idealized projections of the pyrolusite (upper figure) and ramsdellite (middle figure) structures onto (001), and an analogous projection of the structure of nsutite (lower figure). Nsutite contains irregular alternations of ramsdellite domains linked by pyrolusite domains (arrows) (after Giovanoli et al., 1967).

Ring or framework structures

The structure of a-MnO_2 and minerals of the hollandite group, which is shown in Fig. 7-4 is based on the ramsdellite structure (Byström and Byström, 1950). The $[MnO_6]$ octahedra again share edges and form double chains running along the c-axis. The octahedra of the double chains share corners with adjacent double chains to give a three dimensional framework with pseudo-tetragonal symmetry (Fig. 7-4). This produces a large cavity which accomodates H_2O as well as the large monovalent and divalent cations, such as Ba^{2+}, Pb^{2+}, K^+, and Na^+ in hollandite, coronadite, cryptomelane, and manjiroite, respectively. Each large cation is surrounded by eight oxygen atoms situated at the corners of a slightly distorted cube (e.g., Ba—O distances in hollandite are 2.74 Å to these eight oxygens) and four other oxygens (Ba—O = 3.31 Å) at the corners of a square at the same level along the c-axis as the large cation. The c_o dimension (2.86 Å) again represents the Mn—Mn internuclear distance. Disordering of K^+, Ba^{2+}, etc. and H_2O occurs in the cavities, which are probably no more than half-filled, otherwise unfavorable cation repulsions would occur when Ba^{2+}—Ba^{2+}, K^+—K^+, etc. pairs are as close as 2.86 Å (Byström and Byström, 1951). On the other hand, significant amounts of H_2O, K^+, NH_4^+, etc., are necessary to prevent collapse of the structure of synthetic a-MnO_2 (Butler and Thirsk, 1952), otherwise submicroheterogeneities are formed in which regions of pyrolusite and ramsdellite are interdispersed and coexist with regions of a-MnO_2 in the same crystal. Natural hollandites and cryptomelanes,

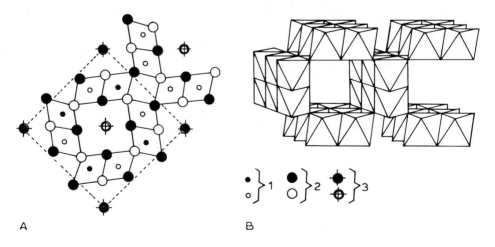

A   B

Fig. 7-4. The hollandite structure-type. A. The structure of hollandite (or α-MnO_2) projected onto (001); white circles, atoms at zero level; black circles, atoms at level $c/2$ (after Byström and Byström, 1950). B. The framework structure of hollandite showing tunnels between double chains of edge-shared $[MnO_6]$ octahedra parallel to c (after Clark, 1972).
1 = Mn; 2 = oxygen; 3 = Ba, K, Pb, Na, or H_2O.

however, appear to be stable to very high temperatures because they retain the Ba^{2+} and K^+ ions. A significant property of the sieve-like structure of hollandite is that it displays pronounced cation exchange properties, which is an important feature noted later when considering element uptake by manganese nodules. In order to maintain a charge balance in the structure accomodating the large exchangeable cations (Ba^{2+}, K^+, Na^+, Pb^{2+}), the linked [MnO_6] octahedra must contain a proportion of the manganese ions in oxidation states lower than Mn(IV) (e.g., Mn^{2+}, Mn^{3+}). This is reflected in the average metal—oxygen distances of tbe [MnO_6] octahedra, 1.98 Å, which is significantly larger than the mean Mn—O distances in pyrolusite (1.88 Å) and ramsdellite (1.89 Å).

Akaganéite, β-FeOOH, is isostructural with the hollandite group (Fig. 7-4). Its structure accommodates H_2O molecules and OH^-, Cl^-, F^-, $SO_4{}^{2-}$, $NO_3{}^-$ ions in the large cavities (Keller, 1970; Giovanoli, 1972; Feitknecht et al., 1973). Synthetic β-FeOOH has spindle-shaped crystals which appear under the electron microscope to be built up of parallel needles or rod-like sub-crystals packed into an orthogonal array. The rods are typically 60 Å square in section and up to 6,000 Å in length. When assembled into regular parallel bundles, crystals about 600 Å wide with tapered ends are observed. Evidence suggests that these rods are hollow with internal diameter being about 30 Å (Gallagher, 1970). The significance of this structure type in relation to todorokite is discussed later.

Psilomelane has a structure related to that of hollandite (Wadsley, 1953). It consists of *treble* chains of MnO_6 octahedra joined by double (ramsdellite-like) chains to form a series of tunnels or tubes running in the direction of the b-axis (Fig. 7-5). The b_o dimension (2.88 Å) of psilomelane, therefore, corresponds to c_o of pyrolusite, ramsdellite, and hollandite. The tunnels are occupied by Ba^{2+}, K^+ ions and H_2O molecules. Thus, the psilomelane structure bears resemblances to the hollandite structure, to which it decomposes at high temperatures. Psilomelane also has cation exchange properties and requires some of the manganese to be in oxidation states lower than Mn (IV) in order to balance the charge of the exchangeable large cations (Ba^{2+}, K^+). The psilomelane structure differs from the pyrolusite, ramsdellite, and hollandite structures described earlier by having *three* distinct octahedral sites. Two of them (the M1 and M3 sites) each have average metal—oxygen distances of 1.91 Å, which is significantly smaller than that (1.99 Å) of the third site (M2), indicating that the M2 site is the one enriched in the lower-valence cations. No known iron oxyhydroxide phase is isostructural with psilomelane.

Layer structures

The crystal structure of chalcophanite, $ZnMn_3O_7 \cdot 3H_2O$, consists of single sheets of water molecules between layers of edge-shared [MnO_6] octahedra, with Zn atoms located between the water layer and oxygens of the [MnO_6]

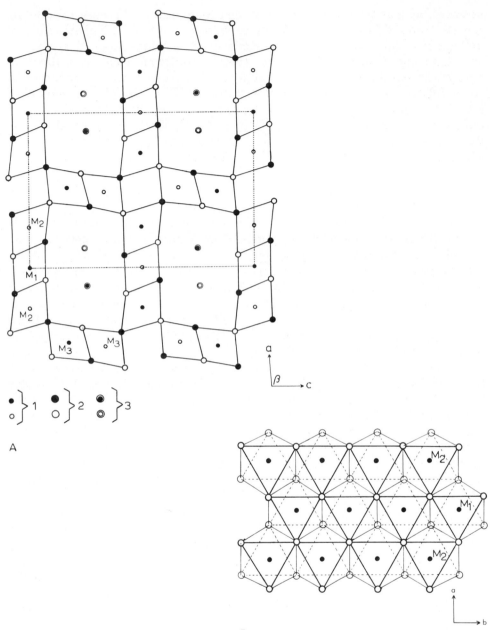

Fig. 7-5. The psilomelane structure. A. Projection onto (010), showing four linked tunnels. The unit cell is shown by broken lines. Open circles, atoms at zero level; shaded circles, atoms at level $1/2\ b$ (after Wadsley, 1953). B. The triple chains of edge-shared $[MnO_6]$ octahedra parallel to b.
$1 = Mn; 2 = $ oxygen; $3 = $ Ba, K, or H_2O.

layer (Wadsley, 1955). The stacking sequence along the c axis is thus $- O -$ Mn $- O - Zn - H_2O - Zn - O - Mn - O -$ (Fig. 7-6) and the perpendicular distance between two consecutive MnO_6 layers is about 7.17 Å. The water molecules are grouped in open double hexagonal rings, while vacancies exist in the layer of linked $[MnO_6]$ octahedra, so that six out of every seven octahedral sites are occupied by manganese. Each $[MnO_6]$ octahedron shares edges with five neighboring octahedra and is adjacent to a vacancy. The Zn atoms are located above and below the vacancies in the manganese layer and are coordinated to three oxygens of the $[MnO_6]$ layer (Fig. 7-6). Each Zn atom completes its coordination with three water molecules so as to form an irregular coordination polyhedron. The chemical compositions of natural chalcophanites differ significantly from the ideal formula $Zn^{2+}Mn_3{}^{4+}O_7 \cdot 3H_2O$. Not only is the water content variable, but there is a deficiency of Mn^{4+} ions and the number of cations usually exceeds 4 per formula unit. These trends indicate that some Mn^{2+} replaces Mn^{4+} in the linked octahedra, accounting for the larger average Mn—O distance of 1.95 Å. Additional cations also occur in interstitial positions between the H_2O layers and oxygens of the MnO_6 layers.

Lithiophorite, $(Al, Li)MnO_2(OH)_2$, also has a layer structure (Wadsley, 1952) in which layers of edge-shared $[MnO_6]$ octahedra alternate with layers of $(Al, Li)(OH)_6$ octahedra. The stacking sequence along the c axis is: $- O - Mn - O - OH - (Al, Li) - OH - O - Mn - O -$, and two consecutive $[MnO_6]$ layers are about 9.5 Å apart. Vacancies occurring in the sheets of linked $[MnO_6]$ octahedra in the chalcophanite structure are not characteristic of the linked $[MnO_6]$ octahedral layers of lithiophorite, although vacancies may exist between the sheets. Similarly, the ordered vacancies characteristic of the gibbsite, $Al(OH)_3$, structure (Megaw, 1934) are not found in the layer of linked $[(Al, Li)(OH)_6]$ octahedra in lithiophorite. On the other hand, substitution of Mn^{2+} for Mn^{4+} in the $[MnO_6]$ layers is required to maintain charge balance. Thus, Giovanoli et al. (1973b) in an X-ray and electron diffraction study of a synthetic lithiophorite formulated it as $[(Mn_5{}^{4+}Mn^{2+}O_{12})^{2-}(Al_4Li_2(OH)_{12})^{2+}]$.

A layer structure also occurs in the synthetic phases $CdMn_3O_8$ and Mn_5O_8 (Oswald and Wampetich, 1967). There are vacancies in the edge-shared $[MnO_6]$ containing Mn^{4+} ions (average Mn^{4+}—O distance = 1.87 Å). The Mn^{2+} and Cd^{2+} ions lie above and below the empty Mn^{IV} sites, and are six-coordinated by oxygens forming a distorted trigonal prism (average Mn^{2+}—O distance = 2.21 Å).

Crystal structure of the birnessite group

Although the crystal structure of a natural birnessite has not been determined, information has been derived for synthetic compounds of the birnessite group from electron diffraction measurements (Giovanoli et al., 1969, 1970a, b; Giovanoli and Stahli, 1970). Platelets of synthetic

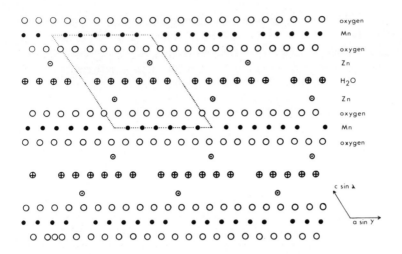

A

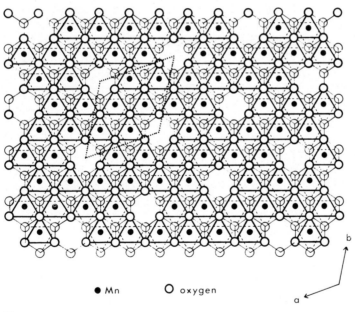

● Mn   ○ oxygen

B

Fig. 7-6. The chalcophanite structure (after Wadsley, 1955). A. Projection along the *b*-axis. Vacancies in the Mn layers define the rhombus unit cell. Note that one out of every seven Mn position is a vacancy. B. The edge-shared [MnO_6] layer viewed normal to the basal plane. The vacant octahedral sites at the origin are at the corners of a rhombus outlining the plane of the Mn atoms. Note that each Mn atom is adjacent to a vacancy.

$Na_4 Mn_{14} O_{27} \cdot 9H_2 O$ and $Mn_7 O_{13} \cdot 5H_2 O$ were deduced to have structures modelled on that of chalcophanite. Thus, sheets of water molecules and hydroxyl groups are located between layers of edge-shared $[MnO_6]$ octahedra separated by about 7.2 Å along the c-axis. One out of every six octahedral sites in the layer of linked $[MnO_6]$ octahedra is unoccupied, and Mn^{2+} or Mn^{3+} ions are considered to lie above and below these vacancies. These low-valence manganese ions are coordinated to oxygens in both the $[MnO_6]$ layer and the $(H_2 O, OH)$ sheet. The position of sodium in the intermediate layer is uncertain. Projections of the lattice of $Na_4 Mn_{14}$ $O_{27} \cdot 9H_2 O$ are shown in Fig. 7-7.

The structures proposed for synthetic birnessites contrast with those originally proposed for "manganous manganite" or "7 Å manganite". All the synthetic products were prepared from oxidized aqueous Mn $(OH)_2$ suspensions. The structure of pyrochroite, Mn $(OH)_2$, (compare Fig. 7-8A) consists of layers of edge-shared $[Mn (OH)_6]$ octahedra, in which each octahedron shares edges with six neighbouring octahedra to form a two-dimensional layer. Feitknechtite (β-MnOOH) has a similar structure (Meldau et al., 1973). Successive layers, 4.74 Å apart, are held together by hydrogen bonding. It was suggested (Feitknecht and Marti, 1945; Buser et al., 1954) that the structure of synthetic "manganous manganite" (birnessite) consists of layers of $[Mn^{IV} O_6]$ octahedra interdispersed with layers of relatively unoxidized Mn^{II} $(OH)_2 \cdot 2H_2 O$, giving the prominent reflections in X-ray powder diffraction patterns. The structure proposed by Giovanoli et al. (1969, 1970a, b), however, indicates that discrete pyrochroite layers are, in fact, not retained in birnessite. In more strongly oxidized synthetic samples, the Mn^{2+} ions no longer exist to bind successive $[Mn^{IV} O_6]$ layers together. Instead, the layers of linked $[Mn^{IV} O_6]$ octahedra are randomly orientated, and constitute the δ-MnO_2 phase which does not give basal X-ray reflections.

It is noteworthy that the habits of natural and synthetic birnessites correlate with the proposed layer structure. Thus, electron micrographs of natural birnessites (Brown et al., 1971; Sorem and Fewkes, Chapter 6, this volume, Fig. 6-7; Woo, 1973) and synthetic birnessites (Giovanoli et al., 1969, 1970a, b) show crystals with platey and lamellar habits, which differ significantly from the acicular habit of todorokite.

Deductions on the todorokite structure

The crystal structure of todorokite has not been determined, mainly because single crystals suitable for a structural analysis have not been found. As noted earlier, there is evidence to suggest that the phase in manganese nodules called "10-Å manganite" (Buser, 1959) corresponds to the terrestrial mineral todorokite. Buser and Grütter (1956) suggested that the "manganite" (todorokite) phase in manganese nodules is related structurally to lithiophorite, because prominent d-spacings around 9.6 Å and 4.8 Å for

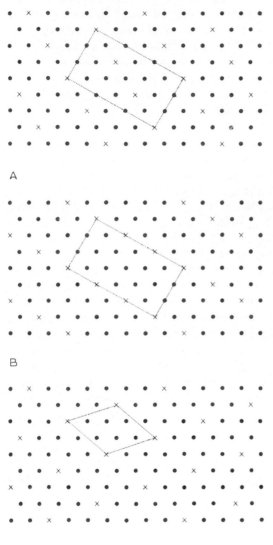

A

B

C

Fig. 7-7. Projections of the structure proposed for synthetic birnessite, Na_4Mn_{14} $O_{27} \cdot 9H_2O$ (after Giovanoli et al., 1970a, b). The projections show Mn atoms only in the basal planes at zero level (A) and at about 7.2 Å along c (B). The different locations of vacancies in levels A and B necessitate doubling of the c_o parameter. A comparable projection for chalcophanite is shown for reference (C).

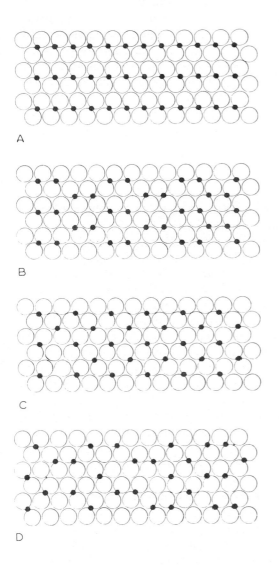

Fig. 7-8. Hexagonal close-packed oxygen layers and cation sites in the structures of: A. $Fe(OH)_2$; B. goethite; and C. δ-FeOOH. A schematic disordered structure for the $FeOOH \cdot xH_2O$ phase is shown in D.

manganese nodules corresponded to similar basal spacings in the X-ray powder diffraction pattern of lithiophorite. As a result, there is a widely held view (Mero, 1962; Arrhenius, 1963; Giovanoli et al., 1973a) that lithiophorite is the structural model for one of the dominant manganese oxide phases in manganese nodules. However, synthesis experiments of lithiophorite by Giovanoli et al. (1973b) showed that substantial amounts of Na^+ or other cations cannot be substituted in the lithiophorite structure. They concluded that lithiophorite differs fundamentally from the "buserite" (todorokite) group, the only similarities being the basal reflections around 9.52 Å and 4.74 Å and the Mn—Mn distances within the linked [MnO_6] octahedra.

A structural correlation of "10-Å manganite" (todorokite) with lithiophorite is unsatisfactory when crystal habits of terrestrial todorokites and lithiophorites are compared. Crystals of lithiophorite consist of laminae showing one perfect cleavage parallel to (001). This habit correlates with the layer structure determined for lithiophorite. Todorokite specimens, on the other hand, consist of fibrous aggregates of small needle-shaped crystals, resembling many specimens of cryptomelane, hollandite, akaganéite, and psilomelane. Furthermore, electron micrographs of todorokite (Straczek et al., 1960; Hariya, 1961) show that the crystals consist of narrow lathes or blades elongated along one axis (parallel to b) and showing *two* perfect cleavages parallel to the (001) and (100) planes. Minerals of the hollandite and psilomelane groups also show two perfect mutually perpendicular cleavages at right angles to the elongation of the acicular crystals. These resemblances suggest that todorokite has a crystal structure resembling those of hollandite and psilomelane and not that of lithiophorite. This correlation is further born out when comparisons are made between the cell parameters summarized in Table 7-I. If todorokite does have a framework structure analogous to those of hollandite or psilomelane, then it is possible that Mn^{2+} and Mn^{4+} ions occur together in chains of edge-shared [MnO_6] octahedra which are stacked in such a manner so as to be able to accommodate large cations and H_2O molecules.

The problems regarding the todorokite structure and other manganese (IV) oxide phases are succinctly summarized by Wadsley (1963): "The wide variety of ill-defined manganese dioxides of variable composition and unknown structure ... often containing substantial quantities of water as well as small amounts of foreign ions, may be more closely related to the two barium-containing tunnel structures (hollandite and psilomelane) than to the dimorphs of manganese dioxide with which they are often compared. If water alone is present in the tunnels, the diffraction patterns of the host phases will alter considerably. Disordered phases of no fixed composition or structure could arise from the irregular intergrowth of one tunnel compound with the other, a likely correspondence in view of the exact correspondence of the two in certain directions and of the presence of the residual fragments of the (pyrolusite) structure in both."

The structure of hydrated ferric oxyhydroxide polymer

Natural and synthetic FeOOH·xH$_2$O specimens usually give poor X-ray and electron diffraction patterns. However, several broad and weak bands have been measured in the patterns of a variety of samples (Van der Giessen, 1966; Towe and Bradley, 1967; Okamoto et al., 1972; Feitknecht et al., 1973), including notable peaks around 2.55 Å and 1.47 Å. These features resemble d-spacings observed at 2.40 Å and 1.42 Å in δ-MnO$_2$, the discrepancies being caused by the smaller ionic radius of Mn^{4+} (0.54 Å) compared to Fe^{3+} (0.645 Å), and correspond to reflections from the (10$\bar{1}$0) and (11$\bar{2}$0) planes of a hexagonal close-packed oxygen framework.

The structure of FeOOH·xH$_2$O may also be correlated with those of goethite (a-FeOOH) and Fe (OH)$_2$. As discussed earlier, goethite is isostructural with ramsdellite and consists of linked double-chains of edge-shared [Fe (O, OH)$_6$] octahedra (Fig. 7-2). The structure of Fe (OH)$_2$ (amakinite), on the other hand, as well as isostructural pyrochroite (Mn (OH)$_2$), consists of two-dimensional layers of edge-shared [Fe (OH)$_6$] octahedra. Another way of viewing these two structures is in terms of a hexagonal close-packed oxygen framework in which the Fe^{3+} or Fe^{2+} cations are distributed in an ordered array among the octahedral sites (Fig. 7-8A and B). Another iron (III) oxyhydroxide polymorph, δ-FeOOH, (Okamoto, 1968) also has an ordered distribution of Fe^{3+} ions among the octahedral sites (Fig. 7-8C), but there is disorder of O^{2-} and OH$^-$ ions. The structure of hydrated ferric oxyhydroxide (FeOOH·xH$_2$O) is believed to contain Fe^{3+} ions in the octahedral sites of a hexagonal close-packed oxygen network (Towe and Bradley, 1967; Okamoto et al., 1972; Feitknecht et al., 1973), but the Fe^{3+} ions are almost randomly distributed in the octahedral sites with but a slight degree of order among them (Fig. 7-8D). The very small crystallites observed in electron micrographs, typically 30—100 Å in diameter, indicate that long-range ordering extends over only 5—10 unit cells.

Crystal chemistry of manganese nodules

A large number of elements are enriched in manganese nodules relative to their abundances in seawater or "average" crust. They include the transition metals Mn, Fe, Co, Ni, and Cu, as well as Zn, Mo, Ba, and Pb. Many of these metals are also enriched in terrestrial manganese oxide minerals, or, along with Na, K, and Ca, form discrete phases with manganese (Table 7-I). In order to understand the relative enrichment of these elements in hydrated manganese oxide or iron oxyhydroxide phases, it is necessary first to establish the oxidation states of each element and stabilization energy of each cation. The transition metal and B subgroup elements of the periodic table occur in a range of oxidation states on the earth's surface. Known valencies of elements enriched in manganese nodules are summarized in

TABLE 7-XIII

Ionic radii of certain metals enriched in manganese nodules

Metal: Valence	Mn	Fe	Co	Ni	Cu	Zn	K	Na	Ba	Ca	Mg	Pb	Mo	Ti
M^{4+}	0.54											0.775	0.65	0.605
M^{3+}	0.65	0.645	0.525*	0.56*										0.67
M^{2+}	0.82	0.77	0.735	0.70	0.73	0.745			1.36	1.00	0.72	1.18		
M^+					0.96		1.38	1.02						

*Low spin electronic configuration.

Table 7-XIII, together with their common ionic radii (Shannon and Prewitt, 1969). Crystal field stabilization energies and radius ratio criteria result in transition metal ions favouring octahedral sites in oxide structures (Burns, 1970). The larger cations K^+, Ba^{2+}, Pb^{2+}, Ca^{2+}, and Na^+ predominate in sites with higher coordination numbers, such as the cavities found in the hollandite and psilomelane structures (Fig. 7-5).

Since manganese nodules form in oxidizing environments, cations with higher valencies predominate over reduced oxidation states. Indeed, results of Mössbauer studies described earlier show that iron occurs principally as Fe^{3+} ions in manganese nodules, while ESR and ESCA measurements indicate the presence of only small amounts of Mn^{2+} ions. Spectroscopic techniques have not been developed to the extent of being able to identify the oxidation states of Co, Ni, and Cu. As a result, interelement relationships derived from microprobe measurements (Burns and Fuerstenau, 1966; Cronan and Tooms, 1968; Friedrich et al., 1969; Sano and Matsubara, 1970; Ostwald and Frazer, 1973) have been used to deduce the valencies of these elements in manganese nodules. Most nodules show strong positive correlations between Mn, Ni, Cu, and Zn, suggesting that the divalent cations Ni^{2+}, Cu^{2+}, and Zn^{2+} substitute for Mn^{2+} in todorokite and birnessite host phases. Some nodules show a correlation between Mn and Co (Cronan and Tooms, 1968; Ostwald and Frazer, 1973), suggesting the presence of Co^{2+} ions. Other nodules show a well-defined apparent Fe—Co correlation (Burns and Fuerstenau, 1966; Sano and Matsubara, 1970), which has been interpreted as Co^{3+} substituting for Fe^{3+} ions in the $FeOOH \cdot xH_2O$ phase (Burns, 1965). Thermodynamic arguments (Goldberg, 1961a; Sillèn, 1961; Burns, 1965) support the existence of Co (III) in the marine environment. It is noteworthy that Co^{3+} ions have a low-spin configuration in oxide structures, and that the ionic radius of Co^{3+} (0.525 Å) is considerably smaller than Fe^{3+} (0.645 Å) so that cobalt may not be tolerated in the $FeOOH \cdot xH_2O$ phase. In fact, the mineral heterogenite, CoOOH, is not isostructural with any of the

FeOOH polymorphs (Hey, 1962; Strunz, 1970). On the other hand, the radius of Co^{3+} is similar to that of Mn^{4+} (0.54 Å). Therefore, the strong fractionation of cobalt into synthetic, terrestrial, and marine manganese oxide phases (Fukai et al., 1966; McKenzie, 1970, 1972; Loganathan and Burau, 1973) may be the result of Co^{3+} substituting for Mn^{4+} ions in the $[MnO_6]$ octahedra (Burns and Burns, 1974), leading to a very high crystal field stabilization energy for cobalt (Burns, 1970). It has also been suggested (Burns and Fyfe, 1967; Burns, 1970) that low-spin Ni^{4+} ions may be formed in strongly oxidizing environments and be stabilized in the octahedral Mn^{4+} sites. The Mn^{3+} and low-spin Ni^{3+} ions are relatively unstable, but may be stabilized in very distorted octahedral sites (Burns, 1970).

In summary, the most likely cations occurring in manganese nodules are Mn^{4+}, Fe^{3+}, Co^{3+}, Ni^{2+}, Cu^{2+}, and Zn^{2+}, with smaller proportions of Mn^{2+}, Mn^{3+}, and Co^{2+}, and negligible amounts of Fe^{2+}. The uptake of cobalt, nickel, and copper into the host ferromanganese oxide phases of manganese nodules may be interpreted as follows. The birnessite phase, by analogy with chalcophanite (Fig. 7-6), accommodates Ca^{2+}, Mn^{2+}, and perhaps, Na^+ ions between the sheets of $H_2O + OH^-$ groups and layers of edge-shared $[MnO_6]$ octahedra. The divalent cations enriched in manganese nodules probably enter predominantly into these between-layer sites of birnessite. However, by analogy with chalcophanite which must accommodate some Mn^{2+} ions in the Mn^{4+} sites, a proportion of the divalent cations in birnessite may be present in the $[MnO_6]$ layer. The Co^{3+} ions are much more readily accommodated in the octahedral Mn^{4+} sites or vacancies in the $[MnO_6]$ layers of birnessite *and* the disordered δ-MnO_2 phase.

Although the todorokite structure is unknown, this mineral does appear to contain essential Mn^{2+} ions, and probably has some form of linkage with edge-shared $[MnO_6]$ octahedra, possibly resembling the framework structures of hollandite and psilomelane. Therefore, the divalent cations probably replace the Mn^{2+} ions in sites which may bear similarities to the M2 octahedra of the psilomelane structure (Fig. 7-5). Small amounts of Co^{3+} may also replace Mn^{4+} ions in the $[MnO_6]$ octahedra. The uptake of large cations such as Ba^{2+}, K^+, and perhaps Pb^{2+} into the nodules suggests the presence of a phase similar to hollandite and psilomelane having large cavities in a framework of edge-shared $[MnO_6]$ octahedra. Some of the Pb, as well as Ce and Mo, may occur as tetravalent cations substituting for Mn^{4+} in the linked $[MnO_6]$ octahedra.

Nucleation and authigenesis of manganese nodules

There is abundant evidence summarized in the last section that crystal structure is a major factor in controlling the uptake of metals into manganese nodules. The crystalline ferromanganese oxide phases contain sufficient coordination sites to accommodate favourably and stabilize the transition metal ions Mn^{4+}, Mn^{3+}, Fe^{3+}, Co^{3+}, Co^{2+}, Ni^{2+} Cu^{2+}, and perhaps Ni^{4+}. The question arises whether structural factors control the growth of the host phases. Recently, hypotheses on the mechanism of nucleation and authigenesis of many nodules (Brown, 1971; Burns and Brown, 1972; Burns and Burns, 1975; Burns et al., 1974) have been proposed and are summarized as follows.

Orange-brown coatings of iron oxyhydroxide phases are frequently observed on substrates such as foram tests, coral, sharks' teeth, palagonite, phillipsite, and other secondary phases associated with altered volcanic rocks. The chemical constituents of these substrates contain anion radicals derived from weak acids (e.g., carbonates, phosphate, silicates), suggesting that $FeOOH \cdot xH_2O$ was precipitated when locally high pH's were produced in seawater trapped in cavities during solution and hydrolysis of the substrate. After the precipitation of $FeOOH \cdot xH_2O$ reaches an advanced stage of deposition, incipient brown-black coatings of manganese (IV) oxide become visible (Burns and Brown, 1972). Electron microprobe traverses across the boundary between the manganese oxide growth and the nucleus frequently show coronae of iron preceding the manganese deposit (Burns and Brown, 1972). Such evidence for the growth of manganese (IV) oxide on the $FeOOH \cdot xH_2O$ layer, together with the fact that manganese nodules always contain major proportions of manganese and iron, suggests that there are crystallographic relationships between the Mn and Fe phases which allow contemporaneous or cyclic structural intergrowths.

The tabulations of oxide minerals of manganese and iron in Tables 7-I and 7-VIII show the existence of isostructural minerals of Mn and Fe, notably ramsdellite and goethite, hollandite group and akaganéite, and probably δ-MnO_2 and $FeOOH \cdot xH_2O$. In such pairs of minerals having identical crystal structures, or layers of atoms in common in a specific crystallographic plane, the possibility exists for orientated intergrowths of one crystalline phase on the other, provided there is a similarity of lattice parameters or interplanar spacings. This phenomenon is termed epitaxy. An excellent example of expitaxial growth is ramsdellite (MnO_2) crystallites which can be seen by SEM to grow in crystallographic continuity with goethite (a-$FeOOH$) (Finkelmann et al., 1972, 1974). There is less than 10% mismatch cell parameters between the smaller ramsdellite unit cell and the goethite structure (Tables 7-I and 7-VIII). This mismatch is sufficiently small to permit epitaxial intergrowth of the two phases perpendicular to the layers of hexagonally close-packed oxygen atoms.

Although δ-MnO_2 and $FeOOH \cdot xH_2O$ consist of very small, disordered crystallites, both phases contain cations in octahedral sites of hexagonally close-packed oxygen layers. Therefore, these two phases in manganese nodules are highly susceptible to epitaxial intergrowths, which probably initiates nucleation and leads to the intimate association of manganese and iron oxide phases. This intergrowth of δ-MnO_2 and $FeOOH \cdot xH_2O$ inhibits the formation of the ordered layer structure of birnessite. However, local enrichments of divalent cations (Ca^{2+}, Ni^{2+}, Cu^{2+}, Mn^{2+}) break the sequence of oscillatory intergrowths of δ-MnO_2 and $FeOOH \cdot xH_2O$ and lead to the development of birnessite or todorokite crystallites.

CONCLUDING REMARKS AND SUGGESTIONS FOR FURTHER MINERALOGICAL STUDIES

Several crystalline phases have now been identified in manganese nodules. The predominant minerals are manganese (IV) oxides related to the terrestrial minerals todorokite, birnessite, and nsutite. Disordered phases showing only short range crystallographic order include δ-MnO_2 and hydrated ferric oxyhydroxide polymer (incipient goethite). The most common accessory minerals are goethite, quartz, feldspar, clays (montmorillonite, illite), and zeolites (phillipsite). While electron microscopy techniques have revealed the presence of euhedral crystals of many of these minerals, the principal method for identifying the constituent minerals has been X-ray powder diffraction analysis, using both diffractogram and photographic methods. This technique has severe limitations for the intimately intergrown assemblages of cryptocrystalline minerals and materials amorphous to X-rays (Sorem, 1972; Sorem and Foster, 1972b), and therefore cannot be used for quantitative analyses of mineral proportions. Uncritical use of X-ray diffraction techniques in the past may have led to erroneous reports of minerals in manganese nodules. Ambiguities over some mineral identifications are suggested by recent evidence showing that certain phases are unstable when manganese nodules are exposed to the atmosphere and are out of contact with seawater (Brown, 1971, 1972; Burns et al., 1974). Indiscriminate sample preparation and failure to remove effects of preferred orientation may also lead to misidentifications, such as δ-MnO_2 in X-ray patterns of orientated todorokite crystallites (Faulring, 1962). Various techniques have been suggested to extract the minute minerals and to prepare samples in order to improve the quality of X-ray diffraction patterns of minerals from manganese nodules (Brown, 1972; Sorem and Foster, 1973). It is recommended that future X-ray diffraction measurements be made using Mn-filtered Fe-K_α radiation. This source not only provides the simplest method of reducing acute fluorescence effects when Co and Cu radiations are used, but it increases the precision of lines at large d-spacings (e.g., 9.5—10 Å) which are essential for accurate phase identifications in manganese nodules. Debye-Scherrer cameras equipped with Gandolphi attachments are recommended to minimize the adverse effects of preferred orientation (C. Frondel, personal communication, 1974). Giovanoli et al. (1973a) have recommended using focussing Guinier-De Wolff X-ray cameras and Mo X-radiation with densitometer profiles of acquired patterns for increased resolution of diffuse, broad peaks found at smaller d-spacings for the minerals of manganese nodules. Selected area electron diffraction patterns, both spot and ring forms, constitute an important technique in the future for obtaining lattice symmetries and spacings of crystalline phases too small to give X-ray patterns. However, great care must be exercised to avoid deterioration of the sample *in vacuo* under the electron beam.

While Mössbauer spectroscopy has not provided definitive identifications of the iron-bearing minerals, nevertheless, the technique has demonstrated that the Fe^{3+} state predominates in manganese nodules. ESR and ESCA spectroscopies have revealed the presence of some Mn^{2+} ions in the predominantly Mn (IV) minerals. Oxidation states of other elements have been inferred from synthesis products and interelement relationships with the major Mn and Fe cations. Thus, the presence of such ions as Mn^{3+}, Co^{3+}, Co^{2+}, Ni^{2+}, Cu^{2+} Pb^{4+}, Ba^{2+}, Ca^{2+}, and K^+ are implied in the manganese (IV) oxide phases. Crystal structure data are scanty for explaining the locations of these cations in the structures of host phases in manganese nodules. In birnessite, most of the divalent cations substitute for Mn^{2+} between the layers of edge-shared $[MnO_6]$ octahedra, while Co^{3+} (and perhaps Mn^{3+} and Ni^{4+}) are very readily accommodated in the vacancies and Mn^{4+} sites of the linked $[MnO_6]$ octahedra in both birnessite and δ-MnO_2. Unfortunately, no satisfactory crystal structure data are available for todorokite, and are urgently needed for a more complete interpretation of the crystal chemistry of manganese nodules. The habits of terrestrial todorokites are suggestive of structural relationships similar to those found in the hollandite group and psilomelane. These structures contain not only sites for accommodating divalent and trivalent transition metal ions but also larger cavities for holding large cations also enriched in manganese nodules.

CHAPTER 8

RATES OF ACCRETION

T.L.KU

INTRODUCTION

The growth rates of manganese nodules provide a critical insight into the processes which govern their formation. During recent years this problem has received increasing attention, largely as a result of the expanded capability in the field of isotope geochemistry. This chapter outlines the principles of applying radiometric techniques to the study of deep-sea nodule formation, discusses the major results and their interpretation, and indicates some important limitations in the methods used. Findings from the radiometric dating studies are compared with those obtained by non-radiometric evidence. A brief account is also given to the problem of dating shallow-water continental margin manganese nodules by radiometric means.

RADIOMETRIC METHODS

The age or rate of accretion of manganese nodules can be determined by dating the nuclei around which the ferromanganese oxide layers accumulate, or by assessing the age difference among the successive layers of these oxides within the nodule. As the age of a nodule is generally defined to be the time when the first oxide layer begins to form, the "nucleus dating" approach accordingly gives only a maximum age for the nodule. The mean accumulation rate derived from such an age is therefore a minimum. The second approach hinges on the assumption that the radioactive nuclides are incorporated into the nodule layers at a constant rate when the latter are deposited from seawater.

To allow accurate determination of the amount of daughter nuclides generated, one common rule in choosing a radioisotope for dating is that the magnitude of its half-life of decay should be comparable to the age of the object, or to the time-constant of the process involved. In this respect, only the following four isotopes have been found to have a major application in nodule dating: ^{230}Th, ^{231}Pa, ^{10}Be, and ^{40}K. As an exception, one additional isotope, ^{238}U, has proved useful. Although this isotope has an extremely long spontaneous fission half-life, the great sensitivity in the method of detecting its fission "daughters" enables it to cover a wide range of ages.

Uranium-series disequilibrium: ^{230}Th *and* ^{231}Pa

The principles of using the daughter products of the uranium radioactive decay series as the basis of age determination methods may be summarized as follows. The processes of weathering, transport and deposition lead to a separation among various elements represented in the decay series. A newly formed authigenic mineral such as a manganese nodule will therefore show deviations from the equilibrium parent-to-daughter isotopic ratios. If the mineral constitutes a closed system, then the departure from equilibrium will gradually disappear with time at a rate depending on the half-lives of the isotopes involved.

Among the radioactive daughters of ^{238}U and ^{235}U, only ^{234}U (half-life = 2.48 · 10^5 years), ^{230}Th (half-life = 75,200 years), ^{226}Ra (half-life = 1,622 years) and ^{231}Pa (half-life = 34,300 years) have sufficiently long half-lives to be of value in most geochronological studies. The use of ^{234}U and ^{226}Ra is relatively limited; one reason is that these two isotopes are more susceptible to post-depositional migration (Kröll, 1955; Ku, 1965). We shall first consider ^{230}Th in the ^{238}U series and ^{231}Pa in the ^{235}U series.

The uranium content of the ocean is rather uniform (about 3.3 μgU/l) due to the formation of strong complexes such as $UO_2(CO_3)_3^{4-}$. By contrast, ^{230}Th and ^{231}Pa which would grow into equilibrium with this oceanic uranium are readily hydrolyzed and precipitated to the ocean floor. ^{230}Th and ^{231}Pa are therefore enriched with respect to their uranium parents in recently deposited deep-sea sediments and manganese nodule layers, whereas in seawater they are greatly depleted. Since the "excess" (or U-unsupported) ^{230}Th and ^{231}Pa in a sediment will disappear in a predicted logarithmic manner, their distribution gives rise to a potential dating tool. If, for example, there is an excess of 20 units of ^{230}Th when a sediment is originally deposited and if at present one finds only 10 units of excess, then the sediment must be one half-life old (i.e., 75,200 years, see Fig. 8-1). Similarly, an excess of 5 units of ^{230}Th corresponds to an age of 150,400 years and so on. The same logic applies to the excess ^{231}Pa. The method assumes that the initial ^{230}Th excess is constant. In other words, the ^{230}Th concentration observed for the uppermost layer is assumed to be the initial concentration of this isotope at all points within the deposit. The method has a built-in cross-check in that it predicts an exponential decrease of ^{230}Th with depth. A variation of the method is to use ratios such as $^{230}Th/^{232}Th$ and $^{230}Th/^{231}Pa$ instead of ^{230}Th concentrations. The assumption will then be that these ratios in freshly deposited sediments remain constant with time. A discussion of the implications of such a variation is beyond the scope of this chapter, but in most cases the variation will alter the age calculations only to a minor degree.

The first measurements of the radioactivities of manganese nodules were made early in the century by Joly (1908). He noted high contents of ^{226}Ra

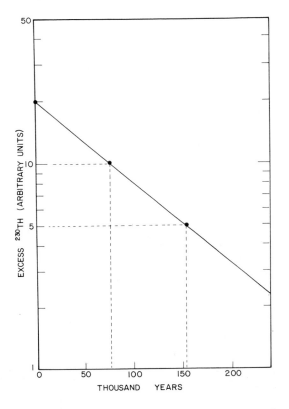

Fig. 8-1. Change with time of the amount of excess^{230}Th in a marine sediment or manganese nodule.

in nodules taken from the *Challenger* and *Albatross* stations in the equatorial south Pacific. Subsequent workers including Iimori (1927), Kurbatov (1936), Pettersson (1943), Von Buttlar and Houtermans (1950), while confirming Joly's observation, also found a decrease of ^{226}Ra from the outer layers towards the central parts of the nodules. By assuming that this ^{226}Ra was "unsupported" by its parent ^{230}Th, the rates of radial growth of the deep-ocean nodules were estimated to be 0.7 to 65 mm/10^3 years (Pettersson, 1943, 1955; Von Buttlar and Houtermans, 1950). Later, it was pointed out by Goldberg and Arrhenius (1958) that the assumption that ^{226}Ra was derived directly from seawater (and hence unsupported) could well be erroneous. Because of the very high ^{232}Th contents of the nodules, these authors suggested that ^{230}Th must also have been effectively collected from the seawater by these concretions along with ^{232}Th. Taking the extreme case that all the ^{226}Ra in the nodules is ^{230}Th-unsupported (it takes only about 8,000 years for ^{226}Ra to grow into equilibrium with ^{230}Th, and hence the observed ^{226}Ra gradient actually reflects that of ^{230}Th), Goldberg and Arrhenius suggested that the actual rates of deposition were about a factor of 50 (the ratio of ^{230}Th to ^{226}Ra half-lives) lower than the previous estimates, i.e., 0.01 to 1.3 mm/10^3 years. This suggestion has subsequently been supported by direct measurements of ^{230}Th in later investigations.

Although no detailed radiochemical data were presented by Goldberg, he went on to deduce from the ^{230}Th distribution that the growth rate of a nodule from the North Atlantic is about 100 mm/10^6 years and that of the "Horizon" nodule from the North Pacific is about 10 mm/10^6 years (Goldberg, 1961a, b, 1963b). Subsequently, much more detailed analytical work on the uranium-series nuclides has been carried out (Bender et al., 1966; Ku and Broecker, 1967, 1969; Barnes and Dymond, 1967; Bhat et al., 1970). An observation common to all the investigators is the very high concentration of ^{230}Th in the surface layers, which diminishes exponentially toward the central portions of the nodules. In most cases the unsupported ^{230}Th disappears completely before reaching a depth of 5 mm below the surface. A similar sharp decrease in concentration with depth is observed for ^{231}Pa (Ku and Broecker, 1967, 1969). Fig. 8-2 shows data reported by Ku and Broecker (1967) for a concretion (V—21—D2) from the North Pacific. Despite the differences in their respective half-lives, the three isotopes, ^{234}U, ^{230}Th, and ^{231}Pa, all show an exponential decrease with depth, giving concordant rates of about 4 mm/10^6 years.

Two points need to be clarified here with regard to the plots of Fig. 8-2. First, due to the recoil phenomenon in alpha decay, fractionation between the two uranium isotopes, ^{238}U and ^{234}U, does occur in nature. The so-called "natural Szilard—Chalmers effect" (Dooley et al., 1966; Cherdyntsev, 1971, p. 80) makes the ^{234}U atoms more susceptible than the ^{238}U atoms to be leached out of a mineral during weathering. The activity ratios of ^{234}U/^{238}U in natural waters are therefore larger than the equilibrium value of unity and in seawater the ratio has a uniform value of 1.15 (Thurber, 1962; Koide and Goldberg, 1965). This ratio is found in the surface of the North Pacific nodule studied by Ku and Broecker. As shown in Fig. 8-2C, the growth rate of 4.6 mm/10^6 years is derived on the assumption that the ^{234}U/^{238}U ratio of the oceanic uranium has been constant at 1.15 for the last several hundred thousand years. Secondly, there is a ± 20% uncertainty in the estimate of "depth" of the analyzed layers within the nodule. These depths should be measured along the direction of radial growth. In practice, thin layers of nodule material are scraped successively from the nodule surface following its general curvature. The thickness (and hence depth) of each layer was then calculated from the nodule's mean density, the area scraped, and the weight of the material removed. In their estimation of the thickness of the layer scraped, Ku and Broecker (1967, 1969) and Bender et al. (1966) used a nodule density value of 2.49 g/cm^3 (Mero, 1965a, p. 135). However, recent measurements (T. O'Neil, personal communication, 1973) on sixteen specimens from five of the stations reported by the above-

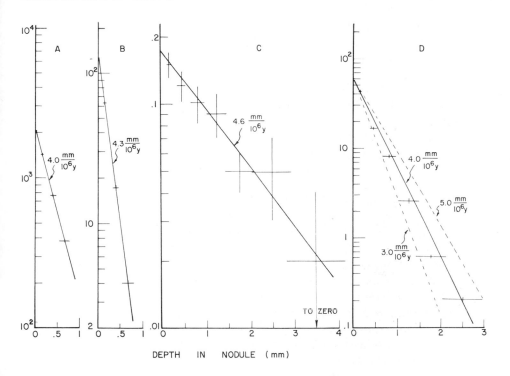

DEPTH   IN   NODULE (mm)

Fig. 8-2. Depth distribution of radioactivity and the accumulation rates derived therefrom in a manganese nodule (V-21-D2) from the Pacific. Adapted from data of Ku and Broecker (1967). The vertical and horizontal lines denote probable-error limits. Units for the Y-axis:
A. Excess ^{230}Th (dpm/g) = dpm ^{230}Th/g $-$ dpm ^{234}U/g.
B. Excess ^{231}Pa (dpm/g) = dpm ^{231}Pa/g $-$ dpm ^{235}U/g.
C. Excess ^{234}U/^{238}U = (dpm ^{234}U $-$ dpm ^{238}U)/dpm ^{238}U.
D. Excess ^{230}Th/^{232}Th = (dpm ^{230}Th $-$ dpm ^{234}U)/dpm ^{232}Th.
The two dashed lines are drawn to show the "sensitivity" in the rate estimation.

mentioned workers indicate that the water-free densities of the nodules range from 1.80 g/cm^3 to 2.11 g/cm^3, with a mean value of 1.96 g/cm^3. This value is about 20% lower than Mero's estimate, although these measurements do support his statement that nodules from a given deposit (station) have a rather uniform density. If the 1.96 g/cm^3 value is adopted instead of 2.49 g/cm^3, then the reported rates are about 20% too low. On the other hand, as pointed out by Ku and Broecker (1969), because of the unevenness of nodule surfaces, layers could not be strictly peeled off along the growth or isochron surfaces. This could have caused a mixing of material across such surfaces, giving rise to too small a radioactivity gradient, and hence too high a measured growth rate. As the two effects (underestimating nodule density vs. "mixing" of layers) tend to cancel each other, the quoted 20% error is considered to be realistic. Uncertainties in radioactivity measurements are much smaller than those of the depth assignments (Fig. 8-2); hence the reported rates should be accurate to within at least a factor of two, if not ±20%.

The problem of assessing radioactivity gradients across the growth layers on a submillimetre scale as discussed above has been met with success in a study by Heye and Beiersdorf (1973). These investigators used a nuclear emulsion technique which is an elaboration of the method employed by Von Buttlar and Houtermans (1950). A nuclear emulsion plate is exposed to a cross-section of a nodule for 30 days and afterwards the alpha-particle traces are counted under the microscope. The alpha traces, due to the decay of ionium (^{230}Th) can be distinguished from those due to uranium and thorium decay. By this means, the distribution of ^{230}Th in small intervals (0.1 − 0.2 mm) parallel to the internal growth structure of the nodule can be "mapped" out.

Although attempts have been made to date nodules using a variety of radiometric methods, the published nodule growth rates have so far come largely from the uranium-series methods (Table 8-I). The above example as depicted by Fig. 8-2 probably represents the most detailed analyses on a single specimen. Studies on other samples were less elaborate in terms of number of measurements made on each specimen and in terms of "layer depth" assignments except for those measured by the nuclear emulsion technique. With this in mind, one may be impressed to note from Table 8-I the relatively narrow range (2—8 mm/10^6 years) in which the majority of the values fall.

Since the upper age limits for the use of ^{231}Pa, ^{230}Th, and ^{234}U are about 200,000 years, 400,000 years, and 1 million years, respectively, in general only the near-surface 2—10 mm layers are datable by these nuclides.

The ^{10}Be method

^{10}Be is the longest-lived (half-life = 1.5 · 10^6 years) radioisotope produced by the interaction of cosmic rays with atmospheric nuclei, mainly nitrogen and oxygen. Being non-volatile, this species is washed out of the atmosphere by rain and is incorporated mainly in deep-sea sediments in a time very short compared to its half-life. The assumptions underlying the ^{10}Be dating technique, pioneered by Goel et al. (1957) and Merrill et al. (1960), are analogous to those of the ^{230}Th and ^{231}Pa methods. It is assumed that: (1) the concentration of ^{10}Be or the specific activity ^{10}Be/^{9}Be in freshly formed deposits has remained constant with time at any point on the ocean floor; and (2) ^{10}Be does not migrate within the deposits. Because of its extremely low abundance, e.g., 1—10 dpm (disintegration per minute) per kg of sediments (Goel et al., 1957; Merrill et al., 1960), ^{10}Be is difficult to measure precisely. With the advancement of low-level beta-counting techniques in recent years, however, tangible results seem to demonstrate that the amount of ^{10}Be in pelagic sediments approximates to that predicted from the rate of ^{10}Be production in the atmosphere. The results also show that ^{10}Be concentration in sediments diminishes with depth at rates comparable to what is expected from the decay of ^{230}Th (Amin, 1970).

Analyses of ^{10}Be have been made on three manganese nodules (Somaya-julu, 1967; Krishnaswami et al., 1972). A rather steep decrease (by a factor of 5 to 20) of ^{10}Be contents from surface to a depth of about 20 mm is observed. Only in one of the nodules are measurements made on three layers and they show an exponential decrease of ^{10}Be with depth. The rates of accumulation of the three nodules are reported at 0.8 ± 0.1, $3.8 \pm ^{1.8}_{1.0}$, $1.8 \pm ^{0.5}_{0.2}$, all in mm/10^6 years. In two of the cases, ^{230}Th analyses have also been made, giving rates of comparable magnitude (Table 8-1). It should be noted that the ^{10}Be rates are calculated over the outermost 10—20 mm of the nodules while the ^{230}Th rates cover only the outermost 2—3 mm.

The K—Ar method

Potassium—argon dating has been applied to volcanic minerals and glass which are often found as nuclei in manganese nodules. The works of Evernden et al. (1964) and Dymond (1966) appear to demonstrate the following: (1) volcanic minerals and glass, especially those of ash layers, form without incorporating significant quantities of ^{40}Ar; (2) minerals such as anorthosite, amphibole and plagioclase, and glass retain argon very well at the low temperatures encountered in the deep sea; and (3) the ^{40}Ar produced in the mineral by decay of ^{40}K can be reliably distinguished from that absorbed from the atmosphere.

Barnes and Dymond (1967) measured K—Ar ages of volcanic minerals and glass shards separated from the nuclei of three nodules from the Pacific and obtained ages ranging from 2 to 29 m.y. By assuming a uniform growth for the entire thickness (varying between 5 mm and 100 mm for the three nodules) of the ferromanganese oxide layerings around the nuclei, growth rates of the order of 1—4 mm/10^6 years were calculated. These rates are lower limits, because in the calculation the time intervals between the volcanic episodes and the beginning of the oxide coating were unknown and assumed to be negligibly short. In one of the nodules, DWHD 47, two different minerals with different K and radiogenic ^{40}Ar contents were analyzed. Both analyses give concordant rates of 0.5—1 mm/10^6 years, suggesting that the minerals were possibly formed free of "frozen-in" initial ^{40}Ar. The same nodule also shows a rate of 6 mm/10^6 years for the outermost 3 mm crust based on excess ^{230}Th.

TABLE 8-I

Published growth rates of pelagic nodules based on radiometric methods

Sample	Latitude	Longitude	Depth (m)	Growth rate (mm/10^6 yr.)	Method	Reference[1]
North Atlantic:						
C58—100	30°57'N	65°47'W	4,800	4	^{230}Th,^{231}Pa	(1)
A266—41[2]	30°59'N	78°15'W	830	<2	^{230}Th,^{231}Pa	(1)
G74—2374[2]	30°31'N	79°01'W	876	<2	^{230}Th,^{231}Pa	(1)
G74—2384[2]	30°53.5'N	78°44'W	843	<2	^{230}Th,^{231}Pa	(1)
Lusiad AD4	6°03'N	32°22'W	1,020	8—10	^{230}Th,K-Ar	(2)
BP1[2]	—	—	—	~0[3]	^{230}Th	(13)
South Atlantic:						
V16—T3	13°04'S	24°41'W	4,415	3	^{230}Th,^{231}Pa	(1)
North Pacific:						
FanBd—20	40°15'N	128°27'W	4,500	3	K-Ar	(2)
Horizon	40°14'N	155°05'W	5,500	2.5	K-Ar,^{230}Th	(2,3)
V21—D2	35°54'N	160°19'W	5,400	4	^{230}Th,^{231}Pa, ^{234}U	(1,4)
V21—71a	27°54'N	162°31'E	5,870	2.5	^{230}Th	(1,5)
V21—D4b	14°25'N	145°52'W	4,618	3	^{230}Th,^{231}Pa	(1)
6A	19°39'N	113°44'W	4,000	4	^{230}Th,^{231}Pa	(1)
Carr 5	9°26.5'N	113°16.5'W	3,700	17—24	^{230}Th,^{234}U	(2)
MP 26	19°N	171°W	1,464	10	^{230}Th	(2)
Zetes—3D	40°16'N	171°20'E	3,000	0.8—2.3	^{10}Be,^{230}Th Fission track	(6,7)
Tripod—2D	20°45'N	112°47'W	3,000	4	^{10}Be,^{230}Th	(6,7)
Dodo—9D	18°16'N	161°50'W	5,500	2.1	^{230}Th, Fission track	(6,7)
Dodo—15—1	19°23'N	162°20'W	4,160	1.8	^{10}Be	(8)
Wah 24F—8	8°18'N	153°03'W	5,143	10(3 nods.)	^{230}Th	(9)
V21—116	19°34'N	134°30'E	5,826	28	Fission track	(10)
2P—52	9°57'N	137°47'W	4,930	7.3	^{230}Th	(11)
Sta. 3996	4°56'N	135°29'E	—	33	^{230}Th	(12)
Sta. 3782	23°55'N	173°39.9'E	—	40	^{230}Th	(12)
South Pacific:						
V18—D32	14°18'S	149°32'W	2,000	3.1	^{230}Th	(1)
V18—T119a	12°27'S	159°25'W	5,000	6	^{230}Th,^{231}Pa	(1)
V18—T119b	12°27'S	159°25'W	5,000	3	^{230}Th,^{231}Pa	(1)
DWHD—47	41°51'S	102°01'W	4,240	1—6	K-Ar,^{230}Th	(2)
DW72	21°31'S	85°14'W	920[4]	18	^{230}Th	(2)
2P—50	13°53'S	150°35'W	3,695	1	^{230}Th	(11)
TF—1, TF—2	13°52'S	150°35'W	3,623	1(2 nods.)	^{230}Th	(7)
E24—15	35°48'S	134°50'W	4,696	6.4	^{230}Th	(14)
D023G	near Tuamoto I.		1,600	3.5	^{230}Th	(16)

Sample	Latitude	Longitude	Depth (m)	Growth rate (mm/10^6 yr.)	Method	Reference[*2]
Pacific:						
J1	—	—	—	10.7—15.1	^{230}Th	(13)
H1d	—	—	—	6.8	^{230}Th	(13)
G1d	—	—	—	3.6—10.3	^{230}Th	(13)
Indian Ocean:						
V16—T19a	29°52'S	62°36'E	4,500	2.8	^{230}Th	(1)
V16—T19b	29°52'S	62°36'E	4,500	2.9	^{230}Th	(1)
V16—T19c	29°52'S	62°36'E	4,500	2.3	^{230}Th	(1)
M1	east of Madagascar		—	~0[*3]	^{230}Th	(13)
Dodo—66a	19°56'S	100°E	—	5.5	^{230}Th(^{226}Ra)	(15)
Antarctic:						
E17—36	55°S	95°W	4,700	3	^{230}Th,^{231}Pa	(1)
E5—4	60°02'S	67°15'W	3,475	4—19	^{230}Th	(14)

[*1] (1) Ku and Broecker (1969)
   (2) Barnes and Dymond (1967)
   (3) Goldberg (1963)
   (4) Ku and Broecker (1967)
   (5) Bender et al. (1966)
   (6) Krishnaswami et al. (1972)
   (7) Bhat et al. (1973)
   (8) Somayajulu (1967)
   (9) Somayajulu et al. (1971)
 (10) Shima and Okada (1968)
 (11) Krishnaswami and Lal (1972)
 (12) Nikolayev and Yefimova (1963)
 (13) Heye and Beiersdorf (1973)
 (14) Kraemer and Schornick (1974)
 (15) Bhandari et al. (1971)
 (16) Boulad et al. (1975)
[*2] Nodules from the Blake Plateau, hence not strictly pelagic.
[*3] Not presently growing.
[*4] Reported depth may be in error.

Fission track method

This relatively new method (Fleischer et al., 1965) utilizes the spontaneous fission of ^{238}U. The rather large energy released in the process causes the recoiled fission fragments to damage the crystal lattice of a U-containing mineral and leave behind them a zone of weakness, or "tracks". Upon suitable etching of the mineral surface, each such track becomes visible under the microscope. Although the half-life of this fission process is long ($\sim 10^{16}$ years), the fact that the record of each fission event is observable makes this method sensitive enough to determine ages as young as a few thousand years, depending on the uranium contents of the dated material. For instance, for a natural glass sample containing 10 ppm uranium, the youngest age datable by fission track counting will be about 30,000 years. For low uranium material (such as basic volcanics) of relatively young age, this technique can be tedious. It gives large counting errors due to very low track densities.

The fission track age obtained for a volcanic glass found in the centre of a North Pacific nodule (Shima and Ikada, 1968) is $(0.5 \pm 0.2) \cdot 10^6$ years, yielding an average growth rate of 28 mm/10^6 years (Table 8-1) for its 1.4 cm outer shell. Like the K—Ar method, this rate should be a lower limit. Aumento (1969) measured fission track and K—Ar ages for basalts from the Mid-Atlantic Ridge and estimated the accumulation rates of the ferro-manganese coatings of the basalts to be 1.6—4.1 mm/10^6 years. In a recent paper, Krishnaswami et al. (1972) indicated that the fission track data on the basaltic nuclei of two nodules are in essential agreement with the ages of the nodules deduced from ^{10}Be and ^{230}Th dating.

RADIOMETRIC DATING OF SHALLOW-WATER, CONTINENTAL MARGIN NODULES

Because of the proximity of continental sources for iron and manganese and the possible diagenetic remobilization of manganese within the sediment column (Lynn and Bonatti, 1965; Manheim, 1965), nodules from near-shore, shallow-water environments are thought to form more rapidly than their deep-ocean counterparts. Recent radiometric data based on uranium-series disequilibrium have shown the difference in accretion rates between the shallow- and deep-water nodules to be more than two orders of magnitude (Ku and Glasby, 1972). The principles of Ku and Glasby's age computation are somewhat different from those for the deep-sea cases mentioned above. As shown in Table 8-II, the major difference lies in the fact that, instead of having U-unsupported ^{230}Th and ^{231}Pa, in the shallow-water environment these two isotopes are deficient with respect to their uranium parents in newly formed nodule layers. With the passage of time, decay of the uranium

TABLE 8-II

Comparison of radiochemical data between continental-margin and deep-ocean nodules[*1]

Sample locality	Water depth (m)	Depth in nodule (mm)	$\dfrac{230\text{Th}[*2]}{234\text{U}}$	$\dfrac{231\text{Pa}[*2]}{235\text{U}}$	Max. age (10^3 yr.) calculated from	
					$\dfrac{230\text{Th}}{234\text{U}}$	$\dfrac{231\text{Pa}}{235\text{U}}$
Jervis Inlet, British Columbia	339	3.9—5.5	0.095±.006	0.19±.02	11±1	10±1
Northwest Indian Ocean	3794	0.55—0.85	2.42±.14	1.90±.19	—	—

[*1] Data from Ku and Glasby (1972).
[*2] Activity ratios.

will gradually replenish the ^{230}Th and ^{231}Pa reservoirs. The rate of replenishment will be such that half of the residual deficiency of ^{230}Th disappears each 75,200 years (for ^{231}Pa, each 34,300 years) until equilibrium (activity ratios ^{230}Th/^{234}U and ^{231}Pa/^{235}U both equal to 1.00) is established. The ages are then computed from the extent of the residual deficiency of ^{230}Th and ^{231}Pa, as in the dating of marine carbonates (Barnes et al., 1956), rather than from the residual excess of the two radionuclides as described previously and illustrated in Fig. 8-1.

The ages listed in Table 8-II (corresponding to a accretion rate of about 0.4 mm/10^3 years) are calculated by assuming that the newly formed nodule acts as a closed system. It may well be that not all the ^{230}Th and ^{231}Pa are produced in situ as assumed; some may be due to precipitation from the overlying waters and/or contamination from detrital minerals in the oxide layers. The ages obtained are, therefore, at best *maximum* values. Nodules from fresh-water lakes have also been dated radiometrically using the decay of unsupported Ra^{226} (Krishnaswami and Moore, 1973). These nodules show even higher accretion rates of the order of 2 mm/10^3 years.

EVIDENCE FROM NON-RADIOMETRIC TECHNIQUES

Evidence on nodule growth rates based on techniques other than the radiometric methods is relativley more abundant for shallow- than for deep-marine environments. In many cases, maximum age limits are placed by the Pleistocene sea-level rise. For example, Manheim (1965) cites the post-Wisconsin uplift in the central and northern Baltic Sea to place the maximum ages for the nodules occurring in the area. Grill et al. (1968a, b) suggest a maximum age of 12,000 years for the Jervis Inlet (British

Columbia) nodules, as these investigators consider the nodules too friable to withstand glacial transport. Similar conclusions are drawn for the Loch Fyne nodules in glaciated valleys of Firth of Clyde (Glasby, 1970). Extreme cases for very rapid accumulation are, for instance, the iron-oxide coating found on two ∿50-year old artillery shell fragments in the San Clemente Basin off southern California (Goldberg and Arrhenius, 1958) and the man-made steel objects encrusted with ferromanganese deposits collected by divers off Oahu, Hawaii (Andrews, 1972). The above-mentioned examples point to the high and variable growth rates (millimetres to tens of centimetres per thousand years) for the shallow-water deposits. This is not unexpected considering that iron and manganese accumulation should to a first approximation be related to the proximity of sources for these metals.

For the deep-sea nodules, one "field" observation has been the large quantity of sharks' teeth or whales' earbones forming nodule nuclei which are sometimes found at a single site. Such concentration of large vertebrate parts provides a qualitative evidence depicting slow deposition of manganese (Murray, 1876; Murray and Renard, 1891; Joly, 1908). Similar evidence is demonstrated by the presence of cosmic spherules in deep-sea manganese nodules (Murray and Renard, 1884; Finkelman, 1970). Recently there have been several more quantitative non-radiometric approaches. A few of these attempts is described briefly below.

Palaeomagnetism

Crecilius et al. (1973) have made magnetic inclination measurements on 25 nodules of both deep-sea and shallow-water origin. They found that the nodules contain natural remanent magnetization (NRM) of the same magnitude as deep-sea sediments. Of the selected seven large deep-sea nodules examined, three showed either no reversals or inconclusive results, four (two of these, Dodo 15 and Horizon, have been dated radiometrically, see Table 8-I) showed magnetic reversals in the deeper layers. Based on the assumption that the magnetization direction in the layers has not subsequently changed since their precipitation and that the reversal found has a minimum age of 0.69 m.y. (the Brunhes–Matuyama boundary), the observed NRM reversals indicate slow overall growth rates of the order of millimetres per million years, as depicted by the radioactive techniques. By contrast, these investigators found no magnetic reversals in three nodules from Jervis Inlet in British Columbia, indicating much more rapid growth of these shallow marine concretions (Grill et al., 1968a, b; Ku and Glasby, 1972).

As noted by Crecilius et al. (1973), the main limitations of the palaeomagnetic technique at present lie in the difficulty of slicing structurally intact layers thin enough to record only one reversal and yet large enough to give adequate signals above background noise of the magnetometer. Failure to observe reversals may sometimes be due to sampling material containing more than one reversal. In addition, the translation of the measured polarity changes into palaeomagnetic reversals is complicated by uncertainties as to the fixation of nodule position on the sea floor (Heye and Beiersdorf, 1973).

Perhaps one piece of exact information from the palaeomagnetic studies is that occurrences of magnetic reversals in a nodule, if established, do indicate that the nodule has grown during at least two magnetic periods, or it is at least 690,000 years old (Brunhes—Matuyama boundary). It is interesting to note that Crecilius et al. (1973) find 130° difference in the NRM direction between a 2—4 mm thick iron-manganese crust and its underlying basalt from the Explorer Seamount (49°N 131°W). This suggests that the Fe—Mn oxides were deposited at some considerable time after the basalt was extruded.

Racemization of isoleucine

Amino acids have been found to undergo racemization in nature and the rates of the reaction, especially those involving isoleucine, are slow enough to be useful in geochronology (Hare and Mitterer, 1967; Hare and Abelson, 1968). The racemization of L-isoleucine produced D-alloisoleucine and the two diastereoisomers are readily resolvable analytically. This renders the measurement of the extent of racemization of isoleucine relatively simple. In addition, the rate constants as a function of temperature can be empirically derived in the laboratory.

Several recent publications have dealt with the use of the ratio of alloisoleucine to isoleucine to date deep-sea sediments and fossil bones (Bada et al., 1970; Wehmiller and Hare, 1971; Bada, 1972). One of the investigators (Bada, 1972) reports that the racemization reaction in bone follows reversible first-order kinetics up to an allo/iso ratio of 1:1 (equilibrium ratio = 1.38) and that the measured allo/iso ratio in a shark's vertebra found in the core of a manganese nodule from Horizon Guyot (19°30'N 168°50'W) is 0.78. Using a half-life of 3.6 m.y. for the racemization reaction at 3°C (bottom water temperature near Horizon Guyot), an age of 8.7 m.y. is obtained for the vertebra core. This gives a minimum accumulation rate of 0.6—1.2 mm/10^6 years for the manganese crust. The ^{230}Th distribution of the manganese crust from the same nodule has also been measured and a rate of 1.2 mm/10^6 years derived (Bada, 1972).

The rate (or half-life) of racemization of isoleucine is strongly temperature dependent, and an uncertainty in the temperature history of the sample of ± 2°C would yield an error of ± 50% in the age estimate (Bada, 1972). However, temperature fluctuations of 2°C or more are seldom encountered in the deep ocean.

Palagonitization

Moore (1966) notes that the outer surface of the basalt lavas dredged from sea floor off the island of Hawaii commonly shows two thin rinds: an inner palagonite (hydrated basalt glass) layer and an outer layer of hydrous manganese oxide. The thickness of the two rinds is roughly proportional to each other. The palagonite forms by inward alteration of basaltic glass. During this alteration, Na, Ca and Mn are lost whereas K, Ti and Fe are gained by diffusional transport of these elements (Moore, 1966). The manganese lost provides a potential source for the manganese which is continually accumulating in the deep sea. The thickness of the palagonite and that of the manganese oxide rinds therefore both increase with age (Moore, 1966; Hekinian and Hoffert, 1975; Morgenstein and Riley, 1975). Using the accumulation rate of 3 mm/10^6 years (Bender et al., 1966) for the manganese oxide rinds, Moore (1966) estimated the ages of the lava samples; these are in general agreement with limits inferred from the available ^{14}C, K—Ar and magnetic reversal data on the Hawaiian volcanics.

From the "manganese dates" and thickness of the palagonite layers, rates of submarine palagonitization of the order of 2 mm/10^6 years are derived (fig. 4 of Moore, 1966). Although subaqueous hydration rate could be a function of composition and ambient temperature, similar rates have been obtained by Morgenstein (1969) from the measurement of palagonite thickness in other sediment and lava samples of known ages. More recently, Morgenstein (1973a) has applied the hydration-rind dating technique to assess the manganese accretion rates on the Waho shelf of Hawaii and found them to be 10—20 mm/10^6 years. Based on the fission-track age measurement on a basaltic rock from the Rift Valley in the Atlantic by Storzer and Selo (quoted in Hekinian and Hoffert, 1975), Hekinian and Hoffert (1975) estimate the accumulation rates of manganese and palagonite encrustation on the rock to be both of the order of 2—4 mm/10^6 years.

EVALUATION OF THE RESULTS

The results presented in Table 8-I show one salient point: the growth of deep-sea manganese nodules is a slow process. The rates of millimetres to centimetres per *million* years are distinctively much lower than the deposition rates of pelagic sediments which are of the order of millimetres to centimetres per *thousand* years (Ku et al., 1968). Of the 62 measurements of different isotopes on 46 nodules (Table 8-I), more than two-thirds give values in the range of 2—8 mm/10^6 years. Such slow accumulation apparently also applied to the formation of iron-manganese crust on deep-sea rocks. In addition to the manganese coatings on basaltic lava samples mentioned earlier, Scott et al. (1972b) report U-series rates of 11 mm/10^6

years for a manganese crust on a limestone from the Atlantis Fracture Zone and 1.2 mm/10^6 years for a crust on a volcanic ridge obliquely cutting the fracture zone. Moore (1973) shows rates of 1—2 mm/10^6 years for five manganese crusts collected from the Atlantic and Indian oceans, and also obtains a uranium-series age of 9,500 years for a branching coral dredged from the east flank of the Mid-Atlantic Ridge. It has a 0.1 mm thick crust of manganese indicating a minimum Mn accumulation rate of 10 mm/10^6 years. Kraemer and Schornick (1974) also report accumulation rates 5—16 mm/10^6 years for Mn—Fe crusts dredged from the South Pacific Ocean floor. Considering that all these results are derived from a variety of techniques and from different investigators, one has to be impressed by their apparent agreement. Available quantitative or semi-quantitative estimates based on non-radiometric methods certainly also support the low rates delineated by radioisotopic methods.

The age-dating tools discussed above do have some limitations in their applicability, however. Inherent to all the dating methods is the question of "closed system". Cherdyntsev et al. (1971), Lalou and Brichet (1972), and Lalou et al. (1973) have found excess ^{230}Th in the cores of several nodules. Cherdyntsev et al. (1971) also found depletion of ^{230}Th and ^{231}Pa in the volcanic cores of eight nodules. These investigators use the data to indicate young ages or rapid formation of nodules. However, the possibility exists that their specimens represent open-system conditions involving contamination and/or post-depositional migration of isotopes. Recent studies have shown that excess ^{230}Th may be associated with clay material infiltrated along fissures of nodule specimens (Ku et al., 1975). In addition, the excess ^{230}Th in the nodule nuclei (Lalou and Brichet, 1972; Lalou et al., 1973) appears to be a very limited feature at most. While these findings demand further detailed studies, they also point to the desirability of having ages measured by two or more independent methods.

The second limitation is that the "nucleus dating" methods (K—Ar, fission track, isoleucine racemization) give only maximum ages. They provide no information as to whether or not the nodules have grown continuously or whether they are presently growing. Thus, as mentioned before, determination of the age of the nucleus permits the computation of only the *minimum mean* rates of accretion. This limitation is potentially surmountable by dating the oxide layers directly. If it were possible to measure radioactivities continuously along the entire depth of the manganese layers, it would be possible to unravel the complete growth history of the nodule. Significant departures from exponential decay patterns would presumably signify changes in growth rate or growth hiatus in nodule growth. The chief limitation of this approach lies in the resolution of thin isochron layers as well as in analytical precision. Because of the knobby surface and the cusp-like growth structure common to deep-sea nodules, the example shown in Fig. 8-2 probably approaches a limiting case of resolution in the present status of the art. In this case, only those major changes in growth rates over periodicities of more than 100,000 years (corresponding to more than half a millimetre in thickness) would lead to significant departures from the logarithmic line, and hence be detectable.

The third limitation is that the uranium-series methods are only useful for the outer several millimetres of crust, representing at most one million years of nodule history. For layers deeper than 3—4 mm, the question of "continuity" of growth may be more difficult to assess, as the sensitivity of the ^{10}Be method has yet to be improved. In view of a great variety of microstructural and compositional variations within nodules (Sorem, 1973), there is reason to suspect that the growth of a nodule may be intermittent and not remain constant over hundreds of thousands of years. It should therefore be emphasized that the reported growth rates are *average* values taken over a million years (and are usually expressed in the unit of mm/10^6 years).

The observation that concordant results have been obtained for the same nodules using isotopes of different decay half-lives and chemical affinities is important in several aspects. First, it demonstrates that any *extensive* post-depositional remobilization of the radioisotopes involved in age-dating is unlikely to have taken place. Secondly, it may be taken to mean that for these nodules there has been no enormous changes in the rate of formation between the inner and the outer portions. Thirdly, it negates the possibility that the observed rapid drop-off of ^{230}Th and ^{231}Pa activities could be due to process(es) other than radioactive decay. One school of thought (Bonatti and Nayudu, 1965; Arrhenius, 1967; Cherdyntsev et al., 1971; Lalou and Brichet, 1972) suggests that nodules may accrete by a rapid succession of precipitation events, presumably due to submarine volcanic eruptions. This rapid accumulation results in iron—manganese layers with low ^{230}Th. The high surface ^{230}Th concentration is acquired by absorption during subsequent long exposure of the nodule to sea water (presumably only ^{230}Th, but not iron and manganese, is collected at that time). The observed ^{230}Th gradient could then be generated by inward diffusion of ^{230}Th (Arrhenius, 1967) or by the artificial "dilution effect" as a result of the penetration of surface material to greater depth during the layer-peeling analytical procedures. The "dilution effect" has been apparently minimized by the nuclear emulsion approach (Heye and Beiersdorf, 1973). Further-more, if the above suggestion is correct, then it would indeed be fortuitous to have radioisotopes of different decay constants (^{230}Th, ^{231}Pa, ^{234}U, ^{10}Be) penetrate to different depths and create various concentration gradients so as to give comparable apparent growth rates.

It should be pointed out that a main reason for advocating the above-mentioned fast-accumulation hypothesis is the concern that slow-accumulating nodules would inevitably be buried by the faster-depositing sediments surrounding them (Menard, 1964; Bonatti and Nayudu, 1965;

Goodell et al., 1971). However, one may note from the following argument that such a hypothesis still leaves the "burial" problem unsolved. From the oceanic uranium concentration of about 3 μg U/l, it can be estimated that under 1 cm^2 of sea floor (water depth 4 km), there will be about 950 dpm of unsupported ^{230}Th. Ku and Broecker (1969) have shown that the integrated amount of unsupported ^{230}Th in the 16 deep-sea nodules they studied averages about 220 dpm/cm^2. Thus, on the average, the so-called "exposure" time for the nodules to collect ^{230}Th from sea water would be about 28,000 years. During this time about 14 cm of sediments would be deposited (taking 5 mm/10^3 years as an average for sediment accumulation rate). These sediments would have no difficulty in covering the nodules, since the mean diameter of the 16 nodules is only about 4 cm. Some of the nodules have integrated ^{230}Th as high as over 800 dpm/cm^2 (Ku and Broecker, 1969). This would require over 200,000 years of exposure time and such nodules would have encountered about one metre of sediments raining from above during its lifetime.

IMPLICATIONS REGARDING NODULE GENESIS

The variability in the chemical and physical characteristics of marine ferromanganese deposits suggests that they may have multiple origins (Bonatti et al., 1972). The nodular aggregates and crusts on which attention has been focussed here are among the different forms of these deposits. The major mechanisms for their formation that have been proposed include inorganic (hydrogenous) precipitation of manganese from normal sea water (Kuenen, 1950; Goldberg and Arrhenius, 1958), precipitation from volcanic hydrothermal solutions (Murray and Renard, 1891; Bonatti and Nayudu, 1965), and diagenetic concentration of manganese at the seawater—sediment interface (Manheim, 1965; Lynn and Bonatti, 1965). Manganese growth rates associated with these three mechanisms are expected to be somewhat different. The nature of manganese accumulation for the submarine hydrothermal type may be fast and episodic, whereas for the hydrogenous precipitation it should be slow and continuous. The diagenetic deposits will have their manganese accumulated at rates faster than those of the hydrogenous type, and they are mostly to be found in hemipelagic regions where the deposition rate of organic matter is relatively high (Manheim, 1965; Bonatti et al., 1972).

The age-dating results summarized above indicate that the nodules and crusts from the deep sea that have been examined so far are mostly of the slowly accumulated hydrogenous precipitate type. Manganese-rich sediments derived from submarine hydrothermal activities have been found along oceanic ridges such as the East Pacific Rise (Skornyakova, 1965; Boström and Peterson, 1966) and the Red Sea Rift (Miller et al., 1966). Manganese

oxide crusts forming at fast rates of up to mm's/10^3 years as determined by
the uranium-series isotopes have also been reported recently as local
products of hydrothermal activities at oceanic ridge crests (Scott et al.,
1974; Moore and Vogt, 1975). The radiochemical characteristics of these
materials are markedly different from those of deep-sea nodules; in particular the hydrothermal deposits contain very low ^{232}Th and show defici-
encies of ^{230}Th and ^{231}Pa with respect to their uranium progenitors.
Indeed, radiochemical evidence has yet to be presented to show that the
debouching thermal fluids accrete rapidly to form extensive *nodular*
deposits. Even if the ultimate source of manganese may have come from the
ocean-ridge volcanism, a major portion of the metal must escape local
precipitation and be spread widely through the deep ocean.

Insights into the nature of manganese deposition in the ocean can be
gained by comparing the manganese accumulation rates in nodules with
those in sediments. Fig. 8-3 gives such a comparison for samples taken from
different parts of the deep sea. It is clear that manganese nodules grow not
because of their unusual capability to attract available manganese, but rather
because some mechanism prevent objects like sharks' teeth and rock fragments from being covered with sediments (Bender et al., 1970b; Somayajulu
et al., 1971; Kraemer and Schornick, 1974). Given a set of similar conditions

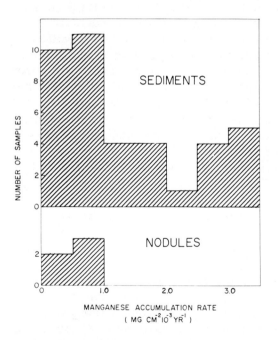

Fig. 8-3. Histogram showing rates of manganese incorporation into nodules and sediments (from Bender et. al., 1970b).

(bottom current, availability of potential nucleating surfaces, etc.), the growth of ferromanganese oxides will depend mainly upon the length of time the nucleus has been in contact with seawater. Thus, Moore (1966) finds no manganese coating on the young lava samples dredged from the active submarine rift zone of Kilauea volcano and Aumento (1969), Glasby (1972b) and Hekinian and Hoffert (1975) note an increase of manganese encrustation on rocks away from the axes of mid-ocean ridges.

Although the various dating methods have their limitations and interpretation of results is not without ambiguity, the present summary does reveal a certain degree of internal consistency in the age data obtained to date. This consistency should provide a boundary condition in the formulation of theories of nodule genesis. The question of whether there exist deep-sea nodules forming at rates orders of magnitude higher than the millimetre-per-million-years figures assembled here is, nonetheless, too important to dismiss without further investigation.

CHAPTER 9

THE FORM OF MANGANESE AND IRON IN MARINE SEDIMENTS

H. ELDERFIELD

INTRODUCTION

The distribution and concentrations of manganese and iron in marine sediments reflects the results of lithogenous, hydrogenous, biogenous, volcanogenic and diagenetic processes operative in the oceans. In certain regions of the sea floor it appears that one process dominates the others in its control over the deposition of manganese and iron whereas elsewhere the geochemistry of the sediment may reflect a complex interplay of several processes. In regions of submarine volcanic activity it is possible that sea water will become locally enriched with iron and manganese and when these elements are removed from solution a metalliferous deposit is produced. Similarly, diagenetic processes can release manganese to the sea-water—sediment interface where its precipitation results in a metal-rich sediment. A component common to both deposits is the fraction of manganese, iron and minor metals which is removed from the overlying water column (in solution and as lithogenous and biogenous particles) by normal sedimentary processes. Hence, it is important to consider sedimentary models for the accumulation of metals from normal sea water since it is upon this background that volcanogenic and diagenetic processes are imposed to produce sediments rich in manganese and/or iron. In addition, these background sedimentary processes can themselves produce metal-rich deposits in certain oceanic regions, and so it is necessary to attempt an evaluation of the conditions under which they are formed so as to interpret more quantitatively the geochemistry of manganese- and iron-rich deposits in regions where casual association would seem to indicate one specific origin. Following this, the roles of submarine volcanism and diagenesis in supplying metals to the upper layers of marine sediments can be viewed in terms of their modification of the sedimentary patterns produced by the deposition of manganese and iron from sea water and its suspended load. The observation of broad-scale compositional patterns in the geochemistry of sediments from the major oceans affords a basis upon which specific models for the accumulation of manganese and iron can be tested. Since regional variations in the geochemistry of Atlantic and Pacific sediments are known in more detail than those from the other oceans a consideration of inter- and intra-oceanic variations in sediment composition will, initially, be restricted to these two areas.

COMPOSITIONAL VARIATIONS IN OCEANIC SEDIMENTS

The distribution of manganese in Atlantic deep-sea sediments (Turekian and Imbrie, 1966) is characterized by relatively high concentrations of manganese in mid-ocean regions, roughly coincident with the Mid-Atlantic Ridge, with decreasing concentrations towards the continents. The region to the west of the ridge at approx. $20°N$ is also rich in manganese. In addition, certain other elements, including copper, cobalt, nickel, zinc and lead, are also enriched in mid-Atlantic sediments as compared with those nearer shore and as compared with shallow-water sediments (Table 9-I). For example, the distribution pattern for cobalt is similar to that for manganese and the belt of cobalt-rich sediments is strongly associated with the region of the mid-Atlantic. Implicit in these observations is the assumption that the elements are associated with the non-carbonate fraction of the sediment with the biogenous component acting merely as a diluent, and for this reason metal contents are expressed on a carbonate-free basis. Bruty et al. (1972) have shown that zinc in the carbonate fraction has some influence on the distribution pattern for this element but that the same general contrast is maintained between deep-sea and nearer-shore regions with or without correction for this factor. Trace element analysis of biogenous carbonate (Chester, in Riley and Chester, 1971) shows that this is true for some other enriched elements. The distribution pattern for iron in the Atlantic is not generally sympathetic with that for manganese. The data of Chester and Messiha-Hanna (1970) indicate some decrease from mid-ocean to nearer-shore regions in parts of the North Atlantic although in the region around $15°N$ there is a latitudinal belt of sediment with a relatively constant iron content stretching from Africa to and across the mid-ocean ridge.

The regional distribution of manganese in Pacific deep-sea sediments (Skornyakova, 1965) shows two important features: (1) there is a marked contrast between the lower manganese concentrations in North Pacific sediments and the higher values in the South Pacific; (2) in the South Pacific, manganese is particularly enriched in sediments overlying the region of the East Pacific Rise. This second feature is also apparent in the distribution pattern for iron but, unlike manganese, the overall pattern shows two latitudinal belts of high iron content, in the North and South Pacific, separated by a belt of lower iron in the region of the Equator. In common with Atlantic sediments, deep-sea sediments from the Pacific are enriched in manganese and certain minor metals as compared with the shallow-water sediment average (Table 9-I). However, Pacific deep-sea sediments are also enriched in these metals as compared with Atlantic deep-sea sediments with, on average, three times more cobalt, manganese and nickel in Pacific sediments.

The sediments associated with the East Pacific Rise are atypical of Pacific deep-sea sediments, containing significantly more manganese and iron and

TABLE 9-I

Fe, Mn and minor metals in some marine sediments

	Near-shore sediments*[1]	Deep-sea sediments		
		Atlantic*[2]	Pacific*[3]	E. Pacific Rise*[4]
Fe	4.83	5.74	6.50	18.0
Mn	850	3,980	12,500	60,000
Cu	48	115	570	730
Co	13	39	116	105
Ni	55	79	293	430
Pb	20	52	162	180
Zn	95	130	200	380
Cr	100	86	130	55
V	130	140	130	450

[1] Data from Wedepohl (1960).
[2] Data from Wedepohl (1960) and Turekian and Imbrie (1966).
[3] Data from Goldberg and Arrhenius (1958), Wedepohl (1960), El Wakeel and Riley (1961) and Landergren (1964).
[4] Data from Boström and Peterson (1969) and Dasch et al. (1971).
Fe in wt.%; other metals in ppm.

with a characteristic trace element assemblage (Table 9-I). Sediments from the crest of the Rise are enriched in elements including copper, arsenic, chromium, zinc, cadmium, barium and vanadium as compared with sediments adjacent to the Rise. In common with Atlantic deep-sea sediments and other Pacific sediments, cobalt and nickel are enriched in East Pacific Rise sediments as compared with shallow-water sediments. However, cobalt and nickel are depleted in sediments from the crest of the Rise as compared with those from the flanks of the Rise. This pattern is significantly different from the overall distribution of these elements in the Atlantic where they decrease away from the Ridge.

THE ACCUMULATION OF MANGANESE AND IRON FROM THE GEOSPHERE

It can be seen from the compositional variations described above that certain regions of the sea floor favour the deposition of sediments rich in manganese and/or iron. Attempts to explain these and similar distribution patterns throughout the world's oceans, and to explain the formation of discrete ferromanganese phases, are now generally based on polygenetic models where factors such as hydrogenous precipitation from sea water, diagenetic remobilization within the sediment column, halmyrolysis of submarine volcanic rocks, and submarine hydrothermal or geothermal activity may, in various combinations, play a part in controlling the overall geochemistry of these deposits. Bonatti et al. (1972) have used these four mechanisms as the basis for a classification of submarine ferromanganese deposits, and suggest sub-divisions into hydrothermal—hydrogenous, hydrogenous—diagenetic etc. where the elements in a particular deposit are supplied from different sources. The successful use of classifications such as this requires an estimate of the supply of manganese and iron from the various geosphere components and a knowledge of the mechanisms by which these elements are deposited on the sea floor. The following sections describe the various geosphere sources of manganese and iron, and the models and mechanisms which have been proposed to explain their accumulation in marine sediments.

Lithogenous material

The various sedimentary mechanisms that have been proposed to explain metal accumulation in marine sediments are based largely on the association of an element with the lithogenous and/or hydrogenous components of the sediment and differences in the rates of accumulation of each component or fractionation of the component in the oceans. The *clay association model* (Turekian, 1965) assumes that the element is simply associated with one major component of the sediment. For deep-sea sediments having a clay/carbonate matrix the biogenous carbonate component acts as a diluent for many of the metals concentrated in the lithogenous clay component, and so the concentration of the element in the carbonate-free sediment simply equals that in the lithogenous component. Since compositional variations are observed in the carbonate-free fraction of sediments it is evident that this model cannot be used to represent the input of manganese etc. to deep-sea areas, and although the model has been shown to apply for the vertical distribution of minor elements at some specific locations (Turekian, 1965, 1967) a mechanism where the elements are partitioned between at least two non-carbonate fractions of the sediment must be invoked to explain regional variations in sediment chemistry. The *differential transport model* (Turekian, 1967) involves such a partition.

The differential transport model assumes that the element is associated with one major component of the sediment (the lithogenous component) and that this component is fractionated during transport in the oceans. This is based on the relative importance of two basic mechanisms of sediment transport in different oceanic regions: (1) bottom transport and deposition; (2) eupelagic deposition; and the different compositions of the sediment carried by each transport mechanism. The lithogenous material that is deposited in pelagic sediments is derived from the upper layers of the water column and so this eupelagic component is likely to be dominant in regions remote from continents, particularly on topographic highs such as the mid-ocean ridge system. In contrast, deep-sea regions adjacent to continents will contain more of the bottom-transported component. If the eupelagic component contains more fine-grained detritus rich in manganese, cobalt, nickel etc. than the bottom-transported component then the enrichment of these metals in mid-oceanic sediments compared with those nearer the continents (Table 9-I) is explained by the differential transport of the former component to mid-ocean regions. The fine-grained particles are considered to be composed of manganese and iron oxides from weathering profiles and clays and other aluminosilicate debris which has sorbed the enriched elements during stream transport. To explain differences between the average composition of Pacific and Atlantic deep-sea sediments (Table 9-I) Turekian argues that a greater proportion of the Atlantic than Pacific ocean floor is composed of near-continental abyssal plain areas where the sediment is dominated by the bottom-transported component.

This theory is supported by a voluminous literature describing the sorption of minor elements on clay minerals in rivers and on freshly-precipitated hydrous oxyhydroxides of manganese and iron, and if this material is differentially transported to mid-ocean areas then the observed distribution patterns may be explained. In addition, Chester and Messiha-Hanna (1970) have shown that the composition of deep-sea sediments, when stripped of sorbed trace elements and ferromanganese oxides by a technique of selective chemical extraction, is similar to that of near-shore sediments (Table 9-II), and so the metal enrichment observed in deep-sea areas can be

TABLE 9-II

Metal contents of the lithogenous material in marine sediments*

	Near-shore sediments	Lithogenous fraction of deep-sea sediments	
		Atlantic	Pacific
Fe	4.8	6.8	7.4
Mn	850	582	743
Cu	48	67	212
Co	13	12	16
Ni	55	63	46
Cr	100	72	91
V	130	120	92

*Based on Chester and Messiha-Hanna (1970); Fe in wt.%; other metals in ppm.

regarded in terms of a significant partition of the enriched elements (manganese, cobalt, nickel etc.) between the lithogenous fraction of the sediment (having a composition similar to near-shore sediments) and the non-lithogenous fraction of the sediment (comprising the sorbed and coprecipitated metals in fine-grained river detritus). In contrast, the metals which are not enriched in deep-sea areas (e.g. chromium and vanadium) are strongly associated with the sediments' lithogenous fraction. Some details of this model have been criticized. Chester (in Riley and Chester, 1971) argues that since it is the pelagic sediments which contain significant amounts of the metal-rich non-lithogenous material then it is difficult to understand why Pacific pelagic clays have a higher content of trace elements than Atlantic pelagic clays. Also, clay minerals presumably must remain deflocculated in order to reach deep-sea areas and hence retain a surface charge and with this the ability to sorb trace elements from sea water so that the differential transport model is, at least, modified by further reactions in the marine environment (Elderfield, 1976a). These criticisms may, to some extent, be minimized (e.g., by arguing that since Turekian (1967) suggests extremely fine particles, $<0.5\mu$m, as carriers of the critical trace elements then compositional differences between pelagic sediments (70% $<2\mu$m) may be a result of different proportions of $<0.5\mu$m metal-rich sediment in different regions; by arguing that the surface charge of pelagic clays ultimately is neutralized by major cations from sea water and not trace elements) but one factor which is an inherent assumption in the differential transport model gives rise to reservations that this model can, alone, explain the geochemistry of manganese etc. in deep-sea sediments. This assumption is that the metal enrichment originally occurs in the riverine environment and that enrichment by removal of elements dissolved in sea water is of little importance. If this is the case then can it be assumed that manganese, cobalt and nickel, for example, are maintained in steady state in the oceans by removal only in manganese nodules and is it simply a coincidence that these same metals are enriched in deep-sea sediments as compared with those nearer shore?

The great difficulty in assessing the differential transport model lies in its treatment of the size of the eupelagically deposited detritus and, ultimately, in our incomplete knowledge of the speciation of manganese and minor metals in sea water. If the stream-supplied, metal-rich, particles are considered to be of colloidal dimensions, say 0.1μm, then the metals become part of the fraction of sea water which is normally defined as being soluble (passing through a filter with a nominal pore size of 0.5μm) and the criticism made above becomes invalid. (See Chapter 10 of this volume for a discussion of the form of manganese and iron in sea water.) If the eupelagic component exists in sea water as extremely fine particles it may be more appropriate to consider this model, along with the models described in the following section, as a theory for the precipitation of "soluble" manganese from sea water rather than for the deposition of lithogenous material.

Sea water

The accumulation of manganese in marine sediments as a result of direct precipitation from sea water was proposed by Wedepohl (1960). In the models described above the metals are assumed to be entirely associated with the lithogenous component of the sediment. In the *trace-element veil model* the elements are partitioned between the lithogenous and hydrogenous components. The model was originally developed as a means of obtaining broad-scale relative accumulation rates of sediments from their trace-element contents and was popularized by Wedepohl who used it as a means of explaining the enrichment of manganese and other metals in Pacific deep-sea clays relative to their Atlantic counterparts. Turekian (1965, 1967) further developed and quantified the model and critically applied it to temporal as well as regional variations in sediment composition. The method assumes that the source of the enriched elements is sea water (Wedepohl considered that they are originally of volcanic origin but this is irrelevant to the application of the model) and that the inter-oceanic variation is caused by the constant supply of these elements to the sediment imposed on the differing clay accumulation rates in the Atlantic and Pacific oceans. Hence, sediments with low rates of clay accumulation have high contents of manganese etc. and vice versa, and so the model also explains the high content of these elements in deep-sea sediments as compared with nearer-shore sediments. The important assumption in the model is that the accumulation rates of the enriched elements from sea water are constant over the world's ocean and, consequently, their residence times must be sufficiently long to prevent localized precipitation taking place.

Turekian (1967) has criticized this model for failing to specify a mechanism for the removal of the metals from sea water. Turekian doubts that clay minerals can act as efficient scavengers because trace-element levels are similar in river and seawater and, hence, the sorptive capacity of clays would be reached during stream supply. Also, biological processes are considered unlikely to concentrate the metals in question. One possible mechanism which has been suggested (Elderfield, 1972a) involves the precipitation of a manganese oxide phase from sea water which scavenges trace elements and so deposits manganese, cobalt, nickel etc. in marine sediments as a "manganese oxide component" compositionally similar to manganese nodules. The "manganese oxide component" may, in fact, be discrete micronodules which have been included in analysed sediment samples or manganese oxides which have precipitated on clays, skeletal debris or other fine-grained material in the sediment. This suggestion is based on measurements of manganese/element ratios in the non-lithogenous fraction of North Atlantic deep-sea sediments which for cobalt, nickel and copper, are similar to those for deep-sea manganese nodules and differ from manganese/element ratios of North Atlatntic near-shore sediments which, for some elements, are similar to those for near-shore nodules (Table 9-III). An additional criticism Turekian (1967) has made of Wedepohl's conception of the trace element veil model is that a constant, independent removal of metals from sea water to marine sediments implies a homogenous distribution of trace elements in sea water. This is not always observed and there is some evidence (Schutz and Turekian, 1965) of a low trace-element concentration barrier between the Atlantic and Pacific oceans. If the removal mechanism involves the precipitation of a manganese oxide component compositionally similar to manganese nodules then the constraints of sea water chemistry described by Turekian apply both to the nodules and sediment, and whilst differences in trace-element concentrations of sea waters obviously are important in relation to rates of metal precipitation, other factors, such as oxygen advection, catalysis, and redox reactions, also are important, particularly for slowly accumulating deposits. Interpretations based solely on distribution coefficients can be misleading. For example, the concentrations of the rare-earth elements in the iron-rich sediments overlying the East Pacific Rise (a rapidly accumulating deposit) show an almost constant partition between sea water and sediment with a distribution coefficient of about 10^6 (Bender et al. 1971) whereas in deep-sea manganese nodules (a slowly accumulating deposit) this simple partition is modified by redox reactions at the sediment—water interface (Glasby, 1973a) resulting in cerium enrichment, and hydrolysis parameters of the rare-earth ions resulting in a more extensive fractionation of the lighter than heavier elements (Ronov et al., 1967).

Chester et al. (1973) have tested the suggestion that the concentration of trace elements by manganese nodules and the enrichment of these elements in slowly accumulating sediments are related processes by comparing the concentrations of manganese, cobalt and nickel in deep-sea clays from the southwest region of the North Atlantic with manganese nodules from this region. Their results (Table 9-IV) show that certain of the sediments (Group A in Table 9-IV) have non-lithogenous contents of these metals ($\sim$ 8% Mn, $\sim$ 0.13% Ni, $\sim$ 0.09% Co) which begin to approximate to the composition of the nodules (see also Glasby, 1975). These studies are consistent with the model for the evolution of deep-sea manganese nodules proposed by Bender et al. (1966) which involves a uniform "rain" of manganese over the entire ocean floor. Bender et al. (1966, 1970a, b, 1971) have shown that the accumulation rate of manganese in deep-sea sediments in more or less constant over the world's oceans, except for the East Pacific Rise, and is similar to that in deep-sea nodules (Table 9-V). Hence, they conclude that no special factors are required for deep-sea nodule deposition and the partition of manganese between the sediment and nodules is linked to the availability of nucleating agents for manganese precipitation. In the North Pacific there is some indication that manganese accumulates in sediments at a rate proportioned to the clay accumulation rate (Bender et al., 1970b). This

TABLE 9-III

Manganese/element ratios for Mn nodules and the non-lithogenous fraction of sediments from deep-sea and near-shore areas
(From Elderfield, 1972a)

	Deep-sea		Near-shore	
	sediment	nodule	sediment	nodule
Cu	33	47	3	1,740
Co	79	68	840	860
Ni	41	33	50	240
Zn	—	47—470	10	31—1,750
Pb	—	190	22	2,700
Cr	47	19,000	6	1,750

TABLE 9-IV

Mn, Co and Ni in manganese nodules and the non-lithogenous fraction of deep-sea clays from the northwest Atlantic Ocean*

		Mn	Co	Ni
Clays	Group A	8.08	0.09	0.13
	Group B	0.72	0.01	0.02
Manganese nodules		11.1—18.9	0.24—0.79	0.16—0.59

*Based on Chester et al., 1973; values in wt.%.

TABLE 9-V

Rates of Mn accumulation in the deep sea*

Mn nodules	0.2—1.0
Pelagic sediments	0.1—3.4
East Pacific Rise sediment	35

*Based on Bender et al., 1970b, 1971; units are 10^{-6} g cm^{-2} per year.

suggests that clay minerals or a related phase may act as nucleating agents for manganese precipitation in the same way as volcanic fragments etc. accumulate manganese in nodules (or that manganese is introduced into the N. Pacific in association with clay minerals) and, as might be expected, no simple model can explain the details of the world-wide deposition of manganese in slowly accumulating sediments. However, the present evidence clearly suggests that compositional variations in deep-sea sediments are related to the removal of metals in the $< 0.5\mu m$ fraction of the overlying water column, and the essentials of the model described above are similar to the trace-element veil model as a simplified representation of the accumulation of manganese and associated minor metals in oxic marine sediments. This model is equally (with the differential transport model) consistent with the results of Chester and Messiha-Hanna (1970; see Table 9-II) in that the constancy of composition of the lithogenous fraction of Atlantic sediments is an expression of the material upon which the "manganese oxide component" is imposed with higher concentrations of this component in deep-sea sediments than in those deposited nearer the continents.

Biogenous material

The role of biogenous material in the marine geochemical cycling of manganese and iron is not understood. Nor is the contribution made by such material to the Mn and Fe contents of marine sediments. The former cannot be estimated until accurate measurements are available of these metals in sea water which can be compared with the distributions of nutrient elements. Some estimate of the contribution of biogenous matter to sediment composition can be made from the levels of manganese and iron in marine organisms (Table 9-VI). It is clear that a significant fraction of the iron in marine sediments overlying regions of high biological productivity may have a direct biogenous origin. For example, opaline silica is accumulating at up to 150 mg cm^{-2} per 1,000 year in the sediments at the north of the East Pacific Rise and if this has an iron content of 3,500 ppm (Table 9-VI) then the accumulation ratio of biogenous Fe (~ 0.5 mg cm^{-2} per 1,000 year) may contribute up to about 15% of the sedimentary iron flux in this region. According to the data in Table 9-VI the manganese content of biogenous material should not contribute significantly to gross compositional variations in the geochemistry of marine sediments.

THE ROLE OF SUBMARINE VOLCANISM

Having described the deposition of manganese, iron and some associated minor metals in marine sediments in terms of their removal from the overlying water column by lithogenous, hydrogenous and biogenous

TABLE 9-VI

Iron and manganese in marine organisms

	Fe	Mn
Dry plant tissues:		
Plankton	3,500	75
Algae	690	53
Dry animal tissues:		
Molluscs	200	10
Crustacea	20	2
Fish	30	0.8
Hard tissues:		
Algae ($CaCO_3$)	3,400	55
Forams ($CaCO_3$)	11,000	—
Forams (SiO_2)	300	8
Corals ($CaCO_3$)	1,500	6.3
Molluscs	900	4.8

*Data from Bowen (1966); units are ppm.

processes it is necessary to consider the role played by submarine volcanism in modifying this simple interpretation. In this context the effect of submarine volcanism is important for two reasons: (1) in regions of submarine volcanic activity metal-rich solutions may be added to sea water and the oxidation of iron and manganese from these solutions will result in the localized deposition of metalliferous sediment, as a direct result of submarine volcanism, having a more rapid rate of accumulation than the sediments adjacent to the metalliferous deposit: (2) together with stream supply, submarine volcanism may act as an important source of certain metals in "normal" sea water which ultimately are precipitated in a slowly accumulating manganese oxide component of the sediment. In this case, the role of submarine volcanism is indirect, with manganese etc. of volcanic and non-volcanic origins being deposited simultaneously, at the same rate, and by similar mechanisms. An assessment of these direct and indirect roles of submarine volcanism in controlling the deposition of manganese and iron in marine sediments requires a knowledge of the geochemistry of metalliferous sediments associated with volcanic features on the sea floor and of the mechanisms for the transfer of elements from submarine basaltic magmas to sea water.

Volcanogenic sediments

Submarine volcanic activity may result in the formations of ferromanganese deposits at the site of the activity. Metalliferous sediments have been found at several, often extensive, sites along the mid-ocean ridge system (Table 9-VII) and the rapid accumulation rates of these deposits has led to suggestions that they are of local (volcanic) origin. Their geochemistry differs significantly from that of the slowly accumulating deposits found in deep-sea areas being composed dominantly of cryptocrystalline iron oxides and Fe smectites with a characteristic assemblage of associated metals (including manganese, copper, chromium, boron, arsenic and mercury) present in much higher concentrations than in normal deep-sea sediments. In the "active ridge sediments" overlying the East Pacific Rise (described earlier) iron and manganese are accumulating at rates on average one order of magnitude higher (and up to 200 times higher) than in surrounding areas of the Pacific (Tables 9-VII). Copper, cobalt and nickel also are accumulating more rapidly than in normal deep-sea sediments but cobalt and nickel, unlike iron, manganese and copper, tend to accumulate at a higher rate in the slowly-accumulating sediments adjacent to the Rise than in the more rapidly accumulating sediments on the Rise crest. Isotopic analysis of East Pacific Rise sediments (Bender et al., 1971; Dasch et al., 1971) suggests that the strontium and uranium present is derived mainly from sea water whereas the lead is mostly of local volcanic origin. In addition to active ridge deposits, where a volcanic origin is suggested by geochemical studies of the sediments, direct evidence that volcanic exhalative processes can give rise to iron and manganese deposits is seen at certain sites where thermal solutions locally enrich natural waters with iron, manganese and certain minor metals (Table 9-VIII).

Supply mechanisms

The various mechanisms that have been suggested for the supply of metals to sea water by submarine volcanism range from general invocations of hypothetical ascending solutions related to magmatic processes to specific models of chemical exchange between basalt and sea water. There is general agreement that some type of "hydrothermal solution" is introduced to sea water and that metal enrichment in sediments is produced by a combination of direct precipitation of metals from this mineralizing solution and coprecipitation and sorption of metals present (at normal levels) in sea water by these precipitates. If the hydrothermal solution is introduced to an oxic sea floor an iron- and manganese- oxide deposit is produced, whereas if the overlying water column is anoxic iron sulphides are formed and manganese precipitation is minimal. The supply mechanism described below ranges from primary, high-temperature volatilization to low-temperature submarine weathering.

TABLE 9-VII

The geochemistry of some metalliferous marine sediments in regions of submarine volcanism

	Fe	Mn	Cu	Co	Ni	Zn	Pb	As	Mo	V	B	Hg
Pacific Ocean: East Pacific Rise[1]												
Aver. concentr.	18.0	6.0	730	105	430	380		145	30	450	500	
Max. concentr.	22.6	8.8	860	96	520	290		400	50	650	690	
Max. accum. rate $(10^{-6}\,g\,cm^{-2}\,yr^{-1})$	410	150	1.48	0.17	0.90	0.50		0.69	0.09	1.12	1.18	
Atlantic Ocean: Mid-ocean ridge 45°N[2]												
Aver. concentr.	8.0	0.41						174				0.414
Max. concentr.	14.1	0.47						361				0.549
Indian Ocean: Red Sea[3]												
(a)	25.9	1.6										
Aver. concentr. (b)	45.0	0.9	1,800	45	25	8,300	155	130	90	90		
(c)	17.0	0.9										
(d)	21.4	27.3										

[1] Data from Boström and Peterson (1969), Bender et al. (1971) and Dasch et al. (1971).
[2] Data from Cronan (1972b).
[3] Data from Bischoff (1969) and Hendricks et al. (1969).

(a) = Fe montmorillonite facies; (b) = amorphous — goethite facies; (c) = sulphide facies; (d) = manganite facies.

Fe and Mn in wt. %; other metals in ppm.

TABLE 9-VIII

The compositions of natural waters in volcanic regions

		Fe	Mn	Si	Zn	As	Pb	Cu
1. Hokkaido, Japan	Okunoju	17	1.03	—	0.45	0.23	—	—
	Ojunuma	44.8	0.5	—	2.1	0.31	0.3	—
2. Deception Is., Antarctica	crater lakes	0.31	2.4	50.7	—	—	—	—
	sea water	0.18	0.13	7.8	—	—	—	—
3. Red Sea brine		81	82	27.6	5.4	—	0.63	0.26
4. Geyser areas associated with volcanism		0.9	2.5	—	—	—	0.02	0.02
5. Acid SO₄ springs associated with volcanism		—	—	—	0.14	1.9	—	0.03

Data from: 1. Wauschkuhn (1973); 2. Elderfield (1972b) — max. concentrations listed;
3. Craig (1969); 4 and 5. White et al. (1963).
Units are mg/l.

Magmatic volatilization

The suggestion that manganese in pelagic sediments originates from submarine hydrothermal springs was made by Gümbel (1878). In their study of the sediments overlying the East Pacific Rise Boström and Peterson (1969) suggested a deep-seated origin for the enriched metals, and Boström and co-workers (e.g., Boström, 1967; Boström et al., 1969, 1971, 1973) have hypothesized that manganese and the other metals which are enriched in "active-ridge sediments" have, at least partly, a juvenile origin and are excess volatiles from the upper mantle. These suggestions are based in part on geophysical evidence (e.g., high heat-flow values on ocean ridges), on the recognition of anomalous ^{3}He in the deep sea (Clarke et al., 1969), on the "well-known relation between basic magmatism and ores of Cu, Ni and Co" (Boström et al., 1973), on the recognition of copper- and iron-sulphides in mid-ocean magmatic rocks (Baturin, 1971; Dmitriev et al., 1971; Moore and Calk, 1971), and because many of the enriched elements are excess volatiles as determined by Horn and Adams' (1966) geochemical balance calculations (e.g., B, As, Cd, Hg) or have very large ionic radii (e.g., Ba, U). Boström et al. (1971) conclude that the volatile elements and those with incompatible ionic radii may be released into hydrothermal solutions during the differentiation of pyrolite into peridotite and basalt. The major problems in assessing theories based on mantle degassing are that no satisfactory means are available of analysing volatiles from cooling magmas, and in the possibility that the hydrothermal solutions are contaminated by, or even formed from, the rocks or sediments through which they travel. The speculation that the metals enriched in a sediment associated with a volcanic feature are derived from some deep-seated volcanic source does little to improve our understanding of sediment genesis. The speculation is usually possible because little is known of the deep-seated source and of the mechanisms by which its volatiles may be delivered to the sediments. Krauskopf (1967) has criticized this approach to "explaining" ore deposition as unsatisfactory since "correct though it may be . . . it merely transfers the area of ignorance from the ore deposit to a magma . . . and our information about magmas is so very limited that we can neither check the hypothesis nor use it to make predictions. Building hypotheses in this manner is a harmless pastime, but it contributes little to our knowledge of ore deposition."

One approach would be to calculate the amount and type of metals that can be transported in a magmatic gas mixture the composition of which is deduced from analyses of volcanic gases. This could be attempted by comparing the maximum volatilities of various metals in equilibrium with the most stable solid phases in a high-temperature gas mixture consisting largely of water vapour, HCl, HF and sulphur gases. However, a large range of values is generated for most metals reflecting the temperatures chosen and the partial pressures of oxygen which are deduced from the stable minerals found in association with intrusive rocks, and there are many limitations to this treatment (e.g., volatility but not solubility is considered, perfect-gas behaviour is assumed, some of the chemical data is unreliable). A rough "volatility series" obtained using this method is: Hg>Sb>As>Bi>Pb>Sn>Mn>Fe>Cd>Zn>Co>Mo>Cu>Ni>Ag>Au, which suggests that if iron and manganese are transported as volatiles in a high-temperature magmatic gas then transport is also possible for most of the metals enriched in active-ridge sediments. However, the amount that can be transported falls off dramatically with decreasing temperature and is unimportant for most metals below $400°C$ (Barnes and Czamanske, 1967). Since the temperature is known at which metal deposition occurs in marine sediments, it is evident that our knowledge of volatilities is useful only in showing that: (1) iron and manganese can, theoretically, be separated from a magma on the basis of volatility and can be transported in significant amounts ($> 10^{-5}$ atm) at high temperatures; (2) some gross fractionation may occur of metals that can be transported with iron and manganese (e.g., As and Pb), and those that cannot (e.g., Ag and Au). The mechanisms that operate to transport metals of magmatic origin in hydrothermal solutions through the intermediate and lower temperature ranges are common to metals in hydrothermal solutions formed at these temperatures by other volcanogenic processes and, as such, are irrelevant to an assessment of the magmatic contribution of manganese and iron to marine sediments.

Metamorphism

The most widely accepted theories for the sources of hydrothermal solutions advocate reactions which may be loosely classified under Krauskopf's (1967) definition of metamorphic processes: "processes which grade into sedimentary reactions at one end of the temperature scale and into igneous reactions at the other". One such reaction (Boström, 1967) involves the magmatic emanations described above not as the source for manganese and iron but as a means of their extraction from basaltic rocks. Using estimates by Jagger (1940) and Rubey (1951) for the composition of gaseous emanations, Boström has shown that even small fractions of the CO_2 and HCl therein can release significant amounts of manganese from lavas. Assuming that 0.5% of the HCl and CO_2 that is added to the oceans by degassing in association with its present volume of water dissolves manganese according to the equations:

$$MnSiO_3 + 2HCl \rightarrow Mn^{2+} + 2Cl^- + SiO_2 + H_2O$$
$$MnSiO_3 + CO_2 \rightarrow Mn^{2+} + CO_3{}^{2-} + SiO_2$$

then about $10^{11}-10^{12}$ g Mn can be added to sea water yearly. If, then, submarine basaltic rocks are envisaged as a sourcy for manganese, it is pertinent to consider the composition of hydrothermal solutions and their ability to extract metals from igneous rocks. In the majority of cases where a hydrous phase transports ore constituents this hydrothermal solution is an alkali—chloride brine (Helgeson, 1964; White, 1968) containing elements in similar proportions to sea water. The possibility that sea water acts to leach metals from lavas is strengthened by considerations of heat flow on oceanic ridges (e.g., Elder, 1965; Deffeyes, 1970; Bonatti et al., 1972) which indicate that convective hydrothermal circulation is required to explain measured heat flow patterns. Since the ocean floor is free from thick sediment in this region and, because of sea-floor spreading, newly-formed oceanic crust can react freely with sea water for about $80 \cdot 10^6$ years it is not unreasonable suppose that some reaction is possible. The computed permeability in layer II of the oceanic crust is consistent with thermal convective flow (Deffeyes, 1970; Lister, 1972) and, since percolation of sea water through cracks, fissures, fractures, joints etc. of ocean-floor lava is probable, there is good reason to believe that leaching may occur to considerable depths within the crust. The rates of decrease of seismic velocities and natural remanent magnetism in layer II with distance from the ridge crest suggest that a layer of basalt about 0.6 km thick is exposed to hydration by sea water and oxidation of ferromagnetic minerals (Cox et al., 1972; Hart, 1973a, b).

One specific mechanism involving basalt—sea-water interaction (Corliss, 1971) has gained some popularity as a model for the origin of metal-bearing submarine hydrothermal solutions. Corliss has observed that the slowly cooled holocrystalline interiors of submarine basalt flows are depleted in several metals, including manganese and iron, as compared with the quenched flow margins. These elements are presumed to be fractionated into residual silica-rich solutions during crystallization of the lava and are ultimately located as residual solid phases along intergranular boundaries. Additional elements are mobilized during the formation of sulphide phases in the siliceous residue and by deuteric alteration of olivine. The interaction of sea water with these residual phases, initially at near-solidus temperatures, mobilizes manganese, iron etc. as chloride complexes to form the hydro-thermal solution which eventually is introduced to the sea floor as a submarine hot spring. Corliss estimated that 10,000 ppm iron and 120 ppm manganese may be released from flow interiors by this mechanism.

If water convection exists beneath the sea floor where the geothermal heat source is present below a thick sediment cover, circulation of heated water allows the leaching of metals from the sediment. This process occurs in the Red Sea where manganese and iron are extracted from evaporites, shales etc. and introduced into axial deeps in a brine with little evidence of a magmatic source for these and other elements (Bishoff, 1969; Craig, 1969).

Additional evidence for subsea-floor metamorphism comes from what can be termed a "reverse uniformitarianism" approach. The recognition of the similarity between metamorphosed ophiolite sequences and mid-ocean ridge metabasic rocks has led to studies (e.g., Spooner and Fyfe, 1973) of ophiolitic rocks which have been used to describe models of chemical mass transfer involving sea water and ocean crust. The evidence from these studies

is that temperatures up to 400°C were developed, at depths ∿ 300 m below the basalt—sea-water interface and that metamorphism took place in a subsea-floor geothermal system where redox conditions allowed leaching of Mn, Fe and other transition metals to form a metalliferous brine which is discharged into sea water to form a metal-rich active ridge sediment. These speculations gain considerable support from recent experimental studies of basalt—sea-water interaction carried out in this laboratory (Table 9-IX) which have demonstrated that manganese, iron, copper and zinc are readily leached at 500 bar and 190°C, conditions which simulate reasonably those of ocean-floor metamorphism.

TABLE 9-IX

Chemical mass transfer during an experimental simulation of basalt—sea-water interaction at 500 bar, 190°C
(Analyst P. Williams)

Time of run (h)	Element in sea water (ppm)				
	Fe	Mn	Cu	Zn	Ni
0	<.05	<.05	0.04	0.07	0.27
50	15.6	4.55	2.03	0.43	0.27
75	25.1	4.65	1.35	0.71	0.30
100	25.0	6.55	1.77	0.58	0.27
150	27.9	—	1.15	0.51	0.25
300	27.4	13.3	0.84	0.58	0.27
Net exchange	+27.4	+13.3	+0.80	+0.51	0

Halmyrolysis

Some early speculations on the role played by low-temperature submarine weathering of submarine basalts as a source of manganese and iron were hampered by the assumption that halmyrolytic leaching takes place only at the sea-water—sediment interface. However, the geophysical evidence of alteration of layer II material (described above) supports the hypothesis that a significant proportion of layer II is a zone of chemical reaction between sea water and ocean crust. If the decreases observed in seismic velocities of layer II with distancs from the mid-ocean ridge are caused by a progressively thickening layer of altered basalt then low-temperature weathering reactions assume some real significance. The reactions proposed by Corliss (1971) might be expected to affect chemical exchange between lavas and sea water when basaltic rocks are erupted onto the ocean floor and during their early exposure to sea water but not when they are 85 million year old.

With the general acceptance of the concept of sea-floor spreading it has been possible to link observations that the major element chemistry of basalts varies systematically with distance from ocean ridges (e.g., McBirney and Gass, 1967) also to the increasing age of the basaltic rocks as oceanic crust moves away from constructive plate margins. One of the more consistently reported trends in basalt chemistry (summarized in Elderfield, 1976a) is of their increasing water content due to the alteration of ferromagnesian minerals to hydrous aluminosilicates. Unfortunately, data on other chemical species are often conflicting because of deuteric and metamorphic alteration of some of the samples studied. The most thorough assessment of chemical exchange between basalt and sea water is that by Hart (1970, 1973a, b) who has combined chemical data on weathering trends with the geophysical evidence described earlier to construct a model for large-scale chemical reactions between sea water and the upper 2—3 km of ocean crust. The overall mass balance for this reaction (Hart, 1973a) is, in units of 10^{14} g/yr.:

187 Fresh Basalt + $0.60K^+$ + $0.82Mg^{2+}$ + $2.07Na^+$ + $7.36H_2O$

$\rightarrow$182 Altered Basalt + $8.06Ca^{2+}$ + $6.90SiO_2$ + $0.5Fe^{2+}$ + $0.079Mn^{2+}$

Hart's model includes an assessment of the roles played by primary and retrograde greenschist metamorphism but these processes have no effect on the mass balance for manganese and no net effect on the balance for iron. Thus, the model predicts that halmyrolysis is responsible for an annual input to sea water of $5 \cdot 10^{13}$ g Fe and $0.8 \cdot 10^{12}$ g Mn. Unfortunately, these values conflict with the results of Hart's (1970) study which indicate a net loss of manganese and iron from sea water. Hart is more confident of the results of the later (1973a) study which includes data from highly altered samples which should emphasize trends obscured using less altered samples, and is consistent with the results of Melson and Thompson (1972) whose comparison of metals in clay mineral products of basalt alteration with the initial glass indicates a loss of manganese to sea water. Similar evidence by Bonatti et al. (1972) points to a depletion in the altered rim of a basaltic glass of more than 800 ppm of the 1,600 ppm manganese in the unaltered nucleus.

THE ROLE OF DIAGENESIS

The role of diagenesis in redistributing manganese in the sediment column is now well documented (e.g., Manheim, 1965; Lynn and Bonatti, 1965; Li et al., 1969; Calvert and Price, 1972). The instability of tetravalent manganese in solid phases under reducing conditions is predicted by simple thermodynamic criteria and the presence on continental margins and marginal sea floors of large volumes of anoxic sediment overlain by a thin oxic layer supports the contention that the surface enrichment of manganese in such regions is caused by the upward migration of Mn^{2+} ions due to a diffusive gradient into the oxic zone where precipitation occurs. Concentration profiles of manganese in the pore waters of sediments (e.g., Li et al., 1969; Bischoff and Ku, 1971) show that such gradients exist (and, in addition, infer the presence of diagenetic Mn carbonate below the oxic-anoxic boundary in the sediment column) and application of advection—diffusion—reaction models to these data (e.g., Anikouchine, 1967; Michard, 1971) may be made to estimate the diagenetic flux. Since the majority of deep-sea sediments are either oxic throughout or have large units of oxic sediment above the anoxic zone it is important to assess the role of diagenesis as a mechanism for manganese enrichment at the sediment surface in such regions. Bender (1971) has done this by using a simple diffusion model to place an upper limit on the diagenetic contribution to the high manganese concentrations found in deep-sea sediments. His conclusion is that manganese cannot diffuse through the large units of oxic sediment found in deep-sea areas rapidly enough to contribute significantly to the observed accumulation rates of Mn at the water—sediment interface.

Thus, a generalization would be that the diagenetic remobilization of manganese over significant thicknesses of sediment column is quantitatively important in near-shore regions but not in deep-sea regions. However, it is incorrect to assume that a sharp contrast exists between the redox characteristics of continental margin sediments and those in adjacent deep-sea areas. For this reason a major problem still exists in our understanding of this diagenetic process in that no accurate estimate is available of the removal rate of Mn from pore water both in the anoxic zone (presumably as Mn carbonate) and more importantly, in the oxic zone (as hydrous Mn IV oxides). This estimation is presently hindered by the lack of precise analytical data for pore waters of oxic sediments and by our incomplete understanding of the kinetics of manganese precipitation in marine sediments. While there may be some justification in assuming that the removal of manganese by rhodochrosite precipitation can be represented by first-order kinetics, the kinetics of oxidation of Mn^{2+} ions is likely to require more detailed experimental study than is presently documented before the diagenetic manganese flux can be quantitatively assessed.

CONCLUSIONS

The previous sections in this chapter have centred on discussions of the sources of manganese and iron in marine sediments and of the mechanisms by which they are added to and redistributed within the sediment column. These discussions have emphasized the precipitation of authigenic phases of

Mn and Fe in oxic marine environments and have neglected to consider the formation of iron sulphides and the detailed characteristics of lithogenous iron-bearing minerals on the grounds that these topics are largely irrelevant to the main theme of this volume. In addition, emphasis has been placed on the geochemistry of manganese and iron rather than their mineralogical form in oxic sediments.

On the basis of the literature which is reviewed here it is obvious that one can make simple statements attributing the Mn and Fe in any marine sediment to one or more of the sources described earlier because of some environmental association, and this approach has many proponents. However, it is more constructive to determine the relative importance of the principal mechanisms by which authigenic manganese and iron phases are supplied to the sediment surface, an exercise which requires the knowledge of more than the metal contents of marine sediments. In the section describing the accumulation of Mn and Fe as geosphere components it was suggested that a simple mechanism is one of manganese precipitation from

TABLE 9-X

Oceanic fluxes of manganese
(From Elderfield, 1976b)

Input	stream supply	0.2	
	submarine volcanism $\begin{cases} \text{magmatic leaching} \\ \text{deutric alteration} \\ \text{halmyrolysis} \end{cases}$	$\begin{matrix} 0.1{-}1.9 \\ 1.1 \\ 0.8 \end{matrix}$	$2.2{-}4.0{\cdot}10^{-12}$ g yr.$^{-1}$
Output	sedimentation	calculated from: 1. comparison of Atlantic and Pacific sedimentation rates (4.1) 2. accumulation rate of non-lithogenous manganese by (a) partition geochemistry data (1.2). (b) shale correction method (7.4). (c) Mn nodule accumulation (6.1). 3. Horn and Adams' (1966) geochemical balance (0.7)	$0.7{-}7.4{\cdot}10^{-12}$ g yr.$^{-1}$ $(0.2{-}2.1{\cdot}10^{-6}$ g cm^{-2} yr.$^{-1}$
Redistribution	in anoxic and 'mildly reducing' sediments		$\sim 330{-}0.1{\cdot}10^{-6}$ g cm^{-2} yr.$^{-1}$
	in oxic sediments		$\sim {<}10^{-12}?{\cdot}1.0^{-6}$ g cm^{-2} yr.$^{-1}$

sea water imposed upon its association with lithogenous material. This can be represented by: $\Sigma F_{Mn} = F_{Mn} + [Mn]_L F_L$, where the total flux of manganese being added to the sediment surface is equated with the flux of manganese from solution (F_{Mn}) and the lithogeneous flux caused by lithogenous material having a manganese concentration of $[Mn]_L$ sedimentary at a rate F_L. Since $[Mn]_L$ is constant for most deep-sea sediments (Chester and Messiha—Hanna, 1970) and if (for the reasons described earlier) F_{Mn} is more or less constant over large areas of the deep-sea then this equation may be used as the basis for simple budget calculations to assess whether the magnitude of the flux from solution is consistent with stream supply being the major source of manganese in the surface layers of deep-sea sediments. Similarly, the advection—diffusion—reaction models (described earlier) may be used to assess the diagenetic flux and its role in controlling the geochemistry of deep-sea and marginal sediments. There is considerable justification in using simplified theoretical models to describe complex natural systems and, in addition, suitable boundary conditions may be inserted into equations describing such models to aid the production of simple mathematical solutions. The problem of assessing the volcanogenic flux is more difficult. Evidence of a physical association between a volcanic feature on the sea floor and a deposit rich in manganese and iron must be regarded as casual rather than causal in the absence of corroborative data. Measurements of a more rapid accumulation of Mn and Fe in the sediment where such an association exists than where it does not taken together with isotopic data of volcanically supplied metals found in association with FeMn-rich sediments (e.g., Bender et al., 1971) are important evidence of the existence of a volcanogenic flux. However, an assessment of its magnitude requires studies of chemical exchange between basalt and sea water (as described earlier) together with careful determination of the sedimentary processes operative in ocean-ridge environments upon which volcanic processes are imposed. In a paper produced concurrently with this review (Elderfield, 1976 b) I have considered some of the problems described here and have attempted to provide some estimates of the fluxes of manganese to the oceans. Table 9-X provides a summary of these estimates. It is clear that sedimentary and diagenetic processes exert an important influence on the manganese content in the majority of marine sediments. However, submarine volcanism, is, by implication, an extremely significant source of manganese in the oceans although it appears that the majority of Mn in marine sediments is not of direct volcanic origin. As diagenesis acts as a secondary process to redistribute manganese added to marine sediments from all sources, so authigenesis acts as a secondary process to redistribute manganese added to sea water by submarine volcanism as well as by stream supply.

CHAPTER 10

MECHANISMS OF REMOVAL OF MANGANESE, IRON AND OTHER
TRACE METALS FROM SEA WATER

J. W. MURRAY and P. G. BREWER

INTRODUCTION

Most trace metals pass relatively quickly through the oceans; that is their residence time is short compared to the major ions or water itself. Sea water contains a complex mixture of organisms and particulate matter, and maintains at least superficial contact with the sediments on a time scale of 1,000 years or less. The varied mechanisms by which metals are removed from sea water reflects this heterogeneity, and it seems probable that no single mechanism is universally dominant. In this chapter the principal mechanisms which have been proposed for the removal of trace metals from sea water are reviewed, and, where the data permit, an attempt is made to quantify their importance.

A comparison of residence times for trace metals in sea water calculated using metal input and removal fluxes gives good agreement (Goldberg and Arrhenius, 1958) suggesting that trace metals are at steady state with respect to a balance between processes adding and removing them from sea water. If no other removal mechanism were operating, the concentration of a metal in sea water would be regulated by the solubility of its least-soluble compound. A comparison of the concentration of Co, Ni, Cu, Zn and Ba in sea water with the solubility of their least-soluble phases is shown in Table 10-I. All of these metals are present in sea water in much lower concentrations than would be predicted by these solubility calculations. (It is worth noting, however, that the sea water concentrations of Cu, Zn and Ba are within a factor of 25 of the predicted solubility concentration and solid solution formation may lower the effective solubility by this amount (Bodine et al., 1965; Stumm and Morgan, 1970, p. 207)). The failure of trace-metal concentrations to be limited by solubility considerations is not a problem of supply. Goldschmidt (1937) noted that the amount of metal weathered from the continents during geologic time far exceeds the quantity present in sea water (see also Rankama and Sahama, 1962, pp. 294—296; Krauskopf, 1956). In order to explain this discrepancy he proposed that metals are removed from sea water by adsorption, particularly on iron-oxide precipitates. Krauskopf (1956) pointed out that besides precipitation and adsorption other processes such as precipitation as sulphides and removal associated with organic matter must also be considered.

TABLE 10-I

A comparison of the metal concentration of sea water with the solubility of least-soluble solids

Metal	Concentration in sea water (gl^{-1})	Solubility concentration* (gl^{-1})	Solid
Co	$.004-0.4 \cdot 10^{-6}$ (a)	$0.5 \cdot 10^{-3}$	$CoCO_3$
Ni	$0.1-2.7 \cdot 10^{-6}$ (b)	$658 \cdot 10^{-3}$	$Ni(OH)_2$
Cu	$0.1-4.5 \cdot 10^{-6}$ (b)	$21.6 \cdot 10^{-6}$	CuO
Zn	$1-10 \cdot 10^{-6}$ (b)	$28.8 \cdot 10^{-6}$	$ZnCO_3$
Ba	$2.6-37 \cdot 10^{-6}$ (d)	$49 \cdot 10^{-6}$ (c)	$BaSO_4$

*Solubility calculations for Co, Ni, Cu and Zn were made using solubility and stability constants from Sillén and Martell (1964).
$pK_{so}(CoCO_3) = 9.63$; $pK_{so}(Ni(OH)_2) = +15.21$; $pK_{so}(ZnCO_3) = +10.84$
$pH = 8$, $CO_3^{2-} = 2 \cdot 10^{-5} ml^{-1}$
$\gamma_{Me^{+2}} = 0.12$ Latimer (1952)
$\gamma_{OH^-} = 0.68$ Burns (1965)
$\gamma_{CO_3^{2-}} = 0.20$ Garrels and Thompson (1962)
(a) Robertson (1970); (b) Spencer and Brewer (1970); (c) Church and Wolgemuth (1972); (d) Andersen and Hume (1968.)

Veeh (1967) and Bertine and Turekian (1972) have proposed that a large fraction of the removal of molybdenum and uranium from sea water may occur in reducing sediments. Although the exact mechanism for the observed enrichment is not clear, sulphide formation is a strong possibility. Veeh (1967) has found a high removal rate of uranium in reducing sediments and calculated that if only 0.4% of marine sediments were reducing they would remove uranium at a rate equal to that supplied to the ocean by streams. Bertine and Turekian (1973) have arrived at a similar value for their molybdenum budget. Both of these estimates are in agreement with the percentage of the ocean (0.3%) that lies under water of high organic productivity plus the area of stagnant basins. For most trace metals the formation of sulphides as a removal mechanism appears less important. Most metals are more evenly distributed throughout marine sediments. Furthermore, most sulphides found in marine sediments are of diagenetic origin. By far the chief source of iron for pyrite formation in most sediments are detrital iron minerals (Berner, 1971). Thus, sulphide formation acts on metals that have already been removed from sea water by other mechanisms. Removal of metals as sulphides can only be important locally and could not account for manganese nodules as they are typically found in oxidizing environments.

The removal of trace elements associated with organic matter may also be important in certain locations. Calvert and Price (1970b) found Cu and Ni concentrations up to 129 ppm and 455 ppm, respectively, in sediments from the southwest African shelf that contained up to 40% organic matter. However, the average deep-sea sediment has only 0.5% organic carbon and metals associated with organic matter, with the possible exception of copper, should make a negligible contribution to the total metal present. If chelation by organic matter were a major influence on the trace-metal content of sediments and nodules, the metal enrichment relative to sea water should reflect the Irving-Williams order:

$$Mn < Fe < Co < Ni < Cu > Zn$$

(The Irving-Williams order is a well-documented sequence of the relative stability of transition metals with organic chelators — see Irving and Williams, 1953; Orgel, 1966, p. 87). Goldberg (1965) has pointed out that the relative concentration factors of metals in marine organisms closely parallel the Irving-Williams order. This suggests that selectivity by marine organisms is initiated by complexing reactions taking place in the body tissues or on their exposed external surfaces. For comparison, the enrichment of metals relative to sea water is shown in Table 10-II for Pacific and Atlantic deep-sea nodules, Pacific seamount nodules, and red-clay sediments. The enrichment relative to sea water for the nodules follows the order:

$$Co \geqslant Mn > Fe > Ni > Cu > Zn$$

and for deep-sea clays:

$$Fe > Mn > Co > Cu \geqslant Ni > Zn$$

These observed orders of enrichment are different from the Irving-Williams order; however, there are other biological pathways besides simple chelation with organic matter that may be important for removal of trace elements from sea water.

MINOR ELEMENTS IN SEA WATER

Estimates of removal processes for minor elements from sea water must be preceded by some information on the dissolved metal concentrations. Unfortunately the data in this field are scant and erratic. Present estimates of the molar concentrations, and presumed chemical species, of the chemical elements in sea water are shown in Table 10-III. These data should be used with caution. Sampling problems for analysis of trace metals in sea water are severe. Robertson (1968) has shown that sea water samples stored in glass or

TABLE 10-II

Enrichment factors

Element	Average sea-water concentration[1]	Pacific deep-sea nodules[2]	Atlantic deep-sea nodules[2]	Pacific seamount nodules[3]	Deep-sea clays[4]
Mg	1.294 (g kg^{-1})	13.0	13.0	11.9	
Ca	0.413 (g kg^{-1})	46.0	65.5	69.0	
Sr	0.008 (g kg^{-1})	$1.01 \cdot 10^2$	$1.12 \cdot 10^2$	$1.60 \cdot 10^2$	22.5
Ba	40 (μg kg^{-1})	$4.5 \cdot 10^4$	$4.25 \cdot 10^4$	$7.22 \cdot 10^4$	$5.75 \cdot 10^4$
Mn	2 (μg kg^{-1})	$1.21 \cdot 10^8$	$0.85 \cdot 10^8$	$0.85 \cdot 10^8$	$3.3 \cdot 16^6$
Fe	10 (μg kg^{-1})	$1.40 \cdot 10^7$	$1.75 \cdot 10^7$	$1.53 \cdot 10^7$	$6.5 \cdot 10^6$
Co	.03 (μg kg^{-1})	$1.17 \cdot 10^8$	$1.03 \cdot 10^8$	$1.42 \cdot 10^8$	$2.1 \cdot 10^6$
Ni	2 (μg kg^{-1})	$4.95 \cdot 10^6$	$2.10 \cdot 10^6$	$1.68 \cdot 10^6$	$1.1 \cdot 10^5$
Cu	2 (μg kg^{-1})	$2.65 \cdot 10^6$	$1.00 \cdot 10^6$	$0.42 \cdot 10^6$	$1.2 \cdot 10^5$
Zn	5 (μg kg^{-1})	$9.4 \cdot 10^4$	—	—	$3.3 \cdot 10^4$

[1] From Spencer and Brewer (1970).
[2] From Mero, quoted in Arrhenius (1963).
[3] From Goldberg (1965), Mero (1965a), Cronan (1967, 1969), and Menard et al. (1964).
[4] From Turekian and Wedepohl (1961).

polyethylene bottles at their natural pH rapidly lose In, Sc, Fe, Ag, U and Co by adsorption on to the walls of the container. Acidified samples stored in polyethylene containers showed minimum loss. Contamination of the sample by the sampler is a frequent source of error. Schutz and Turekian (1965) carried out a major study of the trace-element concentrations of the world's oceans; however, samples were taken with galvanized barrels, epoxy-lined barrels, and some surface samples with the pump used for cooling the hydrowinch. These techniques are clearly unsuitable for certain elements.

The analytical techniques used are not always adequate for the precise determination of very low concentrations. The sea water trace-element intercalibration study, carried out by Brewer and Spencer (1970), revealed large differences between the 26 participating laboratories; only five elements were determined with a coefficient of variation of $< 10\%$, and none of these five were transition metals.

In spite of these problems the data listed in Table 10-III do show general trends, and even though detailed vertical profiles of most elements are missing, conclusions from the abundance can be made.

TABLE 10-III

The abundance of the chemical elements in sea water and estimates of their residence time

Element	Chemical species	Total concentration (molar)	Residence time* (years)
H	H_2O	55	—
He	He(gas)	$1.7 \cdot 10^{-9}$	—
Li	Li^+	$2.6 \cdot 10^{-5}$	$2.3 \cdot 10^6$
Be	$BeOH^+$	$6.3 \cdot 10^{-11}$	—
B	$B(OH)_3, B(OH)_4^-$	$4.1 \cdot 10^{-4}$	$1.8 \cdot 10^7$
C	HCO_3^-, CO_3^{2-}, CO_2	$2.3 \cdot 10^{-3}$	—
N	$N_2, NO_3^-, NO_2^-, NH_4^+$	$1.07 \cdot 10^{-2}$	—
O	H_2O, O_2	55	—
F	F^-, MgF^+	$6.8 \cdot 10^{-5}$	$5.2 \cdot 10^5$
Ne	Ne(gas)	$5.9 \cdot 10^{-9}$	—
Na	Na^+	$4.8 \cdot 10^{-1}$	$6.8 \cdot 10^7$
Mg	Mg^{2+}	$5.3 \cdot 10^{-2}$	$1.2 \cdot 10^7$
Al	$Al(OH)_4^-$	$7.4 \cdot 10^{-8}$	$1.0 \cdot 10^2$
Si	$Si(OH)_4$	$7.1 \cdot 10^{-5}$	$1.8 \cdot 10^4$
P	$HPO_4^{2-}, PO_4^{3-}, H_2PO_4^-$	$2 \cdot 10^{-6}$	$1.8 \cdot 10^5$
S	$SO_4^{2-}, NaSO_4^-$	$2.8 \cdot 10^{-2}$	—
Cl	Cl^-	$5.5 \cdot 10^{-1}$	$1 \cdot 10^8$
Ar	Ar(gas)	$1.1 \cdot 10^{-5}$	—
K	K^+	$9.9 \cdot 10^{-3}$	$7 \cdot 10^6$
Ca	Ca^{2+}	$1 \cdot 10^{-2}$	$1 \cdot 10^6$
Sc	$Sc(OH)_3^0$	$1.3 \cdot 10^{-11}$	$4 \cdot 10^4$
Ti	$Ti(OH)_4^0$	$2 \cdot 10^{-8}$	$1.3 \cdot 10^4$
V	$H_2VO_4^-, HVO_4^{2-}$	$5 \cdot 10^{-8}$	$8 \cdot 10^4$
Cr	$Cr(OH)_3, CrO_4^{2-}$	$5.7 \cdot 10^{-9}$	$6 \cdot 10^3$
Mn	$Mn^{2+}, MnCL^+$	$3.6 \cdot 10^{-9}$	$1 \cdot 10^4$
Fe	$Fe(OH)_2^+, Fe(OH)_4^-$	$3.5 \cdot 10^{-8}$	$2 \cdot 10^2$
Co	Co^{2+}	$8 \cdot 10^{-10}$	$3 \cdot 10^4$
Ni	Ni^{2+}	$2.8 \cdot 10^{-8}$	$9 \cdot 10^4$
Cu	$CuCO_3, CuOH^+$	$8 \cdot 10^{-9}$	$2 \cdot 10^4$
Zn	$ZnOH^+, Zn^{2+}, ZnCO_3^0$	$7.6 \cdot 10^{-8}$	$2 \cdot 10^4$
Ga	$Ga(OH)_4^-$	$4.3 \cdot 10^{-10}$	$1 \cdot 10^4$
Ge	$Ge(OH)_4$	$6.9 \cdot 10^{-10}$	—
As	$HAsO_4^{2-}, H_2AsO_4^-$	$5 \cdot 10^{-8}$	$5 \cdot 10^4$
Se	SeO_3^{2-}, Se^0	$2.5 \cdot 10^{-9}$	$2 \cdot 10^4$
Br	Br^-	$8.4 \cdot 10^{-4}$	$1 \cdot 10^8$

TABLE 10-III (continued)

Element	Chemical species	Total concentration (molar)	Residence time* (years)
Kr	Kr (g)	$2.4 \cdot 10^{-9}$	—
Rb	Rb^+	$1.4 \cdot 10^{-6}$	$4 \cdot 10^6$
Sr	Sr^{2+}	$9.1 \cdot 10^{-5}$	$4 \cdot 10^6$
Y	$Y(OH)_3{}^0$	$1.5 \cdot 10^{-10}$	—
Zr	$Zr(OH)_4{}^0$	$3.3 \cdot 10^{-10}$	—
Nb		$1 \cdot 10^{-10}$	—
Mo	$MoO_4{}^{2-}$	$1 \cdot 10^{-7}$	$2 \cdot 10^5$
Tc			—
Ru			—
Rh			—
Pd			—
Ag	$AgCl_2{}^-$	$4 \cdot 10^{-10}$	$4 \cdot 10^4$
Cd	$CdCl_2$	$1 \cdot 10^{-9}$	—
In			—
Sn	$SnO(OH)_3{}^-$	$8.4 \cdot 10^{-11}$	—
Sb	$Sb(OH)_6{}^-$	$2 \cdot 10^{-9}$	$7 \cdot 10^3$
Te	$HTeO_3{}^-$		—
I	$IO_3{}^-, I^-$	$5 \cdot 10^{-7}$	$4 \cdot 10^5$
Xe	Xe(gas)	$3.8 \cdot 10^{-10}$	—
Cs	Cs^+	$3 \cdot 10^{-9}$	$6 \cdot 10^5$
Ba	Ba^{2+}	$1.5 \cdot 10^{-7}$	$4 \cdot 10^4$
La	$La(OH)_3{}^0$	$2 \cdot 10^{-11}$	$6 \cdot 10^2$
Ce	$Ce(OH)_3{}^0$	$1 \cdot 10^{-10}$	—
Pr	$Pr(OH)_3{}^0$	$4 \cdot 10^{-10}$	—
Nd	$Nd(OH)_3{}^0$	$1.5 \cdot 10^{-11}$	—
Pm	$Pm(OH)_3{}^0$		—
Sm	$Sm(OH)_3{}^0$	$3 \cdot 10^{-12}$	—
Eu	$Eu(OH)_3{}^0$	$6 \cdot 10^{-13}$	—
Gd	$Gd(OH)_3{}^0$	$4 \cdot 10^{-12}$	—
Tb	$Tb(OH)_3{}^0$	$9 \cdot 10^{-13}$	—
Dy	$Dy(OH)_3{}^0$	$6 \cdot 10^{-13}$	—
Ho	$Ho(OH)_3{}^0$	$1 \cdot 10^{-12}$	—
Er	$Er(OH)_3{}^0$	$4 \cdot 10^{-12}$	—
Tm	$Tm(OH)_3{}^0$	$8 \cdot 10^{-13}$	—
Yb	$Yb(OH)_3{}^0$	$3 \cdot 10^{-12}$	—
Lu	$Lu(OH)_3{}^0$		—
Hf		$4 \cdot 10^{-11}$	—
Ta		$1 \cdot 10^{-11}$	—

Element	Chemical species	Total concentration (molar)	Residence time* (years)
W	$WO_4{}^{2-}$	$5 \cdot 10^{-10}$	$1.2 \cdot 10^5$
Re	$ReO_4{}^-$	$4 \cdot 10^{-11}$	—
Os			—
Ir			—
Pt			—
Au	$AuCl_2{}^-$	$2 \cdot 10^{-10}$	$2 \cdot 10^5$
Hg	$HgCl_4{}^{2-}, HgCl_2{}^0$	$1.5 \cdot 10^{-10}$	$8 \cdot 10^4$
Tl	Tl^+	$5 \cdot 10^{-11}$	—
Pb	$PbCO_3, Pb(CO_3)_2{}^{2-}$	$1 \cdot 10^{-10}$	$4 \cdot 10^2$
Bi	$BiO^+, Bi(OH)_2{}^+$	$1 \cdot 10^{-10}$	—
Po	$Po^0, PoO_3{}^{2-}$		—
At			—
Rn	$Rn(gas)$	$2.7 \cdot 10^{-21}$	—
Fr			—
Ra	Ra^{2+}	$3 \cdot 10^{-16}$	—
Ac			—
Th	$Th(OH)_4{}^0$	$4 \cdot 10^{-11}$	$2 \cdot 10^2$
Pa		$2 \cdot 10^{-16}$	—
U	$UO_2(CO_3)_2{}^{4-}$	$1.2 \cdot 10^{-8}$	$3 \cdot 10^6$

*Most recent estimate; data taken principally from Goldberg et al. (1971).

BIOLOGICAL REMOVAL

Certain trace metals are known to be associated, in significant concentrations, with the bulk plankton assemblage (e.g., Vinogradov, 1953; Goldberg, 1965). Estimates of the degree to which minor elements are removed from sea water by plankton are usually obtained from the calculation of enrichment factors. Enrichment factors may be defined as the ratio of the concentration of an element in an organism to that concentration directly available from an organism's environment, e.g., for phytoplankton, directly from sea water. The following elements have been found to be accumulated by phytoplankton, by a factor of at least 10^3, from sea water: Al, As, Ba, Be, Cd, Ce, Cr, Co, Cu, I, Fe, Pb, Mn, Ni, Nb, Pu, Sc, Ag, Zn, Zr (Bowen et al., 1971). The picture is complicated by grazing of the phytoplankton crop by zooplankton, and excretion of the unassimilated material. Kuenzler (1965) has estimated that $> 90\%$ of the flux of iron through the thermocline occurs through sinking of faecal pellets and dead organisms. However, the simple observation that an element is concentrated by plankton does not ensure marked non-conservative behaviour in sea water, nor does it automatically predict that biological transport is quantitatively important in its sedimentary accumulation.

In spite of this large amount of data on trace metals in marine organisms, the role of organisms in transporting metals to the sediments is difficult to evaluate. The exact fraction of the metals that are removed from the surface water and released to the deep water due to breakdown of protoplasm and redissolution of calcareous and siliceous tests is unknown. Broecker (1971), Hurd (1973) and Heath (1974) have estimated that between 90 and 99% of the biogenic opal produced in surface waters dissolves before it is preserved in the sediment column. Similar calculations by Li et al. (1969) indicate that roughly 15% of the $CaCO_3$ produced accumulates in the sediments. These calculations show that an enormous amount of recycling of marine biogenic material with its associated trace metals occurs before incorporation into the sediments. In spite of this high turnover of biogenic material there is good evidence that biological processes provide an important mechanism for the removal of some metals from sea water.

The removal and deposition of barium has been related to biological processes. Revelle et al. (1955) pointed out that strikingly high concentrations of barium are found in siliceous sediments. In these sediments barium concentrations reach 1—2%, which is considerably higher than the average value of 0.2% in deep-sea clays (Riley and Chester, 1971, p. 390). Goldberg and Arrhenius (1958) found a strong correlation between organic productivity in the surface layer of the ocean and the barium concentration in the underlying sediments. They found that the barium prevails as barite and to some extent in apatite. Church and Wolgemuth (1972) have calculated that the saturation value for barite at a depth of 5 km and $1°C$ is 44—49 $\mu g/kg$ of barium. The deep-ocean barium concentration is about half this value (Bacon and Edmond, 1972) indicating that most of the world's oceans are undersaturated with respect to barite.

The presence of barite in sediments (Goldberg and Arrhenius, 1958; Church and Wolgemuth, 1972) and the correlations between Ba and Si in the water column (Bacon and Edmond, 1972), in suspended matter (P. G. Brewer, unpubl. data), and surface sediments (Revelle et al., 1955) suggest that Ba is transported downward by siliceous organisms. Bacon and Edmond (1972) have shown, by means of a one-dimensional advection—diffusion model of dissolved Ba, that at a station in the South Pacific, most of the Ba input to the deep water takes place at the sediment—sea water interface and that little or none occurs due to decomposition of organisms in the water column. If most of the dissolution takes place at the sediment—water interface, Ba is released in the surface layers of the sediment whereupon it can be removed by other mechanisms, such as adsorption or precipitation as barite. The barium concentration in interstitial waters appears to approach barite saturation values indicating the importance of barite as a removal mechanism for Ba from sea water. A similar biological regeneration model has been proposed by Greenslate et al. (1973) to explain the high concentrations of Cu and Ni in nodules just north of the Pacific equatorial high productivity zone. In this case the Cu and Ni have been carried to the seabed by $CaCO_3$, which then dissolved leaving the Cu and Ni to be incorporated into manganese nodules by other mechanisms.

Biological processes may also be important for removing copper from sea water. Revelle et al. (1955) found a poor correlation between copper and manganese in Pacific pelagic sediments and suggested there is both a biogenous and a hydrogenous source of copper. The hydrogenous copper was estimated to be a constant factor of the manganese concentration and subtracted from the total copper to give the biogenous contribution. The calculated biogenous copper correlated very well with barium, which was shown in the previous paragraphs to be carried primarily by opal. Turekian and Imbrie (1966) also found a strong correlation between copper and calcium carbonate in Atlantic Ocean sediments, while other investigators have suggested that the biogenous copper is associated with the soft parts of organisms (Revelle et al., 1955; Chester and Hughes, 1969). A clue regarding the biogenic control of Cu has been presented by Boyle and Edmond (1975). They analyzed for Cu in surface samples collected across the Circumpolar Current in the South Pacific. The Cu concentrations reported (1—3 nmole kg^{-1}) are near the low end of the range of recently reported values (1—32 nmole kg^{-1}) (Spencer and Brewer, 1970). There is a strong correlation of Cu with nitrate and phosphate resulting in Cu:N and Cu:P molar ratios of $1.09 \cdot 10^{-3} \, 4$ and $1.7 \cdot 10^{-3}$ suggesting that Cu may be biochemically controlled. There was a co-variance with alkalinity as well; however, this would require that biogenic carbonates contain approximately 60 ppm Cu, which is a factor of 5—6 times greater than reported values.

Spencer and Brewer (1969) have examined the distribution of copper in sea water of the Gulf of Maine and the Sargasso Sea. They found higher Cu values in slope water east of the Gulf Stream, and lower values in the Sargasso Sea. Assuming a Cu:P ratio of $6.5 \cdot 10^{-3} : 1$ in marine phytoplankton, they calculated that, should no resolution occur, the Cu content of surface near-shore waters would be depleted by 20% at the end of growing season. This would indicate a *minimum* residence time of 5 years for Cu in surface waters, with respect to biological removal. Iron is transported in significant quantities by marine organisms (Lowman et al., 1971) and it has been estimated that transport of iron out of the mixed layer by all biological processes would take $\backsim$ 7 years. Since the mixed layer residence times for both Cu and Fe calculated in this way are shorter than the residence time of water itself (ca 30 years), then biological removal is seen to play a dominant role.

Several other biological factors have been proposed for removing trace metals from sea water. Turekian et al. (1973) have recently analyzed the tests of pteropods collected from surface waters for several trace metals. They suggested that metals are introduced into pteropods by occlusion of small (tenths of microns), iron-rich particles that have adsorbed the metals from sea water. They speculate that this process may be representative of all pelagic calcareous organisms, and, if so, it implies that enrichment by adsorption on metal oxides may be an important process for the removal of metals by organisms. Additional evidence for concretions growing within microcavities in plankton skeletal remains, particularly diatom frustules, has been found by Greenslate (1974a). These concretions are mostly manganese oxide ($>$ 98%) and Greenslate infers that nucleation occurs in the micro-cavities and grows progressively into the shell's interior. Another biological factor has been proposed by Arrhenius and Bonatti (1965) who have suggested that phosphatic fish debris concentrate rare earth elements (REE) and uranium and may be a significant vector in transporting REE from the water column to the sediments.

Several other workers have suggested that bacteria play a major role in nodule genesis (e.g., Graham and Cooper, 1959; Ehrlich, 1972). In spite of the occurrence of a large microbial flora on manganese nodules, their metabolic activity is difficult to evaluate because of the large changes in temperature and pressure that occur when samples are collected (Jannasch and Wirsen, 1973). The role of bacteria is further complicated by the coexistence of Mn(II) oxidizers, MnO_2 reducers and organisms inactive to manganese (Ehrlich, 1972). In fact, reducers are more dominant, in terms of absolute numbers, than oxidizers.

It has recently been discovered that organisms other than bacteria live on the surface of manganese nodules (Greenslate, 1974a, b). Some of these organisms are tube-building benthic Foraminifera of the genus *Saccorhiza*. *Saccorhiza* tubes are found intact on the surface of carefully collected and preserved nodules and within the interior of nodules. These tubes are made of ferromanganese micronodules, mineral grains, radiolarian fragments and sponge spicules cemented with iron oxide containing trace metals (Dudley and Margolis, 1974). Other unidentified organisms build structures on nodule surfaces that contain large amounts of Fe and Mn oxides. These biologically derived structures are found in all the nodules which Greenslate (1974b) has examined and may provide a solid superstructure within which Fe and Mn oxides can accumulate. Additional organisms of the genera *Reophax*, *Rhizammina*, *Hormonsina* and *Tolypammina* have been found on nodules collected from depths above the calcium carbonate compensation depth (Dudley and Margolis, 1974). These organisms construct tubes out of foraminifer tests and sponge spicules with a carbonate-rich cement.

The exact role these benthic foraminifers play in removing trace metals from sea water is uncertain; however, because of their universal occurrence, they must be considered in any model to explain the origin and growth of manganese nodules.

EVIDENCE FOR ADSORPTION

Adsorption is one of the mechanisms most frequently proposed to account for the removal of metals from sea water in order to maintain the observed steady state. Even many of the biological mechanisms mentioned in the previous section involve adsorption of metal ions on occluded or incipiently formed iron and manganese oxides (Ehrlich, 1972; Turekian et al., 1973; Greenslate, 1974). Crearer and Barnes (1974) and Glasby (1974b) have concluded that adsorption may be a dominant means of extracting Mn, Fe, and other trace metals from sea water into ferromanganese nodules.

Many workers have observed correlations among certain elements in manganese nodules, in sediments, and in suspended matter. These correlations have been established using statistical techniques, such as simple linear correlation coefficients (e.g., Willis and Ahrens, 1962; Carvajal and Landergren, 1969; Cronan and Tooms, 1969), by more sophisticated multivariant analysis such as factor analysis (e.g., Turekian and Imbrie, 1966; Cronan, 1967; Spencer et al., 1972), and by studying in-situ relationships using the electron-probe X-ray microanalyzer (e.g., Burns and Fuerstenau, 1966; Aumento et al., 1968; Cronan and Tooms, 1968; Friedrich et al., 1969). These correlations are summarized in Table 10-IV. The most frequently observed correlations are of various metals with manganese or iron. This would suggest either a similar source for the elements or that the correlation is due to adsorption onto the oxides of iron and manganese.

Goldberg (1954) was the first to point out interelement relationships in nodules. Using bulk chemical analyses and bivariate plots, he found correlations of nickel and copper with manganese, and of titanium, zirconium and cobalt with iron. Riley and Sinhaseni (1958) suggested that his correlations of iron with zirconium, or of manganese with copper, were not statistically significant. However, later workers (Table 10-IV) substantiated the correlation of manganese with copper.

Goldberg proposed that adsorption onto particulate Fe and Mn species accounted for the covariances. He suggested that the sorption depends on the relationship between the charge density of the adsorbed ion and that of the adsorbing surface. Adsorption will only take place in response to electrostatic forces (i.e., only ions with a charge opposite to the charge of the surface will be adsorbed); those ions with the largest charge densities will be most effectively scavenged. If MnO_2 were negatively charged and iron oxide positively charged, the distribution of adsorbed ions between these two phases should indicate which elements are present in sea water as cations and which as anions. This concept will be discussed later.

TABLE 10-IV

A summary of metal correlations in marine samples*

	Correlations
Nodules	
1. Goldberg (1954)	Fe—Co—Ti—Zr; Mn—Ni—Cu
2. Willis and Ahrens (1962)	Ni—Cu; Fe—Co
	Mn—Fe (negative)
3. Burns and Fuerstenau (1966)	Fe—Co—Ti—Ca
	Mn—Ni—Cu—Zn—Mg—K—Ba—Al
4. Sevast'yanov and Volkov (1966)	Mn—Ni—Co—Cu—Mo
5. Barnes (1967a)	Pb—Co; Co—δMnO$_2$
6. Aumento et al. (1968)	Mn—Co—Ni
7. Cronan (1967)	Fe—Ti—H$_2$O; Cr—detrital
8. Cronan and Tooms (1968)	Mn—Cu—Ni—Ca—K
9. Cronan (1969)	Mn—Ni—Cu—Mo; Co—Pb
10. Calvert and Price (1970a)	Mn—Ba—Co—Ni—Si—Mo
	Fe—As—Pb—Y—Zn—P
11. Brown (1971)	Pb—Co; Mn—Cu—Ni—Co
Sediments	
1. Turekian and Imbrie (1966)	Mn—Co—Ni
	Cu—CaCO$_3$
2. Carvajal and Landergren (1968)	Mn—Co—Ni
3. Cronan (1969)	Mn—Co
4. Watson and Angino (1969)	Fe—Ni; Fe—Co; Co—Ni
5. Boström (1970)	Mn—Cu—Co—CaCO$_3$
6. Calvert and Price (1970b)	Cu—Ni—Pb—Zn—organic matter

*Positive correlations unless otherwise stated.

Adsorption onto iron and manganese oxides was also invoked by Bender et al. (1970b), Dasch et al. (1971), Piper (1972), Dymond et al. (1973) and Sayles and Bischoff (1973) to explain the enrichment of metals in sediments from the East Pacific Rise, and by Aumento et al. (1968) to explain the enrichment of Co and Ni in manganese pavement from the San Pablo seamount. Sevast'yanov and Volkov (1966) used similar arguments to explain the correlation of Ni, Co, Cu and Mo with Mn in Black Sea nodules. They further hypothesized that the large adsorption capacity of solid manganese hydroxide (MnO(OH)$_2$) is due to the acid properties of the solid which will tend to form salts with divalent metals.

A strong correlation of Co and Sb with Mn was observed in particulate matter of the Black Sea (Spencer et al., 1972). Adsorption or coprecipitation of Co and Sb on solid MnO_2 was invoked to explain the associations. Evidence for molybdenum adsorption by MnO_2 in Saanich Inlet was found by Berrang and Grill (1974) and in the Black Sea by Pilipchuk and Volkov (1974).

Riley and Sinhaseni (1958) observed that if the majority of the minor elements in nodules are adsorbed directly from sea water, concentration factors relative to sea water should prove useful in elucidating this mechanism. They suggested that those elements with low enrichment factors (see Table 10—II) are not strongly adsorbed from sea water because they tend to form hydroxides with ionic bonds, whereas those elements that have the highest enrichment factors (Co, Ni, Cu, Zn, and Pb) have high ionic potentials (i.e., charge/radius) that lead to their enhanced adsorption. If adsorption from sea water is the controlling mechanism, then the adsorption selectivity sequence found in the laboratory experiments should compare well with the sequence of concentration factors. The concentration factors (Table 10-II) suggest that the selectivity sequence may be Co > Ni > Cu > Zn > Ba > Sr > Ca > Mg.

Carvajal and Landergren (1969) studied the interrelationships of Mn, Co and Ni in marine sediments and found that there is a strong correlation between these three metals. Their results further indicated that Co was relatively more enriched than Ni. If the removal of these elements from sea water is by adsorption on manganese phases, this implies that Co is more readily adsorbed than Ni.

EVIDENCE THAT SEA WATER IS THE SOURCE OF THE METALS IN MANGANESE NODULES

In addition to evidence that the enrichment of metals in sediments and manganese nodules is by adsorption on iron and manganese oxides, there is also evidence suggesting that sea water is the source of the adsorbed metals. Somayajula et al. (1971) have suggested that the growth of nodules in the central equatorial Pacific is often not continuous, and that nodules stop accumulating when buried. In fact they seem to be growing during only 10% of their lifetime. Based on Be^{10} geochronology, the age of the surface layer of the nodules presently buried is the same as that of the enclosing sediment. The implication is that sea water is the source of the metals, or that the source is due to some mechanism acting only at the sediment—water interface.

Piper (1972) has found that nodules from a water depth of less than 3,500 m exhibit a rare earth element (REE) distribution (exclusive of Ce) similar to that of sea water. This suggests that the REE in shallow water nodules are derived from sea water without appreciable fractionation, possibly by adsorption on iron and manganese oxides. On the other hand, he found that the REE distribution in deep-water nodules ($>$ 3,500 m) was a mirror image of the REE pattern in sea water. There is some similarity in the REE patterns of deep-water nodules and calcite which suggests that the depth dependency in the patterns is due to the dissolution of $CaCO_3$ in the deeper parts of the ocean. After the metals are released by the dissolution of $CaCO_3$ they may be incorporated into the nodules by adsorption onto the iron and manganese phases (see also Greenslate et al., 1973). The inverse relationship between nodules and sea water suggests that the formation of deep-water nodules is directly responsible for the distribution of REE in sea water. Using a smaller set of samples, Glasby (1973a) also concluded that sea water was the source of the REE in nodules. Dymond et al. (1973) have reached a similar conclusion for sediments from the East Pacific Rise but also suggest that the REE may also be controlled by the distribution and relative abundance of fish debris (see also Arrhenius and Bonatti, 1965).

The conclusions of Piper and Glasby are somewhat different from those of A. M. Ehrlich (1968) and Bender (1972) who found that the REE composition of nodules is similar to that of the surrounding sediments. Ehrlich deduced from this that nodules incorporate their REE from sediments via a surface exchange process and not from sea water.

Variations in isotopic ratios can be used to determine the source of metals. For example, the ratio of $^{87}Sr/^{86}Sr$ is about 0.703 in volcanic rocks and 0.709 in sea water. Carbonate shells and manganese nodules have a $^{87}Sr/^{86}Sr$ ratio of 0.709 which suggests strontium incorporation from sea water (Bender, 1972). Similar conclusions have been drawn for metalliferous sediments from the East Pacific Rise (Bender et al., 1971; Dasch et al., 1971). Although in this example the interpretation may be complicated by exchange of strontium between the sediments and adjacent or interstitial sea water (Dymond et al., 1973). The ratio $^{234}U/^{238}U$ has also been used as evidence that the uranium in manganese nodules and East Pacific Rise sediments is derived from sea water (Ku and Broecker, 1969; Dymond et al., 1973). On the other hand, the variation in the isotopic composition of lead is the same in nodules and coexisting sediments (Chow and Patterson, 1962), suggesting that lead in sediments and nodules is precipitated from sea water, or that lead is carried on detritus and incorporated into nodules by some sort of surface transfer or diagenetic process. The lead isotopes in the East Pacific Rise sediments have very different ratios and plot within the range representing volcanic lead suggesting an ultimate magmatic or hydrothermal origin (Dymond et al., 1973).

THE REMOVAL OF IRON AND MANGANESE FROM SEA WATER

The removal of iron and manganese from sea water is dependent upon the forms in which these metals exist. The correct identification of the form of the metals is difficult because both iron and manganese can exist in different oxidation states and can form a variety of solution species and solid phases. For example, the removal mechanism for iron will depend on whether it exists as Fe^{2+}, $Fe(OH)_2^+$, $Fe(OH)_4^-$, colloidal hydrous iron oxide, or as a dissolved complex with another inorganic or organic ligand. The following discussion will concentrate on the mechanisms of removal of these metals, taking into account the variations in speciation that may occur.

Coagulation and sedimentation

Both iron and manganese can exist as colloids or associated with fine-grained clay minerals in sea water. For example, iron may exist as colloidal iron oxide (Harvey, 1937; Kester and Byrne, 1972), manganese as colloidal MnO_2 (Goldberg, 1954), or both iron and manganese may be associated with fine grain-size detrital material (Carroll, 1958, Turekian and Imbrie, 1966; Bender, 1972). The size of these particles can be extremely small which complicates the distinction between particulate and dissolved concentrations. Van der Giessen (1966) has found that synthetic hydrated iron oxide gel particles may be as small as 30 Å. Colloidal particles with sizes approaching 30 Å will be very stable with respect to removal by sedimenation and will remain dispersed for long periods of time unless they agglomerate due to coagulation or formation of faecal pellets (Arrhenius, 1963; Gordon, 1970; Kranck, 1973) into larger aggregates having faster settling velocities. Thermodynamically the large surface area present in a finely dispersed system represents a large amount of free energy which by recrystallization or agglomeration tends to reach a lower value. The lowest energy state is therefore attained when colloid particles have coalesced.

The destabilization or coagulation of natural inorganic colloids is achieved by reducing the potential of repulsion between the particles through compaction of the double layer due to an increase in ionic strength (Hahn and Stumm, 1968), by compensation of the charge on the surface by specific adsorption of oppositely charged species or by the adsorption of organic films which produce a low-energy surface. The rate of coagulation depends on the frequency of collisions and on the efficiency of particle contact.

The type of kinetics depends on particle size and fluid shear. If the fluxes of particles toward each other are controlled by Brownian motion, coagulation is referred to as perikinetic and the decrease in concentration of particles with time follows a second-order rate law (Stumm and Morgan, 1970, p. 494):

$$\frac{dn}{dt} = -\frac{4kT}{3\eta} \alpha_p n^2 \tag{1}$$

where n = number of particles per unit volume; dn/dt = number of particles per unit volume per unit time; k = Boltzmann constant; T = temperature in degrees Kelvin; η = absolute viscosity; and α_p = fraction of collisions leading to permanent agglomeration:

$$\frac{4kT}{3\eta} = 3.41 \cdot 10^{-12} \text{ cm}^3 \text{sec}^{-1} \text{particle}^{-1}$$

For relatively large particles or high fluid shear rates the decrease in concentration is governed by first-order kinetics (orthokinetics):

$$\frac{dn}{dt} = -\alpha_o \frac{4(du/dz)V_m}{\Pi} n \tag{2}$$

where V_m = volume of particles per volume of solution; du/dz = velocity gradient (root-mean-square); and α_o = collision efficiency. It is usually assumed that $\alpha_o = \alpha_p$.

In the interior of the ocean, where the dimension of turbulent eddies is large, we may safely assume that only the Brownian motion of suspended particles is responsible for their collision (Lal and Lerman, 1973). However, at the sea surface, in the bottom boundary layer (where manganese nodules are located) or in estuarine regions it is possible that coagulation is controlled by orthokinetics (eq. 2).

Theories of stability, for example the Derjaguin—Landau—Verwey—Overbeek (DLVO) theory (Verwey and Overbeek 1948), can be applied to calculate the stability of hydrophobic colloids in media of inert electrolytes (Hahn and Stumm, 1970). However, for natural systems, relative values of stability need to be determined experimentally in media that closely resemble the natural systems in chemical composition. This is important because the adsorption of metal ions, which can destabilize colloids at very low concentrations, can restabilize these same dispersions at higher concentrations (Matijevic, 1967; Stumm and O'Melia, 1968). Adsorption of organic matter can also affect the surface charge and stability of natural colloids (Nemeth and Matijevic, 1968). Neihof and Loeb (1974) have suggested that most suspended material in natural waters is coated with organic matter which changes the nature of the surface from hydrophobic to a more hydrophylic character. This would reduce interparticle repulsion. Organic matter may also cause aggregation through bridging (Busch and Stumm, 1968).

Edzwald et al. (1974) have experimentally determined the collision efficiency, α, of the clay size fraction of natural estuarine sediments and found that estuarine water, containing divalent cations, trace metals and organic matter, produced higher α values than NaCl solutions of equivalent ionic strength. Their α values in 0.34 M estuarine water range from 0.15 for montmorillonite to 0.07 for illite. Using these values and the average particle concentration in deep-sea water of $5 \cdot 10^2$ particles cm^{-3} (Eittreim, 1970) and assuming perikinetic coagulation kinetics (eq. 1) the calculated residence time of particles in sea water ranges from 200 to 400 years. If the average depth of the ocean is 4,000 m then the average settling rate is 40 m year^{-1} and, assuming Stokes' settling law, the average particle radius is 1.2 μ. These are all reasonable figures suggesting that coagulation may be an important mechanism for removing suspended matter from sea water.

Unfortunately the importance of coagulation has only been quantified for clay minerals. Field data suggest that considerable variation can occur in the removal of iron and manganese colloids from sea water. Coonley et al. (1971) observed in-situ removal of iron in the Mullica River estuary, presumably due to a sequence of hydrolysis reactions with sea water. However, they did not find unusual amounts of iron in the sediments of the Mullica estuary. It appears that in spite of the increase in ionic strength in the estuary the colloids are either stabilized by organic matter, restabilized by specific adsorption of metal ions or the coagulation is not rapid enough to prevent them from being swept out to sea by the moderately strong tidal currents.

On the East Pacific Rise, extensive iron and manganese-rich sediments have been found (e.g., Boström and Peterson, 1966) and it has been suggested that these sediments result from the precipitation of iron and manganese oxides from hydrothermal solutions emitted from the crest of the ridge. If so, this suggests that iron and manganese oxides are removed from sea water without being transported far from their source resulting in high iron and manganese accumulation rates (Bender et al., 1971; Dasch et al., 1971).

The importance of coagulation for the growth of manganese nodules is uncertain. Margolis and Glasby (1973) have found that most manganese nodules contain micro-laminations which vary in thickness from 0.25 to 10 μm. They suggest that the formation of individual laminae may reflect abrupt changes in nodule growth rate over relatively short periods of time and that this is most likely caused by variations in bottom current velocity and possibly the suspended matter concentration. Fewkes (1973) has observed colloidal sized material both on nodule surfaces and within nodules. The general appearance suggests that microscopic iron and manganese particles have aggregated together forming successively larger clusters which in turn coalesce to form botryoidal masses. With time these isotropic and X-ray amorphous masses recrystallize to form more crystalline, anisotropic material. The same chemical forces that influence coagulation should be important in a particle accretion model, i.e., particle size, fluid shear and efficiency of collision.

Solubility

Under certain conditions both iron and manganese can be supersaturated with respect to solid phases in sea water. The chemistry of aqueous iron involves the +II and +III oxidation states. Iron (II) is thermodynamically unstable over the entire pH range of natural waters in the presence of oxygen and the kinetics of oxidation to iron (III) at the pH of sea water are rapid. Thus, iron in oxygenated sea water is in the +III oxidation state. The calculated solubility of ferric hydroxide in sea water is highly pH dependent with a minimum iron concentration of about $1 \cdot 10^{-8}M$ at pH 8.0 (Kester and Byrne, 1972). This concentration is about ten times less than the observed "dissolved" iron concentration in sea water which suggests that more iron is apparently dissolved in sea water than calculated because it is present as colloidal particles that pass the 0.45-μ filter, or is solubilized by other organic or inorganic ligands, such as carbonate ion, that were not included in the thermodynamic calculations (Rashid and Leonard, 1973). The oxide phases that would form when Fe(III) precipitates from natural waters have been described by several authors (Stumm and Morgan, 1970, p. 527; Hem, 1972; Chapter 7, this volume).

Thermodynamic considerations predict that manganese should exist in the +IV oxidation state in sea water. In this oxidation state manganese is extremely insoluble as MnO_2 ($K_{so} = 10^{-56}$; Charlot and Bezier, 1957) and the literature on the mineralogy of the resulting precipitates is extremely confusing (see Chapter 7 for a review). A variety of manganese oxide minerals have been found in nature and much of the confusion that exists is the result of comparing these minerals with synthetic counterparts prepared in the laboratory. Unfortunately, consideration of the thermodynamic data alone for the manganese system can be misleading. Processes may well be rate-limited since the kinetics of oxidation of Mn(II) to Mn(IV) are so slow. Most experimental work has tended to support the concept that the predominant form of manganese in sea water is Mn^{2+} (Stumm and Brauner, 1975). The upper limit on the solubility of Mn^{2+} in sea water is probably controlled by the formation of the solid phase $MnCO_3$. The calculated solubility will therefore be a function of total CO_2 and pH, and in sea water will lie in the range 1—5 m-mole Mn/ℓ (Bischoff and Sayles, 1972). The total Mn concentration in normal ocean water is in the range 0.1—1 μg-at.ℓ^{-1} (Spencer and Brewer, 1970). Sea water is therefore far from being saturated with respect to $MnCO_3$, although there is evidence that saturation may be reached in some interstitial waters (Li et al., 1969; Bischoff and Sayles, 1972).

The existence of the Mn^{3+} ion in sea water has been a frequent topic for speculation. The Mn^{3+} ion has four 3d orbital electrons and can assume either a high spin, $(t_{2g})^3(e_g)^1$, or a low spin, $(t_{2g})^4$, configuration. The competition between these two states renders the ion unstable with regards to both oxidation to Mn(IV), $(t_{2g})^3$, which has a high crystal field stabilizing energy (CFSE) and Mn^{2+}, $(t_{2g})^3(e_g)^2$, which has zero CFSE but has a stable electronic configuration (Burns, 1970, p. 12). The Mn^{3+} ion is therefore unstable in water and disproportionates according to the following reaction (Stumm and Morgan, 1970, p. 525).

$$2 \text{ Mn}^{3+} \text{ (aq)} + 2 \text{ H}_2\text{O}(\ell) = \text{Mn}^{2+} \text{ (aq)} + \text{MnO}_2 \text{ (s)} + 4 \text{ H}^+ \text{ (aq)};$$
$$\Delta G^\circ_{298} = -26 \text{ kcal. mole}^{-1}$$

Strong complexing agents, such as pyrophosphate, might tend to stabilize Mn in the +III oxidation state, however, these species are not present in natural waters. The existence of Mn(III) has been used to explain some of the variations in manganese oxide mineralogy (Giovanoli et al., 1969; Hem, 1972). It is important to note that the examples were Mn(III) has been invoked can also be explained as a combination of Mn(II) and Mn(IV) (Bricker, 1965). Morgan (1967) found that, when starting with a given concentration of Mn^{2+}, simply varying the kinetics of oxidation will produce oxide phases with different structures. For this reason it is misleading to attempt to calculate the equilibrium manganese concentration from knowledge of the oxide phase present (e.g., Bischoff and Sayles, 1972) since rate considerations may have played a dominant role in the formation of the phase. It is also possible that variations in the kinetics of oxidation may explain the variations that have been observed with depth in the oxidation grades (Manheim, 1965) and mineralogy (Barnes, 1967a) of manganese nodules.

In conclusion, it appears that iron, as iron (III), may be removed from sea water by rapid precipitation, while for manganese precipitation may be rate-limited. It is clear that the kinetic considerations are important for understanding the removal of iron and manganese from sea water.

Kinetics of oxidation

At equilibrium, both Mn(II) and Fe(II) are unstable over the entire pH range of natural waters in the presence of oxygen. However, because these two metals have such different rates of oxidation in nature (Krauskopf, 1957) they are easily separated.

The rate of oxidation fo Fe(II) in solutions with pH $\geqslant 5.5$ is first-order with respect to the concentration of Fe(II) and O_2 and second-order with respect to the OH$^-$ ion. The oxidation kinetics follow the rate law (Stumm and Lee, 1961):

$$-\frac{d [\text{Fe(II)}]}{dt} = k [\text{Fe(II)}] [\text{OH}^-]^2 P_{O_2} \tag{3}$$

where $k = 8.0 (\pm 2.5) \cdot 10^{13} \text{ min}^{-1} \text{ atm}^{-1} \text{ mole}^{-2}$ at 20°C. Catalysts, such as other metal ions, i.e., Cu^{2+}, Co^{2+}, solid surfaces, and complex forming anions, HPO_4^{2-}, can increase the reaction rate significantly (Singer and Stumm, 1970). The oxidation of Fe(II) is very rapid above pH 6 and there is a 100-fold increase in the rate of reaction for a unit increase in pH. In the absence of organic matter, the half-time for Fe(II) oxidation at pH 7.0 is 4 min (Stumm and Lee, 1961). From these data it seems likely that any ferrous iron introduced into oxygenated sea water above pH 6.0 will oxidize and coagulate almost instantaneously.

Organic compounds, such as tannic acid, complex Fe(II). Reported equilibrium constants for Fe(II)—tannic-acid complexes range from $9.5 \cdot 10^3$ (Theis and Singer, 1974) to $6.3 \cdot 10^5$ (Schnitzer and Skinner, 1966) and tannic acid and other organic compounds can slow the oxidation of Fe(II). Morgan and Stumm (1964) have pointed out that some organic substances, especially those that contain hydroxy and/or carboxylic groups (e.g., phenols, tannic acid, etc.) can both reduce and oxidize ferrous iron. This apparent contradiction can be described by the following reaction sequence:

Fe(II) + 1/4 O_2 + organic compound $\rightarrow$ Fe(III) – organic complex
Fe(III) – organic complex $\rightarrow$ Fe(II) + oxidized organic compound
Fe(II) + 1/4 O_2 + organic compound $\rightarrow$ Fe(III) – organic complex

Under the proper conditions, a relatively high steady-state concentration of Fe(II) can be maintained as long as the organic material is not fully oxidized. The ferrous—ferric couple is acting merely as a catalyst for the oxidation of organic matter by oxygen. A similar reaction sequence which would be plausible from free energey considerations could also be written for manganese. Unfortunately, little is known about the effect of organic compounds on the kinetics of oxidation of Mn(II) to Mn(IV) or reduction of Mn(IV) to Mn(II) (Giovanoli et al., 1969). Provided the rates are sufficient, this reaction sequence might provide an explanation for the small amounts of Mn(II) found in sea water in spite of the strong thermodynamic preference of Mn(IV) (Stumm and Brauner, 1975) and the small amounts (0.5%) of Fe(II) found in manganese nodules (Burns and Brown, 1972).

The kinetics of manganese (II) oxidation have been studied extensively by Hem (1963) and Morgan (1964). The reaction is autocatalytic, second-order with respect to the OH^- ion concentration, and first-order with respect to P_{O_2} and the Mn(II) concentration. Morgan (1964) proposed the following sequence:

$$Mn(II) + O_2 \xrightarrow[=]{slow} MnO_{2\,(s)}$$

$$Mn(II) + MnO_{2\,(s)} \xrightarrow[=]{fast} (Mn(II) \cdot MnO_2)_{(s)}$$

$$(Mn(II) \cdot MnO_2)_{(s)} + O_2 \xrightarrow[=]{slow} 2MnO_{2\,(s)}$$

which is governed by the following rate law:

$$\frac{-d\,[Mn(II)]}{dt} = k_o\,[Mn(II)] + k\,[Mn(II)]\,[MnO_2] \qquad (4)$$

where $k = k'\,[P_{O_2}]\,[OH^-]^2$.

Morgan verified this reaction scheme experimentally and found that the relative proportions of Mn(IV) and Mn(II) in the solid phase depend strongly on the pH and other variables affecting the kinetics. Extrapolating his rate law to sea water he estimated that at the pH and average temperature of sea water it would take roughly 10^3 years to oxidize 90% of the Mn, provided it was present in the reduced form.

Brewer and Elert (unpubl. data) have studied the kinetics of oxidation of Mn(II) in artificial sea water. Their experiments confirmed the reaction mechanism given by Morgan (1964) and led to a preliminary numerical value for the rate constant. They found that:

$$\frac{d\ Mn(II)}{dt} = k\ [Mn(II)]\ [MnO_2]\ [OH^-]^2\ [O_2] \tag{5}$$

then:

$k = 1.4 \cdot 10^{11}\ l^4$ moles^{-3} ml^{-1} min^{-1} at 25°C

where the O_2 concentration is expressed in ml l^{-1} and the Mn(II) removed rate in moles l^{-1} min^{-1}. This rate constant includes activity coefficients. Although some doubt must be attached to the applicability of this rate constant, it can be tested in various natural situations. Spencer and Brewer (1971) and Brewer and Spencer (1974) have described the redox cycling of manganese in the Black Sea. The conditions there are approximately: temperature = 8.5°C; pH = 7.7; O_2 = 0.2 ml^{-1}; MnO_2 = 10^{-6} mole l^{-1}; Mn^{2+} = 10^{-6} mole l^{-1}.

Assuming that the rate constant doubles for a temperature increase of 10°C (Morgan 1964) an oxidation rate of $1.2 \cdot 10^{-5}$ moles l^{-1}year^{-1} can be calculated. Spencer and Brewer suggested an advective velocity of 0.5 m year^{-1}, and an effective oxidation zone of 10 m. Assuming that the reaction goes to completion within this 10 m interval, then the kinetic model would predict a manganese removal rate of $\sim$ 330 mg m^{-2}year^{-1}, whereas the advection-diffusion model finds 875 mg m^{-2}year^{-1}. The two sets of data agree to within a factor of 3, which is remarkable considering the necessarily crude approximations which are involved.

We can also use this oxidation rate to calculate the residence time of dissolved Mn in sea water. If we assume that all the particulate Mn present in sea water is present as MnO_2, and the rate constant at 5°C is $0.3 \cdot 10^{11}\ l^4$ moles^{-3} ml^{-1}min^{-1}, then in sea water where: Mn^{2+} = $2 \cdot 10^{-9}\ M$; MnO_2 = $0.2 \cdot 10^{-9}\ M$; OH^- = $10^{-6}\ M$; O_2 = 4 ml l^{-1}.

$$\frac{d\ [Mn(II)]}{dt} = 0.48 \cdot 10^{-19}\ \text{moles l}^{-1}\ \text{min}^{-1}$$

$$= 9 \cdot 10^{-12}\ \text{moles l}^{-1}\ \text{year}^{-1}$$

$$= 9 \cdot 10^{-3}\ \text{nM year}^{-1}$$

Using this rate law we can calculate the expected rate of removal of Mn (II) from sea water by oxidation:

$$= \frac{A}{dA/dt} = \frac{2 \text{ nM}}{9 \cdot 10^{-3} \text{nM year}^{-1}} = 200 \text{ yrs}$$

The ratio of Mn/Al in suspended matter is very close to that of crustal rocks suggesting that most of the Mn is associated with silicates rather than MnO_2. If particulate MnO_2 is 1% of the total particulate Mn the residence time becomes 20,000 years. The residence time of Mn(II) in the water column is therefore very long. The kinetics of oxidation are increased enormously, however, when the water contacts the sediment which contains large amounts of solid MnO_2. Thus Mn(II) introduced by the rivers circulates with the ocean until it comes in contact with the sediments whereupon it oxidizes and precipitates as MnO_2. For this reason the residence time of Mn in sea water is on the same order of magnitude as the mixing time of the oceans ($\sim$ 2,000 yrs).

THE REMOVAL OF METALS FROM SEA WATER BY ADSORPTION ON IRON AND MANGANESE OXIDES

The evidence for adsorption of metal ions on manganese and iron oxides in the marine environment was reviewed earlier (pp. 301—303). Although circumstantial evidence for adsorption is abundant, the hypothesis has yet to be adequately tested. Both a qualitative understanding of the mechanism of adsorption (i.e., the types of reactions that take place on the surface) and a quantitative means of predicting the degree to which adsorption can take place are lacking. Those few models proposed to explain this process (e.g., Goldberg, 1954; Goldberg and Arrhenius, 1958) have been based on little or no experimental evidence and do not adequately explain many of the important observations.

Research on the surface chemistry of hydrous manganese dioxide (e.g., Morgan and Stumm, 1964; Posselt et al., 1968a; Murray et al., 1968; Van der Weijden, 1975) and iron oxide (Parks and De Bruyn, 1962; Atkinson et al., 1967; Dyck, 1968; Breeuwsma and Lyklema, 1971, 1973; Tewari et al., 1972; Kinniburgh, 1973; Van der Weijden, 1975) has been carried out. However, it is difficult to extrapolate these data to explain phenomena in the marine environment. Krauskopf (1956) was the first to attempt to quantify the adsorption mechanism. He performed some simple qualitative experiments in an attempt to evaluate Goldschmidt's (1937) suggestion that trace metal concentrations in sea water are kept below the limiting solubility concentration due to removal by adsorption. MnO_2 was the most efficient of the various adsorbents he tested; however, quantitative interpretation of his data is difficult because he used metal concentrations two or more orders of magnitude higher than found in sea water. Evaluation of the effectiveness of the various adsorbents is also difficult because he used different concentrations and presumably different surface areas for each. He made two important observations, however, that indicate the complexity of the adsorption mechanism. He found that the anion forming elements Cr and W are as effectively adsorbed by MnO_2 as are the cation forming elements Ni and Co, and that elements that possess similar charge densities in sea water frequently exhibit large differences in their degree of adsorption. It had previously been proposed (Goldberg, 1954, 1965) that negatively charged MnO_2 adsorbs only the cation forming elements, while positively charged iron oxide attracts only anions.

Kharkar et al. (1968) carried out adsorption—desorption experiments as part of their study of the supply of metals by streams to the ocean. Metals were taken up on various adsorbents in distilled water, then transferred to sea water and the release of the metals monitored. Unfortunately, the manganese phase they used was βMnO_2 (pyrolusite), which is not found in the marine environment and has very different surface properties from hydrous manganese dioxide (Healy et al., 1966; Jenkins, 1970; Stumm et al., 1970). They concluded that trace elements adsorbed from river water are partially released on contact with sea water because of displacement by magnesium and sodium ions present in sea water in high concentration.

Murray (1974, 1975a,b) has recently completed a comprehensive study of the surface chemistry of hydrous manganese dioxide. These experiments were performed using a synthetically prepared hydrous manganese dioxide which was similar in structure (broad X-ray diffraction peaks centred at 7.4 Å, 2.43 Å and 1.63 Å) and oxidation grade (MnO$_{1.93}$) to manganese phases found in nature. It is referred to here as hydrous manganese dioxide because it was synthetically prepared and because of controversy in the literature (see Chapter 7) regarding the existence of 7 Å manganite. The majority of this discussion will be based on the experimental work using hydrous manganese dioxide.

Electrostatic aspects

The thermodynamics and coordination chemistry of the metal-oxide—electrolyte interface have been described in detail (De Bruyn and Agar, 1962; Atkinson et al., 1967; Bérubé and De Bruyn, 1968; Stumm et al., 1970). What follows is a brief review. Readers are encouraged to refer to the above references for a more thorough discussion.

The origin of the charge at the metal-oxide—solution interface is viewed as a two-step process. The metal oxide hydrates by removing H^+ ions, OH^- ions or water molecules from solution as the metal ions on the surface attempt to complete their coordination. The metal hydroxide groups formed in this manner participate in acid-base reactions to produce a positively or negatively charged surface. The surface metal hydroxide groups can then either dissociate or accept H^+ or OH^- ions from solution:

$$-Me-OH^o + H_2O = -Me-O^- + H_3O^+$$

or:

$$-Me-OH^o + OH^- = -Me-(OH)_2^- \tag{6}$$

$$-Me-OH^o + H_3O^+ = -Me-OH_2^+ + H_2O$$

or:

$$-Me-OH^o = -Me-^+ + OH^- \tag{7}$$

It is impossible to determine experimentally whether the reaction is due to the adsorption of H^+ and OH^- ions or to the dissociation of surface sites. The exact location of the pH of zero point of charge, pH(ZPC), will depend on the relative acidity of the surface groups:

$$Me - OH_2^+ = Me-OH^o + H^+ = Me - O^- + 2H^+$$
$$\quad\quad K_1 \quad\quad\quad\quad K_2 \tag{8}$$

and the pH(ZPC) will lie at the pH value for which an equal number of $Me-OH_2^+$ and $Me-O^-$ groups are present, i.e., when the surface is uncharged. Using reaction 8, the pH(ZPC) can be characterized by:

$$[Me-OH_2^+] = [Me-O^-]$$

and:

$$H_{ZPC}^+ = K_1 \ K_2^{1/2} \quad \text{or: pH (ZCP)} = 1/2(pK_1 + pK_2) \tag{9}$$

where K_1 and K_2 define the acidity constants of the surface sites (Stumm et al., 1970; Schindler and Gamsjäger, 1972). In the absence of specific adsorption, the pH(ZPC) is an intrinsic characteristic of the oxide. There is a large variation in the early values reported for the pH(ZPC) of metal oxides (Parks, 1965) and this variation is probably due to a combination of factors including the state of hydration and the purity and history of the samples. Recently an increased awareness of these problems, together with the use of the potentiometric titration method of pH(ZPC) determination, has produced a considerable number of reliable values. The pH(ZPC) of some metal oxides are compared in Table 10-V.

The location of the pH(ZPC) depends on the acidity of the metal ion and the electrostatic field strength of the solid (Healy and Fuerstenau, 1965; Healy et al., 1966). In general, the pH(ZPC) is inversely proportional to the charge to radius ratio of the metal (Parks, 1965) and the size of the unit cell of the solid (Healy et al., 1966).

TABLE 10-V

Comparison of the pH of ZPC for various metal oxides

Oxide	pH(ZPC)	Reference
WO_3	0.5	El Wakkad and Rizk (1957)
Hydrous manganese dioxide	2.2	Murray (1974)
δMnO_2	1.5	Healy et al. (1966)
Mn(II) Manganite	1.8	Healy et al. (1966)
α-MnO_2	4.5	Healy et al. (1966)
γ-MnO_2	5.5	Healy et al. (1966)
β-MnO_2	7.3	Healy et al. (1966)
SiO_2 (qtz.)	2.0	Li (1958)
(amorphous)	3.5	Bolt (1957)
(amorphous)	1.8—2.0	Healy et al. (1968)
(amorphous)	3.0	Tadros and Lyklema (1968)
(amorphous)	2.0	James and Healy (1972b)
TiO_2	6.0	Bérubé and De Bruyn (1968)
	6.4	Schindler and Gamsjäger (1972)
γ-Al_2O_3	8.5	Huang and Stumm (1973)
Amorphous iron oxide	8.1	Kinniburgh (1973)
α-Fe_2O_3	8.5	Parks and De Bruyn (1962)
ZnO	8.9	Blok and De Bruyn (1970)
Ag_2O	11.2	Parks and De Bruyn (1962)
MgO	12.5	Robinson et al. (1964)

The determination of K_1 and K_2 is best achieved by acid-base (potentio-metric) titration curves. Using these curves the adsorption of the potential determining ions (PDI) can be measured and, assuming that reactions 6 and 7 are valid, calculate the surface charge. In the absence of specific adsorption, the surface charge density, σ_0, is simply related to the adsorption densities by:

$$\sigma_0 = F (\Gamma_{H^+} - \Gamma_{OH^-}) \tag{10}$$

where F is the Faraday constant, Γ is adsorption density in μequiv. cm^{-2}, and σ_0 is in μc cm^{-2}. The pH(ZPC) is identified as the pH of intersection of the titration curves at different ionic strengths. The net potentiometric titration curves for hematite ($\alpha - Fe_2O_3$) indicate a pH(ZPC) of 8.5 (Fig. 10-1) (Parks and De Bruyn, 1962). The titration curves for hydrous manganese dioxide do not intersect (Fig. 10-2) and this is because its pH(ZPC) is at too low a pH (Murray, 1974). Normally, in the absence of specific adsorption, there is good agreement between the pH(ZPC) deter-

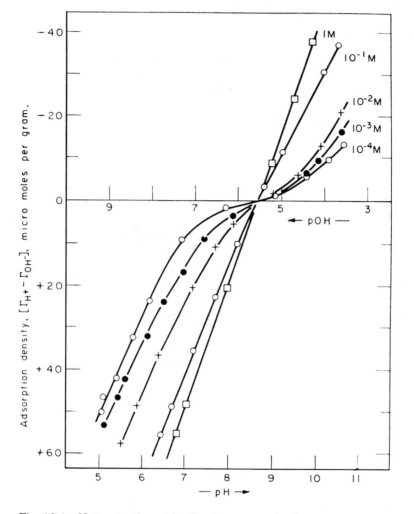

Fig. 10-1. Net potentiometric titration curves for hematite (α—Fe_2O_3) at different ionic strengths. The cross-over point indicates a pH(ZPC) of 8.5 (Parks and De Bruyn, 1962).

mined potentiometrically and the isoelectric point of the solid, pH(IEP), determined by electrophoresis and other electrokinetic measurements.

Using an extrapolation of electrophoretic mobility determinations and Na^+ and K^+ adsorption, the pH(IEP) of hydrous manganese dioxide has been located at pH 2.25 (Murray, 1974). For pH values greater than 2.25, the surface of hydrous manganese dioxide has a negative charge. The overall reaction:

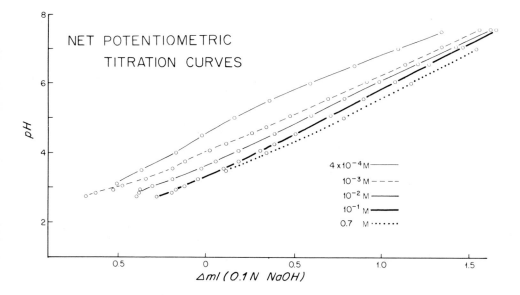

Fig. 10-2. Net potentiometric titration curves for hydrous manganese dioxide at different ionic strengths. The failure of the titration curves to cross is because the pH(ZPC) is less than 3 (Murray, 1974).

$$-\overset{|}{\underset{|}{Mn}}-OH_2^+ \text{ (surface)} = -\overset{|}{\underset{|}{Mn}}-O^- \text{ (surface)} + 2\,H^+$$

has an equilibrium constant of:

$$K_{1\,2} = \frac{[-Mn-O^- \text{ (surface)}]}{[-Mn-OH_2^+ \text{ (surface)}]} \left(\frac{\gamma^-}{\gamma^+}\right) \alpha_{H^+}^{+2} \tag{11}$$

where γ^-, γ^+ are the activity coefficents of the two surface sites, and α_{H^+} is the activity of hydrogen ions. Assuming that at the pH(ZPC) the positive and negative sites are equal and the activity coefficients for the sites are equal, then $K_{1\,2} = 10^{-4.5}$ for $\delta\text{-}MnO_2$. This value compares with a $K_{1\,2}$ of 10^{-17} calculated in a similar manner for $\alpha\text{-}Fe_2O_3$ (Parks and De Bruyn, 1962) and shows that hydrous manganese dioxide attracts protons much less strongly than $\alpha\text{-}Fe_2O_3$.

Using alkalimetric titration curves to obtain the amount of strong base consumed by the surface, surface charge values were calculated that approached $-100\ \mu C\ cm^{-2}$ at pH 8.0 in 0.01 M NaCl. This would be equivalent to an adsorption capacity of $1 \cdot 10^{-5}$ equiv. m^{-2} or $5 \cdot 10^{-6}$ moles m^{-2} of divalent cations. Using a surface area of 260 $m^2 g^{-1}$, which was determined for hydrous manganese dioxide by the B.E.T. method (Murray,

1974), this electrostatic adsorption capacity can be expressed in the same units used to express the ion exchange capacity of clays (mequiv./100 g). The cation exchange capacity of a negative double layer is defined as the excess of counter ions in the double layer which can be exchanged for other cations (Stumm and Morgan, 1970, p. 465). Part of the negative surface charge is also counterbalanced by the exclusion of anions, but this usually compensates for only a small fraction of the total charge (Breeuwsma and Lyklema, 1973). The ion exchange capacity of hydrous manganese dioxide is compared with some clay minerals in Table 10-VI (Grim, 1968, p. 189). At pH8, hydrous manganese dioxide is almost twice as surface-active as the most surface-active clay minerals. The large negative charge produced by the acid-base properties of the surface only explains part of the high adsorption capacity of hydrous manganese dioxide. It cannot account for all the adsorption, because much higher adsorption densities (at least $1 \cdot 10^{-5}$ moles m^{-2}) have been found for transition metals (Murray, 1975a, b).

TABLE 10-VI

Comparison of ion exchange capacity of hydrous manganese dioxide with some clay minerals

Material	Ion exchange capacity (mequiv./100g)
Kaolinite	5—15
Illite	10—40
Chlorite	10—40
Smectite	50—150
Hydrous manganese dioxide	260 (at pH 8.0)

Charge density

The Gouy—Chapman theory (see Stumm and Morgan, 1970, p. 458) predicts that multivalent ions are concentrated in the double layer to a much larger extent than monovalent ions. On the basis of this theory, it is frequently proposed (e.g., Goldberg and Arrhenius, 1958; Goldberg, 1965) that the degree of adsorption is a function of the charge density of the ions. However, using the radii of the unhydrated ions to calculate charge density may be unrealistic because of the large energy of hydration of some ions. When radii of the hydrated ions are used to calculate ionic potential, Ba^{2+} has a larger value than Mg^{2+} (Table 10-VII), and this might explain its greater adsorption. But, if charge density were the controlling factor, it would also be difficult to explain why Co^{2+} adsorbs much more strongly than Ni^{2+}

TABLE 10-VII

Charge densities

Metal ion	Crystallo-graphic radius*[1]	Hydrated radius*[2]	Ionic charge density	Hydrated charge density
Mg^{2+}	0.65	4.23	3.08	0.473
Ca^{2+}	0.99	4.15	2.02	0.482
Si^{2+}	1.13	4.16	1.77	0.481
Ba^{2+}	1.35	4.03	1.48	0.496
Mn^{2+}	0.80		2.50	
Co^{2+}	0.72		2.78	
Ni^{2+}	0.69		2.90	
Cu^{2+}	0.96		2.08	
Zn^{2+}	0.74	4.4	2.70	0.454

*[1] From Pauling, 1960, p. 518.
*[2] From Robinson and Stokes, 1959, p. 126.

(Murray et al., 1968; Murray, 1975a). Both ions have similar radii and solution chemistry. Obviously more controls the adsorption of metal ions than simply coulombic attraction and charge density.

Specific chemical aspects

Goldberg (1965) suggested that if MnO_2 were negatively charged and iron oxide positively charged, the distribution of adsorbed ions between these two phases should indicate which elements are present in sea water as cations and which as anions. Those metals that exist in sea water as cations would adsorb on MnO_2, and those as anions would adsorb on hydrous iron oxide.

The surface of iron oxide has a pH-dependent surface charge (Parks and De Bruyn, 1962; Breeuwsma and Lyklema, 1971); however, the pH(ZPC) of colloidal iron oxide in sea water is uncertain (Harvey, 1937). Goldberg (1965) and Goldberg and Arrhenius (1958) have stated that iron oxides in sea water are positively charged. However, experimental determinations of the pH(ZPC) of various modifications of iron oxide indicate that it can range from pH 6 to 9 depending on the mode of formation and subsequent aging. In general, samples with more ordered structures have a more acid pH(ZPC) (Schuylenborgh and Arens, 1950). Most of the values for goethite, the most common form of iron oxide in sediments, fall between pH 6 and pH 7. A summary of the literature values of the pH(ZPC) of iron oxides is shown in Table 10-VIII.

TABLE 10-VIII

Literature values for the pH(ZPC) of synthetic iron oxides

Phase	Investigator	pH(ZPC)
$\alpha Fe_2 O_3$	Korpi (1960)	9.04 ± 0.05
(hematite)	Parks and De Bruyn (1962)	8.4 ± 0.1
	Albrethson (1963)	8.7 ± 0.1
	Onoda and De Bruyn (1966)	8.3
	Breeuwsma and Lyklema (1971)	8.5
$\gamma Fe_2 O_3$	Iwasaki et al. (1962)	6.7 ± 0.2
$\alpha FeOOH$	Schuylenborgh and Arens (1950)	5.9 to 7.2
(goethite)	Flaningham (1960)	6.1 ± 0.1
	Iwasaki et al. (1960)	6.7 ± 0.2
	Lengweiler et al. (1961)	6.7
	Atkinson et al. (1967)	7.55 ± 0.15
$\gamma FeOOH$	Schuylenborgh and Arens (1950)	5.4 to 7.3
(Lepidocrocite)	Iwasaki et al. (1960)	7.4 ± 0.2
Amorphous iron hydroxides	Hazel and Ayres (1931)	8.6
	Mattson and Pugh (1934)	7.1
	Schuylenborgh and Arens (1950)	8.5
	Kinniburgh (1973)	8.1

In addition, such a classification is complicated by the fact that adsorption can be due to more than electrostatic forces. For example, phosphate, arsenate and selenite are known to adsorb on negatively charged iron oxide surfaces (Hingston et al., 1967, 1971) and MoO_4^{2-} on negatively charged MnO_2 surfaces (Chan and Riley, 1966a). Clearly, it can not be stated that metals associated with MnO_2 exist as cations and that metals associated with iron oxide exist as anions in sea water.

The interaction of metal ions with metal oxides is best explained in terms of both chemical and coulombic attractions (Grahame, 1947; Stumm et al., 1970; James and Healy, 1972c). The total change in standard free energy of adsorption, $\Delta \bar{G}^o_{ADS}$, is the sum of the change in total specific (chemical) adsorption energy, ϕ, the change in electrochemical (coulombic) adsorption energy, $ZF\Psi_\delta$, and the change in secondary solvation energy (eq.12):

$$\Delta \bar{G}^o_{ADS} = \Delta G^o_{COUL} + \Delta G^o_{SOLV} + \Delta G^o_{CHEM} \tag{12}$$

where Ψ_δ is the potential drop at the surface, F is the Faraday, and Z is the charge. Experimentally, it is very difficult to separate the energy of adsorption into its chemical, electrostatic and hydration components. The specific adsorption contribution is estimated by measuring the adsorption that takes place at the pH(ZPC) (Grahame, 1947; Murray et al., 1968). The change in secondary solvation energy can be estimated using a modification of the Born equation but only after making some complicated assumptions and estimates of the dielectric constant in the interfacial region (James and Healy, 1972c). the coulombic term can be estimated near the pH(ZPC) by calculating the surface potential using the Gouy—Chapman model (Kruyt, 1952). Otherwise the coulombic attraction can be estimated by calculating the surface charge from potentiometric titration data (Murray, 1974). Ionic species adsorbed in response to coulombic attraction alone obviously cannot adsorb in amounts greater than the equivalents of surface charge.

The experimental results presented by Murray (1975a, b) indicate that the alkali metal ions have no specific adsorption contribution, thus they only react electrostatically with the surface. Na^+ and K^+ do not adsorb on hydrous manganese dioxide below the pH(ZPC). As a group, the transition metal ions react more strongly with hydrous manganese dioxide than do the alkaline earths. Within each group there is a well defined selectivity sequence. Among the alkaline earths, Ba interacts more strongly than Mg, and among the transition metals, Co interacts more strongly than Ni. The selectivity sequence for all of the metals studied is:

$$Na = K < Mg < Ca < Sr < Ba < Ni < Zn < Mn \leqslant Co$$

The ions that are most easily adsorbed by increasing the pH are the most difficult to desorb by decreasing the pH. This is a reflection of the specific adsorption potentials that increase from the alkali earths to the transition metals (Murray, 1975a).

Other metal oxides that may also be important for removing trace metals from sea water have strong specific adsorption of cations. Kinniburgh (1973) has found that freshly precipitated iron oxide gel (pH(ZPC) = 8.1) exhibits the following selectivity sequence, with all metals having some specific adsorption:

$$Pb > Cu > Zn > Ni > Cd > Ba > Ca > Si > Mg$$

He found slightly different results for freshly precipitated aluminum oxide gel (pH(ZPC) = 9.4):

$$Cu > Pb > Zn > Ni > Co > Cd > Mg > Ca > Sr > Ba$$

In other studies Grimme (1969) found a sequence of $Cu > Zn > Co > Mn$ for goethite (α-FeOOH) and Taniguchi et al. (1970) a sequence of $Zn > Cu > Ni = Co > Mn$ for silica gel. All of these selectivity series are different from the selectivity series for hydrous manganese oxide and also from the order of enrichment relative to sea water of trace metals in marine manganese nodules (Table 10-II). The most striking difference is the position of cobalt. Cobalt adsorbs most strongly on hydrous manganese oxide and is greatly enriched in manganese nodules but shows no unusually large attraction for iron, silica or aluminum oxides.

Specific adsorption is the most frequent explanation for the adsorption selectivity sequence on metal oxides. Unfortunately, the chemical processes controlling it are not well understood. Since specific adsorption of cations invariably seems to involve proton exchange, a stoichiometric reaction model has been developed (e.g., Kurbatov et al., 1951; Stumm et al., 1970; Schindler and Gamsjäger, 1972). For example, comparison of adsorption data and potentiometric titration curves of hydrous manganese oxide in the presence of metal ions (Murray, 1975a) indicates that the metal ions can penetrate increasing amounts from Mg^{2+} to Co^{2+} into the compact part of the double layer to react with protonated sites on the hydrous manganese oxide surface. Any ion that enters the compact layer can modify the double layer in such a way that the charge of the diffuse layer may become reduced or even reversed. This reaction involves the replacement of a proton on the surface by a divalent metal ion on an equal molar basis as shown by reaction 13:

$$\begin{array}{ccc} | & & | \\ - Mn - OH^{\circ} + Co^{2+} = & - Mn - O - Co^{+} + H^{+} \\ | & & | \end{array} \qquad (13)$$

It is possible that the metal ions are associated with more than one surface group. However, Stumm et al. (1970) used a formation curve (a plot of ligand number versus the ligand concentration) to demonstrate that for Ca^{2+} only monodentate associations are formed. A similar reaction was found by Huang and Stumm (1973) for metal interaction with $\gamma-Al_2O_3$ and by Duval and Kurbatov (1952) and Kinniburgh (1973) for metal interaction with hydrous iron oxide.

A second important model that has been developed to explain adsorption data is the James and Healy model (James and Healy, 1972a, b, c). This model was developed to rationalize the considerable evidence available that cation adsorption is closely related to hydrolysis and precipitation (e.g., Dugger et al., 1964; Grimme, 1969; Kinniburgh, 1973). The James and Healy model suggests that for metal-ion adsorption on most metal oxides (especially those with a low dielectric constant) there is a very large positive, and therefore unfavourable, change in solvation energy that tends to prevent metal-ion adsorption (James and Healy, 1972c) (eqn. 12). For these metal oxides adsorption is very small in the low pH range but tends to increase rapidly over a very narrow pH range (James and Healy, 1972a) and this is correlated with a charge reversal of the surface (James and Healy, 1972b). This critical pH range is much lower than the pH at which precipitation or hydrolysis formation would be expected in bulk solution. According to the James and Healy model the high electric field at the surface reduces the dielectric constant of the interfacial medium well below the value for bulk aqueous solution. As a result the solubility product of the metal hydroxide will be smaller and thus more insoluble at the solid-solution interface. The model explains the large increase in adsorption and consequential charge reversal as due to induced precipitation at the solid—solution interface. On hydrous manganese oxide the interaction is more complex. Loganathan and Burau (1973) maintain that during the adsorption process charge is conserved. When a metal ion is adsorbed, equivalent amounts of protons and structural cations are released. In their model of the surface, a metal ion that exhibits specific adsorption must penetrate into the structure of the hydrous manganese dioxide and replace structural Mn(II) or Mn(III). Murray (1975a, b) demonstrated that this mechanism is much less important than suggested by Loganathan and Burau. In the most extreme case, only 10% of the cobalt adsorbed could be accounted for by manganese released to solution. Futhermore, electrophoresis experiments using Co(II) and Mn(II) demonstrated that charge was not conserved during adsorption (Murray, 1975b). The interaction of metal ions with hydrous manganese oxide appears to exhibit features of both the stoichiometric and the James and Healy model. There is a stoichiometric proton—metal-ion exchange reflected by charge reduction at low pH values and a sharp increase in adsorption and consequent charge reversal over a narrow pH range.

In addition to these models that have been proposed for the surface it appears that the selectivity is due to more than cation solution chemistry or to bulk properties of the solid. Surface coordination and constraints of the metal oxide structure must also be considered. An example of this is the strong enrichment of cobalt in manganese nodules and strong adsorption of cobalt on hydrous manganese oxides that has been explained by the oxidation of Co(II) to Co(III) (Goldberg, 1961a; Burns, 1965; Murray et al., 1968). It has been suggested that during adsorption Co^{3+} ions penetrate into the solid phase and replace structural Mn^{2+} or Mn^{3+} (McKenzie, 1970; Loganathan and Burau, 1973). Unfortunately the ionic radius of Co^{3+} (0.525 Å) (Shannon and Prewitt, 1969) in the low spin state (the stable configuration of Co^{3+} in oxide structures) is considerably smaller than Mn^{2+}(0.82), Mn^{3+} (0.65), or Fe^{3+} (0.645 Å) (see Table 7-VII). So that Co^{3+} would not be easily substituted for Mn^{2+} or Mn^{3+} in hydrous manganese oxide, or for Fe^{3+} in hydrous iron oxides. On the other hand, the Co^{3+} radius is similar to Mn^{4+} (0.54 Å). Burns and Burns (Chapter 7, this volume) have therefore suggested that Co^{3+} may substitute for Mn^{4+} ions in the edge-shared (MnO_6) octahedral that form the basis of many manganese (IV) oxide mineral structures. An equivalent site is not available in the iron oxide structure and this probably explains why the affinity of Co(II) for hydrous manganese oxides is so much larger than for other hydrous metal oxides. This also suggests that the large enrichment of cobalt in marine manganese nodules is by adsorption of Co^{2+} on hydrous manganese oxide where some of it can be oxidized to Co^{3+} and substituted for Mn^{4+} ions in (MnO_6) octahedral.

DISCUSSION

It is clear from previous research that the variation in composition of manganese nodules is related to the site of formation and is a function of such variables as sedimentation rate, oxidation and reduction processes in underlying sediment, growth rate, and mineralogy. But clearly, if adsorption is the controlling enrichment mechanism, the selectivity sequence of the hydrous manganese dioxide adsorption experiments should compare well with the enrichment relative to sea water. In both cases, the sequence Co > Ni > Ba > Sr > Ca > Mg is observed. It is not necessary to involve the influence of organic compounds, either as coatings on the surface (Neihof and Loeb, 1972, 1974) or as organo—metallic complexes (Price and Calvert, 1970; Glasby and Hodgson, 1971) in order to explain the enrichment.

It was proposed in the Introduction that adsorption by MnO_2 is an important mechanism for controlling the trace metal concentration in sea water. This proposal can now be tested using the data for cobalt adsorption by hydrous manganese dioxide (Fig. 10-3) (Murray, 1975b) and data for cobalt adsorption by hydrous ferric oxide (Kurbatov and Wood, 1952) and by illite (Chester, 1965). The experimental adsorption data for hydrous ferric oxide and illite were recalculated to moles of cobalt adsorbed per square metre using a surface area of $100 \text{ m}^2 \text{g}^{-1}$ for each phase (Atkinson et al., 1967; Grim, 1968) and plotted in Fig. 10-3. The hydrous ferric oxide and illite experiments were performed in $0.034 \, M$ $NH_4 Cl$ and filtered sea water, respectively.

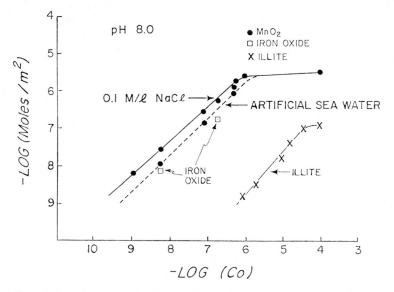

Fig. 10-3. Adsorption isotherm of cobalt on hydrous manganese dioxide, iron oxide and illite.

The slope of the Co—illite adsorption isotherm is parallel to the Co—MnO_2 isotherm but is lower by a factor of about 1,500. The points for cobalt adsorption by hydrous ferric oxide are about 2 to 3 times lower than the Co—MnO_2 adsorption data in 0.1 M NaCl.

For the purpose of comparison, it will be assumed that illite represents an average clay mineral. In order for clay minerals and MnO_2 to remove equal amounts of cobalt from sea water by adsorption, the area of clay removed by sedimentation must be 1,500 times greater than the area of MnO_2 removed. The average accumulation rate of Mn in sediments is 1.3 mg cm^{-2} per 1,000 years (Bender et al., 1970b), 70% of which is present in the hydrogenous rather than the lithogenous fraction (Chester and Messiha—Hanna, 1970). The accumulation rate of hydrous manganese is therefore 0.9 mg cm^{-2} per 1,000 years. Using a surface area of 260 m^{2}g^{-1} for MnO_2 (Murray, 1974), the surface area of manganese removed by sedimentation is 0.37 m^{2}cm^{-2} of sediment per 1,000 years.

The average clay accumulation rate is 0.5g cm^{-2} per 1,000 years (Bender et al., 1970b). Assuming an average surface area of 100 m^{2}g^{-1}, the surface area of clay minerals removed is 50 m^{2}cm^{-2} of sediment per 1,000 years. Therefore, the surface area of clay removed is only 135 times greater than that of manganese dioxide. If MnO_2 adsorbs 1,500 times more cobalt than clays, this implies that roughly one order of magnitude more cobalt is removed by MnO_2 than clay minerals.

The role of adsorption by hydrous iron oxide is more difficult to evaluate. The two experimental points indicate that iron oxide adsorbs about 2 to 3 times less cobalt than MnO_2 at comparable ionic strength. The total iron concentration in marine sediments is much higher than the manganese concentration; however, most of the iron (80%) is tied up in detrital phases (Chester and Messiha—Hanna, 1970). Thus, the amounts of hydrogenous iron and manganese oxides are roughly equal. If these phases had the same surface area, MnO_2 would remove about 2 to 3 times more cobalt from sea water than iron oxide.

The available experimental adsorption data suggest that manganese and iron oxides can remove considerably more metal from sea water by adsorption than clay minerals and that manganese oxide is about 2 to 3 times more effective than iron oxide. This suggests that the adsorption mechanism is possibly the most important on the cobalt concentration in sea water.

CHAPTER 11

ECONOMIC ASPECTS OF NODULE MINING

J. L. MERO

INTRODUCTION

Although deep-sea manganese nodules were discovered over a century ago by scientists of the HMS *Challenger* expedition, few analyses of the nodules for the economically significant elements such as nickel, copper, cobalt and molybdenum were made in those early days and no consideration was given to these deposits as a possible commercial source of metals until the early 1950's when the mining of the nodules was advocated as a possible source of manganese (Mero, 1952). As the result of a haul of nodules taken in relatively shallow water (900 m) about 370 km east of Tahiti on the western edge of the Tuamoto Plateau during the 1957—58 International Geophysical Year (the nodules contained about 2% of cobalt, a valuable metal at that particular time), a study was initiated by the Institute of Marine Resources of the University of California to determine if it might be economic to mine and process the nodules for their cobalt, nickel and copper contents. The results of that study were favourable with respect to the technical and economic factors involved in mining and processing the nodules. All the research and development in this matter dates from the release of the report describing the results of that study (Mero, 1958). To the present time (1975) over $150 million* has been expended in the exploration of the nodule deposits and in the development of mining and processing systems. An additional $100 million is to be spent in the next few years in these activities. The lead in the development of the nodules as a commercial resource has been taken by such groups as Kennecott Copper Corporation, Deepsea Ventures, International Nickel Company, Inc. and the CLB Consortium, a group of about 20 major mining companies from six countries. However, considerable interest in the nodules has been expressed by at least 30 companies and governmental agencies from such countries as Western Germany, Great Britain, Japan, Canada, the U.S.S.R., France, Australia and New Zealand. Numerous academic studies concerning the nodules are being conducted at universities and governmental research agencies in many industrial nations and it appears that within the next five to ten years, assuming no political and/or legal interferences, the nodules

---

*All values in U.S. dollars

should be in full-scale, economic production as a valuable source of important industrial metals.

Some of the studies concerning the economics of mining and processing the nodules indicate that the nodules promise to be a much less expensive and essentially pollution-free source of metals (Mero, 1972), in which case the nodules can be considered as a revolutionary source of industrial metals. Other investigators, however, indicate that the production costs of metals from the nodules will be similar to those of present land sources of metals in which case the deep-sea deposits can be considered only as an alternative source of metals (Rothstein and Kaufman, 1973).

In addition to being an apparently economic and essentially inexhaustible source of metals, the nodules, because of their very large specific surface area (of the order of 100 to 300 m^2/gm of matrix material) and their porosity (of the order of 50 to 60% pore space) (Fuerstenau et al., 1973), are indicated to be valuable in gas absorption (Zimmerly, 1967) and catalytic applications (Weiss, 1968). The nodules have also been shown to be very efficient collecting agents for heavy metals in the purification of crude oil before refining (Weiss and Silvestri, 1973). As well as serving as a source of metals, the nodules may therefore be potentially useful in a variety of other industrial applications.

A number of sources on information are available describing the nodule deposits of the ocean floor in some detail (Murray and Renard, 1891; Agassiz, 1960; Bezrukov, 1960; Skornyakova and Zenkevitch, 1961; Skornyakova et al., 1962; Mero, 1965a; Cronan, 1967; Meylan, 1968; Glasby, 1970; Horn, 1972; Frazier and Arrhenius, 1972; Morgenstein, 1973b). These sources cover theories of formation, known occurrences, chemical and mineralogical composition, surficial concentrations and other associated data bearing on the economic geology of these deposits. These aspects are covered in other chapters. However, some of the information which has a bearing on the economics of the exploration and mining of nodule deposits is included here.

GEOLOGICAL CONSIDERATIONS

Deep-sea manganese nodules are found in all the oceans and many of the seas of the world. The dominant environmental conditions associated with nodule formation are a relatively oxidizing environment, which promotes the precipitation of manganese from seawater, and a low rate of sedimentation which prevents the slowly forming nodules from being buried. Nodule deposits will form in depths of a few metres of water and within a few kilometres of shore, given the proper chemical and physical environment. Although many factors may be involved in controlling the rate of formation of the nodules and their ultimate composition, concentration and size, these other factors do not appear to play a dominant role in the actual formation of the nodule deposits per se. Manganese deposits appear in many forms on the ocean floor, as crusts on rock outcrops, as coatings on blocks of pumice or other loose rocks, as fillings in coral debris, etc. At the present time, economic interest is being shown only in the deposits of the nodules which are most commonly found as loose-lying, roughly spherical concretions at the surface of the soft seafloor sediments. The nodules range in colour from light brown to earthy black, are friable, with a hardness that does not exceed three or four on the Mohs scale, and have a bulk density varying from 2.0 to 3.1. They generally vary from 0.5 to 25 cm in diameter but, on an ocean-wide basis, appear to average about 4 cm in diameter. The size range and distribution of the nodules are important in economic considerations for several reasons. A deposit of the nodules may be of very high surface density (that is the surface of the sea floor may be totally covered by closely packed nodules) but, if the nodules are of a small size (less than about 0.5 cm), the surficial concentration will be relatively low (of the order of 5 kg/m^2 of sea floor or less). Since the mining equipment will be required to deliver the nodules to the surface at rates in the order of 5,000 tons per day, the gathering head would be required to cover about 2 km^2 per day assuming a 50% recovery efficiency in such low concentration deposits. Using a 10 m wide gathering device, the collector head would therefore have to traverse the ocean floor at a velocity of about 10 km/h (5.4 knots). A more practical velocity for such devices would be in the order of 2 km/h. The larger-sized nodules also have the practical mechanical advantage in the ease with which they can be gathered from the sediments on which they are resting. In general, the larger the nodule, the easier it will be to winnow from the associated sediment.

A number of specific shapes of the nodules have been recognized by several investigators and attempts have been made to categorize the morphology of the nodules (Grant, 1967; Meyer, 1973). An interesting fact concerning the morphology of the nodules is that the grade of the nodules appears to vary with the nodule shape (Meyer, 1973). A rough estimate of the grade of a nodule deposit can therefore be made simply by recognizing its morphological classification; this may be possible by viewing the deposits with an underwater television system.

In some deposits the nodules vary considerably in size and appearance while in other deposits the nodules may exhibit a strong group resemblance and may show a relatively small size range. In some cases, the external form of the nodule depends on the shape of the nucleus. In other cases, the nucleus is not a single body, but several. When these nuclei are close to one another, the growing nodules may coalesce to form a single, slab-like nodule with several protrusions. In many areas, however, slab-like objects associated with the nodules are generally blocks of pumice thinly coated with manganese and iron oxides. The chemical nature of the nucleus does not seem to affect the deposition of manganese or the composition of the manganese and iron phases of the nodule. The nuclei may be carbonates, phosphates, zeolites, clays, remains of biota, silicates or various forms of altered and unaltered silica. Any hard object seems to be able to serve as a nucleus for the formation of manganese nodules. The most commonly observed are shards of highly altered pumice, the alteration probably taking place while the nodule is growing.

The hydrochloric acid-insoluble fraction of the nodules, which ranges from about 2 to 40% and averages about 25% of the bulk weight of the nodules, is practically free of the heavy metals that are characteristic of the acid-soluble fraction. The hydrochloric acid-insoluble fraction consists principally of clay minerals together with lesser amounts of quartz, apatite, biotite and sodium and potassium feldspars (Riley and Sinhaseni, 1958). These materials are generally very fine-grained and intimately mixed in the matrix of the nodule such that it appears that it would be impractical to attempt any physical separation of these gangue minerals before chemical processing of the nodules.

In addition to the matrix-included gangue minerals, any practical mining device will recover such extraneous materials as erratic boulders or other rocks, blocks of pumice, sharks' teeth and cetacean earbones; these materials can be mechanically separated from the nodules at the mine site. Such materials, in most instances, will not constitute more than about 5—10% of the total material dredged. Varying amounts of the clay or other sediments on which the nodules are resting will also be recovered; however, these materials are very fine-grained and can be screened from the nodules on the ship.

Although an understanding of the mineralogy of the nodules is important in any economic study, this subject is covered in some detail in other chapters of this volume (cf. Chapter 7). It is important to note, however, that the mineralogical or chemical character of the nodules does vary from location to location in the ocean and that changes in these characteristics do affect the processing of the nodules. Thus, the chemical character of the nodules is another factor to be considered in any nodule exploration programme. Until recently it had been assumed that the nickel, cobalt and copper elements in the nodules were present in solid solution as replacements for manganese and iron in the crystal structure of these elements. Recent work at the University of California at Berkeley has indicated that the bulk of the nickel, cobalt and copper are probably contained in the nodules as ions loosely attached to the surfaces of the manganese and iron crystallites (Fuerstenau et al., 1973). This observation is interesting from a processing standpoint as it indicates that these elements can be stripped from the nodules without putting either the manganese or iron into solution. This study also indicates that cobalt seems to be part of the iron crystal structure in certain nodules. To achieve high recovery efficiencies of this element, the iron crystal structure must therefore be disrupted. The percentage of the cobalt associated in this way in the nodules appears to vary from place to place. Thus, the chemical structure of the nodules is one more factor involved in determining the economic value of the deposits.

Surface density considerations

So far, surface densities of nodules ranging up to about 30 kg/m^2 of sea floor have been found at a number of locations centred around 20°N 114°W (Fig. 11-1). It should be noted, however, that the concentration of nodules in any given area can, and usually does, vary considerably over rather short lateral distances of a few kilometres or, in some cases, a few metres (Kaufman, 1974). Figs. 11-2 and 11-3 illustrate the great changes which can take place in the character of the nodule deposits over a lateral distance of

Fig. 11-1. A sea-floor photograph taken in 3,778 m of water at 20°00'N 113°57'W. The photograph covers a sea-floor area about 160 cm by 160 cm. The nodules average about 6 cm in diameter and the coverage would be about 80%. The concentration estimate is 30 kgm/m^2. (Photograph by N. Zenkevitch, Institute of Oceanology, Moscow, U.S.S.R.)

Fig. 11-2. A sea-floor photograph taken in 4,960 m of water at 12°55.8'N 141°32.6'W, showing a 90% coverage of nodules. The nodules show two distinct size groupings, one averaging about 6 cm and a second, which is more abundant, averaging about 2 cm in diameter. Possibly the nodules represent two families which started growing at different times due to variations in the physiochemical environment or sediment erosional patterns. The estimated concentration is about 25 kg/m^2. The white patches of sediment are probably weathered blocks of pumice which drifted to this location from volcanic eruptions before becoming water logged and sinking to the ocean floor. Smaller patches of sediment covering the nodule bed may be generated by burrowing animals on the sea floor. (Photograph by J. E. Andrews, University of Hawaii.)

about 90 km. Equally large changes are known to take place over lateral distances of a few metres. In some locations, however, such as the area around 20°N 114°W or in an area about 280 km north of Tahiti, a high and uniform surficial concentration of the nodules is maintained over many tens of kilometres and, possibly, hundreds of kilometres.

An overall oceanic average concentration of the nodules in economic grade deposits would be in the range of 5-20 kg/m^2. Within individual deposits, such as the 20°N 114°W field, the deposit may contain as much as one billion (10^9) tons of the nodules in a 100,000 km^2 area.

Fig. 11-3. A sea-floor photograph taken in 4,885 m of water at 13°18.0′N 140°45.4′W, showing manganese crusts, possibly covering soft sediments but, more probably, covering an outcrop of hard rock. Taken about 90 km from the location of Fig. 11-2, this photograph illustrates the extreme changes which can occur in the nodule deposits over relatively small lateral distances. Such changes are known to occur over distances of a few tens of metres. (Photograph by J. E. Andrews, University of Hawaii.)

Much of the information concerning the continuity of manganese nodule deposits is available as the result of extensive exploration activities conducted by the *Valdivia* research group of West Germany. In one of that group's publications, the percent of ocean floor area covered by manganese nodules is shown for approximately a 2,000,000 km² area of the Pacific Ocean between the Clarion and Clipperton fracture zones in a region centred about 2,000 km southeast of Hawaii (Schultz-Westrum, 1973). On the basis of about 1,000 sample points, including some 200 photographic stations, a map was prepared showing the percent of sea floor covered by nodules. While the percent of the ocean floor covered by nodules is an interesting statistic and is of some general value in economic considerations, it is of little help in calculating the concentration of nodules in weight per unit area of the ocean floor, which, from an economic standpoint, is a more important statistic, unless the size distribution of the nodules is also known for the specific area. Using the coverage data given by Schultz-Westrum (1973) and, assuming an average nodule size of 4 cm and an average coverage of nodules in the area surveyed of 20%, the tonnage of nodules contained in this area can be estimated to be about 15 billion metric tons. An average coverage of 20% of 4-cm diameter nodules of bulk nodule density of 2.0 would yield a surficial concentration of nodules of about 10 kg/m^2 or about 10,000 metric tons per square kilometre. In some 15 areas covering a total of about 100,000 km^2, the surface coverage is indicated to be in excess of 50%. The area of which this coverage data is given is also that of relatively high-grade nodules.

Compositional considerations

If the dollar values of the metals contained in the manganese nodules of the Pacific Ocean, based on the following metal prices, $1.00 per percent manganese*, $7.90 per kilogram of cobalt, $4.00 per kilogram of nickel, and $1.80 per kilogram of copper, are plotted on a map and points of equal dollar value contoured (Fig. 11-4), it can be readily seen that the high-grade nodule deposits as outlined by the $120 of contained metals per dry weight ton of nodules, tend to be concentrated in a band extending eastward from about longitude 117°W between the 5°N and 10°N lines of latitude to about 143°W where the band broadens considerably to between 5°N and 20°N extending eastward to about the 120°W line of longitude. This high-grade area in the North Pacific appears to cover about 6,000,000 km^2.

---

*In the case of manganese, the cost of the ore is quoted in U.S. dollars per long ton unit, which is U.S. dollar per percent of contained manganese in the ore.

Fig. 11-4. A map of the Pacific Ocean showing the dollar value of the metals contained in a dry-weight ton of manganese nodules taken from various locations throughout this ocean. The dollar-value numbers of the various samples are calculated on the following basis: $2.20 per percent of manganese; $7.90 per kilogram cobalt; $4.00 per kilogram of nickel; and $1.80 per kilogram of copper. The regularity of the lines connecting points of equal dollar value would tend to indicate that additional extensive high-grade zones of nodules are unlikely to be found in this ocean. Local points showing dollar-value figures in excess of $200 per ton are of nodules showing very high cobalt assays. Such nodules are taken from the tops of seamounts and do not represent mineable deposits due to their limited extent and the roughness of the topography on which they are found. This map clearly reveals the high-grade regions of nodule deposits in the Pacific Ocean, i.e., those areas enclosed by the $120 per dry-weight ton of nodules contours.

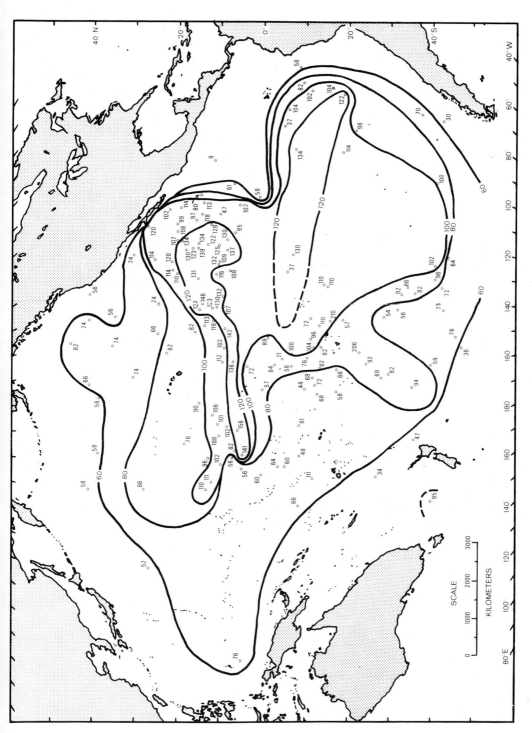

In the South Pacific there appears to be a second high-grade area, covering about 6,800,000 km^2 of ocean floor. This area, however, is highly speculative being based on only three sample points with nodules containing greater than $120 of metals per ton of nodules. This South Pacific high-grade area may be a reflection of the same processes forming the high-grade area of the North Pacific. It is unlikely that the two high-grade areas would merge in the equatorial region because of the high organic productivity of these waters and consequent high calcareous sedimentation rates of this region.

Because of the relatively uniform distribution of the nodule dollar values, as indicated in Fig. 11-4, it is suspected that other high-grade areas of large lateral extent in the North or South Pacific Ocean are unlikely to be found. Smaller high-grade areas, covering a few tens of thousands of square kilometres, however, are very possible, especially in the basin and range areas of the west-central Pacific Ocean. Due to a paucity of data in the southeast-central Pacific it is rather difficult to predict what may occur in this area. Because of the rather high sedimentation rates throughout the area, it is not expected that extensive deposits of high concentration will be found there; however, that does not preclude the occurrence of occasional deposits containing high-grade nodules.

Those areas in the Pacific Ocean north or south of about the 40° lines of latitude are unattractive for mining operations, not only because of the general low-grade of the nodules in these areas but also because of the relatively high percentage of ice-rafted gangue materials found intermixed with the nodules. The weather in these areas is also not favourable for extended mining-vessel operations.

The western section of the North Pacific high-grade area may not be continuous as indicated in Fig. 11-4. Each individual sample point, especially in the westernmost region, may be an indication of a separate nodule deposit of limited extent within a separate valley. West of about 160°W, the floor of the Pacific Ocean is characterized by numerous mountain ranges and valleys. The separate valleys may contain nodule deposits and the character of the deposits, especially as concerns composition, may change markedly from one valley to the next.

The highest grade of nodules so far discovered occurs at a location of 6°N 170°W where the nodules assayed 2.3% Cu, 1.9% Ni, 0.2% Co and 35% Mn on a dry weight basis (Mero, 1965a). The nodules from this deposit, however, appear to be of a small diameter. Thus, although the coverage of nodules on the ocean floor is high at this locality, the surficial concentration may nonetheless be too low for economic mining. Further, little is known of the ocean floor topography in this specific area and the topography may be too rugged to permit efficient operation of mining equipment.

In that section of the high-grade area east of about the 150°W, the ocean floor is relatively devoid of high mountains or mountain ranges and the nodule deposits can be expected to be more uniform in grade and concentration. The nodule deposits can also be expected to be of much greater lateral extent in this area. The deposits may even be continuous throughout this area, although varying on average by a factor of 3 or so in surficial concentration and by a factor of about 2 in grade. Large patches of sea floor in this area will be devoid of surficial nodule deposits as the result of local high sedimentation rates. In addition, at least 20% of the area will be occupied by seamounts or abyssal hills showing manganese crusts or having slopes too steep to permit the operation of mining equipment.

Table 11-I lists the metal content of a representative number of samples throughout the high-grade region of the North Pacific and, where known, the surficial concentration of the nodules at the same or a nearby point. It should be noted that these are assays of a cross-section of a nodule or of the entire nodule from the various locations. Analyses of bulk samples of the nodules from any given location, including the detrital materials which may be dredged along with the nodules, generally tend to show assays about 10—20% lower than that of an entire nodule itself from the same dredge haul. In many cases, however, these detrital or gangue materials can be physically separated from the nodules on the mining vessel. The fine-grained detrital minerals included in the nodule matrix itself cannot be physically separated from the manganese-iron minerals of the nodules. The detrital minerals in the nodule matrix are, of course, included in the assays shown in Table 11-I. In general, the water content of freshly recovered nodules is in the range of 25—35% of the total weight of the nodule. Much of this physically entrained water will evaporate if the nodule is left exposed in air for any length of time. For this reason nodule assays are generally normalized to a dry weight basis with the drying taking place at 110°C to constant weight. Once dried, the nodule material must be kept in water-tight containers for the dried nodule will rapidly absorb moisture from the atmosphere.

Associated sediments

Manganese nodules are found associated with practically all types of pelagic sediments and, in special circumstances, with terrigenous sediments. In general, the nodules are associated most commonly with those pelagic sediments that are deposited the least rapidly. In fact, a principal environmental condition for the formation of deep-sea manganese nodules seems to be a very slow rate of formation of the associated sediments relative to the growth rate of the nodules. Micronodules, a common constituent of all sea-floor sediments formed under oxidizing conditions, are most probably the result of embryo nodules being buried in the sediments. Nodules are therefore most commonly found associated with the red clay type of pelagic sediment. Calcareous sediments appear, in general, to inhibit the formation of extensive beds of nodules, possibly because of the rapid rate of deposition

TABLE 11-I

Metal contents (%) of a representative sampling of manganese nodules from the North Pacific high-grade region

Station:	RC10-91	Chub 5	Amp 3P	DwBd 2	Msn 153	Msn 148	Wah 24	Msn K	Averages
Lat. (N):	12°16'	15°00'	15°04'	10°26'	13°07'	9°06'	8°20'	6°03'	
Long. (W):	120°10'	125°26'	125°05'	130°38'	138°56'	145°18'	153°00'	170°00'	
Depth (m):	4,471	4,380	4,500	4,890	4,927	5,400	5,143	5,400	
Chemical analysis (dry-weight basis)									
Manganese	31.5	27.8	23.2	28.0	29.0	30.9	24.9	35.2	28.8
Iron	5.1	12.2	6.0	9.4	5.9	6.2	7.3	6.4	7.3
Cobalt	0.16	0.47	0.23	0.32	0.38	0.31	0.25	0.19	0.29
Nickel	1.68	1.25	1.91	1.54	1.75	1.79	1.86	1.87	1.71
Copper	1.40	1.02	1.24	1.50	1.56	1.50	1.65	2.30	1.52
Contained metal values ($/ton)*	125.90	122.90	128.60	130.50	144.40	141.70	135.30	153.00	135.50
Sea-floor concentration (kg/m²)	—	—	15	15	8	10	—	5	

*At the following metal values: $1.00 per percent of manganese; $7.90 per kg of cobalt; $4.00 per kg of nickel; and $1.80 per kg of copper.

of these sediments on the ocean floor. An interesting association of the nodules occurs in conjunction with siliceous pelagic sediments, a prime example of which is the high-grade band of nodules lying generally between the Clarion and Clipperton fracture zones. The very high porosity of these siliceous sediments may be responsible for the diagenetic processes at the seafloor—seawater interface which enhances the grade of the nodules in these regions (Horn et al., 1973a).

Motion of the superjacent water seems to be a common environmental characteristic in areas where nodules are forming. Ripple and scour marks are frequently seen in the fine sediments on which the nodules rest. To support growth of the nodules, these sea-floor water currents sweep the precipitating manganese and iron colloidal particles into contact with nuclei and they help to maintain an oxidizing regime on the ocean floor. Should a reducing regime develop as a result of stagnating water in an area where nodules have formed, any existing nodules would probably dissolve. The velocity of these ocean-floor currents is probably not in excess of 0.2 km/h, except near seamounts or in constricted deep-sea valleys where such currents may exceed a velocity of 2 km/h.

Nodules at depth in the sediments

Micronodules are frequently found distributed throughout the bulk of sediment cores taken in deep-sea sediments. In only a few instances do these micronodules constitute more than about 5% of the bulk of the material in the cores. Nodules greater than a centimetre in diameter are frequently found at discrete horizons in the sediment cores; however, such nodules are not found as frequently as they are found at the surface of these cores. Generally, the gravity cores do not penetrate the sediment column more than about a metre and this relatively shallow sampling of the sediments may erroneously favour this conclusion. The nodules found to date at depth in the sediments should not increase the estimated amount of nodules in the Pacific Ocean by more than a factor of 10. In any case, it probably would not be economic to work through, or dredge, a large bulk of sediment to recover those nodules which are buried in the sediment. At the present time, economic calculations concerning the mining of ocean-floor nodules should therefore consider only those nodules which are found at the surface of the sea-floor sediments.

Compositional zones of the manganese nodules in the Pacific Ocean

Chemical analyses have been made on about 700 samples of nodules from the world's oceans, about 500 of which are of nodules from the Pacific Ocean. Mero (1965a) lists assays for 16 elements from about 200 of these locations; Cronan (1967) lists assays for 11 elements from about 125 locations; and Horn et al. (1973a) list analyses for 7 elements from 605 locations in the oceans, 434 of which are from the Pacific Ocean, 94 of which are from the Atlantic Ocean and 77 of which are from the Indian Ocean. When the assay data are plotted on a map of the Pacific Ocean, definite regional variations in the composition of the nodules are noticeable. Fig. 11-5 shows the indicated boundaries of the compositional regions noted. While this chemical compositional distribution of the nodules is of some geochemical importance it is also of economic significance for it will allow the mining of a composition of the nodules with a mix of metals which conforms best with the demand for those metals. Since the mining facility is not fixed in space in the ocean as on land, the operator can move his operation from place to place as the market demand for different metals changes.

The nodules in the areas labeled A in the map of Fig. 11-5 are characterized by manganese/iron ratios generally less than one whereas a general ocean-wide ratio for these metals is generally near two. The iron/cobalt ratios are on average higher than those of the other regions and range as high as 520. These high-iron regions generally lie along the continents and are probably the result of less oxidizing conditions than found in deeper areas of the ocean. Iron is preferentially precipitated relative to manganese in less oxidizing environments.

The three areas in the eastern Pacific are characterized by very high manganese/iron ratios. These ratios range from 12 to 50 and average 23. There are apparently transitional zones between the high-manganese regions and the other compositional regions in which manganese nodules possess compositional characteristics of both regions. Thus, the nodules from transition zone, $BC-2$, assay high in manganese, nickel and copper. The nodules in the B regions seem to be forming very rapidly. The very low content of nickel, cobalt, copper and lead of the nodules of the B regions is probably an indication of the short time span between precipitation of the manganese from solution in seawater and the agglomeration of the manganese sols at the sea floor into the nodules.

The areas farthest removed from land, both continental and island, seem to be regions of relatively high nickel—copper nodules. These regions, labeled C in Fig. 11-5, are, by far, the dominant compositional regions in the Pacific. The manganese/iron ratios of the C regions are relatively stable, ranging from 1 to 6 and averaging 2.1. The copper assays of the nodules from these

---

Fig. 11-5. Map of the Pacific Ocean showing the major compositional regions of manganese nodules. The A regions contain nodules relatively enriched in iron; the B regions contain nodules enriched in manganese; the C regions contain nodules enriched in nickel and copper; and the D regions contain nodules enriched in cobalt. Transitional zones in which the nodules show dual compositional characteristics are indicated by two letters.

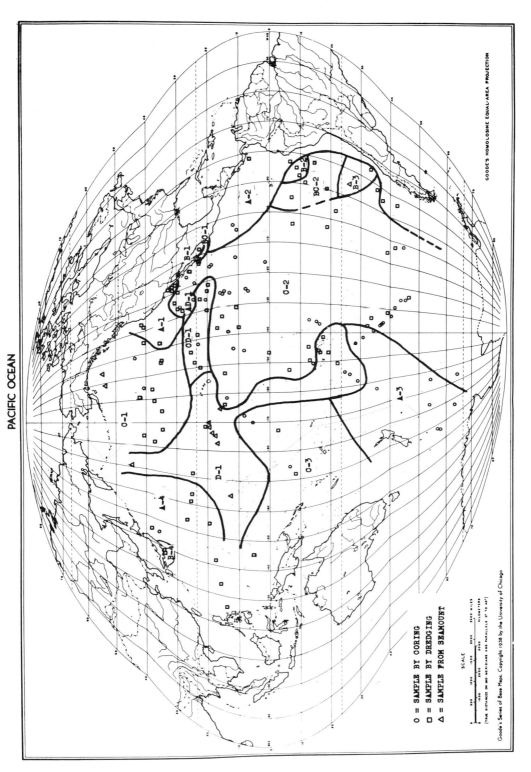

PACIFIC OCEAN

GOODE'S HOMOLOSINE EQUAL-AREA PROJECTION

O = SAMPLE BY CORING
□ = SAMPLE BY DREDGING
△ = SAMPLE FROM SEAMOUNT

SCALE

Goode's Series of Base Maps. Copyright 1936 by the University of Chicago

regions show a greater range in value than do the nickel assays. On the average, the copper content of the nodules tends to increase near the Equator. It is thought that biologic agencies are responsible for fixing at least part of the copper in the manganese nodules in the high productivity zone near the Equator. Greenslate et al. (1973) have proposed a biological mechanism for the equatorial regions whereby the precipitating remains of planktonic organisms incorporate and transport to the sea floor large quantities of transition metals such as copper and nickel.

Centred on topographic highs in the central part of the Pacific Ocean are two regions in which the nodules assay relatively high in cobalt. Eleven samples of nodules from these two regions average 1.2% Co, with the highest value being 2.6% Co. The manganese/iron ratios of these regions vary less than in any of the other regions, ranging from 0.9 to 1.8 and averaging 1.3. The iron/cobalt ratios are also more stable in these regions, ranging from 8 to 39 and averaging 22. The source of the cobalt in these nodules may be the volcanic rocks with which they are associated. However, these shallow areas are also generally highly oxidizing environments, largely due to high currents which sweep over the tops of the seamounts. This condition may be primarily responsible for fixing cobalt in the nodules of these areas (Burns and Fuerstenau, 1966).

In the east-central North Pacific is a zone of relatively high cobalt—nickel—copper nodules; this zone, labeled CD—1, more or less coincides with the high-grade zone of the North Pacific.

Most probable area for nodule mining operations

Considering all factors, it would appear that the high-grade area of the North Pacific would be that most likely in which commercial mining operations will be initiated. Most of the commercial nodule exploration work, with the exception of that conducted by CNEXO in the South Pacific in the areas near Tahiti and the Marquesas Islands, has so far been undertaken in this area. The continuing exploration work of the Kennecott and Deepsea Ventures groups is being concentrated in this region. Not only are extensive deposits of sufficient grade, concentration and continuity found in this area but also the weather is relatively favourable during most of the year. In addition, the deposits are relatively close to North America where a major market exists for the metals to be produced from the nodules.

While extensive deposits of the nodules have been found in the Atlantic and Indian oceans, those deposits tend to be of a considerably lower grade, on the average, than the deposits of the Pacific Ocean, with respect to nickel and copper which are the most important economic metals in the nodules. This lower metal content is probably a reflection of the relatively high rate of continental runoff per unit area received by the Atlantic and Indian oceans and the resulting dilution of the nodules with clastic materials. As grade is a very important factor concerning the economics of nodule mining and processing, all of the commercial attention on the nodule deposits has so far been concentrated in those deposits in the eastern Pacific Ocean.

Tonnages of metals in the nodule deposits of the high-grade area

Assuming an average concentration of 9 kg/m^2 of ocean floor over the 6,000,000 km^2 within the North Pacific high-grade area, a reserve of about 38 billion tons of dry nodules averaging 29% Mn, 0.3% Co, 1.7% Ni and 1.4% Cu would be indicated. The water content of the nodules is assumed to be 30%. Such a reserve would contain about 11.0 billion tons of manganese, 115 million tons of cobalt, 650 million tons of nickel, and 520 million tons of copper. These nodules also contain an average of about 0.14% Zn and 0.06% Mo and the deposit will probably be an important source of these two metals as well.

EXPLORATION OF DEEP-SEA NODULE DEPOSITS

In general, there are two basic techniques of exploring for deep-sea nodule deposits. One technique employees the use of a winch and cable with which an underwater camera or television can be lowered to view the nodule deposits or a dredge bucket, coring device, grab or other sampling device can be lowered to the ocean floor to secure samples of the nodules. In efficient systems, nodule samples and photographs can be obtained on the same lowering. With television systems, a continuous record can be made by the use of video-tape recorders. A second technique of sampling the deposits employs free-fall devices which are dropped into the ocean and sink to the ocean floor at a rate ranging from 35 to 70 m per minute depending on the ratio of the displacement of the flotation element to that of the submerged weight of the ballasting elements. At the ocean floor the free-fall devices, depending on their design, will take a grab sample of the nodules or bottom sediments over an area ranging from 0.1 to 0.2 m^2, release their ballast, and return to the surface, again at rates varying from about 35 to 70 m per minute. Instead of a grab sampler, a camera, corer, water sampler or other data-gathering device can be mounted on the free-fall vehicle. In some designs, both a camera and grab are mounted on the same vehicle.

Deep-sea television systems can yield high-quality data concerning the continuity of the deposits both with respect to surface density and size range of the nodules. Such high-quality data are not attainable with other sampling techniques. In addition, the television systems can be used to monitor current velocities, to determine changes in sea-floor elevation and the direction of maximum slope, to gather information on obstructions which may hinder the operation of mining equipment and to detect sudden changes in the bottom topography all of which are important factors in studying any deposit prior to mining. The German AMR Group have mounted their television camera on a sled which can be towed just above the ocean floor at relatively high velocities of up to about 2 knots. Using a low light level camera, it is possible to view the ocean floor from distances as much as 15 m. On board ship, the German system employs a very elaborate array of signal processing equipment with which it is possible to determine the maximum and minimum sizes of the nodules being viewed as well as the average nodule size automatically. The equipment will also determine the percent coverage which combined with the average nodule size can yield a number concerning the surface density of the nodules.

Still cameras are lowered to the ocean floor via a wire line and, through the use of acoustic devices, can be held just above the ocean floor to take as many as 1,000 separate photographs per lowering while the ship is drifting. Still cameras can also be designed with bottom contact switches so that the camera will record only a single photograph each time it is dropped to the sea floor.

Sampling devices consist of rock dredges, dredge buckets designed specifically for the gathering of the nodules, sediment corers, grabs of various types, .and spade corers. The spade corers are generally designed to take a square core, 20 by 20 cm to 30 by 40 cm in lateral dimensions and to a depth into the sediments ranging from 20 to 50 cm. These devices are particularly useful in securing undisturbed samples of the nodules and sediments for various measurements such as the bearing strength of the sediments as well as for securing accurate data on the concentration of the nodules at the point sampled. These devices are also useful in gathering data concerning the nodule—sediment interface. At present, the best possible combination for the delineation of nodule deposits (especially in detailed surveys of a specific deposit) probably consists of a television system and a spade corer. The exploration of nodule deposits with these devices is, however, slow and time-consuming and therefore expensive per unit of area covered. Regional surveys can be made at considerably less cost through the use of free-fall devices.

Free-fall devices designed specifically for nodule deposit studies were initially developed by Kennecott Copper Corporation (Schatz, 1971) and subsequently by CNEXO and other groups. These devices consist of a light-weight frame in which is mounted one or more glass spheres to provide buoyancy, a grab or camera with which samples or photographs of the nodules can be obtained, weights to carry the device to the ocean floor and which are released at the ocean floor to allow the device to return to the surface, various tripping devices to fire the camera and flash and/or close the grab, and flags, radio beacons and flashing lights to facilitate relocation of the device on its return to the surface. A number of these devices can be dropped at one station or on a predetermined pattern to provide a relatively high degree of certainty concerning the surface density, size distribution and grade of the nodules in a specific area. The success ratio in recovering these devices is generally greater than 90% and, with care, can approach 100%. A major advantage of the free-fall devices is that they can be easily deployed from a small vessel and their initial capital investment is relatively low. In addition, devices can be deployed while the ship is occupied with other survey activities whereas such dual activities are difficult to accomplish with wire line systems.

In any nodule exploration programme, a determination of the sea-floor topography, the average depth and relief of the area and the slope of the abyssal hills is important. In general, standard deep-sea echo sounders are used in making these determinations. Towed vehicles carrying various acoustic devices close to the ocean floor will yield a much higher degree of resolution than those acoustic systems utilizing surface vessel mounted acoustic transducers. Some attempt has been made to use high-resolution, high-frequency and narrow-beam, echo sounders to obtain a better degree of topographic resolution than the standard 12 kHz echo sounders provide. High-frequency seismic probes have also been used to try and differentiate those areas which contain nodule deposits and those areas consisting of rock outcrops or barren sediments. Side-scan sonars mounted on deep-towed vehicles appear to have shown some success in outlining barren patches within nodule deposits (Lonsdale, 1974).

Positioning of the samples and sampling devices, especially at the ocean floor is important. In general, satellite navigation systems are now used for vessel positioning. Under good control, positioning within about 0.2 km can be secured with these systems. Arrays of acoustic transducers have been employed at the ocean floor to position sampling devices at the ocean floor very accurately with reference to the emplaced acoustic transponders. Highly accurate surveys of deposits can therefore be made; however, the position of the deposit area with reference to some benchmark on land is still only as good as the surface positioning system itself.

CONSIDERATIONS IN THE SELECTION OF MINING SITES

There are many factors involved in the calculations used to determine the economic value of a deposit of manganese nodules. The more important of these are the metal contents of the nodules, areal extent of the deposit, surface density of the nodules per unit area of sea floor, continuity of the deposits with respect to metal content and size, size distribution of the nodules within the deposit, water depth, distance to port or process facility, topography of the ocean floor in the deposit area, current velocities throughout the water column, physical characteristics of the associated sediments, frequency and distribution of obstructions to mining systems within the deposit, ease with which the nodules can be processed, and weather. Of these considerations, as long as certain minimum standards are met with reference to all the other factors, the dominant factor in determining the economics of mining any specific deposit in most cases is the metal content of the nodules. Because of this consideration, it is the nodule deposits of the Pacific Ocean and more specifically the deposits between the Equator and the 20°N and between the 110° and 180°W which are of greatest interest at the present time. In general, if the grade of the nodules exceeds 2.8% Ni + Cu + Co on a dry weight basis, if the average surficial nodule concentration exceeds 5 kg/m^2 of sea floor, if the slope of the sea floor does not exceed about 10%, if the percentage of gangue materials mined with the nodules does not exceed about 20% and if the weather of the area permits at least 250 days per year of operations, the deposit would be considered to be of economic grade.

Kaufman (1974) outlines some of the considerations involved in determining the acceptability of a nodule deposit for mining and concludes that, for the mining system envisioned by his company, the deposit should assay at least 20% Mn, 1.0% Ni, 0.8% Cu and 0.2% Co. In addition, the surficial concentration should be greater than 5 kg/m^2. Some of the factors noted by Kaufman (1974) which affect the size of an area needed to contain a mineable deposit are: about 15—25% of the deposit area will consist of obstructions which must be avoided by the mining system; the nodule recovery efficiency of the mining systems is not likely to exceed 60% of the nodules actually swept over by the system; and, the percent of the area which can actually be covered by the mining system nodule gathering head at the sea floor due to lack of control in positioning this device accurately is not likely to be more than about 65%. Considering all of these factors, Kaufman develops a formula which can be used to determine the total area required to contain a mine; that is, an area with a sufficient quantity of nodules of sufficient grade to pay the operating costs of the venture and generate sufficient profits to amortise the capital investment of the venture as well as pay an acceptable rate of return on the investment.

MINING MANGANESE NODULES

Although many techniques of mining the deep-sea manganese nodules have been proposed (Mero, 1958), work is presently proceeding on two types of systems, a hydraulic suction dredge and a mechanical cable-bucket dredge. Three forms of the hydraulic system are under consideration, one powered by centrifugal dredge pumps, another powered by air injected into the pipeline and a third being a combination of the other two.

Essentially the hydraulic system, as shown in Fig. 11-6 consists of a length of pipe which is suspended from a surface float or vessel; a gathering head, designed to collect and winnow the nodules from he surface sediments and feed them to the bottom of the pipeline while rejecting oversized material; and some means of causing the water inside the pipeline to flow upward with sufficient velocity to suck the nodules into the pipeline and move them to the surface of the ocean. In 1970, Deepsea Ventures Inc., a subsidiary of Tenneco Corp., successfully tested an air-lift dredge of the general design as described by Flipse (1969) at a depth of about 760 m on the Blake Plateau, a 250,000 km.^{2} area off the southeastern coast of the United States (LaMotte, 1970). In that test, some 60,000 tons of the nodules were reported to have been recovered at rates varying from 10 to 50 tons/h through a 24-cm diameter pipeline. As the nodules from the Blake Plateau are generally of low grade, most of these recovered nodules were simply dumped back into the ocean. Plans are now being made to extend that system of dredging to depths of 4,500 m.

In general the estimated capital costs of the hydraulic systems of mining the nodules at average annual production rates of one million tons of nodules are between $30 million and $100 million. The indicated production costs of moving the nodules from the ocean floor to the surface with these systems is estimated to be between $10 and $20 per ton of nodules produced.

The second general type of system planned for the mining of the nodules on a full-scale basis is the Continuous Line Bucket (CLB) System. This system consists essentially of a surface vessel, a loop of cable to which dredge buckets are attached at 25—50 m intervals, a traction machine on the surface vessel capable of moving the cable such that the buckets descend to the ocean along one side of the loop, skim over the sea floor on the bottom side of the loop to gather the nodules, and return to the surface on the third side of the loop. This system of dredging the nodules in illustrated in Fig. 11-7. So far this system of recovering the nodules has been tested in a series of experiments, first in 1,500 m of water in 1968, then in 3,600 m of water in 1970 (Masuda et al., 1971), and finally on a full-scale basis in 4,700 m of water in 1972. Because of its great simplicity, the capital costs of the CLB System are relatively low, in the order of $10 million for a two million ton per year system, exclusive of the cost of the surface vessel, which system would be capable of recovering the nodules from any depth up to about 5,500 m of water. The estimated operating costs of this system, including the cost of a chartered surface vessel, are now about $5 per ton of nodules recovered. The CLB System can be mounted on practically any type of vessel capable of crossing the open ocean and of carrying a total load of about 3,000 tons. In addition to its simplicity, the CLB System incorporates a very high degree of flexibility in being able to work in deposits of the nodules of varying concentration and size range, over relatively great sea-floor topographic relief and in a range of sediment bearing strengths and characteristics. A given CLB System can be easily modified to operate in any

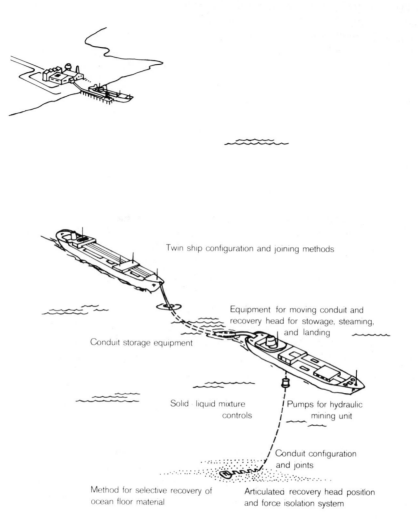

Twin ship configuration and joining methods

Equipment for moving conduit and
recovery head for stowage, steaming,
and landing

Conduit storage equipment

Solid · liquid mixture
controls

Pumps for hydraulic
mining unit

Conduit configuration
and joints

Method for selective recovery of
ocean floor material

Articulated recovery head position
and force isolation system

Fig. 11-6. Two types of deep-sea hydraulic dredges proposed for the mining of manganese
nodules. The system above is that proposed by the Deepsea Ventures Group. It is
supported by a surface vessel and is powered by an air-lift pump with air being injected
into the pipeline at about 35% of the dredging depth. The system on the right, proposed
by J. L. Mero, is supported by a submerged float to prevent surface wave motions from
being transmitted to the dredge pipeline. A centrifugal dredge pump of several stages is
located inside the main flotation tank at about 15% of the dredging depth. The entire
dredge rotates around its long axis with collecting heads fanning out at the seafloor so
that the dredge can cut a very wide swath throughout the nodule deposit with a low
lateral motion of the dredge as a whole. This avoids high energy inputs to move the
dredge at high velocities over the deposit.

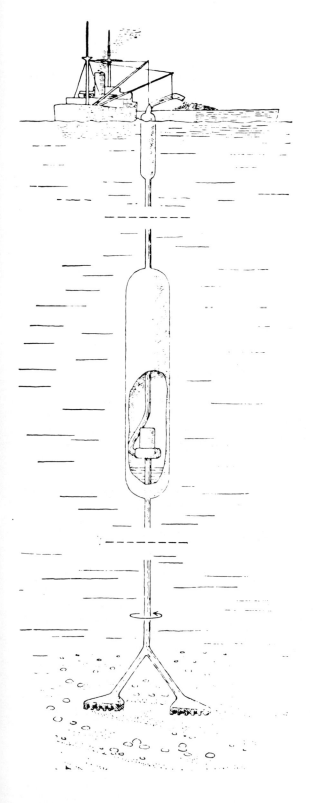

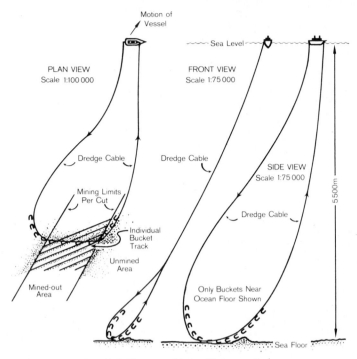

Note: Cable Diameter and Buckets Not Drawn to Scale

Fig. 11-7. A diagram illustrating the design and operation of the Continuous Line Bucket (CLB) System for mining deep-sea nodules as proposed by Y. Masuda. Various forms of this system such as a two-ship system have been developed in order to achieve better control over cable separation and therefore width of the bucket cut at the sea floor.

depth of water and all parts of the system which operate submerged are surfaced several times a day for inspection and repair.

Transport of the nodules to a process facility

There is some indication that the manganese nodules could be processed for their metals aboard a vessel at sea (Mero, 1972). Because of the rather large-power requirements for such a system and the general unavailability of cheap power at sea, it is more likely that the nodules will be transported to some port where inexpensive hydro-electric power, or power produced from natural gas, is available. In general, large-bulk ore carriers can transport ore materials for costs in the range of $0.0003 to $0.0005 per ton—km depending on the size of the carrier and the amount of automated loading and unloading equipment available. It can be expected that the distance from the mining site to a process facility will range from about 2,000 to

10,000 km. Since the carrier must return to the mine site empty, the cost of transporting the nodules can be expected to fall in the range of $3 to $11 per ton depending largely on the size of the transport vessel. This cost estimate assumes a $1 per ton charge is involved in handling the nodules on shore.

As the nodules are of a relatively low density and are easily crushed, it can be expected that they will be pumped from the mining vessel to the carrier and from the carrier to the process plant allowing for high filling and unloading rates. Because of the relatively low capital investment involved in the mining system itself, the CLB System could be installed on an ore carrier and simply shut down while the carrier is returning to port. It is estimated that it will require about six hours to install a CLB System at sea for mining operations from the carrying vessel and a similar amount of time to bring the system inboard when the carrier was filled. At a 15,000 ton/day production rate it would require about seven days to fill a 100,000 ton carrier.

Hydraulic systems, on the other hand, require a relatively large amount of time to install and remove from the sea and because of their rather high capital investment, it is necessary to keep the system working for as much of the time as possible. The hydraulic system therefore requires a surface vessel to act only as the mining system platform and the recovered nodules will have to be transferred to transport vessels at sea.

The economics of mining and processing the nodules

Most estimates of the cost of mining the manganese nodules fall in the range of $5 to $20 per ton of nodules in an operation producing at least one million tons of dry weight nodules per year (Mero, 1958, 1965a, 1972; Clauss, 1972; Kaufman and Rothstein, 1973). The cost of transporting the nodules to a process site and shore handling should fall in the range of $3 to $11 per ton. The cost of processing the nodules is expected to fall in the range of $10 to $20 per ton (Mero, 1968, 1972; Meiser and Müller, 1973). Using the high values of these estimates, it would appear that the cost of mining, transporting and processing a ton of the nodules should not exceed $50 per ton. With manganese nodules averaging 30% Mn, 0.3% Co., 1.5% Ni, and 1.2% Cu and assuming an overall recovery factor of 90% of the contained metals in the nodules, the gross value of the metals recovered from a ton of the nodules would be about $114 using metal prices as indicated in Table 11-I. Each ton of nodules would thus produce a gross operating profit of about $64. Assuming overhead costs of $6 per ton and an effective tax burden of 48%, the net profit on each ton of nodules mined would be about $30. The net profit on a 2 million ton/year operation would therefore be about $60 million. The estimated total capital investment in such an operation, including the mining, processing and transport systems, can be expected to be about $250 million or the rate of net return would be about

24% per annum which is an acceptable figure for a venture carrying the very high risks associated with deep-ocean mining.

Assuming the rate of return on the invested capital desired is 20% per year, it would leave about $10 per ton of the net profits to be applied against repayment of the capital investment. An amortisation tonnage would therefore be about 30 million tons of nodules. At an average concentration of 10 kg/m^2 of wet nodules (concentration estimates are generally determined on a basis of wet weight of nodules which will, on the average contain about 30% of water as the nodules come from the sea) or 7 kg/m^2 of dry nodules, the amortisation tonnage of 30 million tons of dry-weight nodules would cover an area of about 4,300 km^2 of sea floor which is about 0—1% of the total area within the high-grade region of the North Pacific Ocean. In actual practice, it may be assumed that about 25—50% of the nodules in any given deposit area would not be recoverable due to local obstructions in the deposit area which must be circumvented by the mining system and due to inefficiencies of the mining system in the recovery of all the nodules in the deposit (Kaufman, 1974). Consequently, the area required to contain an amortisation tonnage can be calculated to be of the order of 10,000 km^2.

The cost of surveying 30 million tons of dry-weight nodules (about 43 million tons of wet-weight nodules), assuming free-fall sample stations are located at 1—km intervals over a 10,000 km^2 area and that television scans are made along parallel lines 1 km apart, would be about $5 million. This estimate assumes that some 40 free-fall stations can be made per day and that the television scans are made at an average velocity of 1 knot. The estimate also assumes that the ship operating costs are $3,000 per day and that there will be a 50% overall loss of ship time due to returns to port for resupply, weather, mechanical difficulties, etc. The $5 million estimate also includes $2.3 million for survey equipment, data reduction and various overhead expenses of operation. Such an operation, covering some 10,000 km^2 would require a period of about 900 days total time assuming at least 450 days are occupied in actual survey activities. At a $5 million total exploration expenditure, the cost per ton of nodules surveyed would be about $0.17 per ton.

In calculating the future potential value of metals produced from manganese nodules, it would not be wise to include any value for manganese as that metal will be produced in quantities much greater than any reasonable future estimate indicates the market will be able to absorb. In addition, cobalt would be produced in quantities greatly in excess of projected demands so the future value of this metal should be calculated as equal to that of nickel for which metal cobalt can substitute in many industrial applications. Basing the economics of mining the nodules on these criteria, it can be anticipated that, at a recovery efficient of 90% of the contained metals in nodules containing 0.3% Co, 1.5% Ni, and 1.2% Cu, the value of metals recovered from a ton of the nodules would be $81. Using the high values for mining, transport, processing and overhead expenses and again assuming an effective 48% tax burden, this would allow an indicated net profit of $16 per ton or $32 million per year on a 2 million ton per year operation. The rate of return on a $250 million capital investment would be about 13% per year which is about the minimum any mining company could accept for a high-risk venture. A 15% depletion allowance would raise the net profit to about $46 million for a rate of return of about 18% per year.

ENVIRONMENTAL CONSIDERATIONS IN NODULE MINING

Both tests of deep-sea nodule mining systems so far conducted (i.e., the Deepsea Ventures' test of the air-lift hydraulic system in 1,000 m water depth on the Blake Plateau in 1972 and the CLB System in 4,100 m water depth in 1972) have been monitored for possible environmental effects to the ocean. These environmental studies were conducted under the auspices of the U.S. National Oceanographic and Atmospheric Administration. The results of those studies have failed to detect any substantial deleterious effects to the ocean environment (Roels et al., 1973). While it can be expected that small amounts of sediment and gangue materials will be raised to the sea surface in the mining of nodules, either with the hydraulic systems or the CLB System, and that the mining operator will want to separate the gangue materials and return them to the sea floor at the mining site, they can be returned via a standpipe which exists below the thermocline so that no sediments or nutrient-rich bottom waters appear at the surface of the ocean. The sediment and other gangue materials would resettle to the ocean floor and should cause no discernible disturbance to the existing ocean environment. No foreign materials would be added to the seawater in these operations. From theoretical considerations and from actual observations of full-scale mining tests, no damage or altering of the marine environment should occur in the mining of the nodules. Mining operation will occupy only a very small percentage of the total oceanic area, in the order of 0.0001% of the total area of the Pacific Ocean alone, so that even if any adverse effects should occur they will be confined to relatively small areas.

There may be some disturbance of the biota at the ocean floor in nodule mining operations; however, the areas where the nodules are found are the great deserts of the world as far as macroscopic life forms are concerned. Any bacteria or other life forms which may be destroyed in any mining operation can be replaced by populations in the adjacent unmined areas since the mining system will not affect almost 75% of the deposit area. The sea floor itself should be relatively little disturbed by the mining of the nodules as the object in any mining operation will be to gather only those nodules lying loosely at the surface of the sediment and to disturb the seafloor sediment as little as possible. It would be uneconomic to lift gangue sediment and material to the surface.

On the other hand, the mining of the manganese nodules may allow the shutting down of many of the large sulphide mines on land. These mines do cause considerable visual pollution of the landscape and very considerable pollution of the atmosphere in the release of large amounts of sulphur dioxide and other gases in the processing of these ores. Since manganese nodules will be processed with some form of hydrometallurgical process in which the reagents will be largely recovered and recycled, there should be no measurable pollution from this operation. Present indications are that the extraction of metals from manganese nodules requires about 50% of the energy required to produce an equivalent amount of such metals from land sources. In addition, the nodules, which possess large specific surface areas and which are known to be highly efficient absorbers of sulphur dioxide from gas streams (Zimmerly, 1967), may be used in the stacks of existing power plants or other industrial flues to purify the gases presently emitted to the atmosphere by these facilities. The mining of deep-sea nodules should therefore prove, on all accounts, to be a considerable net gain to the present continental environments.

SUMMARY

The major considerations concerning the economic geology of the deep-sea manganese nodules are metal content, surface density and size range of the nodules, continuity of the deposits, sea-floor topography and character of the associated sediments. Although manganese-iron oxide precipitates can be found in many forms on the ocean floor only the loose-lying nodules found at the surface of the soft pelagic sediments are of commercial interest at the present. Environmental factors controlling the commercial recovery of deposits include distance to port, water depth, density of obstructions within a deposit area, water current velocities through the water column, weather, etc. Of all the factors involved in determining the economic value of any given deposit of the nodules, given certain minimum standards concerning the other factors, the most important ones appear to be metal content of the nodules, surface density, continuity and areal extent of the deposits. A minimum grade of nodules, at the present time, would be one containing about 2.8% Ni + Cu + Co on a dry-weight basis. The minimum economic concentration of the nodules appears to be about 5 kg/m^2 of sea floor and the minimum economic size of a deposit should be one containing about 30 million recoverable tons of dry-weight nodules. As only 25—50% of the nodules in any given deposit is estimated to be recoverable, such a deposit is estimated to cover an area of about 10,000 km^2.

Two types of deep-sea dredges are presently under development for the mining of the manganese nodules, a deep-sea hydraulic dredge and a mechanical cable-bucket system. Both systems appear to offer some advantages with the hydraulic system appearing to be advantageous in the mining of a specific deposit for which it is designed while the cable-bucket system appears to be somewhat more flexible in working in a variety of deposits, topographic environments and water depths.

The cost of surveying a deposit of the nodules covering 10,000 km^2 is estimated to be $5 million. The cost of mining, transporting and processing a ton of the nodules is estimated to be between $25 and $50 per ton of nodules at production rates in excess of 1 million tons per year. The value of the metals recoverable from a ton of the nodules in the high-grade region of the North Pacific Ocean would be in the range of $80 to $140 per ton, depending on whether manganese is included as a product of any value or not and depending on the value accorded to cobalt. Total capital investments in a system to handle 2 million tons of the nodules per year are estimated to be in the range of $200 to $250 million and the rate of return on a nodule-mining venture is indicated to be in the range of 13 to 24% per annum.

Environmental studies conducted in conjunction with deep-sea tests of the two types of mining systems presently indicate that substantially no environmental damage will be done in the mining of the deep-sea nodules. Because of the nature of the deposits and the way in which they can be mined, the manganese nodules presently appear to be a relatively pollution-free and energy-saving source of a number of industrially important metals.

CHAPTER 12

EXTRACTIVE METALLURGY

D. W. FUERSTENAU and K. N. HAN

INTRODUCTION

Ocean-floor manganese nodules have recently attracted considerable interest as a promising source of such important metals as nickel, copper, cobalt, manganese and many more. A recent report (Rothstein and Kaufman, 1974) has indicated that the rate of increase in world demand for these elements ranges from 5 to 6% each year. As the land-based reserves of these elements are limited, research and development on economical ways of recovering these elements from deep-sea nodules are exceptionally timely.

The concentrations of different elements in the nodules vary from one location to another. The average concentration of some of the more abundant species in Pacific Ocean nodules, expressed as percentage on an air-dried basis, is the following (Mero, 1965a); Mn, 24; Fe, 14; Si, 9.4; Al, 2.9; Na, 2.6; Ca, 1.9; Mg, 1.7; Ni, 0.99; K, 0.18; Ti, 0.67; Cu, 0.53; Co, 0.35; Ba, 0.18; Pb, 0.09; Sr, 0.081; Zr, 0.063; V, 0.054; and Mo, 0.052. However, a nodule deposit that might be considered for mining could assay as high as 2.3% Cu, 1.9% Ni, 0.2% Co and 36% Mn on a dry-weight basis.

There have been numerous studies in the literature concerning the ultimate origin of the nodule constituents. However, the way in which the metals are accreted into the nodules is not yet fully understood and, consequently, the extraction strategy of these elements from the nodules can hardly be predetermined. Most of the extraction studies presented in the literature are, therefore, in their early stages of demonstrating the feasibility of various processing methods, of finding the important reaction rate parameters, and of formulating mechanisms of reaction rates for extraction of individual elements from the nodules. However, to date information has only been published on studies of the extraction of metals from deep-sea nodules.

In this chapter the possible states of existence of the metals in the nodule will first be discussed in light of the observed data revealed in the literature and in terms of various thermodynamic and kinetic principles. The main objective will be to compile information available on the extraction of the metals from deep-sea nodules, and to interpret the extraction behaviour of various elements in terms of the chemical and physical nature of the nodules.

PHYSICAL AND CHEMICAL CHARACTERISTICS OF NODULES

Nodules from several different locations have been studied in detail, and their locations and designations are given in Table 12-I. Chemical analyses and some physical characteristics of these nodules are given in Tables 12-II and 12-III (Han, 1971; Hoover, 1972). As can be seen in Table 12-III, the nodules are highly porous with a porosity of about 60%, and relatively light with an apparent density of about 1.4. Fig. 12-1 presents the size distribution of pores in the various nodules, as determined by mercury porosimetry (Han, 1971). One of the outstanding characteristics of manganese nodules is their high water content as shown in Table 12-IV (Han, 1971).

As can be seen from Fig. 12-1, about 80% of the total pore volume occurs as pores between 0.01 and 1.0 μm in diameter. Because of these extremely fine pores, nodules contain considerable water at room temperature and atmospheric pressure due to capillary condensation. The results of thermal gravimetric analyses (TGA) show that nodules lose their weight continuously upon heating from room temperature to 1,000°C (Fig. 12.2). The fact that nodules contain an appreciable amount of water has important consequences in the processing of nodules by the segregation process. This also contributes to the costs of processes which may involve a drying step. Heat treatment also affects the pore structure in nodules. When nodules are heated, the total porosity of the nodules does not change but the smaller pores coalesce to form larger pores (Han, 1971). Fig. 12-3 shows that the pores begin to coalesce only after heating the nodule above 450°C. If reduction roasting is required in an ammonia leach process, for example, the change in the nature of the pores may affect the kinetics of processing.

TABLE 12-I

Location of the nodule samples

Nodule designation	Latitude	Longitude	Depth (m)
DH-2	21°50′N	115°12′W	3,430
HRS-1	22°N	114°W	3,400
2P-50	13°53′S	150°35′W	3,623
2P-51	9°52′S	145°56′W	4,900
2P-52	9°57′N	137°47′W	4,930
DWHD-16	16°29′S	145°33′W	1,270

TABLE 12-II

Chemical analyses (%) of the nodule specimens

Element	2P-50	2P-51	2P-52	DH-2	DWHD-16	HRS-1
Mn	19.0	16.8	25.6	23.5	24.5	26.5
Fe	17.0	12.5	5.0	15.1	11.5	9.5
Cu	0.2	0.4	1.2	0.5	0.25	1.0
Ni	0.4	0.6	1.5	0.8	0.7	1.2
Co	0.5	0.2	0.2	02.	1.15	0.2
H_2O*	15.5	9.9	10.1	17.1	17.0	14.2

*The water contents were measured after heating the samples at $110°C$ for 24 h.

TABLE 12-III

Physical characteristics of the nodules

	2P-50	2P-51	2P-52	DH-2	DWHD-16	HRS-1
Apparent density (g/cm^3)	1.36	1.54	1.48	1.43	1.40	1.36
Real density (g/cm^3)	3.50	3.20	3.24	3.53	3.54	3.50
Porosity	0.61	0.52	0.54	0.59	0.60	0.62
Specific surface area (m^2/g)	287	197	118	205	210	173

X-ray analyses of these nodules reveal that the manganese oxides in the various nodules are various manganese dioxides such as δ-MnO_2, 7-Å or 10-Å manganite (Chapter 7). Differential thermal analysis (DTA) and X-ray analysis show that the iron oxides in nodules are various ferric oxides, such as a-FeOOH and/or hematite (Buser and Grütter, 1956; Bonatti and Nayudu, 1965; Manheim, 1965; Andrushchenko and Skornyakova, 1969; Han, 1971) or γ-Fe_2O_3 (Smith et al., 1968). More recently Von Heimendahl et al. (1976)

TABLE 12-IV

Percentage loss of water from manganese nodules (on a wet basis) at different temperatures

	2P-50	2P-51	2P-52	DH-2	DWHD-16	HRS-1
110°C	15.5	9.9	10.1	17.1	17.0	14.2
200°C	19.6	12.8	13.4	20.5	19.2	18.1
300°C	23.2	16.7	17.6	23.7	24.1	19.1
400°C	24.9	18.6	19.0	25.7	25.1	21.9

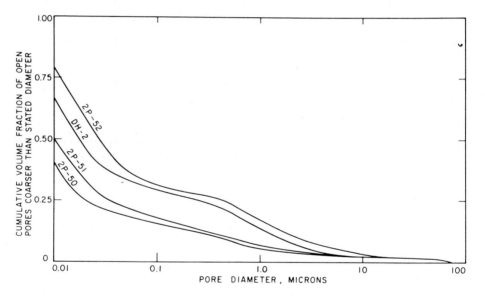

Fig. 12-1. Pore-size distribution of the various nodules as measured by mercury intrusion (after Hoover, 1972).

have carried out a transmission electron microscope study of two different manganese nodule samples and identified the manganese oxides in the nodules with sodium birnessite, todorokite, s-birnessite (Giovanoli et al., 1970a) and hydrohausmanite, and the iron oxides with γ-Fe_2O_3 and ferric hydroxide. They have also observed that the typical particle size of the ferric hydroxides ranges from 30 to 60 Å.

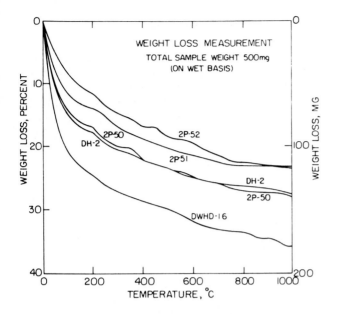

Fig. 12-2. Weight-loss measurement of nodules DH-2, 2P-50, 2P-51, 2P-52 and DWHD-16 at the rate of temperature change, 3°C/min under atmospheric pressure (Han, 1971).

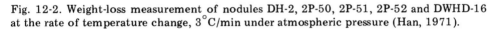

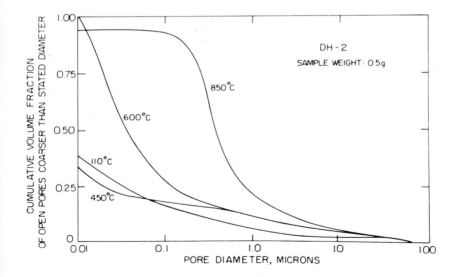

Fig. 12-3. Pore-size distribution of nodule DH-2 at different temperatures, as measured by mercury intrusion (Han, 1971).

FORMATION AND DISTRIBUTION OF METALS IN NODULES

How metals occur and are distributed within nodules is important in a number of extraction processes. The literature indicates that there are two major schools of hypotheses on the origin of deep-sea nodules, namely, the hypothesis of volcanic (Murray and Irvine, 1894; Clarke, 1924; Pettersson, 1945, 1959) and terrestrial sources (Murray and Renard, 1891) and that of diagenetic processes (Lynn and Bonatti, 1965). No matter how these nodule constituent elements arrived in the ocean, it can be shown (Goldberg and Arrhenius, 1958; Han, 1971) that these metal ions suspended in the ocean can be precipitated to form the corresponding stable oxides.

Eh-pH diagrams (Han, 1971) show that divalent ions are oxidized according to the reaction:

$$M^{2+} + nO_2 + 2OH^- \rightarrow MO_{2n+1} + H_2O \tag{1}$$

where Mn^{2+} is a divalent ion, n is a constant: $1/2$ for Mn^{2+}, $1/4$ for Co^{2+} and Fe^{2+}, $1/6$ for Ni^{2+} and O for Cu^{2+}. Fig. 12-4 gives the Eh-pH diagram for Cu, Ni, Co, Fe and Mn at $25°C$ and unit activity of metal ions; only the regions in which the dissolved species are stable are shown. As was mentioned earlier, manganese occurs in nodules as various kinds of MnO_2 and iron as either $FeO(OH)$ or Fe_2O_3. These oxides agree with the equilibrium consideration used in the generation of Eh-pH diagrams. Should the other ions such as Ni^{2+}, Cu^{2+} and Co^{2+} be precipitated or adsorbed on the nodules under the same conditions as manganese and iron, Eh-pH calculations show that the corresponding oxide species are Ni_3O_4, CuO and Co_2O_3 (Han, 1971). The equilibrium concentrations of these ions with respect to their corresponding stable oxides are calculated and tabulated in Table 12-V together with the average concentration of these ions in sea water. These results indicate that under ocean conditions, most metal ions are already saturated with respect to their corresponding stable oxides, and they are subject to precipitation at any time if catalytic media are provided.

The equilibrium analyses made so far are based on ideal solutions. The oceanic environment, especially where nodules form, is quite far from an ideal solution. The ionic strength of the ocean is 0.7, and the average temperature of the ocean bottom is $0°C$. A considerable amount of hydrostatic pressure will also be exerted on the formation of the nodules, since the average depth of the Pacific Ocean, for instance, is known to be 4,280 m (Sverdrup et al., 1946). The effect of pressure and temperature on the equilibrium of the reaction given in eq. 1 has been studied (Han, 1971), and the results are given in Table 12-VI. As can be seen in the table, all the metal ions except manganese are shown to be more stable as ionic species than as oxides at the ocean bottom in comparison with normal room conditions.

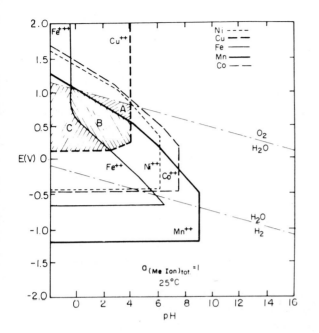

Fig. 12-4. Eh-pH diagrams for the Fe—H_2O, Mn—H_2O, Cu—H_2O, Ni—H_2O and Co—H_2O systems at 25°C. Only the lines bounding the stable ionic species for each metal at unit activity are shown.

TABLE 12-V

The concentration of metal ions in sea water and the corresponding concentration in equilibrium with oxides

Element	Concentr. in sea water (mole/l)*	Concentr. of ions in equilibr. with oxides (mole/l)
Mn	$3.6 \cdot 10^{-8}$	$2 \cdot 10^{-17}$
Fe	$1.8 \cdot 10^{-7}$	$5.2 \cdot 10^{-25}$
Ni	$3.4 \cdot 10^{-8}$	$1.5 \cdot 10^{-8}$
Cu	$4.7 \cdot 10^{-8}$	$7.1 \cdot 10^{-9}$
Co	$1.7 \cdot 10^{-9}$	$2.5 \cdot 10^{-8}$

*Mero, 1965a.

TABLE 12-VI

Ratio of equilibrium constants of $0°C$ and 428 atm to those of $25°C$ and 1 atm

Oxidation reaction	$\dfrac{K_{eq}\ (0°C,\ 428\ \text{atm})}{K_{eq}\ (25°C,\ 1\ \text{atm})}$
Mn^{2+}/MnO_2	1.0
Fe^{2+}/Fe_2O_3	0.132
Co^{2+}/Co_3O_4	0.0039
Cu^{2+}/CuO	0.0043
Ni^{2+}/Ni_3O_4	0.223

In spite of the importance in metallurgical application, the way in which the minor elements are associated with the major ones is not fully understood. On the basis of bulk chemical analysis (Goldberg, 1954; Goodell, 1968) and electron microprobe analysis (Burns and Fuerstenau, 1966), it was found that Ni and Cu positively correlated with Mn in the nodules and, similarly, that Co, Ti and Zn increase with the Fe content. There are several theories on the manner in which these minor elements are present in the nodules. Goldberg (1954), Krauskopf (1956), Burns and Fuerstenau (1966) and Fuerstenau et al. (1973) supported the theory that the metal ions or the metal oxides have been adsorbed on the major constituents of manganese or iron oxides. On the other hand, Buser and co-workers (Buser et al., 1954; Buser and Graf, 1955b; Buser and Grütter, 1956) believe that the minor elements in the nodules substitute for Mn^{2+} in the manganite minerals by an ion exchange mechanism. Burns (1965) suggested the possibility of oxidation of Co^{2+} to Co^{3+} in the sea environment provided the concentration of Co^{2+} exceeds $10^{-8}M$. This reaction is assumed to be catalyzed by $Fe\,(OH)_3$ and is also favoured by $Co\,(OH)_3$ forming a solid solution with $Fe\,(OH)_3$.

These two theories, namely the *adsorption phenomenon* and the *lattice substitution phenomenon*, are the two major postulates on the nature of the occurrence of minor elements in nodules. There are unfortunately no analytical instruments which enable one to prove any of these assumptions. It is, however, possible to deduce the mechanism indirectly by observing the behaviour of dissolution of these minor elements from the nodules under various conditions. For example, if the lattice substitution mechanism is valid, one may expect ion exchange phenomena to control the way the minor elements are leached from the nodules. However, if the metals occur in the nodules due to an adsorption mechanism, the corresponding leaching behaviour of these minor elements will be determined by conditions that promote desorption. On the other hand, if the metals occur only as occluded oxides, conditions for solubilizing the oxides should provide a basis for their extraction. Such an analysis will be introduced in later sections whenever it is appropriate.

PHYSICAL SEPARATION AND CHEMICAL EXTRACTION PROCESSES IN NODULE PROCESSING

As discussed in the previous sections, nodules are composed of conglomerates of colloidal-sized particles. If one could separate the individual manganese-oxide and iron-oxide particles, then a fairly distinct Co separation from Ni-Cu could be effected. However, such methods as flotation, magnetic separation, etc. are effective for particles in the 100 μm size range and not 100 Å range. Thus, methods for processing nodules will have to be based mainly on chemical techniques. Flowsheets will involve initial stages of screening to remove such debris as sharks' teeth, shells, etc., and crushing and grinding for size reduction of the nodules. In the case of processing relatively soft nodules, the size-reduction operation may be carried out simply in hammer mills. Energy consumption for size reduction in nodule processing should not be large; Brooke and Prosser (1969) obtained work indices (i.e., the energy required to grind nodules to 80% minus 100 mesh in size) of 7.7, 13.7 and 8.8 Kwh per ton for nodules 2P-50, 2P-52 and DH-2, respectively.

Processes for the treatment of nodules will be based on decisions as to which metals can be recovered economically. As will be shown in subsequent sections, some processes recover only copper and nickel, others copper, nickel, cobalt and molybdenum, without manganese and iron, still others will also permit manganese recovery, etc. Research on the recovery of economically valuable elements from deep-sea nodules can be classified into two categories, namely, the low-temperature hydrometallurgical processes, using acids, ammonia or reducing agents such as Fe^{2+} or SO_2; and high-temperature pyrometallurgical processes involving various reduction schemes, chlorination, segregation or smelting. Pyrometallurgical processes generally totally alter the physical and chemical structure of the nodules, whereas the performance of hydrometallurgical processes may well depend on the physical and chemical characteristics of the nodules being processed. Some of the leaching processes include a high-temperature reduction treatment to first reduce the metal oxides to metal or to a lower oxidation state prior to the leaching step. Such combined treatments of high and low temperature are classified in this section under their corresponding leaching treatment irrespective of their pretreatment.

Most of the extraction studies in the literature have been concerned with the selective recovery of the minor elements, such as Ni, Cu, and Co. Such processes may be similar to those used for the treatment of other kinds of ores containing minor amounts of these same elements. Selective extraction has advantages over the non-selective extraction of desired metals because extracting small amounts (1—2% weight) of, say, nickel and copper may be economically more promising, selectively extracting may provide a refining step that is simpler and more economical, manganese and iron are still readily available on land, and the physical/chemical characteristics of nodules are such that the selective extraction of minor elements is possible.

ACID LEACHING PROCESSES

Atmospheric pressure processes

There have been a number of studies concerned with the selective leaching of minor elements from nodules with sulphuric acid (Hoover, 1967; Brooke and Prosser, 1969; Fuerstenau et al., 1973; Hubred, 1973; Han and Fuerstenau, 1975a). In general, under the conditions that were used in the acid leaching, it is possible to extract the majority of the Ni and Cu and

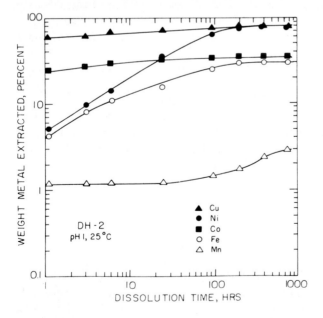

Fig. 12-5. Rate of dissolution of metal ions from a 100 X 200 mesh fraction of DH-2 nodules at pH 1 and 25°C (after Han, 1971).

about half of the Co without destroying much of the major constituents. Typical acid-leaching results, such as those presented in Fig. 12-5 for DH-2 nodule (Han and Fuerstenau, 1975a), demonstrate that the total amounts of Cu, Co, and even Mn dissolved do not depend significantly on time, over a long period. On the other hand, the amounts of Ni and Fe dissolved exhibit a marked time dependence, over a long time period. This is a consequence of the rates of dissolution of Cu and Co being limited by pore diffusion and that Ni being limited by heterogeneous reaction as well as by pore diffusion (Fuerstenau et al., 1973; Han and Fuerstenau, 1973, 1975a). As a result, Cu and Co extractions are markedly affected by particle size, whereas particle size plays only a limited role in Ni dissolution. These phenomena are illustrated in Fig. 12-6, which presents the amount of metal extracted over a wide range of particle sizes for short (1 h) and long (96 h) leaching times. If the particle size of the nodules has been reduced to about 1 μm, the amounts of Cu and Co dissolved are almost independent of leaching time. On the other hand, the amounts of Ni and Fe depend on leaching time, even for 1 μm particles.

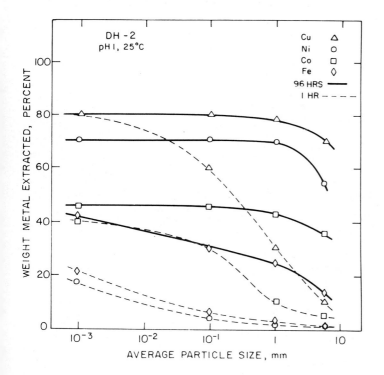

Fig. 12-6. The influence of particle size on the amount of copper, nickel, cobalt and iron extracted from DH-2 nodules for different leaching times.

Figs. 12-7—12-9 present the effect of pH on the room temperature dissolution of Ni, Cu and Co, respectively, from 100 × 150-mesh samples of nodules from five different locations (Fuerstenau et al., 1973). These figures show that the leaching behaviour of Ni and Cu is quite similar for nodules from widely differing locations. On the other hand, cobalt dissolution depends strongly on the nature of the nodule being leached. Co recovery in four of the nodules is low; possibly this is related to the incorporation of the Co into the lattice of the iron minerals in the nodule. In the case of nodule 2P-52, the Fe content of the nodule is low (only 5%) and the bulk of the cobalt is probably associated with the manganese oxide particles, and hence is more readily accessible to the acid.

There are two distinctive responses of Co leaching in an acidic medium. Approximately half of the Co in the nodules is generally leached out easily, while the other half does not react with acid (Han and Fuerstenau, 1975a). This may indicate that the insoluble fraction of cobalt in the nodules exists in a higher oxidation state and/or in lattice association with the major oxide phase, as mentioned in the previous paragraph. It has been observed by electron microprobe analysis (Burns and Fuerstenau, 1966) that the cobalt occurs in the iron oxide phase. However, it should be pointed out that no

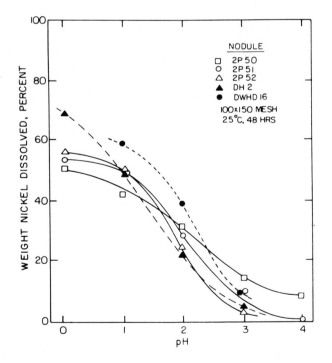

Fig. 12-7. Dissolution behaviour of nickel from various manganese nodules as a function of pH (after Fuerstenau et al., 1973).

relation has been observed in acid leaching between Fe extraction and Co extraction. At pH 0.5, the dissolution of Co did not exceed 45% even though 80% of the Fe was dissolved (Han and Fuerstenau, 1975a).

The temperature-dependency of dissolution is, in general, strongest for Ni, followed by Fe, Mn, Cu and Co in this order (Ulrich et al. 1973; Han and Fuerstenau, 1975a). Copper and cobalt extraction does not vary much with temperature in contrast to nickel extraction. The apparent activation energies of Ni, Cu and Co for a typical nodule are found to be approximately 12, 2.5 and 2.5 kcal./mol, respectively (Han and Fuerstenau, 1975a). The lower values are in accordance with a reaction rate that is limited by pore diffusion.

It should be mentioned that in acid leaching the maximum recovery remains the same irrespective of the particle size of the nodules (Han, 1971). This is an important consequence of the physical and chemical characteristics of nodules in that the pores in the nodules are all open and the minor elements thus have channels to the surface of the nodules. Hence, there is always direct contact established between these species on the surface and the solvent introduced inside the pores.

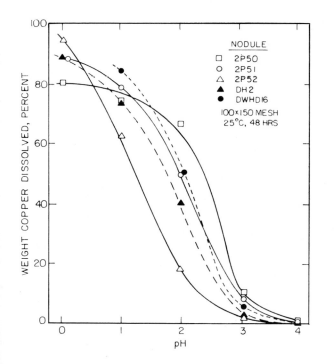

Fig. 12-8. Dissolution behaviour of copper from various manganese nodules as a function of pH (after Fuerstenau et al., 1973).

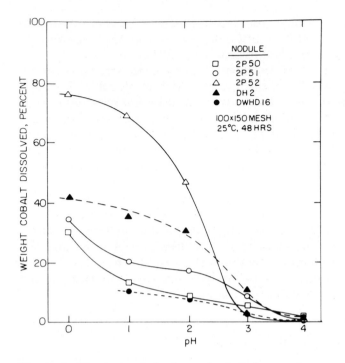

Fig. 12-9. Dissolution behaviour of cobalt from various manganese nodules as a function of pH (after Fuerstenau et al., 1973).

One of the disadvantages of the acid leaching, however, is that there are acid-soluble constituents other than the desired metal compounds in the nodules. These acid-soluble inorganic compounds (zeolites, carbonates, clays) may be roughly equivalent to 20% by weight of the nodules, if one assumes that the equivalent molecular weight of these compounds is 100. Typical results of a room-temperature acid leaching are given in Table 12-VII (Han and Fuerstenau, 1975a), to illustrate metal dissolution and acid consumption.

The selectivity of individual elements in acid leaching can, in general, be achieved by adjusting the temperature, the retention time of the nodules in the reactor and the size of the particles for the given kind and concentration of solvent. Table 12-VIII illustrates the kind of operation and extraction that might be achieved with a two-stage leach conducted at $2°C$ in the first stage and $45°C$ in the second stage. This table shows that a two-stage leach can effect a partial separation, with one leach liquor being (Cu-rich and the other Ni-rich. A similar result can be obtained by a two-stage leach based on

TABLE 12-VII

Summary of metal extraction and sulphuric acid* consumption at 160 h of leaching time
and 25°C for a 100 X 200 mesh sample of DH-2 Nodules at different acid concentrations

| pH | Ni | Cu | Co | Fe | Mn | Acid consumption | | Calculated acid requirements $(cm^3/kg \text{ nodules})$ | Excess acid used $(cm^3/kg \text{ nodules})$ |
						cm^3/kg nodules	kg/kg nodules		
3.0	10	4	14	0.1	0.1	60	0.11	3	57
2.0	44	50	33	9.0	0.9	100	0.18	42	58
1.0	74	80	35	28.0	1.5	160	0.30	112	48
0.5	80	93	45	60.0	2.5	250	0.46	200	50

*Sulphuric acid is expressed as 96.2% H_2SO_4.

TABLE 12-VIII

The results of a process involving leaching a minus-40-mesh sample of nodule HRS-1 at
pH 1 for 90 h at 2°C and with a fresh leach solution for another 90 h at 45°C

| | % recovery: | | | | |
	Cu	Ni	Co	Mn	Fe
First stage leach (2°C)	70	10	16	1.2	21
Second stage leach (45°C)	5	67	5	0.4	10
Combined recovery	75	77	21	1.6	31

differences in leaching kinetics. For example, a Cu-Ni separation can be
made if the nodule sample is first leached for one hour at pH 1, and after
filtering the nodule sample is again leached at pH 1 for a longer time period.
The results of such a process, given in Table 12-IX, yield interesting
separations and recoveries, similar to those obtained in the two-stage
temperature leach.

TABLE 12-IX

The results of a process involving leaching a 100 × 200-mesh fraction of nodule DH-2 at pH 1 and 25°C for 1 h and then with a fresh leach solution for 200 h

| | % Recovery: | | | | |
	Cu	Ni	Co	Mn	Fe
First stage leach (1 h)	60	5	25	1.2	4
Second stage leach (200 h)	20	71	8	0.5	26
Combined recovery	80	76	33	1.7	30

High-temperature acid leaching

Thermodynamic considerations show that metal oxide phases, iron oxide in particular, become increasingly stable in acid medium as the temperature increases (Han and Fuerstenau, 1975b). This is the main reason why high-temperature acid leaching is so successful with lateritic ores. It is also noted that the stability of the oxide phase follows the order Fe>Mn> Ni>Co>Cu as temperature increases. For these reasons, Han and Fuerstenau decided in 1970 to investigate the sulphuric acid leaching of a sample of DH-2 nodules in a titanium-lined autoclave. The leaching experiments were conducted at temperatures ranging up to 220°C and the pressure of oxygen ranging up to 8.06 kgf/cm^2 absolute (at room temperature). The results of these experiments are summarized in Table 12-X. As can be seen from this table, the dissolution of manganese is very sensitive to the pH and can be substantially reduced by applying oxygen under pressure.

The results obtained by Han and Fuerstenau (1975b) are in good agreement with those of Ulrich et al. (1973), who found that about 90% of the Ni and Cu, 40% of the Co, and less than 5% of the Mn and Fe are recovered after 3 h of leaching with 0.3 gram of H$_2$SO$_4$ per gram of nodules at 200°C. On comparison of these results with Eh-pH diagrams, it can readily be seen that the dissolution of Fe and Mn are very sensitive to temperature and to oxygen pressure and that the highest recovery would be expected for Cu because the dissolution of CuO is least sensitive to temperature. The adverse effect of oxygen pressure on Ni and Co dissolution can also be predicted from Eh-pH diagrams.

In conclusion, it was found that high-temperature acid leaching of manganese nodules can be considered successful in view of the recovery, the selectivity, kinetics and the consumption of acid.

TABLE 12-X

The high-temperature leaching conditions and recoveries of metals from 100×200-mesh
DH-2 nodule samples; in each case the reaction time was 60 minutes
(After Han and Fuerstenau, 1975b)

Temp. ($^\circ$C)	Total press. (kgf/cm^2 abs.)	P_{O_2}* (kgf/cm^2 abs.)	pH init.	final	Acid consump. (kg/kg nodule)	% Recovery: Cu	Ni	Co	Mn	Fe
190	10.17	1.24	1.48	1.81	0.45	52	84	35	16	17
220	27.04	3.85	1.48	1.72	0.34	83	85	39	15	13
180	11.58	4.55	1.0	1.04	0.27	93	86	50	23	36
115	7.01	4.55	1.63	1.73	0.09	61	52	10	1.5	3.6
180	21.07	8.06	1.63	1.66	0.045	75	65	26	3	2.3
200	22.83	8.06	1.63	1.64	0.026	90	80	30	5	2
105	22.12	8.06	1.63	1.635	0.01	90	80	30	3.2	2

*At room temperature.

Leaching with reducing acids

There have not been many attempts to extract metals from nodules using
reducing acids, such as sulphurous acid or nitrous acid. This is mainly
because acids would reduce MnO_2, as well as other oxides, and would result
in non-selective leaching. The Eh-pH diagram given in Fig. 12-4 clearly
illustrates how reducing acids function as leaching agents for the various
oxides constituting nodules. By lowering the Eh of the system (making
conditions more reducing), manganese oxides are readily dissolved (Fig.
12-4). By adjusting pH and Eh so that conditions depicted by region B in
Fig. 12-4 are achieved, it should be possible to dissolve the manganese oxides
(plus the oxides of Cu, Ni and Co) and not the iron oxides. Fig. 12-10
presents the results of leaching a minus 48-mesh sample of DWHD-16 nodule
with a solution of 0.2% sulphur dioxide at pH 3. This figure clearly shows
that all four metals are readily dissolved (with the dissolution of Co being
the greatest) with only limited dissolution of Fe. By lowering the pH into
region C of Fig. 12-4, conditions would exist where all five of the oxides
should dissolve in the leach liquor.

In another series of experiments (Han, 1971), hydroxylamine hydro-
chloride was used as the leaching agent at pH 5.5. With this reagent, the

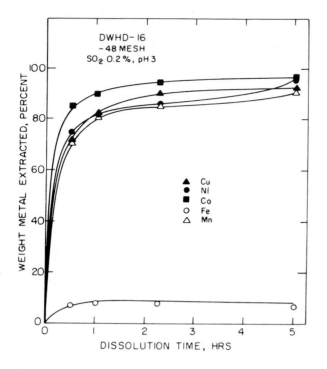

Fig. 12-10. Sulphurous acid leaching of DWHD-16 nodules at 25°C (after Han, 1971).

extraction of Mn, Co, and Ni was complete with only partial dissolution of Cu. Copper extraction was decreased because of a combination of the Eh being too low and the pH somewhat too high.

Mero (1965b) proposed a scheme for the selective separation of Ni and Co from nodules by the reduction leaching of the maganese oxide phase away from the iron oxide phase. The results described in the foregoing paragraphs clearly illustrate that even though the iron oxide phase is not dissolved, all of the cobalt (along with the nickel) is.

Treatment of reducing-acid leach liquors would involve steps for the precipitation of iron and recovery of manganese. For example, the manganese values can be recovered from the leach liquor by the "dithionate process" where calcium dithionate is added to form manganese dithionate and to precipitate calcium sulphate. After the $CaSO_4$ has been removed by filtration, the manganese is precipitated as $Mn(OH)_2$ by adding milk of lime with the resulting precipitated calcium dithionate being recycled. Manganese can also be precipitated as manganese sulphate by evaporation of the solution (Wyman and Ravitz, 1947). The manganese sulphate is then calcined to produce manganese oxides and SO_2.

High-temperature treatment for aqueous leaching

Several techniques are available for pre-treating nodules at high temperature to transform the metals into a water-soluble form. Examples are sulphation roasting or hydrogen chloride roasting.

In 1963, Globus demonstrated that manganese could be extracted from low-grade manganese dioxide ores by roasting the oxide with SO_2 or H_2SO_4 at high temperature. Brooks et al (1970) and Van Hecke and Bartlett (1973) utilized the sulphation technique to extract Ni, Cu, Co and Mn from nodules. After roasting the manganese nodules with either SO_2 or H_2SO_4 at 400—700°C, the product could be leached with water at room temperature. With nodules assaying 1.0% Ni, 0.5% Cu, 0.1% Co, 25% Mn and 10% Fe, in the leaching step they recovered 77% of the Cu, 71% of the Ni, 97% of the Co, 95% of the Mn, and 6% of the Fe (Brooks et al., 1970).

Deepsea Ventures Inc. has developed a process involving a reduction roast of the nodules with HCl (Cardwell, 1973). By controlling the temperature in the roasting furnace, the chlorination of Fe is inhibited in their process. Detailed interpretation of the conditions necessary for the control of the chlorination reactions can be found in the thermodynamic study by Kellogg (1950), as will be discussed briefly in a later section. In the Deepsea Ventures process, the reduction reactions are typified by:

$$MnO_2 + 4\ HCl = MnCl_2 + 2\ H_2O + Cl_2$$

The chlorides can be readily extracted from the roasted product by aqueous leaching methods.

AMMONIA LEACHING

Ammonia leaching with pre-reduction

Ammonia leaching has been widely used as one of the standard hydro-metallurgical techniques for extracting Cu, Ni and Co (Caron, 1924; 1925; Forward et al., 1948; Shimakage et al., 1968, 1969; Kunda et al., 1970). This technique is advantageous over acid leaching in that Cu, Ni and Co form soluble ammonium complexes in alkaline pH ranges in which the acid-soluble constituents do not react. Consequently, ammonia leaching has also been investigated for the extraction of metals from manganese nodules by many investigators (Brooke and Prosser, 1969; Rolf, 1969; Brooks et al., 1970; Brooks and Martin, 1971; Redman, 1972, 1973; Wilder, 1972, 1973; Skarbo, 1973a, b; Han et al., 1974).

Eh-pH diagram for Cu, Ni and Co in the presence of NH_3 is shown in Fig. 12-11 (Han et al., 1974). As can be seen from these, it is theoretically possible to dissolve these oxides under the appropriate conditions of pH and

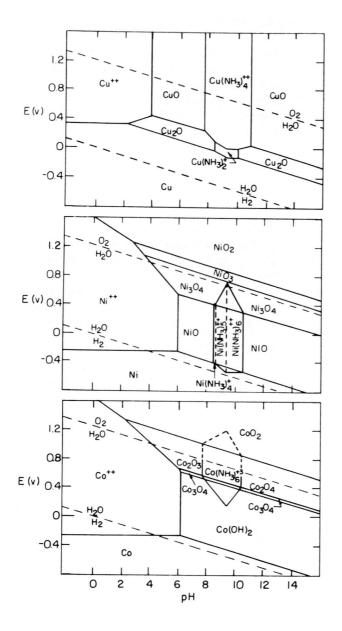

Fig. 12-11. Eh-pH diagrams showing the stable domain of copper—ammonium, nickel—ammonium, and cobalt—ammonium complexes, calculated for unit activity of the ionic species.

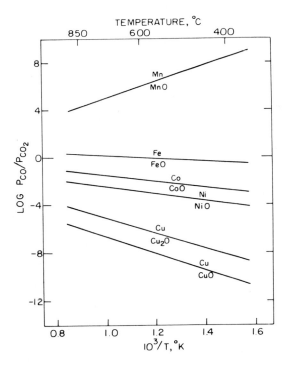

Fig. 12-12. Equilibrium partial pressure ratio of CO/CO_2 versus $10^3/T$ for the reduction of oxides of various metals in nodules (after Hoover, 1972).

Eh. However, in practice the reaction rates of these oxides are very slow. Therefore, most studies that have been carried out using ammonia as the solvent are preceded by a pre-reduction treatment. As shown in Fig. 12-12, Cu, Ni and Co oxides can be preferentially reduced against Mn and Fe oxides and, hence, ammonia leaching of Cu, Ni and Co can be enhanced without introducing Mn and Fe into the solution. It should be noted, however, that the optimum dissolution rate in ammonia solution can be obtained only by proper reduction (Han et al. 1974). For instance, Cu_2O is more easily dissolved in ammonia solution than metallic copper.

Typical results for the ammonia-ammonium carbonate leaching of manganese nodules after pre-reduction is given in Fig. 12-13, which shows the rate of leaching a 48 × 100-mesh sample of DH-2 nodules. Table 12-XI summarizes the results for a 1-h leach of pre-reduced 48 × 100-mesh DH-2 nodules.

These results clearly show that reduction at 400°C is better for copper dissolution than reduction at 600°C, while the recovery of nickel is higher for nodules reduced at 600°C. Cobalt recovery with some nodules is increased by reducing at the higher temperature (e.g., DH-2) but is

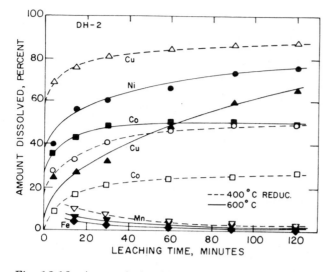

Fig. 12-13. Ammonia leaching of DH-2 nodule samples that had been reduced with a CO—CO₂ gas mixture (60%—40%) at a 200 cc per minute flow rate. The ammonia leaching step was carried out with a mixture of 1.6 M ammonia and 1.6 M ammonium carbonate (after Han et al., 1974).

unaffected with other nodules (e.g., 2P-52). Han et al. (1974) have also shown that the ammonia leaching behaviour is related to the nature of the pores in the various manganese nodules.

It is also interesting to note that, as in the case of the acid leaching, only half of the cobalt in nodules can be recovered by ammonia leaching.

Redman (1972, 1973) reports recoveries of 91% of Ni, 88% for Cu, 72% for Co, and 83% for Mo for nodules leached for 3 h at 25°C with aqueous NH_3/CO_2 after pre-reduction at 600°C. Wilder (1973) proposed a two-step ammonia leaching process of nodules pre-reduced at 800°C. In the first leaching step, which was conducted for 1 h at 25°C with weak ammonia-ammonium carbonate solutions, 0.6% of the Ni and 97% of the Cu were recovered. In the second leaching step, which was conducted for 4 h at 80°C with a strong ammonia-ammonium carbonate solution, 50% of the Ni and 2% of the Cu in the nodules were recovered.

Han et al (1974) also investigated leaching pre-reduced nodules with ammonia-ammonium sulphate solutions. Table 12-XI shows that the recovery of Mn is 40% for pre-reduction at 600°C, probably due to the reaction of MnO with ammonium sulphate.

Skarbo (1973a) demonstrated a process involving simultaneous reduction and leaching with the use of aqueous manganous sulphate together with ammonia-ammonium sulphate. A 4-h leach at 60°C with 5.9 M NH_3, 1 M $(NH_4)_2SO_4$ and 0.5 M $MnSO_4$ recovered 89% of the Cu, 88% of the Ni, 92% of the Co, and 25% of the Mo in a nodule sample.

TABLE 12-XI

Summary of the results of ammonia-ammonium carbonate leaching of a 48 × 100-mesh sample of DH-2 nodules after pre-reduction

Leach system	Reduction temp.	% Recovery:				
		Ni	Cu	Co	Mn	Fe
Carbonate	400°C	43	85	22	4	1
Carbonate	600°C	65	44	50	4	2
Sulphate	400°C	68	95	50	10	0
Sulphate	600°C	74	60	50	42	0

Reduction conditions: 2 roast with a $CO:CO_2$ gas mixture (60%:40%) at 400°C and 600°C.
Leach conditions: 1 h at 25°C with 1.6 M NH_3 and 1.6 M $(NH_4)_2$ CO_3 or with 1.6 M NH_3 and 1.6 M $(NH_4)_2$ SO_4

Ammonia leaching without pre-reduction

Brooke and Prosser (1969) carried out ammonia leaching without pre-reduction using 5 M NH_4OH saturated with $(NH_4)_2 SO_4$ at 110°C for 2 h in an autoclave and obtained recoveries of 41% of the Ni and 77% of the Cu from a sample of DH-2 nodule. In 1970, similar tests were carried out at the University of California using 2 M NH_4OH and 2 M $(NH_4)_2 CO_3$ at 170°C and a typical result is shown in Table 12-XII (Han and Fuerstenau, unpublished data, 1970). As shown in Fig. 12-11, the metal-ammonium complexes are more stable than the corresponding metal oxides under the conditions at which the above tests were performed. The conversion reactions of the metal oxides to the metal-ammonium complexes become more favoured by high temperature. This can be seen more clearly by the results of a recent investigation by Skarbo (1973b), which are also given in Table 12-XII.

SEPARATION OF METALS FROM LEACH LIQUORS

Very little work has been published on the separation of metal values from liquors obtained in the leaching of manganese nodules. However, numerous studies have been carried out for the separation of Cu, Ni and Co from one another in other systems (Queneau, 1961). The adoption of a specific separation process is controlled by the solution conditions which are

TABLE 12-XII

Summary of high-temperature high-pressure ammonia leaching of manganese nodules without pre-reduction

% Recovery:						% Recovery:					
Ni	Cu	Co	Mn	Fe		Ni	Cu	Co	Mo	Mn	Fe
Leaching cond. A*:						Leaching cond. B*:					
80	100	20	0	0		05	95	84	07	—	—

*Leaching conditions A:
1 hour leaching of DH-2 nodules with $2\,M$ NH_3 and $2\,M$ $(NH_4)_2$ CO_3 at $170°C$ and $43.21\ kgf/cm^2$ abs. (Han and Fuerstenau, 1970)
Leaching conditions B-
1 hour leaching of representative nodules with $6\,M$ NH_3 and $6\,M$ $(NH_4)_2$ SO_4 at $250°C$ and $50.25\ kgf/cm^2$ abs. (Skarbo, 1973b)

already predetermined by the extraction method used, that is, whether the system is acidic or alkaline. Various separation approaches for these elements from an aqueous medium have been devised, including selective hydrogen reduction, selective precipitation including cementation, fractional crystallization, ion exchange, and solvent (liquid-liquid) extraction.

The separation of copper in an acidic medium has been carried out successfully in practice. The technique of cementation using scrap iron is a typical example. Copper can also be easily precipitated as CuS by introducing H_2S gas (Warner, 1956; Simons, 1964). The separation of Cu, Ni and Co from one another can also be performed using various solvent extraction techniques (Van der Zeeuw, 1972). Van der Zeeuw was granted a patent for his processing of extracting the copper, nickel, cobalt and iron values employing certain secondary and tertiary carboxylic acids as extraction agents and selectively and sequentially reducing the copper value, the nickel value and finally the cobalt value present in the organic extract to sequentially insoluble metallic precipitates under controlled hydrogen reduction conditions. The removal of iron from a leach liquor containing those metal ions usually occurring in nodules can be accomplished either by hydrolysis under oxidizing environments (Renzoni, 1945) or by adding chemicals that form an insoluble iron complex which can be precipitated out (Mancke, 1956; Kasey, 1971).

Separation of these elements in alkaline solutions, especially in ammoniacal solution, can be carried out in a manner similar to the Sherritt Gordon process (Caron, 1924, 1955; Warner, 1956). The technique of hydrogen reduction has proved to be an economical process (Schaufelberger and Roy, 1955; Schaufelberger, 1956; Benoit and Mackiw, 1961; Mackiw et al., 1962) in the separation of Ni, Co and Cu. The separation of cobalt from nickel can also be achieved by precipitating cobalt preferentially in the presence of ferrous sulphates (Dean, 1959).

Three typical flowsheets for the recovery of the metal values from the leach liquor of three different leaching methods are shown in Figs. 12-14—12-16.

The recovery of metals by ammoniacal extraction of nodules as suggested by Brooks et al. (1970) is shown in Fig. 12-14. In the event that there is appreciable manganese dissolution, this metal is first precipitated as manganese carbonate by heating the leach liquor to 55°C. The Cu and Fe values are then precipitated sequentially as copper sulphide and ferric hydroxide by passing in H_2S and by aeration. The nickel and cobalt carbonate solution remaining after the removal of Mn, Cu and Fe is then heated for recovery of the remaining ammonia and carbon dioxide and to precipitate nickel and cobalt carbonate as a concentrate for further treatment.

Fig. 12-15 is a flowsheet for the treatment of leach liquors produced in the acidic extraction processes proposed by Brooks et al. (1970). After sulphation of nodules, an aqueous leaching operation is performed to extract Cu, Ni, Co, Mn and Fe. The dissolved copper is recovered by cementation with scrap iron. 95% of the Cu is recovered by this conventional operation.

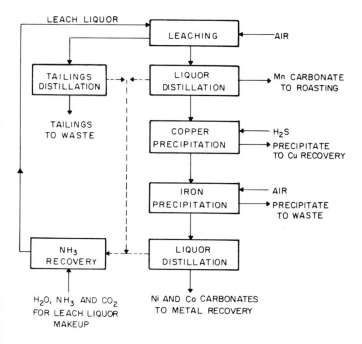

Fig. 12-14. Flowsheet of proposed process for the ammoniacal extraction of metals from nodules (after Brooks et al., 1970).

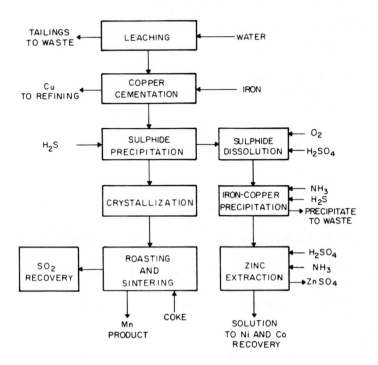

Fig. 12-15. Flowsheet of proposed process for the acidic extraction of nodules (after Brooks et al., 1970).

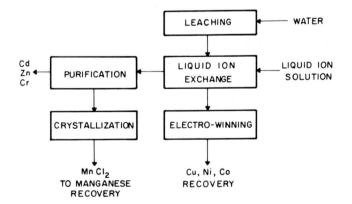

Fig. 12-16. Flowsheet of proposed process for hydrochlorination extraction of metals after the hydrochlorination of nodules (after Cardwell, 1973).

The clarified solution is then treated to remove the Ni, Co and remaining trace of Cu. Over 90% of the Ni and Co and the last trace of Cu were precipitated within 10 min using hydrogen sulphide at 120°C and 4.55 kgf/cm^2 absolute. Less than 1% of the Fe and Mn was precipitated. Manganese sulphate is next crystallized from solution by heating the solution in an autoclave to 200°C (Fuller, 1966). Manganese sulphate monohydrate crystals can then be decomposed to sulphur dioxide for recycling and manganese oxide (Fuller and Edlund, 1966). A similar flowsheet could be utilized for processes involving a direct acid leach or the SO$_2$ reduction leach. The precipitated nickel- and cobalt-bearing sulphides is dissolved in oxygenated and weak sulphuric acid for the recovery of nickel and cobalt using the Sheritt Gordon hydrogen reduction process.

A typical flowsheet for the recovery of the metal values from the leach liquor of water-soluble metal chlorides after hydrochlorination process is shown in Fig. 12-16. The leach liquor is first passed through a series of liquid ion exchange processes where pure aqueous solutions of Cu, Ni and Co are produced. The pure metals in these three separate solutions are then recovered by conventional electrolysis. The manganese chloride solution remaining after Cu, Ni and Co has been separated from it is purified and crystallized out of solution as manganese chloride. The utilization of ion exchange resins for treatment of nodule leach liquors has also been investigated (Iammartino, 1974).

HIGH TEMPERATURE EXTRACTION PROCESSES

Smelting

As discussed in the previous section, a reduction roast is commonly carried out prior to the leaching step in most ammonia leaching operations. There is, however, little information available on other kinds of roasting processes for the treatment of manganese nodules.

Smelting investigations, which have been undertaken both on the laboratory and pilot-plant scale (Beck and Messner, 1970; Vasilchikov et al., 1968), have been aimed at selectively reducing and recovering the primary metals in a metallic product and rejecting the majority of the manganese and iron in a smelting slag. Nodules were preheated at 1,000°C and smelted at about 1,400°C for one hour in the presence of 5% by weight of coke and 5% by weight of SiO$_2$. High recoveries of Cu, Ni, Co, Mo and Fe were obtained. A typical result is listed in Table 12-XIII (Beck and Messner, 1970). The addition of pyrite reportedly improved the cobalt recovery.

The process of the individual metal recovery is largely dependent upon the nature of the metallic products. These could be crushed and ground, followed by a physical or a chemical separation step if necessary and the

TABLE 12-XIII

The results of a smelting process for the treatment of nodules (Beck and Messner, 1970)

% Recovery:					
Cu	Ni	Co	Mo	Fe	Mn
87.2	96.4	92.1	89.3	84.7	3.5

Conditions: Pretreatment at $1,000^{\circ}C$, 1 hour smelting at $1,400^{\circ}C$ with 5% SiO_2, 5% coke, and 5% pyrite added by weight.
Nodule composition (dried): Mn 24.9%, Fe 9.8%, Ni 1.08%, Cu 0.71%, Co 0.15%, Mo 0.05%, SiO_2 13.5%

metallic product would then be treated by a hydrometallurgical and/or pyrometallurgical operation to recover the individual elements. If a leaching process is adopted, then the subsequent processes will be similar to those of low-temperature treatment discussed earlier.

Chlorination processes

Chlorination processes have been recognized as one of the promising extractive methods, especially for the treatment of low-grade ores. The earliest application of this process was in the refining of precious metals, such as gold (Coyle et al., 1966). The application of chlorination metallurgy to processing various kinds of metal oxides has been investigated by a number of researchers (Ketteridge and Wilmshurst, 1964; Amirova et al., 1965, 1966; Lukmanova et al., 1965; Lippert et al., 1969).

The extraction of metals from ocean manganese nodules by a chlorination technique has been reported by Deepsea Ventures Inc. (Caldwell, 1971; Cardwell, 1973). In this process, the chlorination step is performed with excess HCl gas to transform the oxides of Mn, Ni, Cu and Co to the corresponding chlorides. By control of the temperature the chlorination of iron oxide is inhibited. The unique feature of their process is to recover chlorine as a by-product; this has been a major interest to the chemical industry in the past. A detailed study in the mechanism and kinetics of chlorination of copper, nickel and cobalt from nodules has recently been completed by Hoover (1972).

The thermodynamics of the chlorination process has been described in detail by Kellogg (1950). For the five oxides of main interest in nodules, equilibrium diagrams under different partial pressures of oxygen and chlorine were calculated for various temperatures (Hoover, 1972). The diagrams for $400^{\circ}C$ are presented in Fig. 12-17. The small circle in the upper right-hand corner of each of the figures represents oxygen pressures between

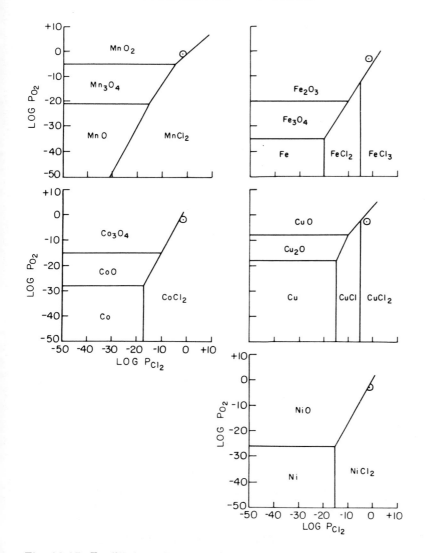

Fig. 12-17. Equilibrium diagrams of oxides under different oxygen and chlorine partial pressures at 400°C (after Hoover, 1972).

0.01 and 0.1 atm, which is the range likely to be of interest in a typical selective chlorination operation. It is clear from these figures that by operating under closely controlled conditions, iron and manganese will not be chlorinated and a selective chlorination of nickel, copper and cobalt oxides might be achieved.

Typical results of the chlorination of DH-2 nodules at 1,050°C as a function of the Cl_2/O_2 ratio is given in Fig. 12-18 (Hoover, 1972). These

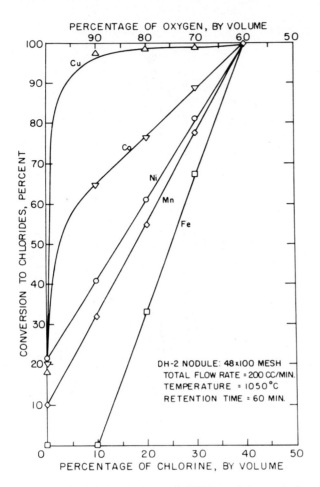

Fig. 12-18. Chlorination of DH-2 nodule as a function of chlorine/oxygen ratio at 1,050°C (after Hoover, 1972).

results show that some degree of separation of the metals as chlorides is possible, with Fe and Mn being the least chlorinated.

An inherent merit of the chlorination process is that selective recovery can also be achieved by vaporization or condensation of these chlorides. As can be seen in Fig. 12-19, the vapour pressures of various metal chlorides differ significantly at a given temperature and may be indicative of the potentially successful transport of a metal chloride.

Segregation

The segregation process is a method of treating oxidized ores in which the ore is heated with reducing agents, generally coke, and halide (chloride) salts.

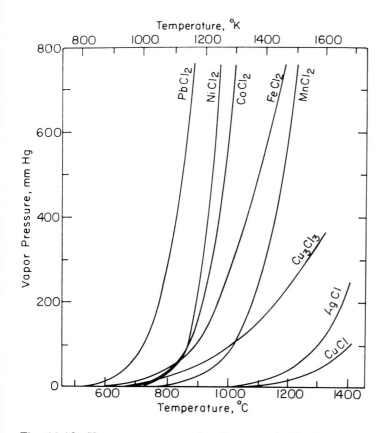

Fig. 12-19. Vapour pressures of various metal chlorides as a function of temperature (after Hoover, 1972).

Solid chloride is added to the ore in order to generate HCl gas, which converts the oxide to chlorides. The varporized metallic chlorides are reduced to metal on the surface of the coke particles. The metal can then be recovered by a subsequent treatment such as screening, magnetic separation, leaching or flotation. The mechanisms of the segregation reactions have been studied by a number of investigators (Rey, 1936; Diaz, 1958; Rampacek et al., 1959; Anonymous, 1959; Rampacek and McKinney, 1960; Martinez, 1967; Dor, 1972; Iwasaki, 1972).

The technique of segregation roasting was recently applied to manganese nodules and proved to be technically feasible for the recovery of Ni, Cu and Co (Hoover et al., 1975).

In this study of segregation processing, the principal variables were the roasting temperature, retention time during roasting, type and amount of chlorinating agent, amount of reducing agent, and nodule particle size. Fig. 12-20 presents a typical result of segregation roasting carried out as a

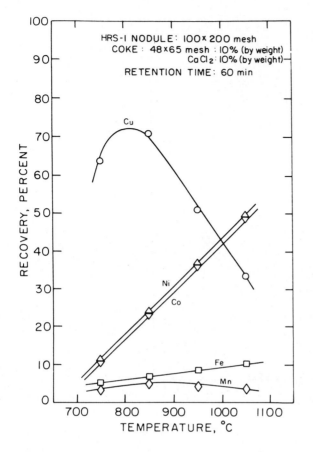

Fig. 12-20. Effect of roast temperture on the recovery of metals on the coke particles, using calcium chloride as the chlorinating agent (after Hoover, 1972).

function of temperature. As can be seen from these results, recoveries are markedly dependent on the temperature and the best overall results for copper were obtained at around 800°—850°C. Nickel and cobalt behave similarly to each other and exhibit an increasing recovery with increasing temperature. The optimum conditions for the segregation roasting of nickel, copper and cobalt were achieved with $CaCl_2$ as the chloride source at a batch retention time of about 2 h; roasting at approximately 850°C yielded the highest recovery for copper and 1,050°C for nickel and cobalt. Under these conditions, the maximum recoveries of 75%, 60% and 60% were obtained for copper, nickel and cobalt, respectively. Similar results were obtained with nodules from a variety of locations in the Pacific Ocean.

SUMMARY

The formation and distribution of some metals in ocean-floor manganese nodules have been discussed in the light of the observed data in the literature and thermodynamic and kinetic considerations of the oxidation of metal ions in the oceanic environment.

There are, in general, two major schools of thought on the mechanism of incorporation of the minor elements such as nickel, copper and cobalt with the major elements such as manganese and iron. One is the lattice substitution mechanism and the other the adsorption mechanism. If the mechanism is lattice substitution, extraction of the metal ions is not possible unless the lattice of the major elements is first broken and exchanged with other ions from the bulk solution. Consequently, the leaching behaviour of minor elements should display a very close relationship with that of major elements. However, it was reported in the literature that the leaching behaviour of minor elements is independent of that of the major elements. Selective extraction of these minor elements against the major is therefore possible.

Extraction of metals from ocean-floor manganese nodules has been divided into two categories, namely, hydrometallurgical processes (with or without high-temperature pretreatment) and pyrometallurgical processes. The hydrometallurgical processes include both acid leaching and ammonia leaching with or without pre-reduction. In some cases, the pre-reduction step involves high-temperature roasting with different types of reagents, e.g., HCl, H_2SO_4, CO/CO_2, etc. Pyrometallurgical processing includes smelting, chlorination and segregation.

Acid leaching shows high dissolution of Cu and Ni (about 80%) without dissolving Mn but dissolves only half of the Co. The selectivity of the desired elements in acid leaching can, in general, be achieved by adjusting the leaching temperature, retention time of nodules in the reactor, the size of nodule particles and the acid concentration. A significant disadvantage of acid leaching is the high acid consumption. A typical leaching experiment at pH 1.0 consumes 0.3 ton of H_2SO_4 per ton of manganese nodules treated. This problem may be overcome by high-temperature leaching in an autoclave. However, the capital cost and the maintenance cost of the reactor will be substantially higher for high-temperature leaching than for low-temperature leaching. Leaching with reducing acids (such as sulphurous acid) yields good extraction of Ni, Co and Mn (over 90%), while that of Cu and Fe is low under the conditions considered. This is the only low-temperature extraction treatment which yields satisfactory recovery of Mn and Co.

Ammonia leaching of nodules after pre-reduction with CO/CO_2 mixtures at high temperatures (500°C—800°C), displays recoveries of Ni, Cu (80%) and reasonable recovery of Co (above 50%). Ammonia leaching can also be carried out at elevated temperature with or without a reductant with satisfactory recoveries of Cu, Ni and Co. However, such a high-temperature, high-pressure operation will require substantial capital investment, as well as operating costs. On the other hand, ammonia leaching in an autoclave will not necessitate the drying costs encountered in the reduction roast.

Smelting of nodules shows satisfactory recoveries of most of metals from manganese nodules. However, this technique would require high capital costs and high operating costs due to high energy comsumption. Further, subsequent operations required for the treatment of the ferro alloy product would be necessary.

Chlorination and segregation roasting processes also yield reasonable recoveries of the metals but generally with low selectivity. There are two ways of controlling the selective extraction of metals in chlorination, namely by selectively chlorinating the elements and by selectively vaporizing and condensing the metal chlorides in the chlorination product.

Unfortunately, there are insufficient data available in the literature at present to permit a meaninful economical assessment of the different processes that might be used for treating nodules. Accordingly, it is difficult at this stage to draw meaningful conclusions concerning the best method for extracting metals from deep-sea manganese nodules. Such an analysis must be left to the commercial organizations working on potential nodule mining and extraction systems. In this chapter, however, a few of the possible extraction methods that can be utilized for the processing of nodules have been summarized and the methods related to the physical-chemical nature of nodules. Only the behaviour of five elements in nodules has been considered in detail, but many of the other numerous elements in manganese nodules would eventually be considered in a commercial nodule processing operation.

CHAPTER 13

ENVIRONMENTAL ASPECTS OF NODULE MINING

A. F. AMOS, O. A. ROELS, C. GARSIDE, T. C. MALONE AND A. Z. PAUL

INTRODUCTION

The mining of manganese nodules from the deep-sea floor has no historical parallel to draw upon for an evaluation of the possible environmental effects. Mining sites will be located thousands of kilometres from land, on the high seas, with 5,000 m of ocean water between the mining platform and the minerals being mined. The complex interactions between atmosphere, ocean water and ocean floor — physics, chemistry, biology and geology — will have to be examined before the effect of nodule mining on the environment can be known. As deep-ocean mining is without historical precedent, this chapter will have to be limited to the consideration of operations that are already in the planning and testing stages rather than the presentation of a general treatise on nodule mining by any method in any ocean. Only those locations where nodule deposits of known commercial value are found will be discussed.

The area of prime interest to the mining industry is the eastern-equatorial Pacific Ocean, specifically the radiolarian (siliceous) ooze sedimentary province (Horn et al., 1972c). Throughout the chapter we will refer to the "nodule zone", defined in Fig. 13-1 as lying between the Equator and the 30°N parallel and extending east to west from the 100°W to the 160°W meridian.

Within the nodule zone, specific sites have been selected for study by a U.S. governmental—academic—industrial panel and will be referenced here (Fig. 13-1). Site *C* has been declared by Deepsea Ventures, Inc. to be their prime site for future mining operations (U.S. Congress, 1975). Several cruises under the auspices of the U.S. Department of Commerce, NOAA, Deep-Ocean Mining Environmental Study (DOMES) project, have already been undertaken to study environmental baseline conditions before mining starts (Table 13-I).

Three basic mining systems will be considered: (1) the continuous-line-bucket (CLB) system (Masuda et al., 1971); the airlift or suction-dredge system using a towed dredge-head (Flipse, 1969); and (3) as (2) above, but using a self-propelled dredging device (Anonymous, 1972).

The environmental baseline conditions existing in the nodule zone will be described, with particular emphasis on those parameters that might be

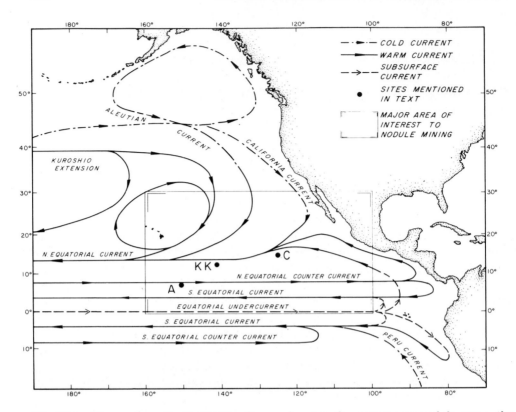

Fig. 13-1. The eastern North Pacific Ocean showing the manganese-nodule zone, the location of test sites mentioned in the text and the surface and subsurface currents. Sources: Sverdrup et al. (1942); Wyrtki (1967).

affected by nodule mining, and then the impact of full-scale mining operations on that environment will be examined. Topics considered to be outside the scope of this chapter are: (1) chemical processing of nodules at sea (not being considered by mining companies in the foreseeable future (Cardwell, 1973)); (2) transportation of nodules from mining site to shore terminals; (3) environmental problems not unique to deep-ocean mining (waste products of extractive metallurgy, tailing disposal sites, increased use of marine terminals, support facilities, transportation); (4) environmental impacts of alternative means of obtaining metal ores; (5) environmental analysis of the utilization of minerals obtained from the marine environment.

It must be emphasized that important data concerning the magnitude and methods of mining operations are known to us only through a few published reports, copies of patents and news releases. Such proprietary information is

TABLE 13-I

Test sites and cruises referred to in text

Test site	Location	Ship	Cruise	Dates	Reference
BP	31°02'N; 78°24'W Blake Plateau	*Deepsea Miner*	—	Aug. 1970	Amos et al., 1972
KK	13°N; 141°W	*Kana Keoki*	72—1	Aug.—Sept. 1972	Roels et al., 1973
A	8°27'N; 151°47'W	*Moana Wave*	74—2	Apr.—May 1974	Amos et al., 1975a,b
C	15°N; 126°W	*Oceanographer*	RP-6-OC -75	Apr.—June 1975	Amos et al., (in prep.)
Geosecs (Profile)	14°3'S; 126°16'W to 28°31'N; 121°30'W	*Melville*	Geosecs leg 10	May—June 1974	Broecker and Mantyla, 1974
Geosecs (Test stn.)	28°29'N; 121°38'W	*Thomas Washington*	—	Sept. 1969	Craig and Weiss, 1970

closely guarded by the mining companies. Critical numbers such as volume of sediment disturbed by the different techniques, and volume of effluent discharged at the sea surface, have been obtained from one U.S. National Research Council report (National Research Council, 1975). Some of the quantities reported in this and other published works are, in our opinion, too speculative.

The *Glomar Explorer* was, until recently, thought to be a vessel constructed exclusively for mining nodules. A great deal of speculation has been published (e.g., Rothstein and Kaufman, 1973; Hammond, 1974) about this system, a major part of which was — it has now been revealed — constructed not to mine nodules but to recover a sunken submarine (U.S. Congress, 1975, p. 32). This situation underscores the difficulties in evaluating the environmental aspects of a mining operation before that operation exists. Yet, the present effort by government, industry and the academic institutions to investigate potential environmental hazards before the onset of full-scale mining can only be regarded as essential. Guidelines for environmentally safe development of deep-ocean mining have been suggested (Roels, 1974; National Research Council, 1975).

ENVIRONMENTAL CONDITIONS IN THE MANGANESE NODULE ZONE

While many physical, chemical and biological measurements have been made in the equatorial Pacific Ocean, detailed measurements in the open ocean areas where nodules will be mined are lacking, particularly in some fields such as nutrient chemistry, phytoplankton productivity and benthic ecology. Deep and bottom measurements in the water column are sparsely located throughout the area. A literature survey of environmental data in the nodule provinces (Amos et al., 1973) provided some of the data presented here. Project DOMES cruises have recently studied the seasonal baseline conditions of the physical, geological, chemical, and biological oceanography at the test sites shown in Fig. 13-1 and Table 13-I. Techniques used and the scope of investigation undertaken have been outlined in Amos et al. (1975a).

The atmosphere

The manganese nodule zone coincides with the N.E. Trade Wind belt in the Pacific Ocean. The mean surface winds over much of the area are from 7 to 10 m/sec (Mintz and Dean, 1952). Between 10°N and the Equator lies the Intertropical Convergence Zone (ITCZ), the narrow region of the "doldrums" that marks the boundary between the N.E. and S.W. trades. The centre of the ITCZ moves from 5°N in February to 10°—12°N in August. The ITCZ lies along the zone of greatest interest to mining and is frequently characterized by cloud-cover accompanied by heavy precipitation. At Sites A and C from April—June, net solar radiation averaged 380 cal. cm^{-2} day^{-1}, compared to 660 cal. cm^{-2} day^{-1} for cloudless days at the same location.

Tropical cyclones occur during the months of May—November but particularly during the period July—September. Usually forming off the west coast of Central America (10°—13°N 90°—110°W), they move west-northwesterly, sometimes reaching as far as Hawaii. Between 1966 and 1973, 65 tropical storms and 54 hurricanes occurred (NOAA, 1974) in the manganese nodule zone. On August 18 1972, six well-developed cyclones and disturbances were photographed by satellite between Hawaii and the Central American coast. The occurrence of such storms will be of concern to mining vessels operating in the area.

The water column

The water depth in the manganese-nodule belt varies from 2,300 to 5,500 m but with an average depth of 5,000 m. High-grade nodules are generally found in water depths between 4,000 and 6,000 m (Mero, 1972) and mining systems will have to operate at these depths. Most of the area consists of a rather flat plain trisected by the east—west trending Molokai, Clarion and Clipperton fracture zones. With the exception of the fracture zones, the abyssal topography shows less than 200 m relief with rather closely spaced abyssal hills (see Chase et al., 1970, for detailed topography). A few widely spaced seamounts are found throughout the area.

Currents

Surface and subsurface currents. A system of zonal currents characterize the surface and near-surface flow throughout the area (Table 13-II). They represent the equatorial extensions of the huge subtropical anticyclonic gyres that dominate the middle-latitude circulation (Fig. 13-1). These large-scale current systems are controlled by the overlying trade-wind system, the density gradients and the Coriolis force due to the earth's rotation. The North and South Equatorial Currents, above and below the Equator, move westward with speeds up to 50 cm sec^{-1}. Between these two currents, north of the Equator, flows the North Equatorial Countercurrent toward the east. The Equatorial Undercurrent flows eastward at speeds up to 150 cm sec^{-1} along the Equator underneath the westerly surface flow. The Eastern Boundary Currents that feed the equatorial current system are the California Current of the North Pacific and the Peru Current in the south. The Peru Current itself does not enter into the nodule zone under discussion. As can be seen from Fig. 13-1 and Table 13-II, the undercurrent also contributes significantly to the total transport of the equatorial flow (Wyrtki, 1967). Some of these currents respond to seasonal influences and show considerable variability in their position, speed, extent and depth.

Bottom currents. Bottom currents in this region distant from continental margins, are generally the result of the slow northward movement of dense, cold water that originally formed at the surface in the Antarctic. Indirect (Heezen and Hollister, 1964) and direct (Isaacs et al., 1966) evidence show that bottom currents in parts of the world's oceans flow at much higher velocities than was previously supposed. Significant oscillations superimposed on the net-drift are caused by the deep ocean's response to tidal forces (Munk et al., 1970; Wimbush, 1972). Peak velocities strong enough to erode and transport sediment have been measured at sites *A* and *C*. The general drift of bottom water in the area is thought to be toward the northeast. This can be inferred from the distribution of potential temperature (Gordon and Gerard, 1970), dissolved oxygen (Wooster and Volkman, 1960) and from the few direct bottom-current measurements made to-date (Reid, 1969; Johnson, 1972). Data from a 14-day record of bottom currents in May and June 1974 at site *A* (Fig. 13-2) show that at 210 m above bottom currents average 2.0 cm sec^{-1} toward ENE and exhibit peak velocities of up to 16.5 cm sec^{-1} at semidiurnal (M2) periods throughout the recording interval. At 1,000 m off bottom the mean current had shifted to due east, about the same average speed with slightly higher peak velocities, while at 2,000 m above bottom the currents may have been below threshold of the current meter (Amos et al., 1975a). A more recent record at site *C* has current measurements at eight levels from 10 to 1,000 m above bottom. The preliminary analysis of the 34-day record during April—May 1975 indicates mean currents to the southwest at all levels, exhibiting semidiurnal peaks of up to 25 cm sec^{-1}. This pattern is consistent with the trends of the bottom

TABLE 13-II

Major currents in the manganese-nodule zone

Name	Where found	Velocity	Depth	Transport (10^{12} cm^3 sec^{-1})	Remarks
North Equatorial Current	8°—30°N across entire region	Westerly; 15 cm sec^{-1}	Surface to >300 m	27	Major surface current in the manganese nodule zone
North Equatorial Counter-current	7°—12°N across entire region	Easterly; avg. 20—30 cm sec^{-1} max. 120 cm sec^{-1}	Surface to 200 m	15	Seasonally variable due to shift in ICTZ; subsurface current (30—<500 m) exists to 85°W
South Equatorial Current	10°S—4°N across entire region	Westerly; 35—45 cm sec^{-1}	Surface to 50 m N. of Equator to 200 m. S. of Equator	49	Seasonally variable N. of Equator; divergence N. and S. of Equator of this current causes equatorial upwelling
California Current	30°—15°N 1,000 km off coast	South, then south-westerly; 15 cm sec^{-1}	Surface to 100—300 m	17	During Feb.—June supplies most of the water to N. Equatorial Current
(Peru Oceanic Current)	30°—5°S	North, then north-westerly	Surface to 700 m	14	Not directly in area of interest, but supplies water to S. Equatorial Current
Equatorial Undercurrent	2°N—2°S across entire region	Easterly; 120—150 cm sec^{-1}	50—300 m	35	Branches N. and S. east of Galápagos Is. to supply water to N. and S. Equatorial Currents

Sources: Cromwell et al., 1954; Montgomery and Stroup, 1962; Wyrtki, 1963, 1967.

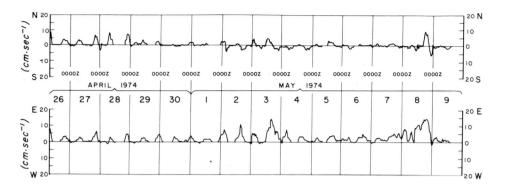

Fig. 13-2. East and north components of current at site A, 210 m above the ocean floor in April and May 1974.

isopleths and with an anticyclonic circulation of bottom water in the northeast Pacific Basin.

Water masses
Water masses, identified by their temperature and salinity (T/S) characteristics, and sometimes by other constituents such as dissolved oxygen or nutrients, are generally formed at the sea surface by interaction with the atmosphere. Away from their sources, the T/S characteristics are modified by spreading and mixing with other water masses to form other water types. In the eastern equatorial Pacific, a complicated series of water masses that are formed outside the area are transported into the region by the circulation system while others are formed within the area itself. Dividing the water column into four vertical domains (surface, subsurface, intermediate, and deep), the water masses and their properties have been summarized in Table 13-III, largely using the system of Wyrtki (1967).

Of greatest importance in our region with respect to the environmental impact of manganese-nodule mining, are the equatorial surface water, the temperate and subtropical subsurface waters, the oxygen minimum layer(s) and the deep (bottom) water. The T/S relationships of the various water masses can be seen at a typical station at site C (Fig. 13-8). Temperature and salinity profiles shown in Fig. 13-4 were made with continuously recording salinity—temperature—depth (STD) sensors (see Amos, 1973, for experimental details).

Temperature
Surface temperatures vary from $18°$ to $27°C$ in the winter and from $20°$ to $28°C$ in the summer (Fig. 13-3). Seasonal upwelling along the Equator and influx of Peru Current water produces a band of cold water (Fig. 13-3) during May to December and a well-defined thermal front is formed

TABLE 13-III

Major water masses in manganese-nodule zone

Name	Core layer in manganese-nodule belt				Where formed	Mechanism of formation
	how identified	range of core parameters				
		salinity (°/oo)	tempera-ture (°C)	depth (m)		
Surface						
Tropical surface water	High temperature; low salinity	33—34	25—28		Tropics north of Equator	Excess of evaporation over precipitation
Subtropical surface water	High salinity; wide range of temperature	35—36.5	18—28	Surface to depth of mixed layer	Centre of S. Pacific anti-cyclonic gyre	Excess of precipitation over evaporation
Equatorial surface water	Medium salinity range; wide range in temperature	34—35	20—28		Equator to 5°N	Mixing of above two, equatorial upwelling, advection of Peru Current water
California Current water	Low salinity; moderate temperature	33—34	15—20		North temperate zones	Extension of Western Boundary Current of N. Pacific gyre
Subsurface						
Temperate sub-surface water	Shallow salinity minimum	34—34.5	12—15	75—150	Coast of California	Minimum formed by seasonal evaporation and warming of surface layer
Subtropical subsurface (North) water	Salinity maximum in upper thermo-cline; high oxygen	34.4—34.8	18—22	30—100	North Pacific anticyclonic gyre	Sinking along density surfaces entrainment by Equatorial Countercurrent
Subtropical subsurface (South) water	Salinity maximum in lower thermo-cline; oxygen minimum	34.6—34.8	11—12	100—>200	South Pacific anticyclonic gyre	Sinking along density surfaces entrainment by undercurrent discharge into N. Pacific by N. Equatorial Current
Oxygen-minimum layer	Extremely low values of dissolved oxygen (<1 ml/l)	Not conservative water mass due to oxygen consumption; includes parts of other subsurface masses		Core: 175—700 upper to lower boundary (<1 ml/l) 100—1400	Eastern equa-torial regions north and south of Equator	Oxygen consumption due to long residence time of water in areas of slow circulation
Intermediate						
North Pacific intermediate water	Deep salinity maximum	34.5—34.55	5—6	600—700	Sub-arctic gyre	Mixing across pycnocline and thermohaline convection at high latitudes
Antarctic intermediate water	Deep salinity minimum	34.5—34.55	5—6	700—900	Sub-antarctic regions	Sinking of Antarctic and Sub-antarctic surface water at polar regions
Deep						
North Pacific deep water	Salinity maximum at ocean bottom	34.68—34.70	0.95—1.2*	Ocean bottom	N. Atlantic, Weddell Sea area	In N. Atlantic and Weddell Sea entering S. Pacific via Antarctic Circumpolar Current

Sources: Sverdrup et al., 1942; Montgomery and Stroup, 1962; Muromtsev, 1963; Reid, 1965; Wyrtki, 1967; Reid and Lynn, 1971.
*Potential temperature.

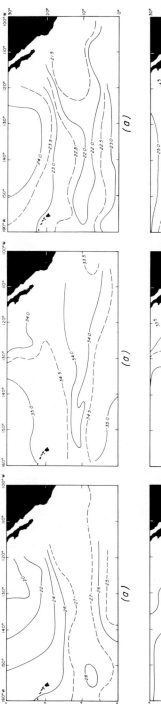
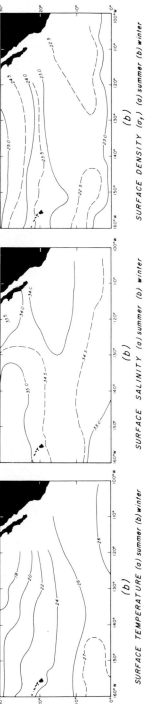

Fig. 13-3. Surface temperature, salinity and density in manganese nodule zone in (a) summer and (b) winter. From Amos et al. (1972), compiled from data archived by NODC up to 1972.

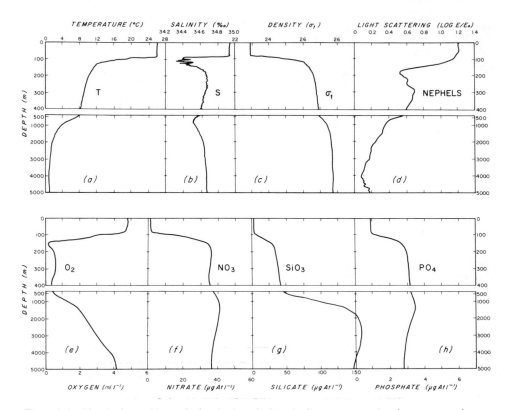

Fig. 13-4. Vertical profiles of physical and chemical parameters in the water column at site A in April and May 1974: (a) temperature, (b) salinity, (c) density (σ_t), (d) light scattering, (e) dissolved oxygen, (f) nitrate, (g) silicate, (h) phosphate; a, b, c, d, are typical continuous profiles at one station; e, f, g, h, are averages of all discrete sample data collected at site A.

(Cromwell and Reid, 1955). Typical vertical temperature distribution (Fig. 13-4a) reveals a homogeneous, mixed layer at the surface 20–100 m thick. Below the mixed layer an intense thermocline is usually found with the temperature dropping to 12°C at 150 m depth. From here to ~1,500 m the temperature decreases monotonically to 3°C. Below this, the decrease is more gradual, reaching a minimum of 1.48°C at 4,250 m. From this depth to the ocean bottom an in-situ temperature increase of a few hundredths of a degree occurs. This increase toward the bottom is an adiabatic increase and in dealing with deep-ocean waters *potential temperature* (θ) is usually used to compensate for pressure effects.

The potential temperature remains virtually constant in the bottom few hundred metres and varies from 0.95°C in the southwest to 1.2°C in the northeast part of our region.

Salinity

The vertical and horizontal distribution of salinity is more complex as the major water masses in the region are identifiable by salinity rather than temperature characteristics. Surface salinities range from $33.5°/oo$ to $> 35°/oo$ throughout the year (Fig. 13-3). Surface waters with salinities $< 34°/oo$ are transported into the equatorial system by the California and Peru currents. The vertical distribution of salinity at site C is shown in Fig. 13-5.

Beneath the mixed layer, and within the thermocline, an alternating series of salinity minima and maxima are often found (see Fig. 13-5). These represent the cores of the various subsurface water masses (Table 13-III). Even in the upper layer of uniform temperature, salinity increases can be found as a result of excess precipitation over evaporation. At the base of the main thermocline, the salinity maximum $(34.6—34.8°/oo)$ of the southern subtropical subsurface water occurs. Below this, the salinity minimum $(34.5—34.55°/oo)$ of the intermediate water is encountered at 600—900 m depth. The region between $8°$ and $30°$ N is where the intermediate waters of the North Pacific Ocean meet and override those of the South Pacific,

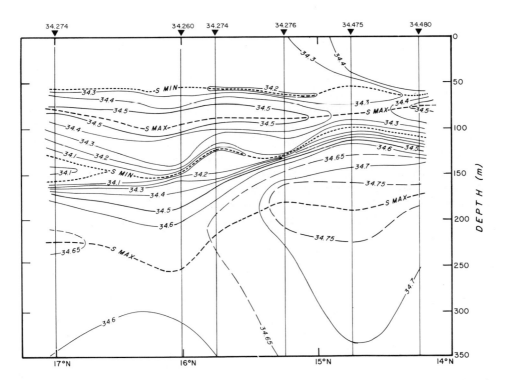

Fig. 13-5. Salinity section down to 350 m at site C in May 1975 from $14°$ N to $17°$ N along $126°$ W. Contoured from STD profile data taken at locations indicated by solid triangles.

sometimes producing a slight salinity maximum between two minima (see section along 160°N in Reid, 1965). From the intermediate salinity minimum there is a gradual increase in salinity to the ocean floor. Bottom values range from 34.68 to 34.70°/oo throughout the entire region.

Sigma-t

The horizontal distribution of σ_t in the region (where σ_t is the conventional way of expressing density in the ocean by the relationship $\sigma_t = 10^3 \cdot (\rho_{s,t,o} - 1)$, where $\rho_{s,t,o}$ = specific gravity at salinity, s; temperature, t; and atmospheric pressure), is largely controlled by the surface salinity (Fig. 13-3).

The vertical density distribution will have a major effect on the fate of mining effluents discharged into the water column. The vertical density field in the nodule belt is much more a function of temperature than salinity because of the small range of salinity with depth (Fig. 13-4c).

Immediately beneath the mixed layer where density is uniform, a strong density gradient (pycnocline) occurs coincident with the thermocline.

Dissolved oxygen

An important feature of the water masses in the nodule zone is the distribution of dissolved oxygen. Oxygenation of ocean waters generally occurs by interaction between atmosphere and ocean at the sea surface. Consequently, the dissolved oxygen content of the surface waters in the region is close to 100% saturation, with values averaging 4.75 ml l^{-1}. Below the mixed layer the oxygen content decreases rapidly to the oxygen minimum layer (Fig. 13-4e, Table 13-III) with values approaching 0 ml l^{-1} (0.08 ml l^{-1} at 125 m at site C). Except for the landlocked Baltic and Black seas, these are the lowest oxygen concentrations found in oceanic waters. But unlike those seas, these oceanic waters produce no hydrogen sulphide (see Grassoff, 1975, p. 565) and the waters are not considered to be anaerobic. The thickness of the layer with values below 1 ml l^{-1} ranges from 100 to 1,400 m.

The mechanisms of formation of this layer are complex and not completely understood (Reid, 1965; Wyrtki, 1967) but basically it is caused by consumption of oxygen beneath areas of high surface productivity off the coasts of Central and South America where the residence time of subsurface waters is long. The strong pycnocline inhibits downward diffusion of oxygenated water from surface layers, yet an equilibrium is maintained by upward diffusion from subsurface layers (Reid, 1965). At both sites A and C, two minima are found, the deeper one at ∿500 m (Amos et al., 1975a). The slight maximum between these minima is probably due to influx of water from the western Pacific, borne by the Equatorial Undercurrent (Wyrtki, 1967).

Beneath the minimum, oxygen increases to 3—4 ml l^{-1} at the bottom, approximately 50% saturation, reflecting the origin of the bottom water, formed at higher latitudes at the sea surface. The change in oxygen gradient near 3,000 m (Fig. 13-4e) may be a residue of the benthic front of the South Pacific described by Craig et al. (1972).

Dissolved nutrients

Vertical profiles of the major dissolved nutrients at site A (phosphate, silicate, and nitrate) are given in Fig. 13-4f,g,h. Typically, the surface waters of the ocean have low concentrations of the inorganic ions of the macro-nutrients (nitrate, nitrite, ammonia, phosphate and silicate) while as a consequence of biological and detrital transports the deeper waters of the world's oceans are enriched with these ions.

The general horizontal and vertical distribution of nutrients in the water column for this area are poorly defined. Montgomery and Stroup (1962) and Reid (1965) have presented comprehensive profiles of phosphate and much has been added to the knowledge of nutrients in the Pacific by the recent GEOSECS profile along 120°W (Broecker and Mantyla, 1974).

Phosphate. Phosphate data are relatively the most abundant compared to other nutrients (see particularly, Reid, 1962; 1965). Surface values range from 0.2 to 0.5 μg-at l^{-1}, remain constant in the mixed layer, increase rapidly to a maximum of $>$ 3 μg-at l^{-1} at between 300—1,200 m, closely paralleling the oxygen minimum and then decline to bottom values of 2.25—2.5 μg-at 1^{-1} (Fig. 13-4h).

The upwelling associated with the departure of the California Current from the California coast is the major ultimate source of phosphate in the manganese nodule belt. At the depths from which upwelled water is drawn ($\sim$200 m) phosphate concentrations are a little over 2 μg-at l^{-1} in the winter (Stefánsson and Richards, 1963). This water, when upwelled, gives rise to a band of nutrient-rich surface water adjacent to the coast in which concentrations are typically in the range 0.25—0.5 μg-at l^{-1}.

Surface phosphate values decline along the course of the equatorial current, but increase to the south to as much as 1 μg-at l^{-1} in the axis of the Equatorial Countercurrent. There is a general decline in surface phosphate towards the north and west to less than 0.25 μg-at l^{-1}.

Nitrogen. Nitrogen as a phytoplankton nutrient is available in at least three forms — nitrate, nitrite, and ammonia; nitrite probably being of least significance. However, the literature is somewhat equivocal concerning the role of organic nitrogen sources in phytoplankton nutrition.

Relatively few data on nitrogen distribution throughout the eastern tropical Pacific are available in the literature, although a number of "mean" values, mostly related to the euphotic zone can be found.

There is a very rapid increase in nitrate from less than 1 μg-at l^{-1} in the mixed layer to over 30 μg-at l^{-1} at 125 m, below which nitrate remains constant or diminishes to a minimum at the same depth as the deeper oxygen minimum (Fig. 13-4f). Maximum values are encountered at 1,000 m (> 45 μg-at l^{-1}) and then decline to 35—37 μg-at 1^{-1} near the bottom.

In a study of organic nitrogen and nitrate along 119°W from 10°S to 20°N, Thomas et al. (1971) report that in the nitrate-rich waters to the south of the Equator (5—10 μg-at l^{-1}) organic nitrogen is lower (6.9 μg-at l^{-1}) and statistically significantly different from the values found north of the Equator (8.2 μg-at l^{-1}) in nitrate-poor waters (NO_3 not detected), and that organic nitrogen is low because the water has recently upwelled in the south. The mean organic nitrogen value in the north (8.2 μg-at l^{-1}) compared with undetectable nitrate and nitrite and very low ammonia concentrations (<1 μg-at l^{-1}) plus experimental enrichment studies, lead Thomas et al. to conclude that the organic nitrogen measured was not a significant phytoplankton nitrogen resource.

Thomas (1970a) divides the eastern tropical Pacific into two general areas: nitrate-rich and nitrate-poor, which are persistent and independent of season. Nutrient-rich water extends from just north of the Equator to 15°S along 105°W and 112°W. Nutrient-poor water extends north and south of this region. The nutrient-poor water contains zero nitrate while the rich water contains up to 7.6 μg-at l^{-1} at 10 m, with a mean of 5.6 μg-at l^{-1}. The high values here and elsewhere along the coast are indicative of upwelling, while further north near the Columbia river Stefánsson and Richards (1963) report that the higher values are correlated with the river water rather than with the ocean. At site $\acute{K}K$ (Roels et al., 1973), there is excellent correlation between nitrate and phosphate profiles: with the exception of the mixed layer and near-bottom samples, the N:P ratio is between 13.6:1 and 14.8:1, strongly supporting the oxidative origin of both nutrients (Redfield, 1942).

Ammonia distribution in 10 m water along 119°W from 10°S to 20°N is described by Thomas and Owen (1971). In this nitrate-rich area, ammonia is the least abundant nitrogen source (<0.5 μg-at l^{-1}), but in the nitrate-poor area, although ammonia is probably even less abundant, it is the major inorganic form of nitrogen. Thomas (1966, 1970b) among others, has suggested that under these circumstances ammonia is in fact the most significant nitrogen source for phostosynthetic organisms. At site KK the ammonia forms a generally featureless profile of values in the range 0—1 μg-at l^{-1} throughout the water column.

Silicate. Silicate has not been considered as either a limiting or important nutrient by many workers and determinations are by no means common. The silicate content of the water column is about 3 μg-at l^{-1} at the surface and within the mixed layer, increases rapidly to 30—40 μg-at l^{-1} at 150 m, remains fairly constant from there to 400 m and then increases steadily down to a maximum of 160 μg-at l^{-1} at about 3,000 m (Fig. 13-4g). Such high values of silicate, greater even than those found in the Antarctic, are indicative of the slow exchange of deep waters of the South and North

Pacific allowing longer time for dissolution of biogenic material (Sverdrup et al., 1942). Below the maximum, silicate content decreases toward the bottom to $\sim$150 μg-at l^{-1}, presumably due to the exhaustion of further soluble silicate material. There is some indication that silicate is regenerated purely by solution effects rather than by oxidative effects as is the case for nitrogen and phosphorus.

Dissolved carbon dioxide, alkalinity

The complexities of the CO_2/carbonate system in the ocean can only be briefly mentioned and the reader is referred to Skirrow (1975) for a detailed treatise. He states (p. 161) that "photosynthesis in the upper layers (of the ocean) leads to the consumption of dissolved carbon dioxide and to the generation of organic matter: decaying organic material falls through the water column and leads to the consumption of dissolved oxygen, to an increase in ΣCO_2 (total carbon dioxide) and to the liberation of phosphate and other nutrients. Dissolution of calcite particles also occurs and is encouraged by the high CO_2 content. This increases the alkalinity and also contributes to the CO_2" The vertical distribution of total alkalinity at site *A* shows an increase from $\sim$2.3 meq. l^{-1} at the surface to a maximum of $>$2.5 meq. l^{-1} at 3,000 m, dropping to 2.45 meq. l^{-1} at the bottom. These results generally agree with the GEOSECS intercalibration station (Takahashi et al., 1970) and with the subsequent GEOSECS Pacific profile (Broecker and Mantyla, 1974). The water depth in the siliceous ooze province is generally greater than the calcium carbonate compensation depth (i.e., the depth at which sedimentation rate of calcareous tests equals the dissolution rate of calcium carbonate in the sea water).

Trace metals

The concentrations of trace metals in sea water are imprecisely known. These metals are in such small concentrations that contamination during the sampling and storage procedures can cause large errors in the final results. Some authorities doubt whether trace metal concentrations can be measured at all using present techniques. Nevertheless, some data are available. Riley and Chester (1971) and Goldberg (1965) present summary data on the concentration of the 75 elements known to occur in the ocean. According to Riley and Chester, iron is present at about 3 μg l^{-1}, while Goldberg states the concentration is 10 μg l^{-1}; the concentration is very variable. Manganese is present at 2 μg l^{-1} and is similarly variable, as is copper at a concentration of about 3 μg l^{-1}. Cobalt concentrations vary widely about 0.1 μg l^{-1}, while nickel is relatively uniform at 2 μg l^{-1}.

Schutz and Turekian (1965) have used neutron activation analysis to determine 18 trace elements in sea water and have concluded that in the central Pacific areas in which Co:Mn and Ni:Mn ratios in manganese nodules are high, are also high in Co and Ni concentrations, possibly of volcanic origin. They also find that Co and Ni concentrations increase with depth under highly productive areas except in regions of persistent upwelling where high steady-state concentrations may be produced. They infer that organic processes are significant in the transport of these two elements.

Lewis and Goldberg (1954) studied the vertical distribution of iron and found little variation of soluble iron, whereas particulate iron varied considerably in the upper 500 m of the Pacific. They attribute this to biological turnover. Measurements on samples collected at site *KK* (Robertson and Rancitelli, 1973) gave Co concentrations oabout 10 times those found at the GEOSECS test station (Spencer et al., 1970) in the northeast Pacific. Zn concentrations were twice as high as those observed in the GEOSECS profile.

Dissolved radon

Radon-222 is a naturally occurring radioactive tracer and is the daughter product of Ra^{226} with a 3.85-day half-life. As the radon diffuses out of the sediments from a constant source (Broecker et al., 1968), the vertical gradient of radon concentration in near-bottom water can be used to compute the vertical coefficient of eddy diffusion. Understanding the near-bottom turbulent boundary-layer processes may be important in evaluating the disturbances caused by mining operations. Comparison of the near-bottom distribution of suspended particles as indicated by nephelo-meter profiles and the distribution of radon suggests that the concentration of particulate materials in bottom waters of the oceans is not solely a function of turbulent processes (Biscaye and Eittreim, 1974).

Preliminary data from the GEOSECS profile along 123°W show "classical" exponential excess radon gradients between 10°N and 20°N (Broecker and Mantyla, 1974) indicating very low rates of vertical diffusion. North and south of this zone the vertical diffusion rate is much higher. Thus, in the zone of greatest interest to the manganese nodule mining industry relatively tranquil conditions exist in the bottom boundary layer. This is quite in agreement with the character of the nephelometer profiles where only a slight increase in light-scattering can been seen on approaching the bottom.

Suspended particulate material

The measurement of present loads of suspended particulate material in the manganese nodule belt is of particular interest in the study of possible additions to this load by mining activities. Particulate material in sea water ranges from colloidal suspensions to organic agglomerates and living plankton (Reiswig, 1972) in the submicron to millimetre size ranges. The total suspended material load may serve to identify a particular water mass such as Antarctic intermediate or Antarctic bottom water (Jerlov, 1968). Qualitative and quantitative measurements can be made by optical methods (light-scattering) or by filtration of sea-water samples and gravimetry or by direct particle counting using a Coulter counter (Sheldon and Parsons, 1967).

Light-scattering. Continuous profiles of light scattering from surface to bottom in the ocean have been made for several years using a nephelometer developed by Thorndike and Ewing (1966). In our investigations in the nodule area we have used (Fig. 13-4d) a modified version of this instrument (Thorndike, 1975) simultaneously with our STD-profiling sensors. GEOSECS has used a laser nephelometer in their profile along $123°W$ (Broecker and Mantyla, 1974).

Although the nephelometer is affected by sunlight in the euphotic zone, the profiles (taken at night) clearly show the high light-scattering of the surface waters. Along the equatorial region two zones of surface water with high particulate concentrations occur centred on $10°N$ and the Equator. These zones are the result of advection of waters formed by upwelling near the Galápagos Islands into the equatorial current system (Jerlov, 1964). These high values remain constant in the mixed layer, then drop rapidly to a minimum coinciding exactly with the oxygen minimum. A maximum in light scattering is found at 300 m followed by a steady decrease to the clearest water in the column at $\sim4,000$ m. The bottom nepheloid layer (Ewing and Thorndike, 1965) is seen below this depth as an increase in light-scattering to the bottom. As there are no identifiable water-mass characteristics associated with the nepheloid layer in the region, the increase in suspended material is caused by turbulent processes within the benthic boundary layer. The values of light-scattering given here as an arbitrary logarithmic ratio have been related to total particulate concentration by Biscaye and Eittreim (1974). At site C, the nephelometer profiles yield values of 40 μg l^{-1} at the surface, 2 μg l^{-1} in the clearest water, and 3 μg l^{-1} at the bottom, using this relationship.

Particulate organic carbon (POC) and nitrogen (PN). The detrital organic carbon fraction of the total suspended load in the open ocean is composed mainly of plant and animal debris derived from the marine food chain and also of some formed in situ through a complex equilibrium between dissolved and particulate organic carbon (Parsons, 1975).

The profiles of POC and PN at site A closely follow each other as a function of depth. High values (100 μg l^{-1} POC, 20 μg l^{-1} PN) at the surface drop rapidly to a minimum at ~100 m. The high surface values probably include a contribution from living planktonic organisms and may be accounted for by the relatively high productivity of the surface waters at site A. This is in contrast to the very low values encountered north of Hawaii (Gordon, 1971) in an area of low productivity. Near the bottom, particularly at some stations, a rapid increase in POC and PN occurs. Gordon (1971) pointed out that POC and PN in the deep water showed temporal changes greater in magnitude than the analytical error of the method.

POC and PN values (Table 13-IV) are higher than Gordon's (1971) data

TABLE 13-IV

Average particulate organic carbon (POC) and nitrogen (PN) at sites A and KK

Depth range (m)	Mean POC (mg m^{-3})	(σ)	Mean PN (mg m^{-3})	(σ)	C:N	Number of samples	Remarks
Site A							
0— 100	48	16	6.6	2.3	7.3	59	Equatorial surface water
100—1,000	28	10	3.4	2.2	8.2	44	Oxygen minimum
1,000—2,000	28	11	2.6	1.9	10.8	17	
2,000—3,000	25	13	2.5	1.2	10.2	13	
3,000—4,000	26	12	3.2	1.9	8.1	12	Deep and bottom water
4,000—bottom	26	15	2.9	2.5	9.0	45	
Site KK*							
0— 100	56	10	6.3	0.8	8.9	10	Equatorial surface water
4,000—bottom	28	12	2.1	2.0	13.3	33	Bottom water

*An insufficient number of samples was collected at site KK to give statistical data at mid-water depths.

north of Hawaii by a factor of 3 in the surface waters and by a factor of 10 in the deep waters.

Primary (phytoplankton) productivity

Primary productivity by phytoplankton is the ultimate source of organic material in the open ocean and is a function of phytoplankton biomass and the specific growth rate of that biomass. Chlorophyll a concentration is an index of phytoplankton biomass and productivity per unit chlorophyll a can be considered an index of specific growth rate. Eppley (1972) has reviewed some of the errors associated with these assumptions which are primarily a consequence of environmentally induced variations in the C:Chl ratio. Since any impact of mining on photic zone primary productivity will most likely involve those factors which affect phytoplankton growth rates (e.g., nutrient supply and light penetration), the following discussion will focus on chlorophyll a-specific productivity.

Phytoplankton distribution and taxonomy. Phytoplankton productivity, biomass and taxonomic composition are poorly documented in the siliceous ooze province. Tropical oceanic phytoplankton communities are dominated by motile phytoplankton species which are small in size and capable of rapid nutrient uptake at low nutrient concentrations (Eppley et al., 1969; Malone, 1971a,b). A summary of phytoplankton species in the Pacific Ocean has been compiled by Allen (1961). Excluding "nanoflagellates", three groups of phytoplankton are important in tropical and subtropical waters: cocco-lithophores, dinoflagellates and diatoms. Hasle (1959) has described the species composition near the Equator ($00°01'N—02°01'N$) at $145°W$. The bulk of the phytoplankton occurred in the upper 150 m and was dominated by species of coccolithophores, diatoms and dinoflagellates, in order of numerical abundance. Semina (1968) found similar results along $174°W$ with peridinians exceeding diatoms in numerical importance north of $3°N$. Small species, with diameters less than $20\ \mu$ made up more than 90% of the total cell numbers along this transect ($37°S—30°N$).

Productivity and chlorophyll a. Surface productivity and chlorophyll *a* concentrations in the eastern equatorial Pacific Ocean average less than $10\ \text{mg C m}^{-3}\ \text{day}^{-1}$ and $0.5\ \text{mg m}^{-3}$, respectively (El-Sayed, 1970; Malone, 1971a). Holmes (1961) reports values of 0.13 to $414\ \text{mg C m}^{-3}\ \text{day}^{-1}$ with a median of $6.1\ \text{mg C m}^{-3}\ \text{day}^{-1}$ for the North Pacific between $0°$ and $25°N$.

Based on limited data, Holmes (1958) suggested that surface levels of productivity and chlorophyll *a* are higher along the northern boundary of the Equatorial Countercurrent (ca. $10°N$) and that this reflects the pattern of thermal ridging along this boundary (Cromwell, 1958).

Water-column levels of primary productivity are reported to be less than $150\ \text{mg C m}^{-2}\ \text{day}^{-1}$ (Ryther, 1969; Koblentz-Mishke et al., 1970). Holmes (1961) summarized productivity measurements in the southeastern Pacific Ocean. Primary production between $0°$ and $25°N$ ranged between 0.01 and $400\ \text{mg C m}^{-2}\ \text{day}^{-1}$ with a median of $0.44\ \text{mg C m}^{-2}\ \text{day}^{-1}$. Based on transects along $119°W$ (Blackburn et al., 1970), $112°W$ (Malone, 1971a) and $155°W$ (Takahashi et al., 1972), water-column levels of chlorophyll *a* rarely exceed $25\ \text{mg m}^{-2}$. The chlorophyll *a* maximum is generally located between 50 and 150 m (Owen and Zeitzschel, 1970; Takahashi et al., 1972) in or near the pycnocline at the base of the mixed layer (Steele and Yentsch, 1960; Hobson and Lorenzen, 1972). The chlorophyll *a* concentrations of these maxima are generally less than $0.25\ \text{mg m}^{-3}$ (Takahashi et al., 1972).

Vertical profiles of phytoplankton productivity and chlorophyll *a* showed little variability between stations at site *A*. Mean values for each light level and depth sampled are given in Table 13-V. Note that the chlorophyll *a* maximum occurred between 30 and 50 m while chlorophyll *a* specific productivity (P/B = mg C per mg Chl *a* per day) peaked between 10 and 20 m. Surface P/B is apparently inhibited by high light intensities (cf., Fogg, 1975).

Phytoplankton productivity and chlorophyll *a* integrated over the photic zone (100% to 1% light depth) ranged from 68 to $155\ \text{mg C m}^{-2}\ \text{day}^{-1}$ and 11 to $21\ \text{mg Chl }a\ \text{m}^{-2}$, respectively (Table 13-VI). Based on these integrated values, P/B ranged from 4.6 to $7.8\ \text{mg C mg Chl }a^{-1}\ \text{day}^{-1}$ with a mean of $6.6 \pm 1.15\ \text{mg C mg Chl }a^{-1}\ \text{day}^{-1}$. Variations in P/B were correlated with variations in incident radiation (Table 13-VI). Using a C:Chl ratio of 98 to

TABLE 13-V

Average productivity and chlorophyll a at site A, April—May 1974

Mean depth (m)	Light penetration (%)	Productivity (mg C m^{-3} day^{-1})	(σ)	Chlorophyll a (mg m^{-3})	(σ)	P/B
2	100	1.1	0.6	0.19	0.04	5.8
9	60	1.8	0.4	0.18	0.03	10.0
20	30	2.1	0.7	0.21	0.03	10.0
32	15	1.4	0.6	0.23	0.05	6.1
51	—	—	—	0.23	0.05	—
72	1	0.8	0.2	0.23	0.06	3.5
84	—			0.21	0.05	
94	—			0.19	0.09	
102	—			0.12	0.10	
121	—			0.04	0.02	
149	—			0.01	0.01	
193	—			0.01	0.00	

TABLE 13-VI

Water column productivity, chlorophyll a and incident radiation at site A, April—May 1974

Date 1974	I	K	Chlorophyll a (mg m^{-2}) 0—150 m	100—1%	Productivity (mg C m^{-2} day^{-1}) 100—1%	P/B
24 April	279	0.07	16.8	14.6	98.4	6.7
26 April	387	0.07	21.4	14.0	100.5	7.2
28 April	439	0.08	34.6	16.0	116.2	7.3
29 April	351	0.08	18.4	10.7	68.4	6.4
1 May	427	0.07	19.2	14.1	87.3	6.2
3 May	255	0.06	23.7	17.4	88.9	5.1
5 May	178	0.06	22.6	17.7	80.6	4.6
6 May	527	0.06	25.9	21.4	154.8	7.2
8 May	559	0.07	27.8	17.4	136.2	7.8
11 May	466	0.07	19.9	14.4	110.0	7.6

Note: I = incident radiation (gcal. cm^{-2} day^{-1}); K = extinction coefficient; P/B = productivity per unit chlorophyll.

convert chlorophyll a to phytoplankton—C (Thomas, 1970b) and the mean P/B of 6.6 mg C mg Chl a^{-1} day^{-1}, carbon-specific phytoplankton growth rates averaged 0.1 doublings per day. A similar calculation using maximum photic zone productivity (400 mg C m^{-2} day^{-1}) and chlorophyll a (25 mg m^{-2}) values reported for the oceanic area between the Equator and 25°N (cf., Holmes, 1961) gives a maximum specific growth rate of 0.2 doublings per day. Thomas (1970a) calculated growth rates of 0.3—0.4 day^{-1} for the nitrate-depleted waters of the eastern tropical Pacific and Eppley et al. (1973) estimated growth rates of 0.2—0.3 day^{-1} for the central gyre of the North Pacific. These results indicate a close coupling between nitrogen flux to the photic zone (primarily in the form of ammonium and urea) and phytoplankton productivity.

Annual cycles of primary productivity and standing crop are poorly documented in subtropical and tropical oceanic waters. Owen and Zeitzschel (1970) observed a significant seasonal variation of phytoplankton productivity in one offshore area (105°—119°W 3°S—15°N) of the eastern tropical Pacific. Average values ranged from 129 mg C m^{-2} day^{-1} in the autumn to 278 mg C m^{-2} day^{-1} in the early spring. The amplitude of the cycle was observed to decrease from east to west, and latitudinal effects were negligible. This cycle is similar, although not as great in amplitude, to that observed at a single station in the northwestern Sargasso Sea (Menzel and Ryther, 1960). Primary productivity ranged from a summer—autumn minimum of 50 mg C m^{-2} day^{-1} to a winter—spring maximum of 830 mg C m^{-2} day^{-1}

The low levels of phytoplankton productivity and standing crop observed throughout most of the eastern tropical Pacific are thought to be caused by low rates of dissolved inorganic nitrogen flux into the photic zone above the thermocline (Dugdale, 1967; Thomas, 1969, 1970a,b). Primary productivity is based mainly on regenerated nitrogen and seasonal fluctuations are probably related to periodic inputs of inorganic nitrogen due to horizontal advection.

Geographic (Holmes, 1958) and seasonal (Blackburn et al., 1970) variations in zooplankton standing stock (displacement volume) in the upper 150 m of the water column are positively correlated with phytoplankton standing crop in the eastern tropical Pacific. Seasonal cycles of chlorophyll a and day zooplankton (105°W to 119°W; 3°S to 15°N) were low in amplitude and nearly identical in phase. Night zooplankton standing stocks ranged from 125 to 150 ml/1,000 m^3 while day zooplankton varied between 75 and 125 ml/1,000 m^3. Stocks generally declined from east to west. Holmes (1958) reports values of 20 to 200 ml/1,000 m^3 in the eastern reaches of the siliceous ooze province pictured by Horn et al. (1972c).

In summary, primary productivity, phytoplankton standing crops and zooplankton standing stocks are low and exhibit little seasonal variability. All the evidence reported to date suggests that epipelagic communities in prospective manganese-nodule mining areas are in approximate steady state, and, as Cushing (1959) has predicted, that phytoplankton and zooplankton productivity are closely coupled.

Other pelagic organisms

Despite the low overall surface productivity in the nodule zone, wide-ranging pelagic animals are found throughout the region, sometimes in large numbers. Typical of these are squid, flying fish, shark, dolphin (fish), birds, porpoises and whales. Many of the pelagic species of birds associate where food supplies are plentiful and a feature of their distribution is a distinct zonation of alternating high and low population densities (Murphy, 1936, p. 89). A surface-marker buoy emplaced at site C for 35 days in the spring of 1975 was "colonized" by four blue-faced boobies (Sula dactylatra) even though very few birds were observed throughout that cruise. With the exception of such temporary effects, it is improbable that mining will have any impact on such wide-ranging species.

The ocean floor

Details of the sedimentology and topography of the siliceous ooze province are given elsewhere in this book (Chapters 2 and 11) and in Horn et al. (1970).

At study sites KK, A, and C, probably typical of the whole region, nodule coverage is extremely variable: from none to dense coverage to pavements on occasional outcrops. Even over small distances, the nodule coverage is often very patchy. Fig. 13-6 shows a nodule field at site C with approximately 2,500 nodules of 4 cm diameter in an area of 4.7 m^2. About 70% of the bottom is covered with nodules in this photograph. Sediment grain size in the radiolarian-ooze/red-clay region is about 1 to 2 microns (Horn et al., 1972c).

Benthic communities

The most direct effect of manganese-nodule mining will be on the bottom-dwelling communities which will be destroyed when buckets and dredge-heads scoop up both nodules and bottom-dwellers from the ocean floor. All of the abyssal fauna live on the sea floor or within the upper few centimetres of the sediment. They range in size from bacteria to large holothurians (Fig. 13-6) up to 0.5 m in length (Heezen and Hollister, 1971). A limited population of free-swimming animals (fish, cephalopods) are present, presumably occurring near the ocean floor where they feed on the abyssal fauna. The larger organisms dwelling on the sediment surface can be studied by underwater photography (Fig. 13-6). Smaller surface-dwellers and sub-sediment dwellers can be studied by dredging or box-coring, while total population densities can be determined by box-coring only.

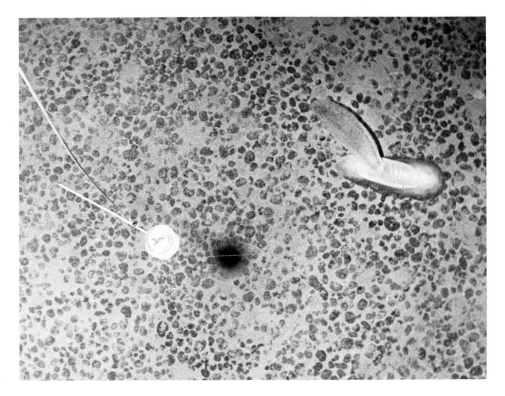

Fig. 13-6. Nodule field at site C with large (0.5 m) holothurian of the genus *Psychropotes*. Suspended compass and vane indicate a bottom current towards 225°. Holothurian with its appendage known as a "sail" is aligned in the direction of the current, a typical stance of some benthic organisms. Nodule size averages 4 cm diameter.

There are very few sources of information on deep-sea benthic communities, particularly in the designated area of interest in the Pacific Ocean. The *Challenger* expedition (1872—76) took two bottom dredges in the region and recovered both nodules and organisms. They are: Sta. 265 (30°22′N 154°56′W; depth 5,395 m) — Porifera (*Ammoconia*), Pelecypoda (*Leda*), Scyphozoa (*Stephanoscyphus* on nodule); Sta. 264 (14°19′N 152°37′W; depth 5,485 m) — Porifera (*Cladorhiza*), Scyphozoa (*Stephanoscyphus* on nodule) (Murray, 1895).

The *Albatross*, of the United States Fish Commission under the scientific direction of Alexander Agassiz (1891) was the next vessel to collect bottom samples in this area. At one station, using a Blake trawl, they collected many manganese nodules, evidence of bottom life "ground to pieces" and some *Webbina*, a small *Voluta*, three species of holothurians, a stem of a *Bathycrinus*, a *Metacrinus*, part of a *Magasella* and some serpulid tubes on nodules. They took less than 10 dredges in this region and the samples were divided into taxonomic groups and sent to various specialists for identification. The following systematists worked on the collections: Dall (Mollusca), Nutting (Alcyonaria), Clark (Crinoidea), Agassiz (Coral and Echinoidea), Cushman (Foraminifera), Richardson (Isopoda), Wilson (Porifera), Eastman (shark teeth and Cetacean bones), and Lütken and Mortensen (Ophiuroidea). Their reports are scattered in the Memoirs and Bulletins of the Museum of Comparative Zoology, Harvard and the Proceedings and Bulletins of the United States National Museum for the years 1892–1930, and their results have not been collated.

Quantitative benthic studies. Subsequent studies of the benthic biota in this region waned until the mining industry began using box-cores to investigate nodule distribution. As their interest has primarily been in the nodules themselves, a systematic study of the benthos in the nodule belt was not begun until the DOMES cruises at sites *KK*, *A*, and *C* (Fig. 13-1).

The quantitative sampling effort on the first two cruises was not successful (Amos et al., 1975a) but at site *C* five replicate 0.25-m^2 box cores were collected at each of seven stations in 4,500 m depth and they represent the most intensive quantitative suite of samples from depths below 3,000 m to date anywhere in the oceans. However, the results will not be known for at least one year. Additional samples will be collected in 1975 and 1976 and a meaningful understanding of benthic community structure should be attained. The only statement possible now regarding the macrofauna is that the following groups are present: Xenophyophoria, Porifera, Cnidaria, Polychaeta, Sipunculida, Tanaidacea, Isopoda, Aplacophora, Pelecypoda, Gastropoda, Crinoidea, Ophiuroidea, Holothuroidea, Bryozoa, Brachiopoda, Ascidacea, Nematoda, Copepoda and Ostracoda. The numerical distribution of the macro- and meio-benthos remains to be determined.

Visual studies of the benthos using bottom photography. In 2,202 bottom photographs taken at site *A* and analyzed for epibenthic megafauna, there were 319 recognizable organisms. They were divided in taxonomic groups as follows: holothurians (198), ophiuroids (30), actinarians (26), porifers (17), asteroids (11), shrimp (11), echinoids (8), pennatulaceans (5), ascidians (2), antipatharians (2), rat-tail fish (2), polychaetes (1), pelecypods (1), ctenophores (1), crab (1), crinoids (1), and unidentified (2). Although these large epibenthononts were sparsely distributed, tracks, trails and faecal piles were common.

The height of the camera off bottom varied, but since almost all of the photographs showed manganese nodules it was possible to determine accurately the area of each frame. The total area photographed was 26,800 m^2, so the average density of the megabenthos was 0.01 organisms/ m^2. Owen et al. (1967) reported values of 1.6, 0.75, and 0.8 organisms/m^2 from a photographic survey at similar depths in the oligotrophic Sargasso Sea. Roels et al. (1973) found faunal densities of 0.02 to 0.05 organisms/m^2 in a photographic survey at site *KK*. Grassle et al. (1975) report densities ranging from 0.03 to 3.55 organisms/m^2 in depths from 500 to 1,830 m in the Atlantic Ocean. It is apparent that the manganese nodule area has a very low epibenthic faunal population density.

The trophic type of the organisms identified is contrary to the views of Sokolova (1972) who stated that suspension-feeders would dominate in oligotrophic waters containing nutrient matter. The most common animal (holothurian) in the photographs from site A was a deposit-feeder. This fact also seems to answer the question asked by Hessler and Jumars (1974) — are megafaunal deposit-feeders absent in the oligotrophic zone? They are not absent, just sparsely distributed, probably due to the low input of food (Rowe, 1971).

IMPACT OF NODULE MINING ON THE MARINE ENVIRONMENT

All of the parameters discussed in the previous section are in some way or another interrelated and to predict completely the effect of a disturbance on this complex environment is not possible at present. The major disturbances that mining will cause and their effect on the marine ecosystem will now be examined.

In each of the three basic mining systems there will be some disturbance of sediments at the ocean floor as the collecting device moves along the bottom. Resuspension of sediment in the near-bottom water will occur. The mining companies will try to avoid transporting metallurgically worthless sediments to the sea-water surface along with the nodules. In practice, it will not be possible to separate all the sediment from the nodules. In the CLB system some of the sediment collected with the nodules in the buckets will wash out as the buckets move through the water column on their way to the surface. The other systems propose to utilize towed or self-propelled bottom-gathering devices connected with hydraulic or airlift pumping systems to transport the nodules to the surface through a pipeline system (Garland and Hagerty, 1972; Welling, 1972). All of these machines have components which contact the ocean bottom and make a first separation of the nodules from the surrounding sediment. This first separation is achieved by a chute with water jets, heavy spring rake tines, a radial tooth roller, harrow blades and water jets, or spaced comb teeth. Many of the machine concepts employ adjustable collecting elements so that changes can be made during the mining operation to accommodate variations in the nodule deposit and sediment characteristics.

After the manganese nodules have been collected from the sea floor, they are transported through the water column to the surface mining vessel either in the buckets of a continuous line dredge or in a water stream through a pipeline. In both modes of transport some or all of the entrained sediment and near-bottom water may be discharged, either at the surface or at intermediate depths in the water column.

The major topics discussed in this section will be the magnitude and effects of disturbances at the ocean floor and within the water column, the fate of mining effluent discharged at the surface and its impact on the ecosystem.

The magnitude of mining operations

To be economically feasible, a mining unit (single mining ship or platform) must harvest at least 5,000 (wet) metric tons per day. With a concentration of 10 kg nodules m^{-2} (National Research Council, 1975), an area of $5 \cdot 10^5$ m^2 day^{-1} must be mined. These numbers are typical of estimates found in the literature on economic aspects of nodule mining. Nodule abundance is frequently quoted in terms of percentage of sediment covered by nodules as revealed by bottom photographs. Although there appears to be some correlation between nodule population densities and size, considerable variation in weight per unit area can be found because of the very nature of the nodule deposit. Nodules are essentially spherical and are found in a single layer on top of the sediment. Assuming the densest possible nodule concentration (diameters of adjacent nodules touching or 91% of the sediment covered), theoretical distributions of 29 to 11,592 nodules m^{-2}, — weighing 243 to 12.1 kg m^{-2}, are computed for nodule populations ranging from 20 to 1 cm in diameter (Table 13—VII). At fixed mining-vessel speeds, from 119,400 to 6,000 metric tons could theoretically be mined per day, or to maintain a 5,000 metric tons day^{-1} yield, the mining-vessel speed would vary from 1.4 to 28 km day^{-1} (Table 13-VII). If the concentration of nodules by weight was a constant 10 kg m^{-2} then the population density would range from 1.2 nodules m^{-2} for 20-cm diameter nodules to 9,549 nodules m^{-2} for 1-cm diameter nodules, while coverage would range from

TABLE 13-VII

Theoretical concentrations of spherical nodules of different diameters

| Diameter of nodule (cm) | Weight of nodule[1] (kg) | Maximum possible nodule concentration[2] | | | | | Concentration of 10 kg m^{-2} | |
		nodules per m^2	kg m^{-2} [1]	mined per day[3] (m.ton$\cdot10^3$)	to mine 5,000 m. tons day^{-1} area[4] ($m^2\cdot10^4$)	distance[5] (km)	nodules per m^2	sedim. cover (%)
1	0.001	11.592	12.1	6.0	41.2	27.5	9.549	75
2	0.008	2.898	24.3	11.9	20.6	13.7	1.194	37
4[6]	0.067	724	48.6	23.9	10.3	6.9	149	18.8
6	0.226	322	72.8	35.8	6.9	4.6	44.2	12.5
8	0.536	181	97.1	47.8	5.1	3.4	18.7	9.4
10	1.047	116	121.4	59.7	4.1	2.7	9.6	7.5
15	3.534	52	182.1	89.6	2.7	1.8	2.8	5.0
20	8.377	29	242.8	119.4	2.1	1.4	1.2	3.8

[1] Wet density of 2 g cm^{-3}; [2] single layer, diameters touching (91.04% coverage); [3] mining cut 32.8 km X 15 m per day; [4] total area covered by 5,000 metric tons; [5] distance travelled by dredge head with 15 m wide mining swath; [6] average nodule diameter in nodule zone.

3.8 to 75% of the sediment surface. This great variability in the abundance of the resource makes it difficult to assess the ultimate magnitude of the mining operation and the disturbance to the environment. Average spherical nodule diameters are given as 2—4 cm (Mero, 1965a, 1972). Note that with the maximum possible nodule coverage a yield of 10 kg m^{-2} is a considerable underestimate. Maximum coverage of nodules with the average diameter of 4 cm would yield 49 kg m^{-2}. The concentration value of 10 kg m^{-2} used in this section would therefore be equivalent to 4- cm diameter nodules covering about 19% of the ocean floor. Larger, more disc-shaped nodules, 5—10 cm in diameter, are also found in the nodule zone (Horn et al., 1973c).

The depth to which a nodule is buried in the sediment will determine how deep a mining device will have to dig (how much sediment is disturbed) to retrieve it. Burial depth may vary from one-third to one-half the nodule diameter (Horn et al., 1973c; Metallgesellschaft AG, 1975).

To assess the environmental aspects of nodule mining, the following quantities are used in this section:

Nodules mined per day	5,000 metric tons (wet weight)
Number of days of active mining per year	300
Nodule size range	1 to 20 cm
Average	4 cm
Shape	spherical
Wet density	2 g cm^{-3}
Nodule concentration (by weight)	10 kg m^{-2}
Mining methods:	continuous-line bucket dredge (CLB) airlift or hydraulic suction dredge

CLB system

Many problems arise in determining the magnitude of CLB mining operations using data published in the literature. CLB buckets used in deep-ocean mining tests (computed from Mero, 1972) had openings of 40 X 125 cm and were 200 cm deep. The total volume was 1 m^3. With the most efficient packing of equal-size spherical nodules in a bucket of this size, from 964 kg of 18.56-cm diameter nodules to 1,373 kg of 1.14-cm diameter nodules could be collected per bucket. The average-sized nodules (here 4.4-cm diameter to fit an exact number of rows in the bucket) would yield 1,299 kg per bucket. To collect 5,000 metric tons of nodules per day, 3,849 buckets would have to be filled. With a 20—50 m distance between buckets (Mero, 1972), the bucket line would have to move at 53—134 m min^{-1}, far faster than the 100 ft. min^{-1} (30.5 m min^{-1}) quoted by Masuda et al. (1971). To pick up 1,299 kg nodules, each 125-cm wide bucket would have to move over 104 m of ocean floor. Thus, if the buckets penetrated 10 cm into the sediment, a total of $5 \cdot 10^4$ m^3 of sediment would be disturbed per day. The width of the mining swath is variously shown as the width (100 m?) of the mining vessel (Masuda et al., 1971), 1,000 m (Mero, 1972), or, with a two-ship operation, 1 nm (1,600 m) (Metallgesellschaft AG, 1975). The mining efficiency of this system may be quite low and the goal of 5,000 metric tons day^{-1} is probably not attainable from a single CLB operation (Hammond, 1974).

Airlift or hydraulic systems

With airlift or hydraulic systems a slurry of nodules, air and/or water are transported to the surface vessel. Several devices have been proposed to collect the nodules on the ocean floor prior to pumping them to the surface (Anonymous, 1972; Metallgesellschaft, 1975). Some of these dredges are towed by the mining ship and some are self-propelled. According to the National Research Council (1975) report, a typical towed device will travel at 38 cm sec^{-1} (32.8 km day^{-1}) and will cut a swath in the ocean floor 15 m wide and 10 cm deep. The National Research Council panel quote a typical self-propelled device as having a track width of 180 cm per side (for the propulsion unit) and a burial depth of 92 cm. It is not clear why this device would need to dig almost 1 m into the sediment, or why the only interaction with the sediment would be where the propulsion unit makes contact with the bottom and not where nodules are removed from the sediment. The total yearly amounts of sediment disturbed are substantial (Table 13-VIII). For comparison, the Mississippi river sediment transfer is estimated to be $2 \cdot 10^8$ m^3 yr^{-1} (Shepard, 1973), only one order of magnitude larger than the figures quoted in the National Research Council report (1975) for a single mining operation. A self-propelled device that may already have been built and tested (U.S. patent 3433 531) consists of a rotating rake and crusher assembly, propelled in a circle about a fixed centre. If this device mined 5,000 metric tons day^{-1}, it would have to sweep round a circle of 400-m radius in one day (mining 100% of the nodules in that circle). The end of the

TABLE 13-VIII

Amount of sediment disturbed at the ocean floor by dredging

Type of dredge	Separation of track-lines (m)	Mining efficiency (%)	Nodules mined per day (metric tons)	Per 300 day year			Suspended solids in bottom 10 m of water (ppm)
				Vol. sediment + nodules removed (m^3)	Vol. sediment alone (m^3)	Vol. water affected (m$^3 \cdot 10^9$)	
Towed dredge	0	100				1.48	9,460
	50	30	4,920	$1.48 \cdot 10^7$	$1.40 \cdot 10^7$	4.91	2,850
	100	15				9.81	1,430
	500	3				48.50	290
Self-propelled dredge*	0	100				1.73	18,440
	50	35	4,600	$3.26 \cdot 10^7$	$3.19 \cdot 10^7$	4.91	6,500
	100	18				9.81	3,250
	500	3.5				48.50	660

*Using a track width of 17.6 m (3.6 m for propulsion device which interacts with ocean floor) and a computed width of 14 m to mine 4,600 metric tons per day.
Source: National Research Council (1975).

sweep arm would be moving at 2.9 cm sec^{-1}. To mine 5,000 metric tons day^{-1} using a 100-m radius circle, 16 such circles would have to be swept per day with the end of the arm moving at 11.6 cm sec^{-1}, not counting any time for moving the ship and setting up between each sweep.

The uncertainties about the magnitude and methods used in the various self-propelled mining systems make it impossible to evaluate accurately the impact of these operations on the environment.

The rest of this section will, therefore, concentrate on the towed airlift dredge system for which reasonable numbers can be found in the literature.

Disturbances on the ocean floor

Sediment plumes

It is obviously in the interest of the mining operation to separate the nodule from the sediment as completely as possible on the ocean floor and to disturb the sediment as little as possible, compatible with efficient collection of the nodules. A cloud of sediment will undoubtedly be stirred up in the near-bottom water layers. Particles of 1—2μ diameter would take from 70 to 250 days to settle 10 m in still water (computed from Gibbs et al., 1971, using a particle density of 2.65 g cm^{-1} and sea water of 0°C, 34°/$_{oo}$). Net bottom currents of 2 cm sec^{-1} are sufficient to keep these particles in suspension even though such currents may not be strong enough to erode them off the bottom under normal conditions (Heezen and Hollister, 1964). Periodic high current peaks in tidally related bottom flow must erode sediment off the ocean floor while turbulent processes occurring at the bottom boundary keep it in suspension. Consequently, mining operations must increase the turbidity of the bottom water and a certain amount of sediment redistribution will take place. In parts of the ocean, strong bottom currents cloud the water sufficiently to reduce visibility in bottom photographs. Curiously enough, when the camera itself contacts the bottom and stirs up a sediment cloud, adjacent photographs taken a few seconds later seldom show evidence of that cloud of material. This may be because such a disturbance stirs up enough sediment to increase the density of the sediment/water mixture and cause the water to sink rapidly. Also, while evidence of previous dredging operations has been observed by the mining companies using underwater television, no resulting turbid water has, to our knowledge, been seen. Clouds of sediment stirred up by artificial disturbances on the ocean floor always appear (in photographs) to be sharply defined and not diffuse. It may be that aggregates and groups of particles that have been cemented by organisms comprise the main bulk of such sediment-clouds and these would not behave like the spherical grains used in determining settling rates. Systems have allegedly been developed that will screen-out 80% to 90% of the sediments at the dredge head (MESA, 1975).

To determine the increase in turbidity of the near-bottom waters during a mining operation, it will be assumed that the sediment disturbed is confined vertically to the bottom 10 m of the water column and laterally to the width of the area being mined (Table 13-VIII). Mining efficiency is defined here as the ratio of ocean floor mined to that left untouched by the dredging device in a given mining pattern.

Almost all studies on the effects of turbidity on organisms have dealt with fresh-water varieties (Wilber, 1971). If the worst-case situation postulated in Table 13—VIII occurs, then the increase in turbidity of the bottom water would be well above the level known to have deleterious effects on fresh-water fish (Herbert and Merkins, 1961). The effects of high turbidity on filter-feeders are the subject of some debate; in fact, large amounts of silt and clay have been shown to have no effect on some oysters (Wilber, 1971). Alternatively, Loosanoff and Tommers (1948) found that very high levels of turbidity inhibit oysters' feeding mechanisms and decrease the pumping rates and that 2,000 ppm turbidity causes 100% mortality in their eggs. Filter-feeders (sestonophages) that are found at the bottom in our region include sponges, crinoids, asteroids, bivalves and tunicates.

The effects of excess turbidity on deposit-feeders (e.g., holothurians) are not known, but presumably they would not affect these organisms as much as the sestonophages. As only relatively surficial sediments will be disturbed, and these will be at or near chemical equilibrium with the overlying waters, little chemical change can be expected. Indeed, the effect of mining will achieve precisely the same kind of turnover that is brought about by benthic in-fauna, but in a rather more rapid and dramatic manner.

The redistribution of manganese-nodule debris can only be looked upon as desirable, since this will provide the necessary nucleation material to allow the continued growth of nodules, thereby ensuring the continued balance of trace metals in the deep waters, and also providing, in the very long term, for renewal of the resource.

It has been argued (Garland and Hagerty, 1972) that the redistribution of sediment on the ocean floor resulting from natural phenomena exceeds by many orders of magnitude, on a world-wide scale, any disturbance caused by all the dredges ever likely to be utilized in deep-sea mining.

One possible effect of disturbing sediment at the ocean floor would be to increase the heat flow into the bottom water. This could give rise to vertical convection currents and horizontal gradient currents if sufficient heat were added to the deep water. If, for example, the heat flow were $0.06°C\ m^{-1}$ and the mining device dug to a depth of 10 cm, with 100% mining efficiency (Fig. 13-8), the temperature increase in the bottom 10 m of the water column would be only $6 \cdot 10^{-5}°C$.

Destruction of benthic communities

While it is unlikely that any mining operation will cover 100% of a given area, it is obvious that sessile organisms which cannot escape the oncoming dredge will be destroyed. Repopulation of the mined-out swath on the ocean bottom from adjacent, untouched zones will undoubtedly occur, although this may take a very long time. Some of the sessile animals have been shown to have an extremely slow reproductive cycle (Turekian et al., 1975). Virtually nothing is known, however, about the life cycles of most of the organisms inhabiting the benthic zone.

Disturbances at the sea surface and within the water column

The major disturbances to the water column caused by mining operations will be related to the types of mining systems used. The CLB system may introduce sediment anywhere into the water column as the buckets are brought up to the mining vessel. Suction-dredge type systems will discharge volumes of bottom water and sediment at or near the sea surface after the nodules have been separated (aboard ship). Quantities of sediment may also remain in the CLB buckets along with the nodules and will either have to be discharged at the surface or transported back to land.

We will first consider the surface-discharge of bottom water and then examine the problem of sediment discharged by both the suction-dredge and CLB-type mining systems.

Surface-discharged mining effluent

Numerous studies have been done on the behaviour of various effluent plumes discharged into rivers and near-shore shallow waters, e.g., turbulent buoyant jets (Fan, 1967; Robideau, 1972), sewage effluent (Brooks, 1960), dredge spoils (Boyd et al., 1972) and (dye) diffusion in the ocean (Okubo, 1971). A selected annotated bibliography of environmental disturbances of concern to marine mining research has been compiled by Battelle Memorial Institute (1971). Most of the papers referenced concern shallow-water problems. Many assumptions have to be made in such models (Neumann and Pierson, 1966, p. 410) which limit their applicability to mining effluent in the open ocean. Some of these assumptions include negligible vertical mixing, negligible mixing in the direction of the current, steady flow rates, the total content of introduced effluent remaining constant as the plume diffuses in the ocean and boundary conditions such as river channels and shallow-ocean bottoms. To our knowledge, no model exists that predicts the fate of deep-ocean mining effluent discharge (such a model, of course, would have its own set of assumptions and limitations). The closest analogs found in the literature are models of buoyant plumes from tall chimneys discharging into the atmosphere in the presence of a horizontal wind. Mining effluent could sometimes be buoyant, but may well be denser than the surrounding ocean, will have a lower temperature, and will be discharged vertically at the surface.* It may also contain sedimentary material which could behave as if it were a dissolved constituent of the effluent or may float or sink independently of the plume.

---

*Mining companies say that they can discharge the effluent anywhere in the water column if necessary, even back at the ocean floor — by which time the effluent will probably be buoyant.

TABLE 13-IX

Properties of deep- and surface-water mixtures

Property	Deep water	Surface water	98.8% surface- to 1.2% deep-water mixture	99.88% surface- to 0.12% deep-water mixture
Salinity (‰)	34.690	34.270	34.275	34.271
Temperature (°C)	1.10	22.35	22.09	22.32
Sigma-t	27.82	23.08	23.13	23.09
Dissolved oxygen (ml l^{-1})	3.75	4.74	4.73	4.74
Phosphate (μg-at l^{-1})	2.45	0.78	0.80	0.78
Nitrate (μg-at l^{-1})	36.50	2.70	3.10	2.74
Silicate (μg-at l^{-1})	155.0	10.3	12.0	10.5

The figure $6 \cdot 10^9$ kg water discharged per 300-day year (National Research Council, 1975) gives a daily discharge rate of $2 \cdot 10^7$ l, or 230 l sec^{-1}. Properties of the bottom water and typical surface water into which it will be discharged are given in Table 13-IX. It is assumed that the water entrained with the nodules is bottom water and does not contain significant quantities of interstitial water which may have a different nutrient concentration. At the moment, the sediment load entrained with the nodules will not be considered. The basic property that will determine the fate of the effluent after discharge is the density of the two water masses. Should the effluent be instantaneously introduced into the water column and should no mixing take place, then the effluent would respond to Archimedian (buoyant) forces and experience an acceleration proportional to the density difference, i.e. $g(\frac{\Delta\rho}{\rho})$.

With a 1.2% deep-98.8% surface-water mixture (Table 13-X), the initial acceleration would equal $4.8 \cdot 10^{-2}$ cm sec^{-2}. With no mixing and a stationary ocean, the parcel of introduced water would ultimately find its way back to its original equilibrium depth — the bottom of the ocean. In reality, mixing will take place. As the parcel sinks, water will be advected in horizontally to replace it and turbulent mixing will take place. The density of the parcel will decrease and the sinking rate will decrease until the process

TABLE 13-X

Concentration of effluent in surface water (%) during a 300 day year mining operation if effluent is confined to area being mined

Depth to which effluent is mixed (m)	Separation in track line (m):		
	50	100	500
	Area of water affected $(m^2 \cdot 10^8)$:		
	4.91	9.81	48.50
1	1.22	0.61	0.12
5	0.24	0.12	0.02
10	0.12	0.06	0.01
25	0.05	0.02	0.005
50	0.02	0.01	0.0025

of molecular diffusion takes over and the parcel will ultimately become indistinguishable from its surroundings.

The effluent is discharged at an initial velocity imparting a momentum that is greater than the buoyant forces alone. Some examples of outfall velocity vs. pipe diameter are given in Table 13-XI. The most probable diameter is between 50 and 75 cm. The separation of the nodules on board the mining vessel will be accomplished by a sieving process, the details of which are not known outside the mining industry, but presumably after processing, the discharged effluent flow in the pipe will be turbulent rather than laminar. It is discharged into a stratified medium that may be flowing with velocities of 50 cm sec^{-1} or more. Based on a detailed analysis of buoyant plumes in the atmosphere, Csanady (1973) considers that heavy plumes such as sewage effluent (and probably mining effluent) may be analyzed in a similar way, although he does not develop the oceanic aspects of his plume theory. An attempt will be made to interpret Csanady's model for oceanic conditions. Initially, the momentum of the plume is additive to the buoyancy forces (in this case, negative buoyancy) and a rapid mixing occurs due to the turbulent flow of the discharge a few pipe-diameters from the source. After this, horizontal momentum of the cross-current is transferred to the plume and it is carried along at a velocity indistinguishable from the surface current. Buoyancy forces still affect its vertical (downward) displacement and mixing continues due to self-generated turbulence caused by buoyancy motions, i.e., entrainment of the ambient water. Next, oceanic turbulence, in the form of Langmuir cells for example, dominates the diminished buoyant forces, leading to more vigorous mixing and break-up of the plume. Finally, the distinct parcels merge and the stage is reached where molecular diffusion occurs and buoyant forces are no longer effective. The above "model" does not consider a stratified ocean.

TABLE 13-XI

Airlift or hydraulic mining systems: residence time of nodules in suction pipe; velocity of effluent outfall*

Suction pipe diameter (cm)	Residence time in pipe for 5,000 m		Outfall pipe diameter (cm)	Velocity (cm sec^{-1})
	(h)	(min)		
10	0	3	10	2,940
20	0	11	20	735
30 **	0	26	30	326
40	0	46	40	183
50	1	11	50 **	118
75	2	40	75	52
100	4	43	100	29

*Assumes: $6 \cdot 10^9$ kg water per 300 day year (National Research Council, 1975)
   $= 2 \cdot 10^7$ litres per day
   $= 231$ litres per second.
**Probable range of pipe diameters used in mining operations.

The stratification of the ocean in our region is characterized by an upper mixed layer 20—100 m thick and a normally intense pycnocline below that. The pycnocline, a region of high stability, will almost certainly act as a barrier to further downward migration of the plume.

Dye-diffusion experiments. By using rhodamine-B dye as a tracer, the development of a plume during an actual mining test was photographed from the air (Amos et al., 1972). Fig. 13-7 shows the dye patch 8 min after injection of the dye into the outboard discharge of the mining vessel, *Deepsea Miner*, in an experiment conducted on the Blake Plateau in August 1970. Some of the processes postulated in the previous section are revealed by the dye patch in this aerial photograph. The plume has expanded rapidly after discharge, is elongated in the direction of the current (although the surface winds of 10 knots were in the opposite direction), shows evidence of having broken up into patches in the region closer to the discharge, and appears more diffuse and probably deeper in the region furthest from the source. Air-charged sedimentary material composed of pteropod tests and made visible by the dye are floating at the surface near the source. These pteropod tests are organizing themselves along convergence boundaries of Langmuir-type cells, and are probably influenced by the surface winds to a greater extent than the body of the dye-patch. Langmuir circulation is an important process in near-surface waters (Assaf et al., 1971) and produces the type of turbulent eddies that will affect the mixing of the plume after its

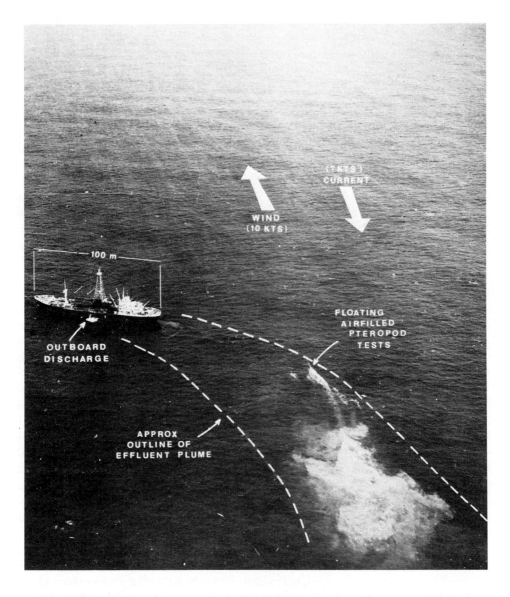

Fig. 13-7. Aerial photograph of dye-marked mining effluent. Photograph taken 8 min after a 3 min injection of dye into outboard discharge of the *Deepsea Miner* during mining tests on the Blake Plateau in August 1970.

buoyancy forces have been dissipated. Fig. 13-7 also reveals two other variables that are not found in the buoyant smoke-stack analogy: (1) the forward motion of the mining vessel, comparable in magnitude to any surface currents, creates, in effect, a line-source for the plume rather than a stationary point-source; and (2) the effluent, free-falling in air before it enters the ocean, creates considerable turbulence at the air—sea interface. Both effects will serve to mix and dilute the effluent more rapidly with the surface water.

Density of effluent. The density difference between effluent and surface water ($\Delta \sigma_t$ = -4.3 at site C) is caused almost entirely by the temperature differential between the two water masses. In Fig. 13-8, showing a typical *T/S* curve at site C, it can be seen that bringing the bottom water to the same temperature as the surface water reduces $\Delta \sigma_t$ to 0.3. Amos et al. (1972) found that considerable warming of the effluent occurred during the transport of nodules from the bottom to the surface. This was apparently due to frictional effects and heat conduction through the pipe. Data on frictional heating generated by nodules and turbulent flow within the pipe have not, to our knowledge, been published. Effects of expanding air used in airlift pumping systems will also affect the temperature in the pipe. Residence time of nodules and water in the pipe (Table 13-XI) will be about 30 min for 5,000 m water depth. During most of that time the pipe will be in contact with ambient temperatures only a few degrees above the initial temperature of the bottom water. Hence heat conduction through the pipe will be minimal. Van Hemelrijck (1973) has computed thermal losses through pipelines used in artificial upwelling projects. For a pipeline off St. Croix, U.S. Virgin Islands, 30 cm in diameter, 2,250 m long with a flow of 65 l sec^{-1}, a temperature increase of 0.77°C would occur. Any temperature increase in the pipe will decrease the density of the bottom water effluent. If the effluent is warmed until it is at the same temperature as the surface water, the density difference between effluent and surface water ($\Delta \sigma_t$) will determine whether the plume will be heavy, buoyant or neutral at the moment of discharge (Fig. 13-9). Large areas in the nodule zone have negative $\Delta \sigma_t$'s where the surface salinities are less than the bottom salinities.

In other areas, if temperature equilibrium were reached before discharge, the plume would be buoyant and remain at the surface.

Dilution of effluent. Any mixtures of bottom water (C) with surface water (A) can be represented by points along the straight line joining C to A on the *T/S* diagram (Fig. 13-8). Because heating will change the density of the bottom water prior to discharge, the ultimate density of any effluent/ surface water mixture can be found by drawing a straight line from A to a point along line C-B representing the increased temperature of the effluent. Dividing this line proportionally will show the density of any ratio of surface water to effluent. Concentrations of effluent in surface water are given in Table 13-X for different mining tracks and mixing depths assuming the

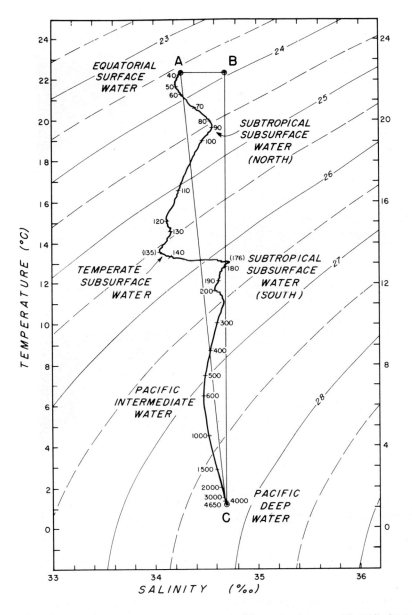

Fig. 13-8. Typical *T/S* curve at site *C*. Water masses are identified at their core-layer values. Depths in metres are marked along the *T/S* curve. Curved lines are isopycnals annotated in σ_t units. A = surface water; C = bottom water; B = bottom water in temperature equilibrium with the surface water. Area with triangle ABC contains all possible mixtures of surface and deep water at any temperature between the two.

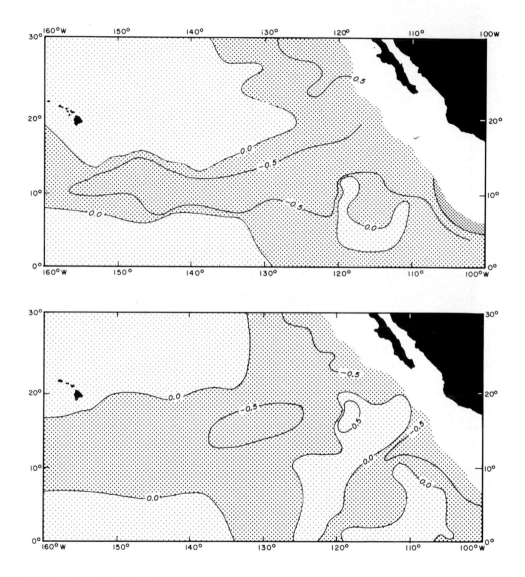

Fig. 13-9. Density difference ($\Delta\sigma_t$) between bottom water and surface water in σ_t units when bottom water is brought to the same temperature as the surface. Top = summer; bottom = winter. Darker shaded areas are where $\Delta\sigma_t$ is negative.

effluent remains within the area bounded by the ship's track. However, complete flushing of the area by prevailing surface currents would take only eleven days along the long axis of a 30 × 150 km mining area with a current speed of 15 cm sec^{-1} (North Equatorial Current). The rapidly expanding plume of Fig. 13-7 had already been diluted to approximately 0.05% of its original concentration 8 min after discharge (the discharge rate during this experiment was half the rate of proposed full-scale operations).

Concentrations of major constituents of the surface water before and after discharge are given in Table 13-IX.

For computations in the following sections, an effluent concentration of 0.12% will be used (discharge rate = $2 \cdot 10^4$ m^3 day^{-1}; plume width = 100 m; plume depth = 5 m).

Transport of effluent by prevailing currents. In much of the region where mining is proposed, the effluent will be transported towards the west by the North Equatorial Current. This current continues westward through 10,000 km of open ocean beyond site C before deflecting northward near the Philippines. The North Equatorial Current transports $6 \cdot 10^{14}$ m^3 of water per 300 day year (Wyrtki, 1967) compared to a mining vessel's $6 \cdot 10^6$ m^3 output. South of 10°N, the North Equatorial Countercurrent would transport mining effluent eastward almost to the coast of Central America where the current deflects north and south to resupply the North and South Equatorial Currents.

Transport of the effluent by currents would make the concentration very patchy. A hypothetical effluent distribution is shown in Fig. 13-10 for a track-line of 32.8 km long with a separation of 100 m and a prevailing current of 15 cm sec^{-1}. Ultimate dilution in one year taken as the ratio of total effluent dumped to the volume of water 5 m deep to be transported past the mining vessel during one year would be 9 ppm.

Surface sediment plume. Of the vast quantity of sediment disturbed by the dredges at the ocean floor (Table 13-VIII) quoted by the National Research Council (1975) report, all but a very small percentage would have to be rejected at the dredge head. According to the National Research Council figures, the quantity of sediment disturbed by a towed dredge would be 2.5 times the volume of *all* the material pumped to the surface and 5.5 times with a self-propelled device. If all this material goes into suspension in the bottom 10 m of the ocean (Table 13-VIII) and the resulting sediment/water mixture is pumped to the surface with the nodules, then the surface water plume will contain 11.5 ppm sediment (towed dredge) and 22 ppm sediment (self-propelled dredge). This is an increase in suspended material of between two and three orders of magnitude over the normal surface-water turbidity. If the sediment screens can remove 80% of the sediment disturbed and redistribute it in the near-bottom water before pumping to the surface, then these quantities would still amount to 2.3 ppm and 4.4 ppm, respectively, for the two systems. These calculations assume that the

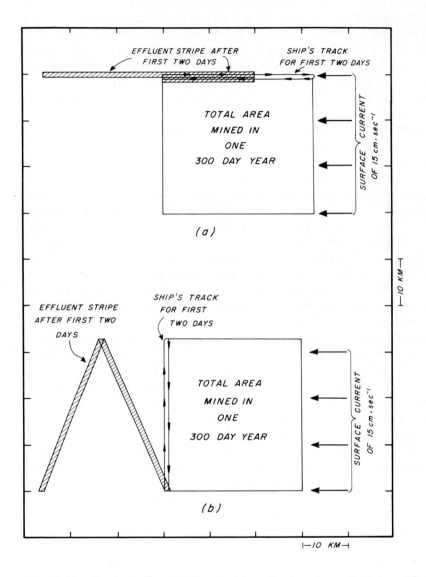

Fig. 13-10. Configuration of effluent stripe after two days of mining with surface current (a) parallel to ship's track, and (b) normal to ship's track. Separation of ship's track and width of effluent stripe are 100 m and are exaggerated 10 times here for clarity of illustration.

sediment remains with the surface plume and is diluted proportionately.

In the upper mixed layer, a region of neutral stability, even a small increase in density may be sufficient to overturn the water down to the thermocline. The increase in density of the sediment/effluent mixture may cause it to sink rapidly even though the settling rate for the individual sediment grain size of $1-2\mu$ is extremely low. If the liquid effluent, when discharged, had the same density as the surface water (1.0235 g cm^{-3}) and contained $1,892$ ppm sediment by volume (80% of $9,459$ ppm screened out at the dredge head), and if the sediment has a wet density of 1.18 (Horn et al., 1972c), then the resulting density of the sediment/effluent mixture would be:

$$\frac{1892}{10^6} \times 1.18 + (1 - \frac{1892}{10^6}) \times 1.0235 = 1.0238 \text{ g cm}^{-3}$$

an increase of $3 \cdot 10^{-4}$ g cm^{-3} over surface water alone.

Using the relationship: $V = \sqrt{2g \, \Delta\rho \, H}$ (Weyl, 1970), where V = velocity in cm sec^{-1}, g = acceleration of gravity, $\Delta\rho$ = density difference in g cm^{-3}, and H = height in cm, the effluent would be sinking at a rate of 17 cm sec^{-1} by the time it was 5 m below the surface. The descending water would not be able to penetrate far into the strong pycnocline where $\Delta\rho$ would diminish to zero very quickly. A fainter cloud of dye can be seen deeper and further from the ship than the main plume (Fig. 13-7). This could be dye-marked sediment that sank more rapidly than the main body of the effluent and was advected further away by the surface current. In the absence of a dye tracer, no increase in turbidity caused by the effluent is visible. Colour photographs taken at the same time (Metallgesellschaft AG, 1975, p.56) reveal a darker blue of the sea surface in the effluent plume. What causes this colour change is not obvious from the photographs. While it is reported that sediment can be seen in the discharge jet of a mining vessel, no cloud of obviously turbid water surrounded the vessel during the Blake Plateau test. Highly porous radiolarian sediment particles may become charged with air during the airlift-pumping process and remain at the surface as the pteropod tests did in the Blake Plateau experiment (Fig. 13-7).

The CLB System's buckets will undoubtedly collect sediment, some of which will be washed out as the buckets are brought to the surface — increasing the turbidity anywhere in the water column. The magnitude of this sediment input in unknown. Dumping of the residue from the buckets after removal of nodules may result in a more concentrated turbidity plume than with the other systems. When tailings from box-cores are hosed over the side of a research vessel, the sediment invariably sinks from view within a few seconds. During a CLB test in the Pacific Ocean in 1972, no increase in turbidity was observed at the surface (Amos et al., 1972).

Impact of surface effluent on the ecosystem

While it is obvious that mining effluent concentrations will be extremely small in the surface waters, it will be necessary to see if such concentrations could have any lasting impact on the ecosystem. It has, for instance, been proposed that the increase in nutrient concentration of the effluent/surface-water mixture might promote phytoplankton growth and create a commercially viable mariculture venture as an off-shoot of ocean mining (see Othmer and Roels, 1973, for a description of a pilot mariculture station using artificially upwelled deep water).

The following chemical/biological changes caused by mining effluent in the surface water will be considered: (1) enhancement of phytoplankton productivity by increased nutrient concentrations in the photic zone; (2) changes in phytoplankton in response to increases in the rate of attenuation of downwelling light caused by an increase in the suspended particulate load; (3) increase in oxygen demand due to (a) increase in organic material from (1) above, and (b) transported benthic biological material as it settles back through the water column; (4) dissolution of siliceous sedimentary materials discharged into surface waters; (5) change in taxonomic composition of the phytoplankton; (6) biological uptake of trace metals.

Nutrient enrichment and phytoplankton productivity. The degree to which productivity is enhanced by the surface discharge of mining effluent will depend on the interaction of several variables: (1) the rate and volume of discharge; (2) the nitrate concentration of the effluent; (3) the turbidity of the effluent and the depth to which it spreads; (4) the rate of dilution by mixing; and (5) the rate of nutrient uptake by the phytoplankters.

Tropical oceanic phytoplankton are characterized by low half-saturation constants for nitrate uptake (Eppley et al., 1969; Eppley and Thomas, 1969; MacIsaac and Dugdale, 1969), and phytoplankton growth rates appear to be nitrogen-limited over much of the tropical Pacific Ocean (cf. Thomas, 1970a,b). These observations and the results of enrichment experiments with near-bottom water collected from a manganese-nodule area in the North Atlantic (Malone et al., 1973b) suggest that any addition of nitrate to the photic zone will stimulate phytoplankton productivity.

Unless the effluent could be contained at the surface in greater concentrations, it is doubtful that the increases in biomass could be of sufficient commercial value to warrant the investment necessary to attempt its harvest. Roels et al. (1975) have proposed utilizing sea-thermal power plant effluent to promote phytoplankton growth in such a contained environment in the open ocean.

Increase in turbidity and phytoplankton productivity. Although there are no data available on the relationship between light extinction and turbidity, it is likely that any increase in suspended sediment load will increase the rate of light attenuation. This increased attenuation rate (higher extinction coefficient) will reduce *P/B* integrated over the mixed layer (as evidenced by the rough correlation between P/B and incident light with a "constant" extinction coefficient). The steady-state productivity calculations will not be affected unless nutrients are removed from the photic zone faster than they can be assimilated by phytoplankton (which is unlikely). Although the presence of suspended sediments appears to reduce the lag time between nutrient enrichment and exponential growth (Malone et al., 1973a) under experimental conditions, it is doubtful whether this could happen in the open ocean.

Increased oxygen demand. A further consideration of the temporary increase of biomass must be the oxygen demand of the organic material during its passage through the food web, and its ultimate decay as detrital material.

While the strong pycnocline inhibits mixing of oxygen from the surface layer into the oxygen minimum layer, discharged organic and photosynthetic material will slowly settle through the water column creating a further demand on this already depleted zone. These waters are already close to anoxic but do not contain measurable sulphide. At an oxygen concentration of about 0.1 mg O_2 per litre, alternative substrates are used for bacterial respiration, characterized by the reactions:

$$2 NO_3^- \rightarrow 2 NO_2^- + O_2$$

and:

$$2 NO_2^- \rightarrow N_2 + 2 O_2$$

When nitrate and nitrite concentrations become sufficiently low, sulphate becomes the substrate for bacterial respiration:

$$SO_4^{2-} \rightarrow S^{2-} + 2 O_2$$

at which point no O_2 remains in the water column (equilibrium for the above reaction lies to the left in oxygenated seawater) and anoxic conditions prevail. However, the intermediate nitrate utilization regime with oxygen present at low concentrations seems to be typical of the Pacific oxygen minimum. Nitrate in the oxygen minimum layer (45 μg-at l^{-1}) contains 67.5 μg-at O_2 l^{-1} (2.16 mg O_2 l^{-1}) which would provide for the oxidation of 0.81 mg C l^{-1} before sulphide formation occurs.

Taking the calculated increase in standing crop of 0.4 μg C l^{-1}, and assuming that this oxidizes to carbon dioxide without transport into other waters, 1.2 μg O_2 l^{-1} will be consumed. To this we must add the oxygen utilisation of transported benthic biological material. Such material includes biogenous sediments and remains of sessile benthic organisms that were caught by the dredging device and brought to the surface. Using undiluted discharge water and sediment from 800 m on the Blake Plateau (Amos et al., 1972), this utilisation was found to be 5.7 mg l^{-1} over a 9 day incubation in the dark at 25°C, the value remaining constant from the fifth day. Assuming a discharge dilution of 0.12%, the utilisation will be 7 μg O_2 l^{-1}. However, we have found experimentally that in undiluted discharge water and sediment incubated in the light, photosynthesis restores oxygen saturation after 8 days, in spite of the fact that organic decay reduces the oxygen content to 20% of saturation after 3—4 days.

To calculate the oxygen demand by transported benthic material, the following assumptions will be made: density of benthic organisms is 0.05 g m^{-2} (Menzies et al., 1973), all of which represents organic carbon; properties of siliceous ooze—carbon content 0.0655% (El Wakeel and Riley, 1961), all of which is oxidizable and none is "refractory"; dry density 2.6 g cm^{-3}; porosity 88% (Horn et al., 1972c); quantity of sediment in effluent 9,460 ppm (Table 13-VIII); dilution of effluent at surface 0.12% (Table 13-IX). The total carbon discharged daily is 24.6 kg contributed by benthic organisms and 38.7 kg contributed by biogenic bottom sediments. The total carbon added to the surface waters is 3.9 mg m^{-3} which would create an oxygen demand of 10.3 μg O_2 l^{-1}. If 80% of the sediment is screened out at the dredge head then the demand becomes only 5.3 μg O_2 l^{-1}. These quantities of oxygen demand are an order of magnitude smaller than the sensitivity of present-day analytical techniques. They are also two orders of magnitude less than the oxygen available from nitrate reduction in the O_2 minimum layer so that the possibility of sulphide formation in this layer as a result of mining effluent is remote.

Dissolution of sedimentary material. The discharged sediments will contain significant quantities of silicon in the form of both clays and biogenous opaline silica. Insofar as these materials have passed through the entire water column and are likely to represent surface sediment which is still being reworked by biological activity, it seems probable that they will be at or near equilibrium with overlying waters. However, as surface waters are generally depleted with respect to both phosphate and silicate, some dissolution is likely to occur, thus ensuring that biological productivity will be nitrogen-limited. These enrichments may also have consequences in terms of species composition.

Changes in taxonomic composition of phytoplankton. The surface-discharge of bottom water could affect the species composition of phytoplankton communities in the receiving water by changing the nutrient content of the water or by introducing new species. If increases in productivity are ultimately nitrogen-limited, compounds such as silicate may accumulate in the photic zone. This could have broad repercussions on species composition, as shown by the effect of silicate enrichment on the relative abundance of diatoms and coccolithophores (Menzel et al., 1963; Malone et al., 1973a). Observations in the North Atlantic (Amos et al., 1972) also indicate that foreign algal species could be introduced into the photic zone by the mining effluent itself. Both of these effects could have an impact on phytoplankton population dynamics as well as on energy flow through phytoplankton-based food chains (Mullin, 1963; Parsons and LeBrasseur, 1970; Malone, 1971b).

Biological uptake of trace metals. The mining systems will undoubtedly result in abrasion of nodule material which will result in the suspension of fine particles of ferromanganese, presenting a large potentially adsorptive/desorptive surface to the water. In the highly oxidizing environment of surface waters it is unlikely that substantial solution of the nodule material will occur, but the possible effects of even a small amount of dissolution, or adsorption/desorption of trace metals, may have far-reaching consequences. It is well known that marine organisms have a need for many of the elements present in nodule material (Fe, Mn, Cu, Co) and that others are extremely toxic (Cd, Pb). In addition, iron and manganese hydroxides are known to be capable of quantitative removal of many metals and organic compounds which may be of biological importance.

Marine organisms concentrate many trace metals by several orders of magnitude relative to their concentrations in the surrounding waters (Vinogradov, 1953; Bowen, 1966; Goldberg, 1967). It is not uncommon for transition metals to be concentrated by factors of 100 to 100,000 (Goldberg, 1965). Numerous trace metals, including Co, Cu, Fe and Mn, are fixed photochemically by phytoplankters and are used either as cofactors for enzymes or as structural components of enzymes. Phytoplankters have been shown to concentrate Co by a factor of $1.5 \cdot 10^3$, Cu by $3.0 \cdot 10^4$ (Lowman et al., 1970) and Ni by $8 \cdot 10^3$ (Nicholls et al., 1959) relative to sea water. An increase in the concentration of trace metals could have several effects, none of which can be predicted on the basis of existing information. Phytoplankton productivity could either be inhibited or enhanced, depending on the concentration of organic chelators, their affinity for specific trace metals, and the degree to which phytoplankton growth is limited by local trace-metal concentrations. Trace metals could also be concentrated by organisms in the receiving waters regardless of whether or not they are required for growth. This could lead to growth inhibition at any given trophic level depending on the magnitude of uptake and the degree to which trace metals are concentrated up the food chain. While these factors are unlikely to limit primary production, they may importantly influence the species composition and growth rates. Further, prolonged mining activities in an area may have consequences in terms of accumulation and biological amplification of toxic metals in the higher trophic levels of the food webs.

CONCLUSIONS

The disturbances caused by deep-ocean mining of ferro-manganese nodules will generally be extremely small compared to natural, large-scale processes of oceanic circulation and sediment redistribution by turbidity currents. If mining effluents are discharged at the surface of the ocean they will remain within the upper mixed layer but will mix rapidly with the surrounding waters and be transported out of the mining area by prevailing equatorial surface currents. Such currents will transport the effluent far from any continental regions, dilution of the effluent will be considerable and the ultimate concentrations of dissolved effluent constituents will be barely measurable. Nutrient enrichment and increased phytoplankton productivity may occur locally on a small scale but would not be sufficient to warrant any commercial mariculture projects unless the effluent was deliberately contained and concentrated. The unlikely possibility remains that dormant phytoplankters transported from the ocean floor could result in changes in the local phytoplankton community structure, with possible consequences in the food web.

At the ocean floor, benthic communities will be destroyed by the dredging operation, but as mining will only be about 35% efficient, repopulation from adjacent untouched areas will occur. The rate at which this repopulation will take place is unknown.

The greatest unknown and the greatest potential hazard is the behaviour and effects of sediment plumes at the bottom of the ocean, within the water column and at the surface. A considerable amount of sedimentary material will be stirred up at the ocean floor. The rate at which this material resettles is not known, but indications are that it will be more rapid than settling times of up to 25 days per metre based on individual grain sizes of the material. Turbulent processes in the bottom boundary layer are very low, so that material will not be stirred up high into the water column. Steady bottom currents of ~ 2 cm sec^{-1} will transport material away from the mining zone.

Toxic trace metals may be leached out from the stirred-up sediment, or from interstitial waters, and can be concentrated by some organisms or may inhibit the growth of others. The increase in particulate material in the surface plume will be several times normal levels in the equatorial region, even with the large dilution expected and a screening-out of 80—90% of the sediments prior to pumping to the surface. Increased turbidity will lower the penetration of sunlight into surface waters and may cause changes in primary productivity. Oxidizable sedimentary material settling down through the water column will increase insignificantly the oxygen demand in the oxygen-minimum layer.

We have not considered here that any metallurgical extraction processes will be carried out at sea. Reagent transportation costs will be prohibitive for economical at-sea processing. Should this become feasible in the future, then the discharge of waste chemical products into the ocean would be far more hazardous than the discharge of mining effluent alone.

Important areas of investigation still to be undertaken are: development of a model to predict the fate of discharged mining effluent; experiments to determine the settling rates of sediments discharged at the surface and disturbed at the ocean floor; continued investigation into such baseline conditions as trace-metal distribution in the water column, composition of interstitial water that may contribute to the effluent; distribution, density and reproductive rates of benthic fauna. Furthermore, some of the predicted changes can only be verified by monitoring actual mining operations or pilot tests so that processes potentially hazardous to the marine environment can be detected and recommendations made for their elimination.

CHAPTER 14

LEGAL ASPECTS OF NODULE MINING

F. M. AUBURN

INTRODUCTION

In examining international law aspects of the commercial exploitation of manganese nodules it is essential to emphasize that there is very little existing law applicable to the deep seabed. Just as plate tectonics has been a focus for the physical sciences in the ocean, so seabed law has been one of the central points of international legal writing in recent years. A large volume of literature has already been created and it must be quite clearly pointed out that much of the discussion concerns proposed and not existing law (e.g., Auburn (1973a), Sisselman (1975), for review).

It would appear that the exploitation of manganese nodules is near and therefore a legal framework for operation is essential regardless of whether the operations will be economically profitable or carried out for political security or other non-profit reasons (Drechsler, 1973). Rules developed for manganese-nodule mining in deep water will directly affect a number of law-of-the-sea issues such as petroleum, Red Sea Brines, seabed nuclear power stations, scientific research, Mid-Atlantic Ridge projects and other peaceful uses of the sea. As United States oceanographers have learnt from the current debate on the freedom of scientific research, the scientist must take an active part in policy making and the formation of international law. With respect to manganese nodules it would appear that a further step may be required — the acquisition of a detailed knowledge of the international-law issues in order to guide the formation of seabed law.

In traditional international law the exploitation of the bed of the sea was essentially ancillary to uses of the land. Precedents such as the Ceylonese pearl fisheries, harvesting of bêche de mer under the Queensland Pearl Shell and Bêche-de-Mer Fisheries (Extra Territorial) Act 1888 or coal mining off Cornwall (Hurst, 1923—1924) can be found, and have been relied upon. But such examples are of little direct utility in regard to the large-scale mining of manganese nodules in deep water.

International law emphasizes horizontal zones such as the territorial sea, contiguous zones and fishery zones. It has proved difficult to avoid the resulting problems. So the negotiation of the Seabed Disarmament Treaty met with serious difficulties arising from basic disagreement on the breadth of the territorial sea. Proposals, such as the Draft Ocean Space Treaty of

Malta (1971), have been put forward to completely restructure this system, but there does not appear to be any present likelihood of their acceptance. New concepts of coastal-zone jurisdiction are constantly emerging. Some recent examples are the Canadian Arctic Waters Pollution Prevention Act 1970 enacting a hundred-mile pollution zone, and the Mexican proposal for a patrimonial sea. Any international regime for the seabed must take into account the effect of such assertions of jurisdiction by coastal states, together with the possibility of claims to mining sites by states. It is conceivable that manganese-nodule mining could be conducted under at least four legal regimes — territorial sea, exclusive economic zone, continental shelf, or complete international control.

The necessity for close co-operation between the oceanographer and the international lawyer is further emphasized by the emergence of vertical zones, superimposed on the existing horizontal divisions. The resulting multiplicity of legal orders, complicated by problems of boundary demarcation (Brown, 1969), is not conducive to clarity or certainty in the law. Vertical boundary questions have only recently become the subject of detailed study. For ocean minerals a particular practical difficulty will arise due to the differing regimes of the seabed and the superjacent waters. Vertical boundary questions are well illustrated by the Red Sea Brines which may be subjected to the differing regimes of the high seas, seabed, continental shelf or a coastal-state claim independent of the continental shelf doctrine.

It is difficult to separate law-of-the-sea issues. The Icelandic fishery zone relies, in part, upon Latin American 200-mile jurisdictions which are in turn related to the breadth of the territorial sea (Auburn, 1972a). The coastal-state claims affect the seabed regime. Such interrelations are reinforced by the views of many states that the issues can only be negotiated at the Law of the Sea Conference in the form of package deals. So the United States is prepared to accept a 12-mile territorial sea as part of a treaty provided it gains acceptance of free transit through international straits. Among other factors relevant to any proposed legal regime for manganese nodules it is necessary to take into account the desires and needs of the land-locked, shelf-locked and mineral producing states (Friedheim, 1969). Development of seabed law involves the utilization of a number of disciplines and the comprehension of various categories of national requirements. It is for this reason that multinational and multidisciplinary conferences have made an important contribution to the detailed discussion of seabed law. Particular mention must be made of the annual meetings of the Law of the Sea Institute of the University of Rhode Island, and the Pacem in Maribus convocations.

In the discussion of manganese-nodule law the physical scientist must accept the use of terms in international law, a usage which sometimes differs from his own. So the geological continental shelf is already much smaller than the legal continental shelf. Similar terminological differences may well arise in regard to the "seabed" and "subsoil".

CONTINENTAL SHELF AND EXCLUSIVE ECONOMIC ZONE

The regime of the continental shelf is a central issue in the law relating to manganese nodules. The continental-shelf concept, in its present form, originated in a Proclamation of President Truman in 1945. The United States government was faced with two conflicting interests. On the one hand, as a major naval power, it opposed any form of sovereignty over the high seas and feared that even a limited jurisdiction might be extended in future to exclude United States warships from coastal areas. On the other hand, growing United States power demands and the development of offshore-drilling technology indicated the desirability of the assumption of jurisdiction over petroleum reserves off the coast of the United States, but beyond its 3-mile territorial sea. The Proclamation therefore stated that the United States regarded the natural resources of the subsoil and seabed of the continental shelf beneath the high seas but contiguous to the coasts of the United States as appertaining to it and subject to its jurisdiction and control. But the character of the high seas above and the right of their free and unimpeded navigation was in no way affected.

The Truman Proclamation set off an international reaction which had not been fully foreseen. The Proclamation had not given any clear guide as to the seaward boundary of the continental shelf, but an accompanying statement suggested a water depth of 600 ft. Some states claimed to this depth, some made no specific depth limitation and others, especially those lacking a continental shelf in the geological sense, claimed sovereignty up to 200 miles offshore. United States opposition to unilateral 200-mile claims is considerably weakened by the unilateral nature of the Truman Proclamation, as the State Department has realized (Stevenson, 1972).

Between 1950 and 1956 the International Law Commission prepared drafts of what subsequently became the Continental Shelf Convention 1958. As the concept was of such recent origin the Commission was compelled to develop the law, as well as codifying it. Article 1 of the Convention defined the continental shelf as the seabed and subsoil of the submarine areas adjacent to the coast but outside the area of the territorial sea to a depth of 200 m or, beyond that limit, to where the depth of the superjacent waters admits of the exploitation of the natural resources of the areas. Islands are also to have continental shelves. The coastal-state's interest, described as "sovereign rights", exists for the purpose of exploring and exploiting the natural resources. The rights are independent of actual development of the resources or of any occupation or proclamation. Natural resources include minerals and certain living resources. The sovereign rights do not affect the legal status of the superjacent waters as high seas.

The Convention came into force in 1964, six years after it was finally drafted. It has been ratified or adhered to by more than 40 states. A number of other states have adopted the principle by state practice. However, there is a considerable divergence on important questions such as the method of delimiting boundaries between adjacent states.

The complexity of manganese-nodule law was considerably increased by the judgments of the International Court of Justice in the *North Sea Continental Shelf Cases* in 1969. The Cases concerned the question whether the equidistance method of delimitation of boundaries between adjacent states contained in Article 6(2) of the Convention could be invoked by The Netherlands and Denmark against West Germany which was not a party to the Convention. A number of observations by the Court on the nature of the continental-shelf concept are of relevance in the present context. The continental shelf constitutes a "natural prolongation" of the land territory, and an inherent right. As a matter of normal topography the greater part of a state's continental shelf will be nearer to its own coasts than to any other state's coasts. The notion of adjacency so constantly employed in continental-shelf doctrine from the start only implies proximity in a general sense. The natural prolongation principle is more fundamental than that of proximity. Articles 1—3 of the Convention which define the shelf and the nature of the sovereign rights of the coastal state, whilst preserving the freedom of the high seas, were regarded in 1958 as reflecting or crystallizing rules of customary international law. "The institution of the continental shelf has arisen out of the recognition of a physical fact and the link between this fact and the law remains an important element for the application of its legal regime." The principle applied is that the land dominates the sea. The Court emphasized that the problem before it was strictly that of boundary delimitation and the other questions relating to the general regime of the continental shelf were examined for that purpose only.

The Continental Shelf Convention is clearly the starting point for any attempt to ascertain the seaward boundary of the shelf. A very extensive literature examining the genesis of the Convention has turned into a "ritualistic fencing" between narrow-shelf and wide-shelf proponents (Krueger, 1970). There are a number of possible interpretations of Article 1 of the Convention (Brown, 1971). The 200-m limit, of itself, cannot be applied today. To accept it would be to ignore the exploitation provision of Article 1 and state practice in offshore licensing which has gone far beyond this depth, some examples being New Zealand (1,000 m), United States (1,000 m) and Canada (3,700 m) (Auburn, 1972b). Another test in Article 1, that of adjacency, is now to be regarded as less fundamental than the natural prolongation principle. Adjacency is further weakened by such instances as the partition of the North Sea (Auburn, 1973b) and of the Timor and Arafura seas between Indonesia and Australia in October 1972.

It is submitted that the sole practical test available today under the Convention is that of exploitation. It would appear to be accepted that exploitation refers to the capabilities of the most advanced state. Therefore the fact that a given coastal state cannot exploit its own offshore areas does not debar its sovereign rights to the fullest extent. Utilizing the dicta previously quoted from the North Sea Cases it is arguable that the continental shelf extends to "at least the landward portion of the continental rise" (National Petroleum Council, 1969) or to the continental slope (Jennings, 1969). It may be concluded that the seaward boundary of the shelf, as defined by contemporary international law, now lies at an uncertain depth. This depth is well below 200 m, perhaps below 1,000 m and may extend to the boundary separating the continental rise from the abyssal plain. The complete division of the ocean floor between coastal states already envisioned in theory (World Lake Concept, 1971) has not yet come to pass, but a number of extensive semi-enclosed seas have been allocated, or may be allocated in the future in this manner. These include the North Sea, the Baltic Sea, the Persian Gulf and the Red Sea.

The primary relevance of the continental-shelf concept for manganese nodules lies in the boundary question. Nodules outside the territorial sea will be either on the continental shelf (or the coastal state's exclusive economic zone) or on the seabed beyond national jurisdiction. Both areas are beneath the high seas which are open to the navigation and other uses of all nations. The major resource to be found in the bed of the sea between 200 m and the abyssal plain will presumably be petroleum, and this is the main focus for the very heated debate on the continental-shelf/seabed boundary. Manganese nodules present additional problems in so far as they are presently recoverable at water depths of several thousand metres far offshore. It would be most difficult, if not impossible, to devise a viable permanent legal regime for manganese nodules which does not directly affect the future regime of the entire seabed. The further significance of the continental-shelf doctrine lies in the analogies which will be drawn from it, for application to the seabed in regard to conflicting uses, environmental protection, scientific research and other matters on which past experience may prove of use.

General support has emerged at the Law of the Sea Conference for an exclusive economic zone of up to 200 miles, measured from the baselines for determining the territorial sea. The zone would offer a compromise between the advocates of a 200-mile territorial sea and the major naval powers. The freedoms of navigation, overflight, laying of submarine cables and pipelines would be given to all states. On one formula the freedoms would be unrestricted, and on another they would be subject to interference by the coastal state with "reasonable justification". One view would include fish in the zone's resources, and there are also proposals to give the coastal state total control over scientific research. As with the continental shelf, the primary relevance of this concept for the manganese-nodule industry lies in the boundary between the zone and the international seabed area. Taking into account very extensive proposals for drawing straight baselines for the territorial sea and enclosing ocean archipelagos, the zone could cover nodule areas. In such a case the coastal state would have total control over mining.

It is possible that the coastal state's rights could extend beyond the zone.

UNITED NATIONS

The initiation of practical work on the commercial recovery of manganese nodules and the increasing interest of the petroleum industry in oil in deep waters led to fears that the entire ocean-bed would ultimately be partitioned. This feeling gained strength from the flexibility of the exploitability criterion and state practice such as the Australian Petroleum (Submerged Lands) Act 1967 making provision for jurisdiction over large areas of the seabed to a water depth of more than 6,000 m. In 1967 Ambassador Arvid Pardo of Malta raised the question of the future of the seabed in the General Assembly of the United Nations. The General Assembly appointed an Ad Hoc Committee on the Peaceful Uses of the Seabed and the Ocean Floor beyond the Limits of National Jurisdiction. The Committee became permanent and was given the task of preparing for the third Law of the Sea Conference. The very extensive working papers and discussions of the Committee, other United Nations organs and the Conference provide a basis for development of the law of the sea. Although the origin of the Committee lay in the views of a number of developing nations that the seabed was a potential source of revenue which could be utilized through an international agency, the Committee examined a very wide range of other law-of-the-sea problems. Ultimately, the list of subjects and issues approved by the Committee in 1972 covered 25 main headings and more than 60 sub-headings, but was stated to be not necessarily complete. A number of these matters, such as innocent passage through straits, exclusive fishery zones beyond the territorial sea, the rights of land-locked countries, archipelagos and zones of peace and security are controversial. Each one could be the subject of a separate conference.

The seabed regime which will emerge from the Conference, provided agreement is reached, will represent a compromise on three levels. Firstly, it will be necessary to reconcile the widely divergent views on the regime of the seabed. Secondly, a particular state may agree to provisions which it does not want in order to achieve its goals in other areas. For instance the United States will be prepared to make concessions in order to reach international agreement on a right of free transit through international straits. Thirdly, each state will have to reconcile its own conflicting internal interests. Thus, in May 1972, the United States and Brazil signed an agreement regulating shrimping by United States nationals off Brazil's coast and well beyond 12 miles offshore, providing a significant departure from previous United States firm opposition to Latin American offshore claims. The possibility that the United States will insist on free transit through international straits "as the price of its participation in a new oceans regime" has already been discussed

(Knight, 1972). Any international agreement on the seabed will therefore represent a series of compromises and, like the Continental Shelf Convention, leave major problems unsolved.

An attempt has been made on a number of occasions by developing states to take effective and immediate action. In 1969 the General Assembly decided, by Resolution 2574D, that pending the establishment of an international regime states and persons were bound to refrain from all activities of exploitation of the resources of the area of the seabed beyond the limits of national jurisdiction. No claim to any part of that area or its resources should be recognized. Of the 62 states voting in favour only two (Finland and Sweden) were developed. The 28 negative votes included the United States, Soviet Union and the United Kingdom (Auburn, 1971b). A proposal on similar lines was submitted by Kuwait to the Seabed Committee in March 1972, but not passed by the General Assembly. In May 1972 the United Nations Conference on Trade and Development (UNCTAD) passed a resolution calling for an end to all activities aimed at commercial exploitation of the seabed area beyond the limits of national jurisdiction. The State Department's view of the 1969 Moratorium Resolution was that as the General Assembly has no legislative powers, its resolutions being only recommendatory, it did not anticipate efforts to discourage United States nationals from manganese-nodule exploration plans. In Malta's view the Moratorium was meaningless as there has been no definition of the boundary of the seabed beyond national jurisdiction. The Moratorium has received the support of a number of states at the Conference and demonstrates the stresses to which any international regime for manganese nodules will be subjected. It will not be easy to regulate and encourage a large-scale ocean mining industry if the interests of the regulators are so conflicting.

It would appear that the manganese-nodule miner may well be under a considerable degree of regulation by a body representing countries having very diverse interests, including those entirely opposed to nodule mining for economic reasons. On the presumption that the eventual international treaty regime is based upon drafts already presented to the Seabed Committee and the Conference, considerable difficulties can be forecast for the nodule miner.

Agreement on seabed law in the United Nations is summarized in Resolution 2749 of the General Assembly, passed in 1970. The seabed, ocean floor and subsoil beyond the limits of national jurisdiction and their resources are declared to be the common heritage of mankind. The area shall not be subject to appropriation by any means by states or persons, natural or juridical, and no state shall claim or exercise sovereignty or sovereign rights over any part of the area. Nor shall rights be acquired which are incompatible with the international regime to be established on the principles of the Resolution. Whilst the Resolution is on one view purely political, it has received strong support at the Conference and will presumably form the basis for the future treaty regime.

The key concept "common heritage of mankind" has been the subject of considerable discussion since 1967. It would appear that there is general support for the principle but a very wide range of disagreement over its content, as evidenced by the differing roles given to the International Authority in various drafts. The view that the seabed and its resources are not subject to national appropriation has gained wide acceptance, but its actual implementation in the crucial case of manganese nodules remains to be seen. Although the continental-shelf/seabed boundary has not yet been agreed upon, it may be presumed that many of the promising nodule sites, at 4,000 m or deeper, will be part of the seabed. It would therefore be surprising if the nodule miners could accept a regime without strong transitional guarantees.

The United Nations seabed discussions have hardly been a model of efficiency. On current tendencies the eventual seabed regime will be far from the clear legal framework required to encourage the large-scale exploitation of minerals. The United Nations has never participated in the regulation of so large an industry and its past failures (due in a large degree to the recalcitrance of its members) hardly augur well for an International Seabed Authority modelled upon the United Nations or its instrumentalities. On current United States policy this may occur, but it does not appear to have deterred investors from expanding large sums on nodule research, mining prototypes and systems. The progress made by United Nations organs so far is not inconsiderable. The law of the sea, and the seabed in particular, is one of the few major political issues centred on the United Nations. There is an area of the bed of the ocean which cannot be acquired by states. Non-participants, especially developing countries, shall gain some benefit from ocean mining. An international seabed convention containing some type of international regime is to be drafted.

Progress of the Law of the Sea Conference in 1975 on seabed issues may be briefly summarized. The boundary of the international seabed area was still a contentious issue, but there appeared to be little substantial opposition to a 200-mile exclusive economic zone giving the coastal state, at the least, the rights presently enjoyed under the continental-shelf doctrine. The fundamental conflict between the developed states having the technological capability of exploitation, and the developing states remained. The former sought an International Authority having restricted powers, the latter favoured full control by the Authority. Developed states favoured laying down detailed rules for ocean mining, regarded as essential by the industry for planning, security of tenure, and financing. The contrary view would leave all but the broadest principles to the Authority. The effect of nodule mining on the economics of developing states was the subject of much debate. The United States argued that the continuing increase in demands for raw materials would offset possible losses, whereas mineral-producing nations took a much more sombre view. Various mechanisms were suggested to compensate for such economic losses. The failure of the Conference to reach final agreement on ocean minerals in 1975, after so many years of preparation, suggests that any convention ultimately agreed upon may be extremely difficult to operate from the viewpoint of industry.

At the conclusion of the 1975 meeting of the Conference, a single negotiating text of a Seabed Convention was presented by the Chairman of the First Committee of the Conference as a basis for negotiation. In view of the stress laid by the President of the Conference on the informality of the text which does not represent any accepted compromise, it would be premature to place any reliance on this document. But, its existence and the fact that it proposes compromises on a number of major issues suggests that the Conference has realized the urgency of its task.

ISLANDS AND SEAMOUNTS

The general location of commercial manganese-nodule deposits is public knowledge (Horn et al., 1972c, 1973c). It is usually, and perhaps correctly, accepted that the precise location of the twenty-year mine sites is known to particular companies and a well kept proprietary secret. It is only possible to discuss geographical criteria on the basis of open publications. The part of the Pacific Ocean referred to in this manner contains numerous small islands, reefs and completely submerged features such as banks and seamounts. It is therefore necessary to examine the extent to which states may rely upon claims over such features to extend their jurisdiction over nodule sites. Small isolated islands and banks have already proved of considerable legal interest. Rockall, a tiny rock in the North Atlantic which can barely supply nesting space for a few birds, provides the United Kingdom with a claim to a very large continental-shelf area whose resources are as yet unknown. Columbia and Nicaragua have disputed sovereignty over Quita Sueño, Roncador and Serrana banks 260 km out to sea and covered by the sea at high tide. On the purely hypothetical national lakes concept nodule mine sites would be allocated between the United States, Mexico, Ecuador, France, New Zealand and "Undivided" (due to disputes concerning the sovereignty of a number of islands). Such a final division is unlikely to occur, but the possibility of claims to sovereign rights over nodule sites cannot be ignored.

Article 10 of the Convention on the Territorial Sea and the Contiguous Zone 1958 defines an island as "a naturally-formed area of land, surrounded by water which is above water at high tide". The High Seas Convention 1958 provides that no state may validly purport to subject any part of the high seas to its sovereignty. The Continental Shelf Convention contains no definition of an island, although it states that islands have their own continental shelves. Provision is made for installations and other devices for exploration and exploitation, but this hardly covers existing natural features and may not comprehend artificial islands, as distinguished from installations such as drilling rigs. Therefore, current continental-shelf law does not provide a clear solution to the question whether partly submerged features and artificial islands have appurtenant shelves* and similar problems may well arise in the seabed.

The legal status of artificial islands has already been raised. In 1971 Belgium informed the Seabed Committee that it had received a plan for an artificial port 27 km offshore. There are a number of other current projects for artificial islands and structures. Some instances will illustrate the problems. In a case before the Fifth Circuit of the United States Court of Appeals in 1970 two enterprises planned to build island states, to be named Grand Capri Republic and Atlantis, Isle of Gold, on Triumph Reef and Long Reef less than 2.5 km outside United States territorial waters. The reefs are completely submerged at mean high water. The Court held that the Secretary of the Army could prevent building work on the reefs under the Outer Continental Shelfs Lands Act. The government did not claim sovereignty over the reefs but rested jurisdiction on the protection of the coral as a natural resource of the shelf. A more extreme case, which did not reach the courts, arose from an attempt to found the new state of Abalonia on Cortes Bank in less than 15 m water, 220 km off San Diego. The Secretary of the Army intervened on the basis that Cortes Bank was part of the continental shelf of the United States (Stang, 1968). The dividing line between jurisdiction and sovereign rights in such cases is of little practical relevance. It is notable that in the case of the Triumph and Long reefs the activity restrained was not within the sovereign rights over the continental shelf.

In January 1972 the Ocean Life Research Foundation proclaimed the establishment of the Republic of Minerva on the isolated North and South Minerva Reefs, south of Fiji and Tonga, announcing that no part of the reefs had been above sea level at high tide and that the founders of the Republic had raised parts of the Reefs above this level. In June 1972, having itself built artificial islands on the Reefs, Tonga claimed them as Teleki Tokelau and Teleki Tonga, with 12-mile territorial seas. This precedent is of particular interest. For the purposes of the Territorial Sea Convention an island is a naturally-formed area of land above water at high tide. The Minerva case underlines the absence of a legal regime for artificial islands. As Tonga claims a territorial sea for the reefs it might in future claim a continental shelf. It is hardly necessary to emphasize the implications of such claims in respect of manganese nodules.

One commentator's reaction to the Minerva episode was that "the occupation of the seamounts appears to have begun". But the utilization of seamounts, instanced by the implanting of scientific instruments on Cobb Seamount, antedated Minerva. An example of the possible legal implications of artificial structures may be found in the boundary provisions of the United States Draft Seabed Treaty of 1970. The boundaries of the zone were to be drawn by straight lines not exceeding 60 miles in length following the

*Shelves that can reasonably be regarded for legal purposes as belonging to or constituting part of the adjacent land mass.

general line of the 200 m isobath and the (undefined) seaward boundary of that zone. In certain circumstances such lines could reach a length of 120 miles. If such a boundary were drawn for the international seabed area there would be a considerable incentive to examine the question of raising reefs, banks and seamounts to either constitute islands having a continental shelf or to become submarine features from which the extremely lengthy straight lines might be drawn. Apart from Tonga which built up Teleki Tokelau and Teleki Tonga at minimal expense, there are a large number of artificial islands in Melanesia built by indigenes. It is clear that even small developing states could thus assume jurisdiction over large areas of the seabed. With or without artificial structures the straight base lines might be of considerable importance to jurisdictions with manganese nodules beyond the territorial sea at a considerable depth, but capable of being enclosed by such base lines such as the Cook Islands. This could be a means of reconciling the divergent interests of the Federal government and Hawaii in manganese nodules within or close to the Archipelago.

Artificial structures may be of particular interest in the commercially exploitable manganese-nodule province. Mexico has quite a high concentration of large nodules 180 km off its south coast. Apart from its offshore areas it has sovereignty over the Revilla Gigedo Islands, has in the past claimed Clipperton Island, and may have claims to reefs, banks and other sea features. Mexico has recently put forward the concept of the patrimonial sea, an offshore resource zone of up to 200 nautical miles which has been judicially elaborated in the dissenting opinion of Judge Padilla Nervo in 1972 in a preliminary judgment in the Icelandic Fisheries Case at the International Court of Justice. The Mexican doctrine gains support from the rapidly increasing number of claims to jurisdiction beyond 12 miles offshore (Auburn, 1972c). Until the precise location of the twenty-year mines is public knowledge it is impossible to examine the various means by which states may attempt to assert jurisdiction over mine sites. But from the examples of Triumph Reef, Cortes Bank and Minerva Reefs it may be assumed that states may very well make strenuous efforts to do so. So although Clipperton Island was awarded to France in 1931 by Victor Emmanuel III of Italy in arbitration proceedings with Mexico the possibility that Mexico may re-assert its claim to this island and its associated exclusive economic zone and continental shelf must be taken into account if manganese nodules are to be exploited in the adjacent sea.

There are a number of small atolls and islands in or adjacent to the commercial manganese nodule province whose sovereignty is disputed. These include Canton and Enderbury, eleven of the Gilbert and Ellice Islands, four of the Northern Cook Islands, and the Tokelaus (Office of the Geographer, 1971). In all these cases the United States disputes sovereignty with either New Zealand or the United Kingdom. Most of the United States claims are ill-founded but, as they are advanced and maintained by a great power, are still cause for concern. If manganese nodules were exploited in these areas the latent sovereignty disputes might well come into the open. It may be suggested that the paper claims be withdrawn and the substantial claims settled on the precedent of the recent United States action in regard to Swan Island (Honduras) and Quita Sueño, Roncador and Serrana Banks (Columbia and Nicaragua).

Hawaii is a particularly interesting exemplar of the problem associated with islands and submerged features. Each island has, according to current United States law, a 3-mile territorial sea. The status of the manganese-crust deposits in the Kauai Channel and elsewhere within and adjacent to the Archipelago, but 3—5 miles or more outside the territorial sea, in water depths of 750—2,000 m remains unresolved. The latent conflict between Hawaii's interests in jurisdiction over minerals so close to its shores and State Department policy may emerge after the claims of other states to jurisdiction beyond the territorial sea have been resolved. It is difficult to refute the view of Hawaiian authorities that manganese deposits in the vicinity of the Archipelago demand single management for the entire resource.

SCIENTIFIC RESEARCH

Since 1968 increasing attention has been paid to the freedom of scientific research in the oceans by international lawyers and oceanographers. The problems include the conditions for entry to foreign ports, innocent passage through the territorial sea, national defence and many others (Tegger Kildow, 1973). The present discussion will be confined to the continental shelf and the seabed.

Article 3 of the Continental Shelf Convention provides that the rights of the coastal state over the shelf do not affect the legal status of the superjacent waters as high seas. Similar wording has been proposed for the seabed regime by the United States, the Soviet Union, Japan and the United Kingdom. Article 2 of the High Seas Convention states that the freedom of the high seas "comprises, inter alia" the freedoms of navigation, fishing, laying cables and overflight. Though the freedom of scientific research was not specified, it was recognized by the International Law Commission and gains considerable support from subsequent state practice. Controversy to date has centred on Article 5 of the Continental Shelf Convention which provides that the exploration and exploitation of the shelf shall not result in any interference with fundamental oceanographic or other scientific research carried out with the intention of open publication. Under Article 5 (8) the consent of the coastal state shall be obtained in respect of any research concerning the continental shelf and undertaken there. Nevertheless the coastal state shall not normally withhold its consent if the request is submitted by a qualified institution with a view to purely scientific research into the physical or biological characteristics of the shelf, subject to the proviso that the coastal state shall have the right, if it so desires, to participate or to be represented in the research, and that in any event the results shall be published.

These provisions have been under repeated attack, particularly from United States oceanographers, on the grounds that fundamental research is limited, the outer boundary of the shelf is uncertain and it is not clear what types of research are subject to the coastal state's control (Schaefer, 1969). Although it had been suggested that Article 5 (8) be deleted (McDougal and Burke, 1962), there was little evidence of serious problems before 1968. In January 1968 the United States' ship *Pueblo* was captured by North Korea in the course of electronic surveillance. At the time the Defence Department described the ship as an environmental research vessel. There were five civilian sister ships of the *Pueblo* including at least one research vessel, the *Trident*. Although the legal problems concerning the *Pueblo* centred on the question whether or not the ship had been operating in the North Korean territorial sea (Rubin, 1969), it is hardly surprising that a number of states became suspicious of United States vessels engaged in scientific research off their coasts outside the territorial sea. Such fears increased when scientific research was backed by government departments which were required by law to support only research relevant to military needs.

In 1968 a Brazilian decree provided that an application must be filed 180 days before the vessel sailed and contain a detailed description of equipment, personnel and projected activities. In 1969 Canada and the Soviet Union made specific provision for scientific research permits. Efforts by oceanographers to support the freedom of scientific research were not helped by the United States Outer Continental Shelf Lands Act under which scientific research can only be carried out for limited purposes by permit (Krueger, 1968). To remedy this the Council of the National Academy of Sciences proposed an "open waters" policy on the United States continental shelf, without permits, subject to suitable notice, participation, inspection, and publication. The United States oceanographers' campaign for freedom of scientific research has been pursued in the International Oceanographic Commission, the Seabed Committee and the Conference. This effort has suffered from the view stated by at least one leading oceanographic institution that, whilst the procedures for obtaining consent are somewhat time-consuming, they have seldom encountered any problems in obtaining the necessary permission (Woods Hole, pers. comm., 1971) and the conclusion of a prominent scientist that the gravity of the situation had been exaggerated. The evidence adduced to support the claim that a serious problem does exist has not yet been presented in sufficient detail, in regard to the continental shelf and the seabed, nor has there been the requisite analysis of the justifications for refusal of permission. Little attempt has been made to separate the instances of refusal of permission to enter ports and work in the territorial sea (both being within the unfettered discretion of the coastal state) from the instances relating to the continental shelf and the seabed. It is also evident that a considerable amount of recent research has in fact been done by United States vessels on the continental shelves of Brazil, Ecuador and other Latin American states.

Although a major aim of those seeking a wider freedom of scientific research is the deletion of Article 5 (8) it is further argued that a new substantive provision is probably required (Burke, 1970). The legal problems of scientific research on the continental shelf and the exclusive economic zone may therefore be similar to those on the seabed. One of the most favourable and detailed proposals in support of the freedom of oceanic research was to be found in the U.S. Draft Seabed Convention of 1970. Article 24 of the Draft provided that the parties would agree to encourage research and to promote international co-operation by international programmes, effective international publication and strengthening the research capabilities of developing states. It is noteworthy that, despite the current controversy over the freedom of scientific research the Draft did not make detailed provision for it nor specifically state the principle. An explanation may be sought in Article 3 under which the seabed area would have been open to use by all states, without discrimination, except as otherwise provided in the draft. This formulation of United States' seabed policy may be the optimum achievable at the Law of the Sea Conference from the oceanographers' viewpoint, but hardly provides the fully guaranteed freedom which they seek.

Manganese nodules in the International Seabed Area would raise additional problems. The Draft did not define scientific research. The definition of exploration specifically excluded scientific research. As exploration would be licensed, and fees paid, the International Authority would be required to carry out a certain measure of supervision of scientific research in order to ensure that it did not constitute "exploration". As the fees constitute part of the Authority's revenue, supervision would no doubt have to be strict. How the Authority would distinguish between scientific research and exploration was far from clear. Such problems will be further compounded by the great secrecy under which manganese-nodule exploration is conducted. Exploration was defined by the Draft as "any operation which has as its principal or ultimate purpose the discovery and appraisal or exploitation of mineral deposits". If the Office for the International Decade of Ocean Exploration (IDOE) were to undertake a sea-going nodule investigation at the request of the ocean mineral industry for the purpose of examining aspects of nodule chemistry, genesis or distribution, would this constitute scientific research or exploration? From the viewpoint of the industry the purpose would be exploration, but IDOE might well regard the work as scientific research. It may be concluded that the United States Draft was more restrictive in some respects than current law.

However, the Draft may present the most extensive freedom of scientific research likely to be accepted by the Law of the Sea Conference. In 1973 the United States submitted a proposal that scientific research in the economic zone would be subject to advance notification to the coastal state, certification that the research would be purely scientific and conducted by a qualified institution, appropriate opportunity to coastal state scientists to participate, sharing of all data and samples, open publication as soon as possible, and assistance to the coastal state in assessing the implications of the data for its interests. In a working paper submitted by Chile and twelve other American states, it was proposed that the International Seabed Authority should itself conduct scientific research on the seabed and also authorize others to do so "provided that the Authority may supervise any research authorized by it".

The United States' Draft contained specific provisions for "deep drilling", defined as any drilling or excavation in the International Seabed deeper than 300 m below the surface of the seabed. For deep drilling for purposes other than mineral exploration or exploitation free permits would have been issued by the Authority and the sponsoring state would certify technical competence and accept liability for any damage. Whilst the intention to safeguard the Deep Sea Drilling Project was clear, it is questionable whether this was in fact achieved. The DSDP has not in the past requested any permits for drilling (M. N. A. Peterson, pers. comm., 1972), but there is at least one recorded instance of problems of freedom of scientific research. The Canadian government denied the right of the *Glomar Challenger* to drill at a depth of 3,650 m at a site off the Grand Banks (Hollick, 1971). The granting of permits, even if they are free, involves some degree of regulation apart from supervision and certification of competence and this would have been performed by the Authority, thus implying the right to refuse permits. In particular the Authority would have had to judge whether or not the proposed drilling posed "an uncontrollable hazard to human safety, property and the environment".

For many oceanographers the present legal status of scientific research in relation to manganese nodules does not appear satisfactory. Article 5 (8) of the Continental Shelf Convention certainly requires much clearer definition of the conditions to be fulfilled. Recent proposals (and the present analysis has concentrated on those proposals giving the widest freedom) are likely to give rise to new problems for scientific research on manganese nodules.

MUNICIPAL LAW

A number of municipal-law matters will require attention. Private international law questions will include that of the classification of the mining plant as an aid to deciding which legal system should be applied to transactions such as mortgages over the plant. In cases of personal accident the same problem will arise. To take the example of a manganese-nodule operation conducted by an international consortium, it is not difficult to envisage a tort action in which there are substantial connecting factors with a number of jurisdictions. Further complexities arise where there is a direct conflict between the applicable systems on levels of compensation or available defences. In one oil-rig case an Englishman was employed by a Dutch company on a rig off the Nigerian coast. An exemption clause in the contract was valid by Dutch law, but not by English law. The English Court of Appeal decided that the contract, which would bar the claim in tort, was subject to Dutch law. But it must be noted that even a clear contractual term indicating the choice of a particular legal system and a particular court, may not be upheld in some jurisdictions.

Provision will have to be made for crimes connected with manganese-nodule operations. If part of an offence is committed on land, and part on the mine site, the accused may be able to plead that, due to statutory or common law limitations on criminal jurisdiction, there has been no offence on land and he is therefore not subject to the court's jurisdiction. So in a recent case the accused wrote a letter to Mrs. X in Frankfurt and posted it in the Isle of Wight. The letter was received in Germany. The question whether this constituted blackmail under the Theft Act 1968 was only resolved after lengthy consideration by the House of Lords.

In view of the significant proportion of the market which will be taken by nodule producers for a number of metals, they will be in considerable danger of criminal and civil anti-trust and related actions throughout the world. The necessity for a comprehensive legal system must be stressed. A recent case illustrates problems of criminal jurisdiction arising during scientific research in an area lacking such a legal system. A civilian was accused, in a United States court, of responsibility for the death of another civilian on Fletcher's Ice Island in the Arctic Ocean in 1970. All those involved were United States citizens on a scientific expedition supported by the government. The defence of lack of jurisdiction was precariously resolved. The Court of Appeals was equally divided on the question, thus sustaining the District Court which had taken jurisdiction, but neither Court detailed the grounds upon which this was done. In Antarctica, held by the United States government to be no-man's land, crimes committed by United States civilians would give rise to similar problems of jurisdiction (Bilder, 1966). It has been suggested that the best advice a lawyer could give to a civilian accused of a criminal offence on Fletcher's Ice Island or Antarctica would be to stay put, with the purpose of selecting the legal system most favourable to himself (Auburn, 1973c). Unless specific provision is made in the laws of all nodule-mining states, a person accused of an offence on a mining plant should obtain independent legal advice before consenting to be removed for trial on land.

Flags of convenience have been utilized for shipping both to avoid domestic costs such as higher seamen's wages and to evade international regulation under treaties adhered to by the state from which the enterprises are in fact (if not in law) carried out (Goldie, 1972). Such responsibility-avoiding devices may be of particular interest to future international nodule-mining consortia which will, in any case, have to select one state as the seat of incorporation of their enterprise. Whether the United States will permit United States companies to participate in a consortium controlled, from a legal point of view, from another country, remains to be seen. Considerations of finance, taxation and strategy may dictate otherwise.

Several governments have demonstrated an interest in the development of ocean mining. In at least one case (the Soviet Union), nodule mining, if undertaken, would be carried out by the government itself. According to the law of a number of states ships of foreign sovereigns are entitled to immunity from suit, even when carrying on commercial transactions. Thus a nodule-mining enterprise in the Pacific, providing it were correctly organized, would enable the Soviet Union to sell nodule minerals directly to a number of states without any liability to be sued in the courts of those countries. This would enable the Soviet Union to dispense with much of the usual insurance cover, providing a substantial competitive advantage. On the other hand such a state enterprise would involve direct international law liability (Bundesverband, 1972). Should the International Seabed Authority be created it could well incur liability for damages even if its activities were limited to regulation. Whether the Authority could be sued in municipal law would depend upon the extent to which it was recognized as a juristic person by that law and upon any specific statutory immunity provisions. It may be presumed that the Authority would require a measure of municipal recognition to enable it to have recourse to municipal courts for urgent environmental action against private nodule miners.

It has previously been pointed out that manganese deposits may provide a source of conflict between Hawaii and the Federal government. Litigation over petroleum rights offshore between state and Federal authorities has a history of a quarter of a century in the United States. Similar problems have arisen in Canada and Australia. Should deposits such as those in the Hawaiian Archipelago prove of commercial interest, the prospect of being involved, albeit indirectly, in prolonged disputes of this nature could be a major factor in the miner's choice between a site close to land and one in mid-ocean.

The relatively small amount of research done to date on the environmental changes resulting from nodule mining has not revealed any potential damage. But the overall effect will only be known when full-scale commercial operations are carried out. Environmental degradation has been inherent in forms of mining undertaken so far on land and must be assumed to be a result of manganese-nodule mining until the contrary is established. Ocean dumping has been the subject of considerable recent debate and at least one developed state has passed legislation which could be interpreted to apply to dumping of nodule wastes by ships carrying its flag on the high seas. There will be a strong temptation for mining enterprises working in mid-ocean, far from conservationists' surveillance, to ignore potential pollution problems in order to reduce costs. It has been argued that the American Mining Congress Draft Bill contains provisions quite inconsistent with any workable rules to prevent large-scale environmental damage (Auburn, 1972c). It may be suggested that, in partial return for municipal legislation of this nature, the manganese-nodule industry must accept effective measures for the enforcement of environmental protection by state agencies.

AMERICAN MINING CONGRESS DRAFT BILL

Discussion so far has been confined to a possible regime laid down in an international treaty, presumably drafted at the Law of the Sea Conference. The alternative to such a treaty is the development of customary international law. A number of factors are involved. On the basis of publicly available information it appears that ocean-mining companies will be capable of full-scale commercial exploitation of Pacific Ocean manganese nodules in the near future. Taking 1978 as the date for signature of a convention embodying a seabed regime it can be argued on the precedent of the 1958 Conventions, that the seabed treaty will not be in force before 1980. Some observers judge the likelihood of agreement by 1980 to be remote. On present trends the seabed convention will have a number of substantial defects. It will be vague on critical matters, due to compromises between irreconcilable views. If there is a strong international seabed authority the developed countries may not ratify the convention, making it of little practical value. Key states, such as the United States, may not ratify a treaty or accept a provisional regime due to differences of opinion between the executive and the legislature, or the federal and state governments. For the ocean miner able to command the required investment capital, the customary international-law approach has several advantages. The miner will develop the law according to his practical requirements, rather than being bound by a regime controlled by states lacking technical expertise. Law would be developed as the need arose, eliminating the laying down of a treaty framework for ocean mining when even the various extraction techniques have not been tested on a commercial scale. Regulatory control would not be in the hands of a United Nations type of international organization subject to the vagaries of international politics. A private-enterprise solution on the lines of the claim to exclusive mining rights filed by Deepsea Ventures with the Secretary of State in November 1974 (Deepsea Ventures, Inc., 1974) and reaffirmed in April 1975 may act as a catalyst to government action. But in traditional international law, only states can assert rights. Therefore, even if a private claim to rights over a mining site were to have some effect in United States municipal law, it is difficult to accept that a claim of this nature has any status in international law.

But a purely customary international-law approach based on such a claim would leave some vital problems unsolved. If, as will be the case with most non-governmental ventures, loan capital is required, the banker will demand that tenure over the mine site be secure. Guarantees will be needed for adequate compensation in the event of an international regime taking over or otherwise seriously reducing the value of the enterprise.

It is to deal with such problems that the American Mining Congress has sponsored a Bill, a recent version of which was introduced in the House of Representatives in 1974 as H.R. 12233 as a framework for manganese-nodule mining by United States private enterprise. The Bill requires detailed analysis because the United States is a leader in nodule research, its enterprises have already spent large sums which the Bill would safeguard, the Bill has been drafted with the support of a number of major companies interested in ocean mining and its supporters have indicated their determination to ensure its passage.

The Bill applies to the "deep seabed", meaning the seabed and subsoil seaward of the continental shelf as defined in the Continental Shelf Convention. United States' nationals could only mine manganese nodules in the deep seabed if licensed under the Bill. Licensees would gain exclusive use of the mine site for a 10-year period. Exploitation would be permitted for 20 years. The minimum expenditure until commercial recovery commences would be 5,900,000 dollars for each block. For manganese nodules a block would comprise 40,000 km^2, reducible to 10,000 km^2 after 10 years. The United States' government would reimburse the licensee for loss or impairment of the investment resulting from any difference between the Bill and any future international regime.

The Bill poses vertical and horizontal boundary problems. There is no definition of the boundary between the continental shelf and the deep seabed. The status of the Kauai Channel deposits would be unclear. The boundary between the "deep seabed" and the overlying waters, the high seas, is not defined. The freedom of the high seas is a cornerstone of present United States policy and will no doubt be clearly preserved in any United States supported ocean mineral regime. The Bill does not make it clear that manganese nodules lying on the seabed are part of the "deep seabed".

The Bill has been presented as an interim regime, and thus within United States ocean law policy as formulated by President Nixon in 1970. Under the Bill the United States government would, for a premium to be determined by the Secretary of Commerce, insure the licensee against damages suffered through the impairment of the insured investment or the removal of hard minerals from the licensed block by any other person against whom a legal remedy does not exist or is unavailable in any legal forum to which the licensee has access. In other words, the government would bear the loss. A licence is valid for a maximum of 20 years from the date commercial recovery commences. The government would also guarantee licensees against any losses due to the establishment of an international regime in so far as the loss was caused by requirements different from the Bill. Taking the estimate of the Japanese Ministry of International Trade and Industry made in January 1973, that 227,000,000 dollars capital is required, the United States government would be under strong pressure to protect its ocean miners, under the Bill, for a lengthy period of time. The liability of the United States government would not be limited to the capital sum, but might well also cover loss of profits. The Bill would therefore provide a regime which would effectively prevent the establishment of any United Nations sponsored manganese-nodule regime affecting United States licensees and differing from the provisions of the Bill for decades to come.

The American Mining Congress Bill may be contrasted with the Basic Conditions presented by the Group of 77 developing countries to the First Committee of the Law of the Sea Conference in 1974. Title to the seabed and its resources would be vested in the Authority. The Authority would exercise direct and effective control over exploration and exploitation at all times. Any contractor's security of tenure would be subject to observance of the provisions of the Convention and the Authority's rules and regulations. The Authority could revise or terminate contracts in case of radical change of circumstances or "force majeure". Any responsibility, liability or risk arising out of the conduct of operations would lie only with the contractor. The contractor would have to transfer technology know-how and data to the Authority on a continuing basis, and train personnel of developing countries. The sole applicable law would be the provisions of the Convention and rules laid down under it. The Basic Conditions provide no details of fees, block size, tenure or choice of court or arbitrator. It is hardly necessary to add that the Conditions outlined would be unacceptable to the drafters of the Bill.

In view of the rapid approach of commercial exploitation there will be increasing pressure in the United States, West Germany and other developed states concerned with ocean mining for such legislation. Clearly nodule exploitation will require a municipal-law framework, whether based upon an international regime, the American Mining Congress Bill or other legislation. The recent establishment of the United States Ocean Mining Administration to plan the licensing of mining operations reflects the lack of progress of the Conference.

CONCLUSIONS

On the presumption that the Law of the Sea Conference reaches agreement, a general outline of future developments may be suggested on the basis of current trends. The 200-mile zone of coastal state jurisdiction has general support, whether in the form of the continental shelf, exclusive economic zone, patrimonial sea or otherwise. Such a jurisdiction will clearly cover the exploration and exploitation of manganese nodules and, taking into account innovations in the drawing of base-lines for the territorial sea, the 200-mile zone would in practice often extend much further than 200 miles from the coast. Some known deposits would therefore fall within the coastal zone. Additional problems arise with isolated islands, atolls and reefs.

It seems to be accepted that the seabed beyond national jurisdiction will be placed under the aegis of an International Authority. A very substantial majority of nations at the Conference support an Authority with wide powers, including direct control of the activities of contractors and mining by the Authority itself. The Authority would consist of a majority of developing countries, some of whom might be concerned at the effect that ocean minerals would have on their own economy. Following the Basic Conditions of the Group of 77 it may be presumed that any private ocean-mining consortium will have substantial difficulties working under a contract with such an Authority. This view is emphasized by the opinion prevalent among developing states that the Authority should take over exploitation itself as soon as it has acquired the requisite skills from the contractors.

Should the concepts put forward by the Group of 77 be incorporated in a Convention, a number of possibilities may be suggested. Ocean miners could exploit the seabed until the treaty came into force, say a period of 4—6 years, in order to recoup their outlay, and then cease production. The ocean-mining industry might persuade governments of developed countries not to ratify the Convention. This alternative has already been canvassed in the United States and could expect some support in the Senate. Miners of nations not parties to the treaty would then request national legislation to protect their interests and proceed to mine outside the provisions of the Convention. They might even decide that the provisions of the Convention were so onerous that they would immediately cease activities.

On present trends it is difficult to foresee any outcome which would satis-fy both the ocean-minerals industry and the majority of participants in the Law of the Sea Conference.

REFERENCES

Aarnio, B., 1918. Om Sjömalmerna i några sjöar i Pusula, Pyhäjärvi, Loppis, Somerniemi och Tammela socknar. Fennia, 41: 1—7.

Agassiz, A., 1891. Three letters to the Hon. Marshall McDonald, U.S. Commissioner of Fish and Fisheries. Bull. Mus. Comp. Zool. Harv., 21: 185—200.

Agassiz, A., 1902. Reports on the scientific results of the expedition to the tropical Pacific, I. Preliminary report and list of stations. Mem. Mus. Comp. Zool. Harv., 26: 1—108.

Agassiz, A., 1906. Reports on the scientific results of the expedition to the eastern tropical Pacific, V. General report of the expedition. Mem. Mus. Comp. Zool. Harv., 33: 1—75.

Alaimo, R., Calderone, S. and Leone, M., 1970. Mineralogia e caratteri genetici degli ossidi di ferro e di manganese nelle concrezioni del flysch numidico siciliano. Atti Accad. Sci. Lett., Palermo, Ser. 4, 30: 3—19.

Albrethson, A. E., 1963. An Electrochemical Study of the Ferric Oxide—Solution Interface. Ph.D. Thesis, Massachusetts Institute of Technology, Cambridge, Mass. (unpublished).

Allen, J. A., 1960. Manganese deposition on the shells of living molluscs. Nature, 185: 336—337.

Allen, M. B., 1961. Our knowledge of the kinds of organisms in Pacific phytoplankton. In: M. S. Doty (Editor), Primary Productivity Measurement, Marine and Freshwater. IGY World Data Center, Natl. Acad. Sci., Washington, D.C., pp. 58—60.

Amin, B. S., 1970. Dating of ocean sediments by radioactive methods. Unpubl. M.S. Thesis, Bombay University, 100 pp.

Amirova, S. A., Pechkovskii, V. V. and Kurmaev, R. Kh., 1965. Chlorination of vanadium trioxide and vanadium spinels in a melt. Zh. Prikl. Khim., 38: 2107—2110 (in Russian).

Amirova, S. A., Pechkovskii, V. V. and Kurmaev, R. Kh., 1966. Effect of the melt on the chlorination of metal oxides. Izv. Vyssh. Ucheb. Zaved. Tsvet. Metall., 9: 62—65 (in Russian).

Amos, A. F., 1973. The deep STD station: techniques for making surface-to-bottom STD profiles and some examples of abyssal microstructure. Proc. 2nd Plessey Envir. Syst. S/T/D Conf. Workshop, San Diego, Calif., Jan. 24—26, pp. 87—101.

Amos, A. F., Garside, C., Haines, K. C. and Roels, O. A., 1972. Effects of surface-discharged deep-sea mining effluent. J. Mar. Technol. Soc., 6(4): 40—45.

Amos, A. F., Garside, C., Gerard, R. D., Levitus, S., Malone, T. C., Paul, A. Z. and Roels, O. A., 1973. Study of the impact of manganese nodule mining on the seabed and water column. In: Inter-University Program of Research on Ferromanganese Deposits on the Ocean Floor — Phase I Report, National Science Foundation, IDOE, Washington, D.C., pp. 221—264.

Amos, A. F., Daubin, S. C., Garside, C., Malone, T. C., Paul, A. Z., Rice, G. E. and Roels, O. A., 1975a. Report on a cruise to study environmental baseline conditions in a manganese nodule province. Proc. Offshore Technol. Conf., Houston, Texas, May 5—8, 1975, 1: 143—158.

Amos, A. F., Amos, L. M. and Paul, A. Z., 1975b. The environmental impact of deep-sea mining. Cruise Rep. Moana Wave Cruise 74—2: April—May 1974, 1116 pp. on Microfiche, NOAA, U.S. Dept. Commerce, Boulder, Colo. 89302.

Andermann, G., 1973. An evaluation of analytical techniques to obtain spectroscopic characterization of molecular properties of manganese nodules. In: Inter-University Program of Research on Ferromanganese Deposits on the Ocean Floor — Phase I Report, National Science Foundation, IDOE, Washington, D.C., pp. 39—43.

Andersen, N. R. and Hume, D. N., 1968. Determination of barium and strontium in sea water. Anal. Chim. Acta, 40: 207—220.

Anderson, B. J., Jenne, E. A. and Chao, T. T., 1973. The sorption of silver by poorly crystallised manganese oxides. Geochim. Cosmochim. Acta, 37: 611—622.

Anderson, C. T., 1942. Laboratory experiments on sulfur dioxide-sulfuric acid leaching of black and brown manganiferous iron ores of the Cuyuna of Minnesota. U.S. Bur. Mines Rep. Invest., 3649: 38 pp.

Andrews, J. E., 1972. Distribution of manganese nodules in the Hawaiian archipelago. Manganese Nodule Deposits in the Pacific Symposium/Workshop Proc., Honolulu, Hawaii, pp. 61—65.

Andrews, J. E. and Meylan, M. A., 1972. Results of bottom photography: Kana Keoki cruise Manganese '72. Hawaii Inst. Geophys. Rep., HIG-72-23: 83—111.

Andrushchenko, P. F. and Skornyakova, N. S., 1965. Composition, structure, and characteristic features of formation of Fe-Mn nodules in the Pacific Ocean. In: D. S. Sapuzhnika (Editor), Manganese Deposits of the Soviet Union. Israel Program for Scientific Translations, Jerusalem, 1970, pp. 101—124.

Andrushchenko, P. F. and Skornyakova, N. S., 1969. The textures and mineral composition of iron-manganese concretions from the southern part of the Pacific Ocean. Oceanology, 9: 229—242.

Anikouchine, W. A., 1967. Dissolved chemical substances in compacting marine sediments. J. Geophys. Res., 72: 505—509.

Anonymous, 1959. USBM Charts Segregation Process. Eng. Min. J., 160: 98—99.

Anonymous, 1972. Mining, bottom crawlers, dredging. Patent Review 7. Underwater J., 4(2): 74—78.

Anonymous, 1974. "Hughes Glomar Explorer" begins sea tests of mining systems. Ocean. Ind., 9(3): 32—34.

Anonymous, 1975. The great submarine snatch. Time Magazine, March 31, pp. 26—30.

Arrhenius, G., 1963. Pelagic sediments. In: M. N. Hill (Editor), The Sea, Vol. 3. Interscience, New York, N.Y., pp. 655—727.

Arrhenius, G., 1967. Deep sea sedimentation: a critical review of U.S. work. Trans. Am. Geophys. Union, 48: 604—631.

Arrhenius, G. and Bonatti, E., 1965. Neptunism and volcanism in the ocean. Prog. Oceanogr., 3: 7—22.

Assaf, G., Gerard, R. and Gordon, A. L., 1971. Some mechanisms of oceanic mixing revealed in aerial photographs. J. Geophys. Res., 76: 6550—6572.

Aston, S. R. and Chester, R., 1973. The influence of suspended particles on the precipitation of iron in natural waters. Estuarine Coastal Mar. Sci., 1: 225—231.

Atkinson, R. J., Posner, A. M. and Quirk, J. P., 1967. Adsorption of potential-determining ions at the ferric oxide—aqueous electrolyte interface. J. Phys. Chem., 71: 550—558.

Auburn, F. M., 1971a. The international seabed area. Int. Comp. Law Q., 20: 173—194.

Auburn, F. M., 1971b. Deep sea mining. Arch. Völkerrechts, 15: 93—96.

Auburn, F. M., 1972a. A New Zealand 200 mile fishery zone. Recent Law, 7: 221—223.

Auburn, F. M., 1972b. The 1973 Conference on the Law of the Sea in the light of current trends in state seabed practice. Can. Bar Rev., 50: 87—109.

Auburn, F. M., 1972c. The Deep Seabed Hard Mineral Resources Bill. San Diego Law Rev., 9: 491—513.

Auburn, F. M., 1973a. Some legal problems of the commercial exploitation of manganese nodules in the Pacific Ocean. Ocean Dev. Int. Law J., 1: 185—200.

Auburn, F. M., 1973b. The North Sea continental shelf boundary settlement. Arch. Völkerrechts, 16: 29—36.

Auburn, F. M., 1973c. International law and sea-ice jurisdiction in the Arctic Ocean. Int. Comp. Law Q., 22: 552—557.

Audley-Charles, M. G., 1965. A geochemical study of Cretaceous ferromanganiferous sedimentary rocks from Timor. Geochim. Cosmochim. Acta, 29: 1153—1173.

Audley-Charles, M. G., 1968. The geology of Portuguese Timor. Mem. Geol. Soc. Lond., 4: 76 pp.

Audley-Charles, M. G., 1972. Cretaceous deep-sea manganese nodules on Timor: implications for tectonics and olistostrome development. Nature Phys. Sci., 240: 137—139.

Audley-Charles, M. G. and Carter, D. J., 1972. Palaeogeographical significance of some aspects of Palaeogene and Early Neogene stratigraphy and tectonics of the Timor Sea region. Palaeogr., Palaeoclimatol., Palaeoecol., 11: 247—264.

Aumento, F., 1969. The Mid-Atlantic Ridge near 45°N; V: Fission-track and ferromanganese chronology. Can. J. Earth Sci., 6: 1431—1440.

Aumento, F., Lawrence, D. E. and Plant, A. G., 1968. The ferromanganese pavement on San Pablo seamount. Geol. Surv. Pap. Can., 68—32: 30 pp.

Baas Becking, L., Kaplan, I. R. and Moore, D., 1960. Limits of the natural environment in terms of pH and oxidation-reduction potentials. J. Geol., 68: 243—284.

Bacon, M. P. and Edmond, J. M., 1972. Barium at Geosecs III in the Southwest Pacific. Earth Planet. Sci. Lett., 16: 50—74.

Bada, J. L., 1972. The dating of fossil bones using the racemization of isoleucine. Earth Planet. Sci. Lett., 15: 223—231.

Bada, J. L., Luyendyk, B. P. and Maynard, J. B., 1970. Marine sediments: dating by the racemization of amino acids. Science, 170: 730—732.

Bancroft, G. M., 1973. Mössbauer Spectroscopy: An Introduction for Inorganic Chemists and Geochemists. McGraw Hill, Maidenhead, 252 pp.

Bandel, K., 1972. Paläokologie und Paläogeographie im Devon und Unterkarbon der zentralen karnischen Alpen. Palaeontographica, 141A: 1—117.

Bandel, K., 1974. Deep-water limestones from the Devonian—Carboniferous of the Carnic Alps, Austria. In: K. J. Hsü and H. C. Jenkyns (Editors), Pelagic Sediments on Land and under the Sea. Spec. Publ. Int. Assoc. Sedimentol., 1: 93—115.

Banning, D. L., 1974. Description of small manganese nodules. Unpubl. Lab. Rep., Dept. of Geology, Washington State Univ., Pullman, Wash. 33 pp.

Barnes, H. L. and Czamanske, G. K., 1967. Solubilities and transport of ore minerals. In: H. L. Barnes (Editor), Geochemistry of Hydrothermal Ore Deposits. Holt, Reinhard and Winston, New York, N.Y., pp. 334—381.

Barnes, J. W., Lang, E. J. and Potratz, H. A., 1956. Ratio of ionium to uranium in coral limestone. Science, 124: 175—176.

Barnes, S. S., 1967a. Minor element composition of ferromanganese nodules. Science, 157: 63—65.

Barnes, S. S., 1967b. The Formation of Oceanic Ferromanganese Nodules. Ph.D. Thesis, Univ. California, San Diego, Calif. (unpublished).

Barnes, S. S. and Dymond, J. R., 1967. Rates of accumulation of ferromanganese nodules. Nature, 213: 1218—1219.

Battelle Memorial Institute, 1971. Environmental disturbances of concern to marine mining research, a selected bibliography. NOAA Tech. Memo. ERL MMTC-3, Marine Minerals Technology Center, Tiburon, Calif., 71 pp.

Baturin, G. N., 1971. Deep-sea sediments of hydrothermal origin. In: L. Zenkevich (Editor), The History of the World Ocean. Nauka, Moscow, pp. 259—277.

Bauwin, G. R. and Tyner, E. H., 1957a. The distribution of non-extractable phosphorus in some grey-brown podzolic, brunizem and planosol soil profiles. Proc. Soil Sci. Soc. Am., 21: 245—250.

Bauwin, G. R. and Tyner, E. H., 1957b. The nature of reductant-soluble phosphorus in soils and soil concretions. Proc. Soil Sci. Soc. Am., 21: 250—257.

Beals, H. L., 1966. Manganese-iron concretions in Nova Scotia lakes. Marit. Sediments, 2: 70—72.

Beck, R. R. and Messner, M. E., 1970. Copper, nickel, cobalt, and molybdenum recovery from deep sea nodules. In: R. P. Ehrlich (Editor), Copper Metallurgy. Proceedings of the Extractive Metallurgy Division Symposium on Copper Metallurgy, American Institute of Mining, Metallurgical and Petroleum Engineers, Denver, pp. 70—82.

Beckwith, R. S. and Reeve, R., 1963. Studies on soluble silica in soils, I. The sorption of silicic acid by soils and minerals. Aust. J. Soil Res., 1: 157—168.

Belov, N. A., Kulikov, N. N., Lapina, N. N. and Semenov, Y. P., 1968. The distribution of iron, manganese and carbonates in deposits of the Arctic Seas. Trudy Arkt. Antarkt. Nauchno-issled. Inst., 285: 67—73.

Bender, M. L., 1971. Does upward diffusion supply the excess manganese in pelagic sediments? J. Geophys. Res., 76: 4212—4215.

Bender, M. L., 1972. Mechanisms of trace metal removal from the oceans. In: D. R. Horn (Editor), Ferromanganese Deposits on the Ocean Floor. National Science Foundation, Washington, D.C., pp. 73—80.

Bender, M. L., Ku, T.-L. and Broecker, W. S., 1966. Manganese nodules: their evolution. Science, 151: 325—328.

Bender, M. L., Broecker, W., Gornitz, V. and Middel, U., 1970a. Accumulation rate of manganese and related elements in the sediments from the East Pacific Rise. EOS Trans. Am. Geophys. Union, 51: 327 (Abstr.)

Bender, M. L., Ku, T.-L. and Broecker, W. S., 1970b. Accumulation rates of manganese in pelagic sediments and nodules. Earth Planet. Sci. Lett., 8: 143—148.

Bender, M. L., Broecker, W., Gornitz, V., Middel, U., Kay, R., Sun, S.-S. and Biscaye, P., 1971. Geochemistry of three cores from the East Pacific Rise. Earth Planet. Sci. Lett., 12: 425—433.

Benoit, R. L. and Mackiw, V. N., 1961. German Patent 1,106,741.

Bentor, Y. K., 1956. The manganese occurrences at Timna' (southern Israel), a lagoonal deposit. In: J. G. Regna (Editor), Symposium Sobre Yacimientos de Manganeso, 4 Asia y Oceania. Int. Geol. Congress, Mexico, 20th. pp. 159—172.

Berner, R. A., 1964. Iron sulfides formed from aqueous solution at low temperatures and atmospheric pressure. J. Geol., 72: 293—306.

Berner, R. A., 1967. Thermodynamic stability of sedimentary iron sulfides. Am. J. Sci., 265: 773—785.

Berner, R. A., 1970. Sedimentary pyrite formation. Am. J. Sci., 268: 1—23.

Berner, R. A., 1971. Principles of Chemical Sedimentology. McGraw-Hill, New York, N.Y., 240 pp.

Bernoulli, D. and Jenkyns, H. C., 1974. Alpine, Mediterranean and Central Atlantic Mesozoic facies in relation to the early evolution of the Tethys. In: R. H. Dott and R. H. Shaver (Editors), Modern and Ancient Geosynclinal Sedimentation, A Symposium. Spec. Publ. Soc. Econ. Paleont. Miner., 19: 129—160.

Berrang, P. G. and Grill, E. V., 1974. The effect of manganese oxide scavenging on molybdenum in Saanich Inlet, British Columbia. Mar. Chem., 2: 125—148.

Berry, L. G. and Thompson, R. M., 1962. X-ray powder data for ore minerals: The Peacock Atlas. Mem. Geol. Soc. Am., 85: 281 pp.

Bertine, K., 1972. Submarine Weathering of Tholeitic Basalts and the Origin of Metalliferous Sediments. Unpubl. manuscript.

Bertine, K. K. and Turekian, K. K., 1973. Molybdenum in marine deposits. Geochim. Cosmochim. Acta, 37: 1415—1435.

Bérubé, Y. G. and De Bruyn, P. L., 1968. Adsorption at the rutile—solution interface, 1. Thermodynamic and experimental study. J. Colloid Interface Sci., 27: 305—318.

Bezrukov, P. L., 1960. Sedimentation in the northwestern part of the Pacific Ocean. Int. Geol. Congr., 21: 39—49.

Bezrukov, P. L., 1963. Studies of the Indian Ocean during the 35th cruise of R.V. "Vityaz". Okeanologiya, 3: 540—549 (in Russian).

Bezrukov, P. L. and Andrushchenko, P. F., 1972. Iron-manganese nodules from the Indian Ocean. Izv. Akad. Nauk SSSR, Ser. Geol., 7: 3—20. (Translated in Int. Geol. Rev., 15: 342—356.)

Bezrukov, P. L. and Andrushchenko, P. F., 1974. Geochemistry of iron-manganese nodules from the Indian Ocean. Int. Geol. Rev., 16: 1044—1061.

Bhandari, N., Bhat, S. G., Krishnaswamy, S. and Lal, D., 1971. A rapid gamma-beta coincidence technique for determination of natural radionuclides in marine deposits. Earth Planet. Sci. Lett., 11: 121—126.

Bhat, S. G., Krishnaswamy, S., Lal, D., Rama and Somayajulu, B. L. K., 1973. Radiometric and trace elemental studies of ferromanganese nodules. Proc. Symp. Hydrogeochem. Biogeochem., I. Clark, Washington, D.C. pp. 443—462.

Bilder, R. B., 1966. Control of criminal conduct in Antarctica. Virginia Law Rev., 52: 231—285.

Biscaye, P. E. and Eittreim, S. L., 1974. Variations in benthic boundary layer phenomena: nepheloid layer in the North American Basin. In: R. J. Gibbs (Editor), Suspended Solids in Water, Plenum, New York, N.Y. pp. 227—260.

Biscaye, P. E., Kolla, V. and Turekian, K. K., 1976. Distribution of calcium carbonate in surface sediments of the Atlantic Ocean. J. Geophys. Res., 81: 2595—2603.

Bischoff, J. L., 1969. Red Sea geothermal brine deposits: their mineralogy, chemistry and genesis. In: E. T. Degens and D. A. Ross (Editors), Hot Brines and Recent Heavy Metal Deposits in the Red Sea. Springer-Verlag, Berlin, pp. 368—401.

Bischoff, J. L. and Ku, T.-L., 1971. Pore fluids of recent marine sediments, II. Anoxic sediments of $35°$ to $45°N$ Gibraltar to Mid-Atlantic Ridge. J. Sediment. Petrol., 41: 1008—1017.

Bischoff, J. L. and Sayles, F. L., 1972. Pore fluid and mineralogical studies of recent marine sediments: Bauer Depression region of East Pacific Rise. J. Sediment. Petrol., 42: 711—724.

Blackburn, M., Laurs, R. M., Owen, R. W. and Zeitzschel, B., 1970. Seasonal and areal changes in standing stocks of phytoplankton, zooplankton and micronekton in the eastern tropical Pacific. Mar. Biol., 7: 14—31.

Blok, L. and De Bruyn, P. L., 1970. The ionic double layer at the ZnO/solution interface, I. The experimental point of zero change. J. Colloid Interface Sci., 32: 518—526.

Bloomfield, C., 1952. Translocation of iron in podzol formation. Nature, 170: 540.

Bodine, M. W., Holland, H. D. and Boresik, M., 1965. Coprecipitation of manganese and strontium with calcite. In: Symposium on Problems of Postmagmatic Ore Deposition, II, Geol. Surv. Czech, Prague, pp. 401—406.

Bolt, G. H., 1957. Determination of the charge density of silica sols. J. Phys. Chem., 61: 1166—1169.

Bonatti, E. and Joensuu, O., 1966. Deep-sea iron deposit from the South Pacific. Science, 154: 643—645.

Bonatti, E. and Nayudu, Y. R., 1965. The origin of manganese nodules on the ocean floor. Am. J. Sci., 263: 17—39.

Bonatti, E., Kraemer, T. and Rydell, H., 1972. Classification and genesis of submarine iron-manganese deposits. In: D. R. Horn (Editor), Ferromanganese Deposits on the Ocean Floor. National Science Foundation, Washington, D.C., pp. 149—166.

Bonatti, E., Zerbi, M., Kay, R. and Rydell, H., 1976. Metalliferous deposits from the Apennine ophiolites: Mesozoic equivalents of modern deposits from oceanic spreading centers. Bull Geol. Soc. Am., 87: 83—94.

Boström, K., 1967. The problem of excess manganese in pelagic sediments. In P. H. Abelson (Editor), Researches in Geochemistry, 2. Wiley, New York, N.Y., pp. 421—452.

Boström, K., 1970. Deposition of manganese rich sediments during glacial periods. Nature, 226: 629—630.

Boström, K. and Peterson, M. N. A., 1966. Precipitates from hydrothermal exhalations on the East Pacific Rise. Econ. Geol., 61: 1258—1265.

Boström, K. and Peterson, M. N. A., 1969. The origin of aluminum-poor ferromanganoan sediments in areas of high heat flow on the East Pacific Rise. Mar. Geol., 7: 427—447.

Boström, K., Peterson, M. N. A., Joensuu, O. and Fisher, D. E., 1969. Aluminum-poor ferromanganoan sediments on active oceanic ridges. J. Geophys. Res., 74: 3261—3270.

Boström, K., Farquharson, B. and Eyl, W., 1971. Submarine hot springs as a source of active ridge sediments. Chem. Geol., 10: 189—203.

Boström, K., Joensuu, O., Valdes, S. and Riera, M., 1972. Geochemical history of South Atlantic Ocean sediments since Late Cretaceous. Mar. Geol., 12: 85—121.

Boström, K., Joensuu, O., Moore, C., Boström, B., Dalziel, M. and Horowitz, A., 1973. Geochemistry of barium in pelagic sediments. Lithos, 6: 159—174.

Boström, K., Kraemer, T. and Gartner, S., 1973. Provenance and accumulation rates of opaline silica, Al, Ti, Fe, Mn, Cu, Ni and Co in Pacific pelagic sediments. Chem. Geol., 11: 123—148.

Boulad, A. P., Condomines, M., Bernat, M., Michard, G. and Allegre, C. J., 1975. Vitesse d'accrétion des nodules de manganèse des fonds océaniques. C.R. Acad. Sci., Paris, 280D: 2425—2428.

Bourbon, M., 1971a. Structure et signification de quelques nodules ferrugineux, manganésifères et phosphatés liés aux lacunes de la série crétacée et paléocène briançonnaise. C.R. Acad. Sci., Paris, 273D: 2060—2062.

Bourbon, M., 1971b. Un example de série pélagique extrêmement condensée, en zone briançonnaise, au Nord-Ouest de Serre-Chevalier (Hautes-Alpes). C.R. Acad. Sci., Paris, 273D: 1899—1902.

Bowden, J. W., Bolland, M. D. A., Posner, A. M. and Quirk, J. P., 1973. Generalised model for anion and cation adsorption at oxide surfaces. Nature Phys. Sci., 245: 81—83.

Bowen, H. J. M., 1966. Trace Elements in Biochemistry. Academic Press, London, 241 pp.

Bowen, V. T., Olsen, J. S., Osterberg, C. L. and Ravera, J., 1971. Ecological interactions of marine radioactivity. In: Radioactivity in the Marine Environment. National Academy of Sciences, Washington, D.C., pp. 200—222.

Bower, C. A. and Goertzen, J. O., 1959. Surface area of soils and clays by an equilibrium ethylene glycol method. Soil Sci., 87: 288—292.

Bowser, C. J., Callender, E. and Rossmann, R., 1970. Electron probe and X-ray studies of ferromanganese nodules from Wisconsin and Michigan. Abstr. Progr. Ann. Mtg. Geol. Soc. Am., 2(7): 500—501 (Abstr.)

Boyd, M. B., Saucier, R. T., Keeley, J. W., Montgomery, R. L., Brown, R. D., Mathis, D. B. and Guice, C. J., 1972. Disposal of dredge spoil: problem identification and assessment and research program development. U.S. Army Wat. Exp. Stn., Technol. Rep. H-72-8, 127 pp.

Boyle, E. and Edmond, J. M., 1975. Copper in surface waters south of New Zealand. Nature, 253: 107—109.

Breeuwsma, A. and Lyklema, J., 1971. Interfacial electrochemistry of hematite (α-Fe$_2$O$_3$). Disc. Farad. Soc., 52: 324—333.

Breeuwsma, A. and Lyklema, J., 1973. Physical and chemical adsorption of ions in the electrical double layer on hematite (α-Fe$_2$O$_3$). J. Colloid Interface Sci., 43: 437—448.

Brewer, P. G. and Ellert, M., 1972. The Kinetics of Manganese Oxidation in Seawater. Unpubl. manuscript.

Brewer, P. G. and Spencer, D. W., 1970. Trace element intercalibration study. Woods Hole Oceanogr. Inst. Rep., 70—62.

Brewer, P. G. and Spencer, D. W., 1974. Distribution of some trace elements in Black Sea and their flux between dissolved and particulate matter. Mem. Am. Assoc. Petrol. Geol., 20: 137—143.

Bricker, O. P., 1965. Some stability relations in the system Mn—O$_2$—H$_2$O at 25° and one atmosphere total pressure. Am. Miner., 50: 1296—1354.

Broecker, W. S., 1971. Calcite accumulation rates and glacial to interglacial changes in oceanic mixing. In: K. K. Turekian (Editor), The Late Cenozoic Glacial Ages. Yale University Press, New Haven, Conn., pp. 239—265.

Broecker, W. S. and Mantyla, A. W., 1974. GEOSECS: Pacific Expedition Preliminary Report, Leg 19, Papeete to San Diego, May 13—June 10, 1974. Unpubl. Rep. Scripps Inst. Oceanogr.

Broecker, W. S., Cromwell, J. and Li, Y. H., 1968. Rates of vertical eddy diffusion near the ocean floor based on measurements of the distribution of excess ^{222}Rn. Earth Planet. Sci. Lett., 5: 101—105.

Brooke, J. N. and Prosser, A. P., 1969. Manganese nodules as a source of copper and nickel — mineralogical assessment and extraction. Trans. Inst. Min. Metall., 78C: 64—73.

Brooks, N. H., 1960. Diffusion of sewage effluent in an ocean current. In: E. A. Pearson (Editor), Waste Disposal in the Marine Environment. Pergamon Press, New York, N.Y., pp. 246—267.

Brooks, P. T. and Martin, D. A., 1971. Processing manganiferous sea nodules. U.S. Bur. Mines Rep. Invest., 7473: 19 pp.

Brooks, P. T., Dean, K. C. and Rosenbaum, J. B., 1970. Experiments in processing marine nodules. Proc. Int. Mineral. Congr., 9th, Czechoslovakia, pp. 329—333.

Brooks, R. R., Presley, B. J. and Kaplan, I. R., 1968. Trace elements in the interstitial waters of marine sediments. Geochim. Cosmochim. Acta, 32: 397—414.

Brown, B. A., 1971. A Geochemical Investigation of Inter-Element Relations in Deep-Sea Ferromanganese Nodules. Ph.D. Thesis, Univ. of Oxford, 293 pp. (unpublished).

Brown, B. A., 1972. A low-temperature crushing technique applied to manganese nodules. Am. Miner., 57: 284—287.

Brown, E. D., 1971. The Legal Regime of Hydrospace. Stevens, London, 236 pp.

Brown, F. H., Pabst, A. and Sawyer, D. L., 1971. Birnessite on colemanite at Boron, California. Am. Miner., 56: 1057—1064.

Bruevich, S. W., 1938. Oxidation-reduction potentials and pH of sea bottom deposits. Verh. Int. Ver. Theor. Angew. Limnol., 8: 35—49.

Brunauer, S., Emmett, P. H. and Teller, E., 1938. Adsorption of gases in multi-molecular layers. J. Am. Chem. Soc., 60: 309—319.

Bruty, D., Chester, R., Royle, L. G. and Elderfield, H., 1972. Distribution of zinc in North Atlantic deep-sea sediments. Nature Phys. Sci., 237: 86—87.

Bryan, W. H., 1952. Soil nodules and their significance. In: M. F. Glaessner and E. A. Rudd (Editors), Sir Douglas Mawson Anniversary Volume. University of Adelaide, Adelaide, pp. 43—53.

Buchanan, J. Y., 1878. Manganese nodules in Loch Fyne. Nature, 18: 628.

Buchanan, J. Y., 1891. On the composition of oceanic and littoral manganese nodules. Trans. R. Soc. Edinb., 36: 459—484.

Buckman, H. O. and Brady, N. C., 1960. The Nature and Properties of Soils. MacMillan, New York, N.Y., 567 pp.

Bundesverband der Deutschen Industrie, 1972. Die Völkerrechtlichen Probleme der Gewinnung Mineralischer Rohstoffe des Meeres aus, der Sicht der Deutschen Industrie. Meeresbergbau und Völkerrecht.

Burckhardt, C. E. and Falini, F., 1956. Memoria sui giacimenti italiani di manganese. In: J. G. Reyna (Editor), Symposium Sobre Yacimientos de Manganeso, 5, Europa. Int. Geol. Congress, 20th, Mexico, pp. 221—272.

Burke, W. T., 1970. Marine Science Research and International Law. Law of the Sea Institute, Kingston, 36 pp.

Burns, R. G., 1965. Formation of cobalt(III) in the amorphous FeOOH·nH_2O phase of manganese nodules. Nature, 205: 999.

Burns, R. G., 1970. Mineralogical Applications of Crystal Field Theory. Cambridge University Press, Cambridge, 224 pp.

Burns, R. G., 1973. The partitioning of trace transition elements in crystal structures; a provocative review with applications to mantle geochemistry. Geochim. Cosmochim. Acta, 37: 2395—2403.

Burns, R. G. and Brown, B. A., 1972. Nucleation and mineralogical controls on the composition of manganese nodules. In: D. R. Horn (Editor), Ferromanganese Deposits on the Ocean Floor. National Science Foundation, Washington, D.C., pp. 51—61.

Burns, R. G. and Burns, V. M., 1975. Mechanism for nucleation and growth of manganese nodules. Nature, 255: 130—131.

Burns, R. G. and Fuerstenau, D. W., 1966. Electron probe determination of inter-element relationships in manganese nodules. Am. Miner., 51: 895—902.

Burns, R. G. and Fyfe, W. S., 1967. Crystal field theory and geochemistry of transition elements. In: P. H. Abelson (Editor), Researches in Geochemistry, 2. Wiley, New York, N.Y., pp. 259—285.

Burns, R. G., Burns, V. M., Sung, W. and Brown, B. A., 1974. Ferromanganese nodule mineralogy: suggested terminology of the principal manganese oxide phases. Abstr. Progr. Ann. Mtg Geol. Soc. Am., 6(7): 1029—1031.

Burns, V. M. and Burns, R. G., 1974. The uptake of cobalt by ferromanganese nodules, soils and manganese (IV) oxides. EOS Trans. Am. Geophys. Un., 56: 1139 (Abstr.).

Burns, V. M. and Burns, R. G., 1976. Evidence of diagenesis and growth in manganese nodules from the North Equatorial Pacific by scanning electron microscopy. Abstr. Progr. Ann. Mtg. Geol. Soc. Am., 8 (8): 796—797.

Burrett, C. F., 1972. Plate tectonics and the Hercynian orogeny. Nature, 239: 155—157.

Busch, P. L. and Stumm, W., 1968. Chemical interactions in the aggregation of bacteria bioflocculation in waste treatment. Environ. Sci. Technol., 2: 49—53.

Buser, W., 1959. The nature of the iron and manganese compounds in manganese nodules. Prepr. Int. Oceanogr. Congr., 1: 962—963.

Buser, W. and Graf, P., 1955a. Differenzierung von Mangan(II)-Manganit und δ-MnO_2 durch Oberflächenmessung nach Brunauer-Emmet-Teller. Helv. Chim. Acta, 38: 830—834.

Buser, W. and Graf, P., 1955b. Radiochemische Untersuchungen an Festkörpern III. Ionen- und Isotopenaustauschreaktionen an Mangandioxyden und Manganiten. Helv. Chim. Acta, 38: 810—829.

Buser, W. and Grütter, A., 1956. Über die Natur der Manganknollen. Schweiz. Miner. Petrogr. Mitt., 36: 49—62.

Buser, W., Graf, P. and Feitknecht, W., 1954. Beitrag zur Kenntnis der Mangan(II)-Manganite und des δ-MnO_2. Helv. Chim. Acta, 37: 2322—2333.

Butler, G. and Thirsk, H. R., 1952. Electron diffraction evidence for the existence and fine structure of a cryptomelane modification of manganese dioxide prepared in the absence of potassium. Acta Cryst., 5: 288—289.

Byström, A. and Byström, A. M., 1950. The crystal structure of hollandite, the related manganese oxide minerals, and α-MnO_2. Acta Cryst., 3: 146—154.

Byström, A. and Byström, A. M., 1951. The positions of the barium atoms in hollandite. Acta Cryst., 4: 469.

Byström, A. M., 1949. The crystal structure of ramsdellite, an orthorhombic modification of MnO_2. Acta Chem. Scand., 3: 163—173.

Caldwell, A. B., 1971. Deepsea ventures readying its attack on Pacific nodules. Min. Eng., 23(10): 54—55.

Callender, E., 1970. The economic potential of ferromanganese nodules in the Great Lakes. Proc. 6th Forum Geol. Ind. Miner. Michigan Geol. Surv. Misc., 1: 55—65.

Callender, E., 1973. Geochemistry of ferromanganese crusts, manganese carbonate crusts and associated ferromanganese nodules from Green Bay, Lake Michigan. In: Inter-University Program of Research on Ferromanganese Deposits of the Ocean Floor — Phase I Report. National Science Foundation, Washington, D.C., pp. 105—120.

Calvert, S. E. and Price, N. B., 1970a. Composition of manganese nodules and manganese carbonates from Loch Fyne, Scotland. Contr. Miner. Petrol., 29: 215—233.

Calvert, S. E. and Price, N. B., 1970b. Minor metal contents of recent organic rich sediments of South West Africa. Nature, 227: 593—595.

Calvert, S. E. and Price, N. B., 1972. Diffusion and reaction profiles of dissolved manganese in the pore waters of marine sediments. Earth Planet Sci. Lett., 16: 245—249.

Calvert, S. E. and Price, N. B., 1977. Geochemical variation in ferromanganese nodules and associated sediments from the Pacific Ocean. Mar. Chem., 5: 43—74.

Cameron, E. N., 1961. Ore Microscopy. Wiley, New York, N.Y., 249 pp.

Cardwell, P. H., 1973. Extractive metallurgy of ocean nodules. Proc. Min. Conv./Environ. Show, Am. Min. Congr., Denver, Colo., Sept. 9—12, 1973.

Caron, M. H., 1924. Recovering values from nickel and cobalt-nickel ores. U.S. Patent, 1,487,145.

Caron, M. H., 1955. Ammonia leaching of nickel sulphide ores: Half century review. Trans. Inst. Min. Metall., 64: 611—616.

Carpenter, R. and Wakeham, S., 1973. Mössbauer studies of marine and fresh water manganese nodules. Chem. Geol., 11: 109—116.

Carpenter, R., Johnson, H. P. and Twiss, E. S., 1972. Thermomagnetic behavior of manganese nodules. J. Geophys. Res., 77: 7163—7174.

Carr, G. L., 1970. Marine Manganese Nodules: Identification and Occurrence of Minerals. M.S. Thesis, Washington State Univ., Pullman, Wash., 101 pp. (unpublished).

Carroll, D., 1958. Role of clay minerals in the transportation of iron. Geochim. Cosmochim. Acta, 14: 1—27.

Carvajal, M. C. and Landergren, S., 1969. Marine sedimentation processes. The interrelationships of manganese, cobalt, and nickel. Stockholm Contr. Geol., 18: 99—122.

Chakravarti, M. N. and Dhar, N. R., 1927. Adsorption of electrolytes by manganese dioxide, and a discussion of the Freundlich adsorption formula. J. Phys. Chem., 31: 997—1033.

Chan, K. M. and Riley, J. P., 1966a. The determination of molybdenum in natural waters, silicates and biological materials. Anal. Chim. Acta, 36: 220—229.

Chan, K. M. and Riley, J. P., 1966b. The determination of vanadium in sea and natural waters, biological materials and silicate sediments and rocks. Anal. Chim. Acta, 34: 337—345.

Charlesworth, J. K., 1957. The Quaternary Era, 2. Edward Arnold, London, 1700 pp.

Charlot, G. and Bezier, D., 1957. Quantitative Inorganic Analysis. Wiley, New York, N.Y., 535 pp.

Chase, T. E., Menard, H. W. and Mammerickx, J., 1970. Bathymetry of the North Pacific. Inst. Marine Resour. Tech. Rep. Ser. TR-6, La Jolla, Calif., 10 charts.

Chau, Y. K. and Riley, J. P., 1965. The determination of selenium in sea water, silicates and marine organisms. Anal. Chim. Acta, 33: 36—49.

Cheney, E. S. and Vredenburgh, L. D., 1968. The role of iron sulfides in the diagenetic formation of iron-poor manganese nodules. J. Sediment. Petrol., 38: 1363—1365.

Cherdyntsev, V. V., 1971. Uranium-234. Israel Program for Scientific Translations, Jerusalem, 234 pp.

Cherdyntsev, V. V., Kadyrov, N. B. and Novichkova, N., 1971. Origin of manganese nodules of the Pacific Ocean from radioisotope data. Geokhimiya, 3: 339—354 (English transl.: Geochem. Int., 8: 211—225.)

Chester, R., 1965. Adsorption of zinc and cobalt on illite in sea water. Nature, 206: 884—886.

Chester, R. and Hughes, M. I., 1969. The trace element geochemistry of a North Pacific pelagic clay core. Deep-Sea Res., 13: 627—634.

Chester, R. and Messiha-Hanna, R. G., 1970. Trace element partition patterns in North Atlantic deep-sea sediments. Geochim. Cosmochim. Acta, 34: 1121—1128.

Chester, R., Johnson, L. R., Messiha-Hanna, R. G. and Padgham, R. C., 1973. Similarities between Mn, Ni and Co contents of deep-sea clays and manganese nodules from the south-west region of the North Atlantic. Mar. Geol., 14: M15—M20.

Chow, T. J. and Patterson, C. C., 1962. The occurrence and significance of lead isotopes in pelagic sediments. Geochim. Cosmochim. Acta, 26: 263—308.

Chuecas, L. and Riley, J. P., 1966. The spectrophotometric determination of chromium in sea water. Anal. Chim. Acta, 36: 240—246.

Chukhrov, F. V., Zvyagin, B. B., Gorshkov, A. I., Yermilova, L. P. and Balashova, V. V., 1973. On ferrihydrite (hydrous ferric oxide). Izv. Akad. Nauk SSSR, Ser. Geol., 4: 23—33 (in Russian). Cf. also C.A. 79: 106696C.

Chun, C., 1908. Manganknollen. In: Wissenschaftliche Ergebnisse der Deutschen Tiefsee-Expedition auf dem Dampfer "Valdivia" 1898—1899, Vol. 10, Jena, pp. 111—114.

Church, T. M. and Wolgemuth, K., 1972. Marine barite saturation. Earth Planet. Sci. Lett., 15: 35—44.

Churchill, R., Simmonds, K. R. and Welch, J. (Editors), 1973. New Directions in the Law of the Sea, 3. Oceana, New York, N.Y., 358 pp.

Clark, G. M., 1972. The Structures of Non-Molecular Solids. Applied Science Publ. Ltd., London, 365 pp.

Clark, J. S. and Brydon, J. E., 1963. Characteristics and genesis of concretionary brown soils of British Columbia. Soil Sci., 96: 410—417.

Clarke, F. W., 1924. The data of geochemistry. Bull. U.S. Geol. Surv., 770: 841 pp.

Clarke, W. B., Beg, M. A. and Craig, H., 1969. Excess ^{3}He in the sea: evidence for ter-restrial primordial helium. Earth Planet. Sci. Lett., 6: 213—220.

Clauss, G., 1972. Theoretical and experimental investigations of deep-ocean mining systems and their economic implications. Int. Ocean Dev. Conf., 2nd, Tokyo, October 5—7, 1972, pp. 1925—1955.

Clute, R. R. and Grant, R. W., 1974. Organic matter and iron-manganese concretions in Chatauqua Lake, New York. Abstr. Progr. Ann. Mtg. Geol. Soc. Am., 6(7): 690 (Abstr.).

Coey, J. M. D. and Readman, P. N., 1973. Characterization and magnetic properties of natural ferric gel. Earth Planet. Sci. Lett., 21: 45—51.

Cole, W. F., Wadsley, A. D. and Walkley, A., 1947. An X-ray diffraction study of MnO_2. Trans. Electrochem. Soc., 92: 1—22.

Coonley, L. S., Baker, E. B. and Holland, H. D., 1971. Iron in the Mullica River and in Great Bay, New Jersey. Chem. Geol., 1: 51—63.

Copeland, L. C., Griffith, F. S. and Schertzinger, C. B., 1947. Preparation of a dry cell depolarizer by air oxidation of manganous hydroxide. Trans. Electrochem. Soc., 92: 127—158.

Corliss, J., 1971. The origin of metal-bearing submarine hydrothermal solutions. J. Geophys. Res., 76: 8128—8138.

Correns, C. W., 1937. Die Sedimente Des Äquatorialen Atlantischen Ozeans. In: Deutsche Atlantische Expedition "Meteor" 1925—1927, Vol. 3. De Gruner, Berlin and Leipzig, pp. 135—298.

Cox, A., Blakely, R. J. and Phillips, J. D., 1972. Geomagnetic reversals and the core-mantle boundary. EOS Trans. Am. Geophys. Union 53: 974 (Abstr.).

Coyle, T. J., Statham, E. F. and Howat, D. D., 1966. Some thermodynamic and kinetic aspects of the refining of gold. J. South Afr. Inst. Min. Metall., 66: 297.

Craig, H., 1969. Geochemistry and origin of the Red Sea Brines. In: E. T. Degens and D. A. Ross (Editors), Hot Brines and Recent Heavy Metal Deposits in the Red Sea. Springer-Verlag, Berlin, pp. 208—242.

Craig, H. and Weiss, R. F., 1970. The GEOSECS 1969 intercalibration station: introduction, hydrographic features, and total CO_2-O_2 relationships. J. Geophys. Res., 75: 7641—7647.

Craig, H., Chung, Y. and Fiadero, M., 1972. A benthic front in the South Pacific. Earth Planet Sci. Lett., 16: 50—65.

Craig, J. D., 1975. The Distribution of Ferromanganese Nodule Deposits in the North Equatorial Pacific. M.Sc. Thesis, Univ. of Hawaii, 104 pp. (unpublished).

Crecelius, E.A., Carpenter, R. and Merrill, R.T., 1973. Magnetism and magnetic reversals in ferromanganese nodules. Earth Planet. Sci. Lett., 17: 391—396.

Crerar, D. A. and Barnes, H. L., 1974. Deposition of deep-sea manganese nodules. Geochim. Cosmochim. Acta, 38: 279—300.

Cressman, E. R. and Swanson, R. W., 1964. Stratigraphy and petrology of the Permian rocks of southwestern Montana. Prof. Pap. U.S. Geol. Surv., 313C: 275—569.

Cromwell, T., 1958. Thermocline topography, horizontal currents and ridging in the eastern tropical Pacific. Bull. Int.-Am. Trop. Tuna Comm., 3(3): 135—164.

Cromwell, T. and Reid, J. L., 1955. A study of oceanic fronts. Tellus, 8: 94—101.

Cromwell, T., Montgomery, R. B. and Stroup, E. D., 1954. Equatorial undercurrents in the Pacific Ocean revealed by new methods. Science, 119: 648—649.

Cronan, D. S., 1967. The Geochemistry of Some Manganese Nodules and Associated Pelagic Sediments. Ph.D. Thesis, Univ. of London (unpublished).

Cronan, D. S., 1969. Inter-element associations in some pelagic deposits. Chem. Geol., 5: 99—106.

Cronan, D. S., 1972a. Composition of Atlantic manganese nodules. Nature, 235: 171—172.

Cronan, D. S., 1972b. The Mid-Atlantic Ridge near 45°N, XVII: Al, As, Hg and Mn in ferruginous sediments from the Median Valley. Can. J. Earth Sci., 9: 319—323.

Cronan, D. S., 1972c. Regional geochemistry of ferromanganese nodules in the World Ocean. In: D. R. Horn (Editor), Ferromanganese Deposits on the Ocean Floor. National Science Foundation, Washington, D.C., pp. 19—30.

Cronan, D. S., 1975. Manganese nodules and other ferromanganese oxide deposits from the Atlantic Ocean. J. Geophys. Res., 80: 3831—3837.

Cronan, D. S., 1976. Manganese nodules and other ferromanganese oxide deposits. In: J. P. Riley and R. Chester (Editors), Chemical Oceanography. Vol. 5. Academic Press, London, pp. 217—263.

Cronan, D. S. and Thomas, R. L., 1970. Ferromanganese concretions in Lake Ontario. Can. J. Earth Sci., 7: 1346—1349.

Cronan, D. S. and Thomas, R. L., 1972. Geochemistry of ferromanganese oxide concretions and associated deposits in Lake Ontario. Bull. Geol. Soc. Am., 83: 1493—1502.

Cronan, D. S. and Tooms, J. S., 1967a. Geochemistry of manganese nodules from the N.W. Indian Ocean. Deep-Sea Res., 14: 239—249.

Cronan, D. S. and Tooms, J. S., 1967b. Sub-surface concentrations of manganese nodules in Pacific sediments. Deep-Sea Res., 14: 117—119.

Cronan, D. S. and Tooms, J. S., 1968. A microscopic and electron probe investigation of manganese nodules from the Northwest Indian Ocean. Deep-Sea Res., 15: 215—223.

Cronan, D. S. and Tooms, J. S., 1969. The geochemistry of manganese nodules and associated pelagic deposits from the Pacific and Indian Oceans. Deep-Sea Res., 16: 335—359.

Csanady, G. T., 1973. Turbulent Diffusion in the Environment. Reidel, Dordrecht, 248 pp.

Cushing, D. H., 1959. The seasonal variation in oceanic production as a problem in population dynamics. J. Conseil, 24: 455—465.

Dale, N. C., 1915. The Cambrian manganese deposits of Conception and Trinity Bays, Newfoundland. Proc. Am. Phil. Soc., 54: 371—456.

Damiani, V., Morton, T. W. and Thomas, R. L., 1973. Freshwater ferromanganese nodules from the Big Bay section of the Bay of Quinte, northern Lake Ontario. Proc. Congr. Great Lakes Res., 12th, pp. 397—403.

Dasch, E. J., Dymond, J. R. and Heath, G. R., 1971. Isotopic analysis of metalliferous sediment from the East Pacific Rise. Earth Planet. Sci. Lett., 13: 175—180.

Dean, J. G., 1959. Process for Separating Cobalt and Nickel from Ammoniacal Solutions. U.S. Patent, 2,915,389.

Dean, W. E., 1970. Iron-manganese oxidate crusts in Oneida Lake, New York. Proc. Conf. Great Lakes Res., 13: 217—226.

Dean, W. E., Ghosh, S. K., Krishnaswami, S. and Moore, W. S., 1973. Geochemistry and accretion rates of freshwater ferromanganese nodules. In: Morgenstein, M. (Editor), Papers on the Origin and Distribution of Manganese Nodules in the Pacific and Prospects for Exploration. An International Symposium organized by the Valdivia Manganese Exploration Group and the Hawaii Institute of Geophysics, Honolulu, July 23 24 25, 1973, pp. 13—20.

Debenedetti, A., 1965. Il complesso radiolariti-giacimenti di manganese-giacimenti piritoso-cupriferi-rocce a fuchsite, come rappresentante del Malm nella formazione dei Calceschisti. Osservazioni nelle Alpi piemontesi e della Val d'Aosta. Boll. Soc. Geol. Ital., 84(1): 131—163.

De Boer, J. H., Tippens, B. C., Tinsen, B. G., Broekhoff, J. C. P., Van der Heuvel, A. and Osinga, T. J., 1966. The t-curve of multimolecular N_2-adsorption. J. Colloid Interface Sci., 21: 405—414.

De Bruyn, P. L. and Agar, G. E., 1962. Surface chemistry of flotation. In: D. W. Fuerstenau (Editor), Froth Flotation. Am. Inst. Min. Metall. Petrol. Eng. Inc., New York, N.Y.

Debyser, J., 1961. Contribution à l'Étude Géochimique des Vases Marins. Soc. Editions Technip., Paris, 249 pp.

Debyser, J. and Rouge, P. E., 1956. Sur l'origine du fer dans les eaux interstitielles des sédiments marins actuels. C.R. Acad. Sci., Paris, 243: 2111—2113.

Deepsea Ventures Inc., 1974. Notice of Discovery and Claim of Exclusive Mining Rights, and Request for Diplomatic Protection and Protection of Investment, by Deepsea Ventures Inc. Filed with the Secretary of State of the United States of America, November 15, 1974, 27 pp.

Deffeyes, K. S., 1970. The axial valley: a steady state feature of the terrain. In: H. Johnson and B. L. Smith (Editors), The Megatectonic of Continents and Oceans. Rutgers University Press, New Jersey, pp. 194—222.

Degenhardt, H., 1957. Untersuchungen zur geochemischen Verteilung des Zirkoniums in die Lithosphäre. Geochim. Cosmochim. Acta, 11: 279—309.

Degens, E. T., Von Herzen, R. P., Wong, H. K., Deuser, W. G. and Jannasch, H. W., 1973. Lake Kivu: Structure, chemistry and biology of an East African rift lake. Geol. Rundsch., 62: 245—277.

De Wolff, P. M., 1959. Interpretation of some γ-MnO_2 diffraction patterns. Acta Cryst., 12: 341—345.

Diaz, C. M., 1958. Mechanism for the Segregation Processes. M.Sc. Thesis, Department of Mineral Engineering, Columbia Univ., 59 pp. (unpublished).

Dietz, R. S., 1955. Manganese deposits on the Northeast Pacific sea floor. Calif. J. Mines Geol.; 51: 209—220.

Dmitriev, L., Barsukov, V. and Udintsev, G., 1971. Rift zones of the ocean and the problem of ore-formation. Proc. IMA—IAGOD Mtg. '70, IAGOD Volume. Society of Mining Geologists of Japan, Tokyo, pp. 65—69.

Doff, D. H., 1970. The geochemistry of recent oxic and anoxic sediments of Oslo Fjord, Norway. Ph.D. Thesis, Univ. of Edinburgh, 345 pp. (unpublished).

Dooley, J. R., Granger, H. C. and Rosholt, J. N., 1966. Uranium-234 fractionation in the sandstone type uranium deposits of the Ambrosia Lake District, New Mexico. Econ. Geol., 61: 1362—1382.

Dor, A. A., 1972. Nickel Segregation. Am. Inst. Min. Metall. Petrol. Eng., Extractive Metall. Div., New York, N.Y., 310 pp.

Drechsler, H. D., 1973. Exploitation of the sea: a preliminary cost-benefit analysis of nodule mining and processing. Marit. Stud. Management, 1: 53—66.

Drosdoff, M. and Nikiforoff, C. C., 1940. Iron-manganese concretions in Dayton soils. Soil Sci., 49: 333—345.

Duchart, P., Calvert, S. E. and Price, N. B., 1973. Distribution of trace metals in the pore waters of shallow water marine sediments. Limnol. Oceanogr., 18: 605—610.

Dudley, W. C. and Margolis, S. V., 1974. Iron and trace element concentration in marine manganese nodules by benthic agglutinated Foraminifera. Abstr. Progr. Ann. Mtg. Geol. Soc. Am., 6(7): 716 (Abstr.).

Dugdale, R. C., 1967. Nutrient limitation in the sea: dynamics, identification and significance. Limnol. Oceanogr., 12: 685—695.

Dugger, D. L., Stanton, J. H., Irby, B. N., McConnell, B. L., Cummings, W. W. and Maatman, R. W., 1964. The exchange of twenty metal ions with the weakly acidic silanol groups of silica gel. J. Phys. Chem., 68: 757—760.

Dunham, A. C. and Glasby, G. P., 1974. Petrographic and electron microprobe investigation of some deep- and shallow-water manganese nodules. N.Z. J. Geol. Geophys., 17: 929—953.

Duval, J. E. and Kurbatov, M. H., 1952. The adsorption of cobalt and barium ions by hydrous ferric oxide at equilibrium. J. Phys. Chem., 56: 982—984.

Dyck, W., 1968. Adsorption and coprecipitation of silver on hydrous ferric oxide. Can. J. Chem., 46: 1441—1444.

Dymond, J. R., 1966. Potassium-argon geochronology of deep-sea sediments. Science, 152: 1239—1241.

Dymond, J., Corliss, J. B., Heath, G. R., Field, C. W., Dasch, E. J. and Veeh, H. H., 1973. Origin of metalliferous sediments from the Pacific Ocean. Bull. Geol. Soc. Am., 84: 3355—3372.

Ealey, P. J. and Knox, G. J., 1975. The pre-Tertiary rocks of SW Cyprus. Geol. Mijnb., 54: 85—100.

Edgington, D. N. and Callender, E., 1970. Minor element geochemistry of Lake Michigan ferromanganese nodules. Earth Planet. Sci. Lett., 8: 97—100.

Edzwald, J. K., Upchurch, J. B. and O'Melia, C. R., 1974. Coagulation in estuaries. Environ. Sci. Technol., 8: 58—63.

Ehrlich, A. M., 1968. Rare Earth Abundances in Manganese Nodules. Ph.D. Thesis, Massachusetts Institute of Technology, Cambridge, Mass., 225 pp. (unpublished).

Ehrlich, H. L., 1963. Bacteriology of manganese nodules, I. Bacterial action on manganese in nodule enrichments. Appl. Microbiol., 11: 15—19.

Ehrlich, H. L., 1964. Microbial transformations of minerals. In: H. Heukelekian and N. C. Dondero (Editors), Principles and Applications in Aquatic Microbiology. Wiley, New York, N.Y., pp. 43—60.

Ehrlich, H. L., 1966. Reactions with manganese by bacteria from marine ferro-manganese nodules. Dev. Ind. Microbiol., 7: 279—286.

Ehrlich, H. L., 1968. Bacteriology of manganese nodules, II. Manganese oxidation by cell-free extract from a manganese nodule bacterium. Appl. Microbiol., 16: 197—202.

Ehrlich, H. L., 1971. Bacteriology of manganese nodules, V. Effect of hydrostatic pressure on bacteria oxidation of Mn^{II} and reduction of MnO_2. Appl. Microbiol., 21: 306—310.

Ehrlich, H. L., 1972. The role of microbes in manganese nodule genesis and degradation. In: D. R. Horn (Editor), Ferromanganese Deposits on the Ocean Floor. National Science Foundation, Washington, D.C., pp. 63—69.

Einsele, W., 1938. Über chemische und colloidchemische Vorgänge in Eisen-Phosphat-Systemem unter limnochemischen und limnogelogischen Gesichtspunkten. Arch. Hydrobiol., 33: 361—387.

Einsele, W. and Vetter, H., 1938. Untersuchungen über die Entwicklung der physikalischen und chemischen Verhältnisse in Jahrzyklus in einem mässig eutrophen See (Sehleinsee bei Langlenargen). Int. Rev. Hydrobiol., 36: 285—324.

Eittreim, S. L., 1970. Suspended particulate matter in the deep waters of the northwest Atlantic Ocean. Unpubl. Ph.D. Thesis, Lamont-Doherty Geological Observatory, Colombia Univ.

Elder, J. W., 1965. Physical processes in geothermal areas. In: Terrestrial Heat Flow. Am. Geophys. Union Monogr., 8: 211—239.

Elderfield, H., 1972a. Compositional variations in the manganese oxide component of marine sediments. Nature Phys. Sci., 237: 110—112.

Elderfield, H., 1972b. Effects of volcanism on water chemistry, Deception Island, Antarctica. Mar. Geol., 13: M1—M6.

Elderfield, H., 1976a. Hydrogenous material in marine sediments; excluding manganese nodules. In: J. P. Riley and R. Chester (Editors), Chemical Oceanography, Vol. 5. Academic Press, London, pp. 137—215.

Elderfield, H., 1976b. Manganese fluxes to the oceans. Mar. Chem., 4(2): 103—132.

Elderfield, H., Gass, I. G., Hammond, A. and Bear, L. G., 1972. The origin of ferromanganese sediments associated with the Troodos Massif of Cyprus. Sedimentology, 19: 1—19.

El-Sayed, S. Z., 1970. Phytoplankton production of the South Pacific and Pacific sector of the Antarctic. In: W. S. Wooster (Editor), Scientific Exploration of the South Pacific. National Academy of Science, Washington, D.C., pp. 194—210.

El Wakeel, S. K., 1964. Chemical and mineralogical studies of siliceous earth from Barbados. J. Sediment. Petrol., 34: 687—690.

El Wakeel, S. K. and Riley, J. P., 1961. Chemical and mineralogical studies of deep-sea sediments. Geochim. Cosmochim. Acta, 25: 110—146.

El Wakkad, S. E. S. and Rizk, H. A., 1957. The polytungstates and the colloidal nature and the amphoteric character of tungstic acid. J. Phys. Chem., 61: 494—497.

Emery, K. O., 1969. A Coastal Pond. Elsevier, New York, N.Y., 80 pp.

Emiliani, C. and Millman, J. D., 1966. Deep-sea sediments and their geological record. Earth Sci. Rev., 1: 105—132.

Eppley, R. W., 1972. Temperature and phytoplankton growth in the sea. Fish. Bull., 70: 1063—1085.

Eppley, R. W. and Thomas, W. H., 1969. Comparison of half-saturation constants for growth and nitrate uptake of marine phytoplankton. J. Phycol., 5: 375—379.

Eppley, R. W., Rogers, J. N. and McCarthy, J. J., 1969. Half-saturation constants for uptake of nitrate and ammonium by marine phytoplankton. Limnol. Oceanogr., 14: 912—920.

Eppley, R. W., Renger, E. H., Venrick, E. I. and Mullin, M. M., 1973. A study of plankton dynamics and nutrient cycling in the Pacific Ocean. Limnol. Oceanogr., 18: 534—551.

Ernst, W. G., 1969. Earth Materials. Prentice-Hall, Englewood Cliffs, N.J., 150 pp.

Evernden, J. F., Savage, D. E., Curtis, G. H. and James, G. T., 1964. Potassium argon dates and the Cenozoic mammalian chronology of North America. Am. J. Sci., 262: 383—393.

Ewing, M. and Thorndike, E. M., 1965. Suspended matter in deep ocean water. Science, 147: 1291—1294.

Ewing, M., Horn, D., Sullivan, L., Aitken, T. and Thorndike, E., 1971. Photographing manganese nodules on the ocean floor. Oceanol. Int., 6(12): 26—27, 30—32.

Fan, L., 1967. Turbulent buoyant jets into stratified or flowing ambient fluids. Rep. No. KH-R-15, W.M. Keck Lab. Hydraulics Water Resources, California Institute of Technology.

Faulring, G. M., 1962. A study of Cuban todorokite. In: M. Mueller (Editor), Advances in X-Ray Analysis. Plenum, New York., N.Y., 5, pp. 117—126.

Faulring, G. M., 1965. Unit cell determination and thermal transformation of nsutite. Am. Miner., 50: 170—179.

Feitknecht, W. and Marti, W., 1945. Über die Oxydation von Mangan(II) Hydroxid mit molekularem Sauerstoff. Helv. Chim. Acta, 28: 129—148.

Feitknecht, W. and Marti, W., 1945. Über Manganite and künstlichen Braunstein. Helv. Chim. Acta, 28: 149—157.

Feitknecht, W., Giovanoli, R., Michaelis, W. and Müller, M., 1973. Über die Hydrolyse von Eisen(III) Salzlösungen, 1. Die Hydrolyse der Lösungen von Eisen(III) Chlorid. Helv. Chim. Acta, 56: 2847—2856.

Fewkes, R. H., 1972. Conglomerate manganese nodules from the Drake Passage. Acta Miner. Petrogr., Szeged., 20: 386 (Abstr.).

Fewkes, R. H., 1973. External and internal features of marine manganese nodules as seen with SEM and their implications in nodule origin. In: M. Morgenstein (Editor), Papers on the Origin and Distribution of Manganese Nodules in the Pacific and Prospects for Exploration. Hawaii Inst. of Geophysics, Honolulu, pp. 21—29.

Finkelman, R. B., 1970. Magnetic particles extracted from manganese nodules: suggested origin from stony and iron meteorites. Science, 167: 982—984.

Finkelman, R. B., 1972. Relationship between manganese nodules and cosmic spherules. Mar. Technol. Soc., J., 9: 34—39.

Finkelman, R. B., Matzko, J. J., Woo, C. C., White, J. S. and Brown, W. R., 1972. A scanning electron microscopy study of minerals in geodes from Chihuahua, Mexico. Miner. Rec., 3: 205—212.

Finkelman, R. B., Evans, H. T. and Matzko, J. J., 1974. Manganese minerals in geodes from Chihuahua, Mexico. Miner. Mag., 39: 549—558.

Flaningham, O. L., 1960. The Electrokinetic Properties of Goethite. Thesis, Mich. Coll. Mining and Technology (unpublished).

Fleischer, M., 1960. Studies of the manganese oxide minerals, III. Psilomelane. Am. Miner., 45: 176—187.

Fleischer, M., 1966. Index of new mineral names, discredited minerals, and changes of mineralogical nomenclature in volumes 1—50 of the American Mineralogist. Am. Miner., 51: 1247—1357.

Fleischer, M., 1971. Glossary of Mineral Species. Mineral Record Inc., 103 pp. (Corrections and additions in: Mineral. Rec., 3: 140—142 (1972); 4: 244—246 (1973).)

Fleischer, M. and Richmond, W. E., 1943. The manganese oxide minerals: a preliminary report. Econ. Geol., 38: 269—286.

Fleischer, R. L., Price, P. B. and Walker, R. M., 1965. Tracks of charged particles in solids. Science, 149: 383—393.

Fleischer, R. L., Jacobs, I. S., Schwarz, W. M., Price, P. B. and Goodell, H. G., 1968. Search for multiply charged Dirac magnetic poles. Gen. Elect. Res. Dev. Cent. Tech. Inf. Ser., 68-C-356: 1—9.

Flipse, J. E., 1969. An engineering approach to ocean mining. Prepr. 1969 Offshore Technol. Conf., 1: 317—332.

Floyd, P. A., 1972. Geochemistry, origin and tectonic environment of the basic and acidic rocks of Cornubia, England. Proc. Geol. Assoc., 83: 385—404.

Fogg, G. E., 1975. Primary productivity. In: J. P. Riley and G. Skirrow (Editors), Chemical Oceanography, Vol. 2. Academic Press, London, 2nd ed., pp. 385—453.

Fomina, L. S. and Volkov, I. I., 1969. Rare earths in iron-manganese concretions of the Black Sea. Dokl. Akad. Nauk SSSR, 185: 188—191 (English transl. pp. 158—161).

Forward, F. A., Samis, C. S. and Kudryk, V., 1948. A method for adapting the ammonia-leaching process to the recovery of copper and nickel from sulfide ore and concentrate. Trans. Can. Inst. Min. Metall., 51: 181—186.

Foster, A. R., 1970. Marine manganese nodules: nature and origin of internal features. M.S. Thesis, Washington State Univ., Pullman, Wash., 131 pp. (unpublished).

Franssen, H. T., 1974. Understanding the ocean science debate. Ocean. Dev. Int. Law, 2: 187—202.

Frazer, J. Z. and Arrhenius, G., 1972. World-wide distribution of ferromanganese nodules and element concentration in selected Pacific nodules. Tech. Rep. Off. Int. Decade Ocean Explor., 2: 51 pp.

Friedheim, R. L., 1969. Understanding the Debate on Ocean Resources. Law of the Sea Institute, Kingston, 65 pp.

Friedrich, G. H., Kunzendorf, H. and Plüger, W. L., 1973. Geochemical investigation of deep sea manganese nodules from the Pacific on board R/V Valdivia — an application of the EDX-technique. In: M. Morgenstein (Editor), Papers on the Origin and Distribution of Manganese Nodules in the Pacific and Prospects for Exploration. Hawaii Inst. of Geophysics, Honolulu, pp. 31—43.

Friedrich, G., Rosner, B. and Demirsoy, S., 1969. Erzmikroskopische und mikroanalytische Untersuchungen an Manganerzkonkretionen aus den Pazifischen Ozean. Miner. Deposita, 4: 298—307.

Frink, C. R., 1969a. Fractionation of phosphorus in lake sediments: analytical evaluation. Proc. Soil Sci. Soc. Am., 33: 326—328.

Frink, C. R., 1969b. Chemical and mineralogical characteristics of eutrophic lake sediments. Proc. Soil Sci. Soc. Am., 33: 369—372.

Frondel, C., Marvin, U. B. and Ito, J., 1960a. New data on birnessite and hollandite. Am. Miner., 45: 871—875.

Frondel, C., Marvin, U. B. and Ito, J., 1960b. New occurrences of todorokite. Am. Miner., 45: 1167—1173.

Fuerstenau, D. W., Herring, A. P. and Hoover, M., 1973. Characterization and extraction of metals from sea floor manganese nodules. Trans. Soc. Min. Eng. AIME, 254: 205—211.

Fukai, R., 1968. A spectrophotometric method for determination of cobalt in sea-water after enrichment with solid manganese dioxide. J. Oceanogr. Soc. Jap., 24: 265—274.

Fukai, R., Huynh-Ngoc, L. and Vas, D., 1966. Determination of trace amounts of cobalt in sea-water after enrichment with solid manganese dioxide. Nature, 211: 726—727.

Fuller, H. C., 1966. Recovery of manganese sulfate crystals from solution by submerged combustion evaporation and by thermal crystallization. U.S. Bur. Min. Rep. Invest., 6762: 30 pp.

Fuller, H. C. and Edlund, V. E., 1966. Decomposition of manganese sulfate by a partial reduction process. U.S. Bur. Min. Rep. Invest., 6794: 18 pp.

Gabe, D. R. and Gal'perina, A. M., 1965. The development of the microzonal mud profile in the absence of microflora. In: B. V. Perfil'ev and others (Editors), Applied Capillary Microscopy. Consultants Bureau, New York, N.Y., pp. 110—116.

Gager, H. M., 1968. Mössbauer spectra of deep-sea iron-manganese nodules. Nature, 220: 1021—1023.

Gallagher, K. J., 1970. The atomic structure of tubular sub-crystals of β-iron(III) oxide hydroxide. Nature, 226: 1225—1228.

Garland, C. and Hagerty, R., 1972. Environmental planning considerations for deep-ocean mining. Prep. 8th Ann. Mar. Technol. Mtg., Washington, D.C.

Garrels, R. M. and Christ, C. L., 1965. Solutions, Minerals and Equilibria. Harper and Row, New York, N.Y., 450 pp.

Garrels, R. M. and Thompson, M. E., 1962. A chemical model for sea water. Am. J. Sci., 260: 57—66.

Garrison, R. E., 1973. Space-time relations of pelagic limestones and volcanic rocks, Olympic Peninsula, Washington. Bull. Geol. Soc. Am., 84: 583—594.

Gattow, G. and Glemser, O., 1961a. Über Manganoxyde VIII. Darstellung und Eigenschaften von Brausteinen, II. (Die γ- und η-Gruppe der Braunsteine.) Z. Anorg. Allg. Chem., 309: 20—36.

Gattow, G. and Glemser, O., 1961b. Über Manganoxyde (IX). Darstellung und Eigenschaften von Braunsteinen III. (Die ϵ-, β-, and α-Gruppe der Braunsteine über Ramsdellit und über die Umwandlungen der Braunsteine.) Z. Anorg. Allg. Chem., 309: 121—150.

Geiger, Th., 1948. Manganerze in den Radiolariten Graubündens. Beitr. Geol. Schweiz. Geotech., 27: 1—89.

Georgescu, I. I. and Lupan, S., 1971. Contributions to the study of the ferromanganese concretions from the Black Sea. Rev. Roum. Geol. Geophys. Geogr. Sci. Geol., 15: 157—163.

Georgescu, I. I. and Nistor, C., 1970. Study of the iron chemical bond by the Mössbauer effect in the ferromanganese concretions of the Black Sea. Rev. Roum. Phys., 15: 819—823.

Georgescu, I. I. and Nistor, C., 1973. A study of the iron chemical bond by the Mössbauer effect in the ferromanganese concretions of the Black Sea. Rapp. Comm. Int. Mer. Medit., 21: 363—365.

Georgescu, I. I., Morariu, M. and Diamandescu, L., 1973. Study of iron-manganese nodules from the Black Sea by Mössbauer spectroscopy. Rev. Roum. Phys., 18: 401—404 (cf. also: C.A. 79: 68820d, 1973).

Germann, K., 1971. Mangan-Eisen-führende Knollen und Krusten in jurassischen Rotkalken der Nordlichen Kalkalpen. Neues Jb. Miner. Paläont. Monatsh., 1971: 133—156.

Germann, K., 1972. Verbreitung und Entstehung Mangan-reicher Gesteine im Jura der Nördlichen Kalkalpen. Tschermaks Miner. Petrogr. Mitt., 17: 123—150.

Ghosh, S. K. and Dean, W. E., 1974. Factors contributing to precipitation of major, minor and trace elements in ferromanganese nodules and associated sediments, Oneida Lake, N.Y. Abstr. Progr. Ann. Mtg. Geol. Soc. Am., 6(7): 751—752 (Abstr.).

Gibbs, R. J., Matthews, M. D. and Link, D. A., 1971. The relation between sphere size and settling velocity. J. Sediment. Petrol., 41: 7—18.

Giese, R. F., Weller, S. and Datta, P., 1971. Electrostatic energy calculations of diaspore (α-AlOOH), geothite (α-FeOOH), and groutite (α-MnOOH). Z. Kristallogr. Kristallgeom., 134: 275—284.

Gillette, N. J., 1961. Oneida Lake pancakes. N.Y. State Conserv., 18: 41.

Giovanoli, R., 1969. A simplified scheme for polymorphism in the manganese dioxides. Chimia, 23: 470—472.

Giovanoli, R., 1972. Discussion of paper of Okamoto et al., 1972: Characterization and phase transformation of amorphous ferric hydroxide. In: J. S. Anderson, M. W. Roberts, and F. S. Stone (Editors), Reactivity of Solids. Proc. I.S.R.S., 7th, Bristol, pp. 341—350; 352—353.

Giovanoli, R. and Stähli, E., 1970. Oxide und Oxidhydroxide des drei- und vierwertigen Mangans. Chimia, 24: 49—61.

Giovanoli, R., Maurer, R. and Feitknecht, W., 1967. Zur Struktur des γ-MnO_2. Helv. Chim. Acta, 50: 1073—1080.

Giovanoli, R., Stähli, E. and Feitknecht, W., 1969. Über Struktur und Reaktivität von Mangan(IV) Oxiden. Chimia, 23: 264—266.

Giovanoli, R., Stähli, E. and Feitknecht, W., 1970a. Über Oxidhydroxide des vierwertigen Mangans mit Schichtengitter, 1. Mitteilung: Natriummangan (II, III)-Manganat (IV). Helv. Chim. Acta, 53: 209—220.

Giovanoli, R., Stähli, E. and Feitknecht, W., 1970b. Über Oxidhydroxide des vierwertigen Mangans mit Schichtengitter, 2. Mangan(III)-Manganat(IV). Helv. Chim. Acta, 53: 453—464.

Giovanoli, R., Feitknecht, W. and Fischer, F., 1971. Über Oxidhydroxide des vierwertigen Mangans mit Schichtengitter, 3. Reduktion von Mangan(III)-Manganat (IV) mit Zimtalkohol. Helv. Chim. Acta, 54: 1112—1124.

Giovanoli, R., Burki, P. and Scheiss, P., 1973a. Report No. 33a: Investigation of manganese nodules. Universität Bern, 12 pp. (English translation of introduction).

Giovanoli, R., Bühler, H. and Sokolaowska, K., 1973b. Synthetic lithiophorite: electron microscopy and X-ray diffraction. J. Microsc., 18: 271—284.

Glasby, G. P., 1970. The Geochemistry of Manganese Nodules and Associated Pelagic Sediments from the Indian Ocean. Ph.D. Thesis, Univ. of London, 674 pp. (unpublished).

Glasby, G. P., 1972a. The mineralogy of manganese nodules from a range of marine environments. Mar. Geol., 13: 57—72.

Glasby, G. P., 1972b. The geochemistry of manganese nodules from the Northwest Indian Ocean. In: D. R. Horn (Editor), Ferromanganese Deposits on the Ocean Floor. National Science Foundation, Washington, D.C., pp. 93—104.

Glasby, G. P., 1972c. The nature of the iron oxide phase of marine manganese nodules. N.Z. J. Sci., 15: 232—239.

Glasby, G. P., 1972d. Effect of pressure on deposition of manganese oxides in the marine environment. Nature Phys. Sci., 237: 85—86.

Glasby, G. P., 1973a. Mechanisms of enrichment of the rarer elements in marine manganese nodules. Mar. Chem., 1: 105—125.

Glasby, G. P., 1973b. Manganese deposits of variable composition from north of the Indian-Antarctic Ridge. Nature Phys. Sci., 242: 106—108.

Glasby, G. P., 1973c. Distribution of manganese nodules and lebensspuren in underwater photographs from the Carlsberg Ridge, Indian Ocean. N.Z. J. Geol. Geophys., 16: 1—17.

Glasby, G. P., 1973d. The role of submarine volcanism in controlling the genesis of marine manganese nodules. Oceanogr. Mar. Biol. A. Rev., 11: 27—44.

Glasby, G. P., 1974a. A geochemical study of the manganese ore deposits of the Harlech Dome, North Wales. J. Earth Sci., Leeds, 8(3): 445—50.

Glasby, G. P., 1974b. Mechanisms of incorporation of manganese and associated trace elements in marine manganese nodules. Oceanogr. Mar. Biol. A. Rev., 12: 11—40.

Glasby, G. P., 1975. Minor element enrichment in manganese nodules relative to seawater and marine sediments. Naturwissenschaften, 62: 133—135.

Glasby, G. P., 1976. Manganese nodules in the South Pacific: A review. N.Z. J. Geol. Geophys., 19: 707—736.

Glasby, G. P. and Hodgson, G. W., 1971. The distribution of organic pigments in marine manganese nodules from the Northwest Indian Ocean. Geochim. Cosmochim. Acta, 35: 845—851.

Glasby, G. P. and Summerhayes, C. P., 1975. Sequential deposition of authigenic marine minerals around New Zealand: Paleo-environmental significance. N.Z. J. Geol. Geophys., 18: 477—490.

Glasby, G. P., Tooms, J. S. and Cann, J. R., 1971. The geochemistry of manganese encrustations from the Gulf of Aden. Deep-Sea Res., 18: 1179—1187.

Glasby, G. P., Bäcker, H., Meylan, M. A., McDougall, J. C. and Singleton, R. J., 1974. Extensive manganese nodule province discovered in the Southwest Pacific near New Zealand. Meerestech. Mar. Technol., 5: 145—147.

Glassley, W., 1974. Geochemistry and tectonics of the Crescent volcanic rocks, Olympic Peninsula, Washington. Bull. Geol. Soc. Am., 85: 785—794.

Glemser, O. and Meisiek, H., 1957. Reine synthetische Braunsteine. Naturwissenschaften, 44: 614.

Glemser, O., Gattow, G. and Meisiek, H., 1961. Über Manganoxyde VII. Darstellung und Eigenschaften von Braunsteinen, 1 (Die δ-Gruppe der Braunsteine). Z. Anorg. Allg. Chem., 309: 1—36.

Globus, A. R., 1963. Method of Producing Manganese Sulfate. U.S. Patent, 3,106,451.

Goel, P. S., Kharkar, D. P., Lal, D., Narsappaya, N., Peters, B. and Yatirajam, V., 1957. The beryllium-10 concentration in deep-sea sediments. Deep-Sea Res., 4: 202—210.

Goldberg, E. D., 1954. Marine geochemistry, I. Chemical scavengers of the sea. J. Geol., 62: 249—265.

Goldberg, E. D., 1959. The processes regulating the composition of sea water. Int. Oceanogr. Congr., Am. Assoc. Adv. Sci., pp. 70—71 (preprint).

Goldberg, E. D., 1961a. Chemistry in the oceans. In: M. Sears (Editor), Oceanography. Am. Assoc. Adv. Sci. Publ., 67: 583—597.

Goldberg, E. D., 1961b. Chemical and mineralogical aspects of deep-sea sediments. Phys. Chem. Earth, 4: 281—302.

Goldberg, E. D., 1963a. The oceans as a chemical system. In: M. N. Hill (Editor), The Sea, Vol. 2. Interscience, New York, N.Y., pp. 3—25.

Goldberg, E. D., 1963b. Mineralogy and chemistry of marine sedimentation. In: F. P. Shepard (Editor), Submarine Geology. Harper and Row, New York, N.Y., 2nd ed., pp. 436—446.

Goldberg, E. D., 1965. Minor elements in sea water. In: J. P. Riley and G. Skirrow (Editors), Chemical Oceanography, Vol. 1. Academic Press, London, pp. 163—196.

Goldberg, E. D., 1967. Review of Trace Element Concentrations in Marine Organisms. Puerto Rico Nuclear Center, San Juan, Puerto Rico, 535 pp.

Goldberg, E. D. and Arrhenius, G., 1958. Chemistry of Pacific pelagic sediments. Geochim. Cosmochim. Acta, 13: 153—212.

Goldberg, E. D. and Koide, M., 1962. Geochronological studies of deep-sea sediments by the ionium/thorium method. Geochim. Cosmochim. Acta, 26: 417—450.

Goldberg, E. D., Koide, M., Schmitt, R. and Smith, R., 1963. Rare earth distributions in the marine environment. J. Geophys. Res., 68: 4209—4217.

Goldberg, E. D., Broecker, W. S., Gross, M. G. and Turekian, K. K., 1971. Marine chemistry. In: Radioactivity in the Marine Environment. National Academy of Sciences, Washington, D.C., pp. 137—146.

Goldie, L. F. E., 1972. Development of an international environmental law — An appraisal. In: Hargrove, J. L. (Editor), Law, Institutions and the Global Environment. Oceana, New York, N.Y., 394 pp.

Goldschmidt, V. M., 1937. The principles of distribution of chemical elements in minerals and rocks. J. Chem. Soc., pp. 655—672.

Goldschmidt, V. M., 1954. Geochemistry. Oxford University Press, London, 730 pp.

Goldsmith, J. R. and Graf, D. L., 1960. Subsolidus relations in the system $CaCO_3$—$MgCO_3$—$MnCO_3$. J. Geol., 68: 324—335.

Goncharov, G. N., Kalyamin, A. V. and Lur'e, B. G., 1973. Iron-manganese concretions from the Pacific Ocean studied by a nuclear,γ-resonance method. Dokl. Akad. Nauk SSSR, 212: 720—723. (Cf. also C.A. 80: 29270a, 1974.)

Goodell, H. G., 1965. Marine geology, USNS "Eltanin", Cruises 9—15. Contr. Sedim. Res. Lab. Fla. State Univ., 11: 196 pp.

Goodell, H. G., 1968. Ferromanganese deposits of the Southern Ocean. Spec. Pap. Geol. Soc. Am., 115: 475—476 (Abstr.).

Goodell, H. G., Meylan, M. A. and Grant, B., 1971. Ferromanganese deposits of the South Pacific Ocean, Drake Passage, and Scotia Sea. Antarct. Res. Ser., 15: 27—92.

Goodier, J. L., 1972. How manganese nodules develop. Oceanol. Int., 7(4): 45—48.

Gordon, A. L. and Gerard, R. D., 1970. North Pacific bottom potential temperature. Mem. Geol. Soc. Am., 126: 23—29.

Gordon, D. C., 1971. Distribution of particulate organic carbon and nitrogen at an oceanic station in the central Pacific. Deep-Sea Res., 18: 1127—1134.

Gorham, E. and Swaine, D. J., 1965. The influence of oxidising and reducing conditions upon the distribution of some elements in lake sediments. Limnol. Oceanogr., 10: 268—279.

Gorshkova, T. I., 1931. Chemical and mineralogical researches of the sediments of the Barents and White Seas. Trudy Gos. Okeanogr. Inst., 2—3: 83—127.

Gorshkova, T. I., 1957. Sediments of the Kara Sea. Trudy Vses. Gidrobiol. Obshch., 8: 68—99.

Gorshkova, T. I., 1966. Manganese in bottom sediments of northern seas of the U.S.S.R. and its biological significance. Proc. All-Union, Res. Inst. Mar. Fish. Oceanogr., 60: 89—102.

Gorshkova, T. I., 1967. Manganese in bottom sediments of northern Seas. In: D. G. Sapozhnikov (Editor), Manganese Deposits of the Soviet Union. Nauka, Moscow, pp. 125—145.

Graham, J. W., 1959. Metabolically induced precipitation of trace elements from sea water. Science, 129: 1428—1429.

Graham, J. W. and Cooper, S. C., 1959. Biological origin of manganese-rich deposits on the sea floor. Nature, 183: 1050—1051.

Grahame, D. C., 1947. The electrical double layer and the theory of electrocapillarity. Chem. Rev., 41: 441—501.

Grant, J. B., 1967. A comparison of the chemistry and mineralogy with the distribution and physical aspects of marine manganese concretions of the southern oceans. Contr. Sediment. Res. Lab. Fla. St. Univ., 19: 99 pp.

Grasselly, G. and Hetenyi, M., 1968. Adsorption properties of some manganese oxides. Acta Miner. Petrogr. Szeged., 18: 85—98.

Grassle, J. F., Sanders, H. L., Hessler, R. R., Rowe, G. T. and McLellan, T., 1975. Pattern and zonation: a study of the bathyal megafauna using the research submersible *Alvin*. Deep-Sea Res., 22: 457—481.

Grassoff, K., 1975. The hydrochemistry of landlocked basins and fjords. In: J. P. Riley and G. Skirrow (Editors), Chemical Oceanography, Vol. 2. Academic Press, London, 2nd ed., pp. 455—597.

Greenslate, J., 1974a. Microorganisms participate in the construction of manganese nodules. Nature, 249: 181—183.

Greenslate, J., 1974b. Manganese and biotic debris associations in some deep-sea sediments. Science, 186: 529—531.

Greenslate, J. L., Frazer, J. Z. and Arrhenius, G., 1973. Origin and deposition of selected transition elements in the seabed. In: M. Morgenstein (Editor), Papers on the Origin and Distribution of Manganese Nodules in the Pacific and Prospects for Exploration. Hawaii Inst. of Geophysics, pp. 45—70.

Griffin, J. J., Windom, H. and Goldberg, E. D., 1968. The distribution of clay minerals in the world ocean. Deep-Sea Res., 15: 433—459.

Grill, E. V., Murray, J. W. and MacDonald, R. D., 1968a. Manganese nodules from Jervis Inlet, a British Columbia fjord. Syesis, 1: 57—63.

Grill, E. V., Murray, J. W. and MacDonald, R. D., 1968b. Todorokite in manganese nodules from a British Columbia fjord. Nature, 219: 358—359.

Grim, R. E., 1968. Clay Mineralogy. McGraw-Hill, New York, N.Y., 596 pp.

Grimme, H., 1969. Die Adsorption von Mn, Co, Cu, and Zn durch Goethit aus verdünnten Lösungen. Z. Pflanzenernähr. Düng. Bodenk., 121: 58—65.

Gripenberg, S., 1934. A study of the sediment of the North Baltic and adjoining seas. Merentutkimuslait. Julk. Skift, 96: 231.

Grütter, A. and Buser, W., 1957. Untersuchungen an Mangansedimenten. Chimia, 11: 132—133.

Gulbrandsen, R. A. and Reeser, D. W., 1969. An occurrence of Permian manganese nodules near Dillon, Montana. Prof. Pap. U.S. Geol. Surv., 650C: 49—57.

Gümbel, C. W., 1861. Geognostische Beschreibung des Bayerischen Alpengebirges und seines Vorlandes, Vol. 1. Perthes, Gotha, 950 pp.

Gümbel, W., 1878. Über die im stillen Ozean auf dem Meeresgrunde vorkommenden Manganknollen. Sitzungsber. Bayer. Akad. Wiss. Math-Phys. K., 8: 189—209.

Hahn, H. and Stumm, W., 1968. Coagulation by Al(III): the role of adsorption of hydrolyzed aluminum in the kinetics of coagulation. Adv. Chem. Ser., 79: 91—111.

Hallberg, R. O., 1972. Iron and zinc sulfides formed in a continuous culture of sulfate-reducing bacteria. Neues Jb. Miner. Monatsh, 11: 481—500.

Hamilton, E. L., 1956. Sunken islands of the Mid-Pacific mountains. Mem. Geol. Soc. Am., 64: 1—97.

Hammond, A. L., 1974. Manganese nodules (II): prospects for deep sea mining. Science, 183: 644—646.

Han, K. N., 1971. Geochemistry and Extraction of Metals from Ocean Floor Manganese Nodules. Ph.D. Thesis, Univ. of California, Berkeley, Calif., 212 pp. (unpublished).

Han, K. N. and Fuerstenau, D. W., 1973. Behavior of metal ions during extraction from ocean floor manganese nodules. Proc. 1st Aust. Heat Mass Trans. Conf., Sec. 6, pp. 41—48.

Han, K. N. and Fuerstenau, D. W., 1975a. Preferential acid leaching of nickel, copper and cobalt from ocean floor manganese nodules. Trans. Inst. Min. Metall., 84C: 105—110.

Han, K. N. and Fuerstenau, D. W., 1975b. Acid leaching of ocean manganese nodules at elevated temperatures. Int. J. Miner. Process., 2: 163—171.

Han, K. N., Hoover, M. and Fuerstenau, D. W., 1974. Ammonia-ammonium leaching of deep-sea manganese nodules. Int. J. Miner. Process., 1: 215—230.

Harder, E. C., 1910. Manganese deposits of the United States. Bull. U.S. Geol. Surv., 427: 1—298.

Harder, E. C., 1919. Iron-depositing bacteria and their geologic relations. Prof. Pap. U.S. Geol. Surv., 113: 1—89.

Harder, H. and Menschel, G., 1967. Quartz formation on the ocean floor. Naturwissenschaften, 54: 561.

Hare, P. E. and Abelson, P. H., 1968. Racemization of amino acids in fossil shells. Carnegie Inst. Wash. Yearbook, 66: 526—528.

Hare, P. E. and Mitterer, R. M., 1967. Non-protein amino acids in fossil shells. Carnegie Inst. Wash. Yearbook, 65: 362—364.

Hariya, Y., 1961. Mineralogical studies on todorokite and birnessite from the Todoroki mine, Hokkaido. Jap. J. Assoc. Miner. Petrol. Econ. Geol., 45: 219—230.

Harland, W. B., Smith, A. G. and Wilcock, B. (Editors), 1964. The Phanerozoic Time-scale, a Symposium. Supplement to Q.J. Geol. Soc., Lond., 120: 458 pp.

Harriss, R. C., 1968. Mercury content of deep sea manganese nodules. Nature, 219: 54—55.

Harriss, R. C. and Troup, A. G., 1969. Freshwater ferromanganese concretions: chemistry and internal structure. Science, 166: 604—606.

Harriss, R. C. and Troup, A. G., 1970. Chemistry and origin of freshwater ferromanganese concretions. Limnol. Oceanogr., 15: 702—712.

Harriss, R. C., Crocket, J. H. and Stainton, M., 1968. Palladium, iridium and gold in deep-sea nodules. Geochim. Cosmochim. Acta, 32: 1049—1056.

Hart, R., 1970. Chemical exchange between sea water and deep ocean basalts. Earth Planet. Sci. Lett., 9: 269—279.

Hart, R. A., 1973a. A model for chemical exchange in the basalt—sea water system of oceanic layer II. Can. J. Earth Sci., 10: 799—816.

Hart, R. A., 1973b. Geochemical and geophysical implications of the reaction between sea water and the oceanic crust. Nature, 243: 76—78.

Harter, R. D., 1968. Adsorption of phosphorus by lake sediment. Proc. Soil Sci. Soc. Am., 32: 514—518.

Hartmann, M., 1964. Zur Geochemie von Mangan und Eisen in der Ostsee. Meyniana, 14: 3—21.

Harvey, H. W., 1937. Note on colloidal ferric hydroxide in sea water. J. Mar. Biol. Ass. U.K., 22: 221—225.

Haskin, L. A. and Gehl, M. A., 1962. The rare-earth distribution in sediments. J. Geophys. Res., 67: 2537—2541.

Haskin, M. A. and Haskin, L. A., 1966. Rare-earths in European shales: a re-determination. Science, 154: 507.

Hasle, G. R., 1959. A quantitative study of phytoplankton from the equatorial Pacific. Deep-Sea Res., 6: 38—59.

Hawkins, L. K., 1969. Visual observations of manganese deposits on the Blake Plateau. J. Geophys. Res., 74: 7009—7017.

Hays, J. D., Saito, T., Opdyke, N. D. and Burckle, L. R., 1969. Pliocene—Pleistocene sediments of the equatorial Pacific, their palaeomagnetic, biostratigraphic and climatic record. Bull. Geol. Soc. Am., 80: 1481—1514.

Hazel, F. and Ayres, G. H., 1931. Migration studies with ferric oxide sols. J. Phys. Chem., 35: 2930—2942.

Healy, T. W. and Fuerstenau, D. W., 1965. The oxide—water interface-interrelation of the zero point of charge and the heat of immersion. J. Colloid Interface Sci., 20: 376—386.

Healy, T. W., Herring, A. P. and Fuerstenau, D. W., 1966. The effect of crystal structure on the surface properties of a series of manganese dioxides. J. Colloid Interface Sci., 21: 435—444.

Healy, T. W., James, R. O. and Cooper, R., 1968. The adsorption of aqueous Co(II) at the silica—water interface. Adv. Chem. Ser., 79: 62—73.

Heath, G. R., 1974. Dissolved silica and deep-sea sediments. Spec. Publ. Soc. Econ.-Paleont. Miner., 20: 77—93.

Heezen, B. C. and Hollister, C., 1964. Deep-sea current evidence from abyssal sediments. Mar. Geol., 1: 141—174.

Heezen, B. C. and Hollister, C., 1971. The Face of the Deep. Oxford University Press, New York, N.Y., 659 pp.

Hekinian, R. and Hoffert, M., 1975. Rate of palagonitization and manganese coating on basaltic rocks from the Rift Valley in the Atlantic Ocean near 36°50'N. Mar. Geol., 19: 91—109.

Helgeson, H. C., 1964. Complexing and Hydrothermal Ore Deposition. Pergamon, New York, N.Y., 128 pp.

Hem, J. D., 1963. Chemical equilibria and rates of manganese oxidation. U.S. Geol. Surv. Water Supply Pap., 1667A: 1—64.

Hem, J. D., 1972. Chemical factors that influence the availability of iron and manganese in aqueous systems. Bull. Geol. Soc. Am., 83: 443—450.

Hem, J. D. and Skougstad, M. W., 1960. Coprecipitation effects in solutions containing ferrous, ferric and cupric ions. U.S. Geol. Surv. Water Supply Pap., 1459: 95—110.

Hendricks, R. L., Reisbick, F. B., Mahaffey, E. J., Roberts, D. B. and Peterson, M. N. A., 1969. Chemical composition of sediments and interstitial brines from the Atlantis II, Discovery and Chain Deeps. In: E. T. Degens and D. A. Ross (Editors), Hot Brines and Recent Heavy Metal Deposits in the Red Sea. Springer-Verlag, Berlin, pp. 407—440.

Herbert, D. W. M. and Merkins, J. C., 1961. The effect of suspended mineral solids on the survival of trout. Int. J. Air Water Pollut., 5: 46—55.

Hering, V. N., 1971. Metalle aus Tiefsee-Erzen. Stahl Eisen, 91: 452—459.

Herzenberg, C. L. and Riley, D. L., 1969. Interpretation of the Mössbauer spectra of marine iron-manganese nodules. Nature, 224: 259—260.

Herzenberg, D. L., 1969. Mössbauer spectrometry as an instrumental technique for determinative mineralogy. In: Mössbauer Effect Methodology. Plenum Press, New York, N.Y., 5: 209—230.

Hessler, R. R. and Jumars, P. A., 1974. Abyssal community analysis from replicate box cores in the central North Pacific. Deep-Sea Res., 21: 185—209.

Hewett, D. F., Fleischer, M. and Conklin, N., 1963. Deposits of the manganese oxides: supplement. Econ. Geol., 58: 1—51.

Hey, M. H., 1962. Cobaltic hydroxide in nature. Miner. Mag., 33: 253—259.

Heye, D. and Beiersdorf, H., 1973. Radioaktive und magnetische Untersuchungen an Manganknollen zur Ermittlung der Wachstumsgeschwindigkeit bzw. zur Altersbestimmung. Z. Geophys., 39: 703—726.

Hingston, F. J., Atkinson, R. J., Posner, A. M. and Quirk, J. P., 1967. Specific adsorption of anions. Nature, 215: 1459—1461.

Hingston, F. J., Posner, A. M. and Quirk, J. P., 1968a. Adsorption of selenite by goethite. Adv. Chem. Ser., 79: 82—90.

Hingston, F. J., Atkinson, R. J., Posner, A. M. and Quirk, J. P., 1968b. Specific adsorption of anions on goethite. Trans. Int. Congr. Soil Sci., 9th, 1: 669—678.

Hingston, F. J., Posner, A. M. and Quirk, J. P., 1970. Anion binding at oxide surfaces — the adsorptions envelope. Search, 1: 324—327.

Hingston, F. J., Posner, A. M. and Quirk, J. P., 1971. Competitive adsorption of negatively charged ligands on oxide surfaces. Discuss. Faraday Soc., 52: 334—342.

Hingston, F. J., Posner, A. M. and Quirk, J. P., 1972. Anion adsorption by goethite and gibbsite, I. The role of the proton in determining adsorption envelopes. J. Soil Sci., 23: 177—192.

Hirst, D. M., 1974. Geochemistry of sediments from eleven Black Sea cores. Mem. Am. Assoc. Petrol. Geol., 20: 430—456.

Hobson, A. and Lorenzen, C. J., 1972. Relationships of chlorophyll maxima to density structure in the Atlantic Ocean and Gulf of Mexico. Deep-Sea Res., 19: 297—306.

Hofmann, H. J., 1969. Attributes of stromatolites. Geol. Surv. Pap., Can., 69—39: 1—58.

Hollick, A. L., 1971. The law of the sea and U.S. policy initiatives. Orbis, 15: 670—686.

Hollister, C. D. and Heezen, B. C., 1967. The floor of the Bellingshausen Sea. In: J. B. Hersey (Editor), Deep-Sea Photography. The Johns Hopkins Oceanographic Studies, 3: 177—189.

Holmes, R. W., 1958. Surface chlorophyll a, surface primary production and zooplankton volumes in the eastern Pacific Ocean. Rapp. P.-V. Reun. Cons. Perm. Int. Explor. Mer, 144: 109—116.

Holmes, R. W., 1961. Summary of productivity measurements in the southeastern Pacific Ocean. In: M. S. Doty (Editor), Primary Productivity Measurement: Marine and Freshwater. IGY World Data Center, National Academy of Sciences, Washington, D.C., pp. 18—57.

Honeyman, D., 1880. Nova Scotian archaeology. Proc. Trans. Nova Scotian Inst. Sci.,
    1: 217—218.
Hoover, M., 1967. Studies on the Dissolution of Copper, Nickel and Cobalt from Ocean
    Manganese Nodules. M.S. Thesis, Univ. of California, Berkeley, Calif., 68 pp. (un-
    published).
Hoover, M., 1972. On the Mechanism and Kinetics of Chlorination of Copper, Nickel
    and Cobalt from Ocean Manganese Nodules. Ph.D. Thesis, Univ. of California,
    Berkeley, Calif., 254 pp. (unpublished).
Hoover, M., Han, K. N. and Fuerstenau, D. W., 1975. Segregation roasting of nickel, cop-
    per and cobalt from deep sea manganese nodules. Int. J. Miner. Process., 2: 173—
    185.
Horn, D. R. (Editor), 1972. Ferromanganese Deposits on the Ocean Floor. National
    Science Foundation, Washington, D.C., 293 pp.
Horn, D. R., Horn, B. M. and Delach, M. N., 1970. Sedimentary provinces of the North
    Pacific. Geol. Soc. Am. Mem., 126: 1—22.
Horn, D. R., Ewing, M., Horn, B. M. and Delach, M. N., 1972a. World-wide distribution
    of manganese nodules. Ocean Ind., 7(1): 26—29.
Horn, D. R., Horn, B. M. and Delach, M. N., 1972b. Distribution of ferromanganese
    deposits in the world ocean. In: D. R. Horn (Editor), Ferromanganese Deposits on
    the Ocean Floor. National Science Foundation, Washington, D.C., pp. 9—17.
Horn, D. R., Horn, B. M. and Delach, M. N., 1972c. Ferromanganese deposits of the
    North Pacific. Tech. Rep. Off. Int. Decade Ocean Explor., 1: 78 pp.
Horn, D. R., Delach, M. N. and Horn, B. M., 1973a. Metal content of ferromanganese
    deposits of the oceans. Tech. Rep. Off. Int. Decade Ocean Explor., 3: 51 pp.
Horn, D. R., Horn, B. M. and Delach, M. N., 1973b. Factors which control the distribu-
    tion of ferromanganese nodules and proposed research vessel's track North Pacific.
    Tech. Rep. Off. Int. Decade Ocean Explor., 8: 20 pp.
Horn, D. R., Horn, B. M. and Delach, M. N., 1973c. Ocean manganese nodules metal
    values and mining sites. Tech. Rep. Off. Int. Decade Ocean Explor., 4: 57 pp.
Horn, M. K. and Adams, J. A. S., 1966. Computer-derived geochemical balances and
    element abundances. Geochim. Cosmochim. Acta, 30: 279—297.
Hrynkiewicz, A. Z., Sawicka, B. D. and Sawicki, J. A., 1970. The Mössbauer effect in the
    Pacific Ocean Fe-Mn nodules. Phys. Stat. Sol., 3: 1039—1045.
Hrynkiewicz, A. Z., Pustówka, A. J., Sawicka, B. D. and Sawicki, J. A., 1972a. Mössbauer
    effect analysis of Fe-Mn nodules from various Pacific Ocean locations. Phys. Stat.
    Sol., 10: 281—287.
Hrynkiewicz, A. Z., Pustówka, A. J., Sawicka, B. D. and Sawicki, J. A., 1972b. Mössbauer
    study of iron-manganese nodules at high temperatures. Phys. Stat. Sol., 9: K159—
    K163.
Huang, C. P. and Stumm, W., 1973. Specific adsorption of cations on hydrous γ-Al_2O_3.
    J. Colloid Interface Sci., 43: 409—420.
Hubred, G. L., 1970. Relationship of morphology and transition metal content of manga-
    nese nodules to an abyssal hill. Hawaii Inst. Geophys. Rep., HIG-70-18:
    38 pp.
Hubred, G. L., 1973. An Extractive Metallurgy Study of Deep-Sea Manganese Nodules
    with Special Emphasis on the Sulfuric Acid Autoclave Leach. Ph.D. Thesis, Univ.
    of California, Berkeley, Calif., 194 pp. (unpublished).
Hurd, D. C., 1973. Interactions of biogenic opal, sediment and seawater in the central
    Equatorial Pacific. Geochim. Cosmochim. Acta, 37: 2257—2282.
Hurlbut, C. S., 1971. Dana's Manual of Mineralogy. Wiley, New York, N.Y., 18th ed.,
    579 pp.
Hurley, R. J., 1966. Geological studies of the West Indies. In: W. H. Poole (Editor),
    Continental Margins and Island Arcs. Geol. Surv. Pap. Can., 66—15: 139—150.

Hurst, C. J. B., 1923—1924. Whose is the bed of the sea? British Yearbook Int. Law, 4: 34—43.

Hutchinson, G. E., 1957. A Treatise on Limnology, Vol. 1. Wiley, New York, N.Y., 1015 pp.

Iimori, S., 1927. Formation of the radioactive manganiferous deposits from Tanokami and the source of manganese in the deep-sea manganese nodules. Sci. Pap. Inst. Phys. Chem. Res., Tokyo, 7: 249—252.

Immartino, N. R., 1974. Metals from Mn nodules. Chem. Eng., 81(25): 52—53.

Innes, H., 1965. The Strode Venturer. Collins, London, 320 pp.

Irvine, R. and Williams, R. J. P., 1953. The stability of transition-metal complexes. J. Chem. Soc., pp. 3192—3210.

Isaacs, J. D., Reid, J. L., Schick, G. B. and Schwartzlose, R. A., 1966. Near-bottom currents measured in 4 kilometers depth off the Baja California coast. J. Geophys. Res., 71: 4297—4303.

Isacks, B., Oliver, J. and Sykes, L., 1968. Seismology and the new global tectonics. J. Geophys. Res., 73: 5855—5899.

Ishibashi, M., 1953. Studies on minute elements in sea water. Rec. Oceanogr. Works Jap., N.S., 1: 88—92.

Ishibashi, M., Shigematsu, T. and Nakagawa, Y., 1953. Quantitative determination of tungsten and molybdenum in sea water. Proc. Pacif. Sci. Congr., 8th, 3: 817—820.

Ishibashi, M., Fujinaga, T. and Kuwamoto, T., 1962. Fundamental investigation on the dissolution and deposition of molybdenum, tungsten and vanadium in the sea. Rec. Oceanogr. Works Jap., Spec. No. 6: 215—218.

Iwasaki, I., 1972. A thermodynamic interpretation of the segregation process for copper and nickel ores. Miner. Sci. Eng., 4(2): 14—23.

Iwasaki, I., Cooke, S. R. B. and Colombo, A. F., 1960. Flotation characteristics of goethite. U.S. Bur. Mines Rep. Invest., 5593: 25 pp.

Iwasaki, I., Cooke, S. R. B. and Yim, Y. S., 1962. Some surface properties and flotation characteristics of magnetite. Trans. AIME, 223: 113—120.

Jaggar, T. A., 1940. Magmatic gases. Am. J. Sci., 238: 313—353.

James, R. O. and Healy, T. W., 1972a. Adsorption of hydrolyzable metal ions at the oxide—water interface, I. Co(II) adsorption on SiO_2 and TiO_2 as model systems. J. Colloid Interface Sci., 40: 42—52.

James, R. O. and Healy, T. W., 1972b. Adsorption of hydrolyzable metal ions at the oxide—water interface, II. Charge reversal of SiO_2 and TiO_2 colloids by adsorbed Co(II). La(II) and Th(IV) as model systems. J. Colloid Interface Sci., 40: 53—64.

James, R. O. and Healy, T. W., 1972c. Adsorption of hydrolyzable metal ions at the oxide—water interface, III. A thermodynamic model of adsorption. J. Colloid Interface Sci., 40: 65—81.

Jannasch, H. W. and Wirsen, C. O., 1973. Deep-sea microorganisms: in situ response to nutrient enrichment. Science, 180: 641—643.

Jedwab, J., 1970. Les sphérules cosmiques dans les nodules de manganèse. Geochim. Cosmochim. Acta, 34: 447—457.

Jedwab, J., 1971. Particules de materière carbonée dans les nodules de manganèse de grandes fonds océaniques. C.R. Acad. Sci., Paris, 272D: 1968—1971.

Jenkins, S. R., 1970. The Colloid Chemistry of Hydrous MnO_2 as Related to Manganese Removal. Ph.D. Thesis, Harvard Univ., Harvard (unpublished).

Jenkyns, H. C., 1967. Fossil manganese nodules from Sicily. Nature, 216: 673—674.

Jenkyns, H. C., 1970a. Fossil manganese nodules from the west Sicilian Jurassic. Eclogae Geol. Helv., 63: 741—774.

Jenkyns, H. C., 1970b. Submarine volcanism and the Toarcian iron pisolites of western Sicily. Eclogae Geol. Helv., 63: 549—572.

Jenkyns, H. C., 1971. The genesis of condensed sequences in the Tethyan Jurassic. Lethaia, 4: 327—352.

Jenkyns, H. C. and Torrens, H. S., 1971. Palaeographic evolution of Jurassic seamounts in western Sicily. In: E. Végh-Neubrandt (Editor), Colloque du Jurassique Méditerranéen. Ann. Inst. Geol. Publ. Hung., 54(2): 91—104.

Jenne, E. A., 1968. Controls on Mn, Fe, Co, Ni, Cu and Zn concentrations in soils and water: significant role by hydrous Mn and Fe oxides. Adv. Chem. Ser., 73: 337—387.

Jennings, R. Y., 1969. The limits of continental shelf jurisdiction: some possible implications of the North Sea case judgment. Int. Comp. Law Q., 18: 819—832.

Jerlov, N. G., 1964. Optical classification of ocean water. In: Physical Aspects of Light in the Sea. Univ. of Hawaii Press, Honolulu, pp. 45—59.

Jerlov, N. G., 1968. Optical Oceanography. Elsevier, Amsterdam, 194 pp.

Jewell, M. E. and Brown, H. W., 1929. Studies on northern Michigan bog lakes. Ecology, 10: 427—475.

Johansen, P. G. and Buchanan, A. S., 1957. An application of the microelectrophoresis method to the study of the surface properties of insoluble oxides. Aust. J. Chem., 10: 398—403.

Johnson, A. H. and Stoke, J. L., 1966. Manganese oxidation by *Sphaerotilus discophorus*. J. Bact., 91: 1543—1547.

Johnson, C. E. and Glasby, G. P., 1968. Mössbauer determination of particle size in microcrystalline iron-manganese nodules. Nature, 222: 376—377.

Johnson, D. A., 1972. Eastward-flowing bottom currents along the Clipperton Fracture Zone. Deep-Sea Res., 19: 253—257.

Johnson, D. G., 1969. Ferromanganese Concretions in Lake Champlain. M.S. Thesis, Vermont Univ., Burlington, Ve. (unpublished).

Joint Committee on Powder Diffraction Standards, 1974. Powder Diffraction File Search Manual, Alphabetical Listing, Inorganic 1974. Also, Sets 1—24 of the Powder Diffraction File (1960—1974).

Joint Committee on Powder Diffraction Stands (JCPDS), 1974. Selected Powder Diffraction Data for Minerals. 1st ed., 833 pp.

Joly, J., 1908. On the radium-content of deep-sea sediments. Phil. Mag., 16: 190—197.

Jones, L. H. P., 1957. The solubility of molybdenum in simplified systems and aqueous soil suspensions. J. Soil Sci., 8: 313—327.

Jones, L. H. P. and Handreck, K. A., 1963. Effects of iron and aluminium oxides on silica in solution in soils. Nature, 198: 852—853.

Jones, L. H. P. and Milne, A. A., 1956. Birnessite, a new manganese oxide mineral from Aberdeenshire, Scotland. Miner. Mag., 31: 283—288.

Jukes-Browne, A. J. and Harrison, J. B., 1892. The geology of Barbados, 2. The oceanic deposits. Q.J. Geol. Soc. Lond., 48: 170—226.

Kanamori, S., 1965. Geochemical study of arsenic in natural waters, III. The significance of ferric hydroxide precipitate in stratification and sedimentation of arsenic in lake waters. J. Earth Sci. Nagoya Univ., 13: 46—57.

Kanwisher, J., 1962. Gas exchange of shallow marine sediments. In: N. Marshall (Editor), The Environmental Chemistry of Marine Sediments. Narragansett Mar. Lab. Occas. Pub., 1: 13—19.

Kaplan, I. R. and Rittenberg, S. C., 1963. Basin sedimentation and diagenesis. In: M. M. Hill (Editor), The Sea, Vol. 3. Interscience, London, pp. 583—619.

Kasey, J. B., 1971. Process for the Selective Separation of Ferric Sulfate from Copper in a Sulfuric Acid Leach Solution. U.S. Patent, 3,586,498.

Kato, K., 1969. Behaviour of dissolved silica in connection with oxidation-reduction cycle in lake water. Geochim. J., 3: 87—97.

Kaufman, R., 1974. The selection and sizing of tracts comprising a manganese nodule ore body. Proc. 1974 Off. Technol. Conf. OTC, 2059.

Kaufman, R. and Siapno, W. D., 1972. Variability of Pacific Ocean manganese nodule deposits. In: D. R. Horn (Editor), Ferromanganese Deposits on the Ocean Floor. National Science Foundation, Washington, D.C., pp. 263—270.

Kausch, P., 1970. Der Meeresbergbau im Völkerrecht. Glückauf, Essen. 144 pp.

Kee, N. and Bloomfield, C., 1961. The solution of some minor elements by decomposing plant materials. Geochim. Cosmochim. Acta, 24: 206—225.

Keller, P., 1970. Eigenschaften von (Cl, F, OH)$_{<2}$ Fe (O, OH)$_{16}$ und Akaganéit. Neues Jahrb. Miner. Abh., 113: 29—49.

Kellogg, H. H., 1950. Thermodynamic relationships in chlorine metallurgy. Trans. Am. Inst. Min. Metal. Petrol. Eng., 188: 862—872.

Kester, D. R. and Byrne, R. H., 1972. Chemical forms of iron in sea water. In: D. R. Horn (Editor), Ferromanganese Deposits on the Ocean Floor. National Science Foundation, Washington, D.C., pp. 107—113.

Ketteridge, I. B. and Wilmshurst, R. E., 1964. Chlorination of ilmenite. Aust. J. Appl. Sci., 15: 90—105.

Kharkar, D. P., Turekian, K. K. and Bertine, K. K., 1968. Stream supply of dissolved silver, molybdenum, antimony, selenium, chromium, cobalt, and cesium to the oceans. Geochim. Cosmochim. Acta, 32: 285—298.

Kim, Y. S. and Zeitlin, H., 1969. The role of iron (III) hydroxide as a collector of molybdenum from sea water. Anal. Chim. Acta, 46: 1—8.

Kindle, E. M., 1932. Lacustrine concretions of manganese. Am. J. Sci., 24: 496—504.

Kindle, E. M., 1935. Manganese concretions in Nova Scotia lakes. Trans. R. Soc. Can., 29: 163—180.

Kindle, E. M., 1936. The occurrence of lake bottom manganiferous deposits in Canadian lakes. Econ. Geol., 31: 755—760.

Kinniburgh, D. G., 1973. Cation Sorption by Hydrous Metal Oxides. Ph.D. Thesis, Univ. of Wisconsin, Madison, Wisc. (unpublished).

Kirkman, J. H., 1973a. Amorphous inorganic materials in three soils formed from loess, 1. Application of selective dissolution techniques. N.Z. J. Sci., 16: 79—93.

Kirkman, J. H., 1973b. Amorphous inorganic materials in three soils formed from loess, 2. Amounts and distribution. N.Z. J. Sci., 16: 95—100.

Klenova, M. V., 1936a. Sediments of the Kara Sea. C.R. Acad. Sci., URSS, 4: 187—190.

Klenova, M. V., 1936b. Ob usloviyakh podvodnogo vyvetrivaniya. Akademiku V.I. Vernadskomu, V.2. Izv. Akad. Nauk SSSR.

Klenova, M. V., 1938. Colouring of Polar Sea sediments. C.R. Acad. Sci., URSS, 19: 629—632.

Klenova, M. V., 1960. Geology of the Barents Sea. Izv. Akad. Nauk SSSR, Moscow, 367 pp.

Klenova, M. V. and Pakhomova, A. S., 1940. Manganese in the sediments of Polar Seas. C.R. Acad. Sci., URSS, 28: 87—89.

Knauss, J. A., 1974. Marine science under suspicion. Mar. Technol. Soc. J., 8: 17—24.

Knight, H. G., 1971. The draft United Nations Convention on the international seabed area: Background, description and some preliminary thoughts. San Diego Law Rev., 8: 459—550.

Knight, H. G., 1972. The 1971 United States proposals on the breadth of the territorial sea and passage through international straits. Oregon Law Rev., 51: 759—787.

Koblentz-Mishke, O. J., Volkovinsky, V. V. and Kabanova, J. O., 1970. Plankton primary productivity of the world ocean. In: W. S. Wooster (Editor), Scientific Exploration of the South Pacific. National Academy of Sciences, Washington, D.C., pp. 183—193.

Korpi, G. K., 1960. Measurement of Streaming Potentials. Thesis, Massachusetts Institute of Technology, Cambridge, Mass.

Koyama, T. and Sugawara, K., 1951. Sulphate coprecipitation in lake and sea water and its concentration in the bottom deposits. J. Oceanogr. Soc. Jap., 6: 190—193.

Kozawa, A., 1959. On an ion-exchange property of manganese dioxide. J. Electrochem. Soc., 106: 552—556.

Kraemer, T. and Schornick, J. C., 1974. Comparison of elemental accumulation rates between ferromanganese deposits and sediments in the South Pacific Ocean. Chem. Geol., 13: 187—196.

Kranck, K., 1973. Flocculation of suspended sediment in the sea. Nature, 246: 348—350.

Krause, D. C., 1961. Geology of the Southern Continental Borderland West of Baja California, Mexico. Ph.D. Thesis, Univ. of California, San Diego, Calif. (unpublished).

Krauskopf, K. B., 1956. Factors controlling the concentrations of thirteen rare metals in sea water. Geochim. Cosmochim. Acta, 9: 1—32B.

Krauskopf, K. B., 1957. Separation of manganese from iron in sedimentary processes. Geochim. Cosmochim. Acta, 12: 61—84.

Krauskopf, K. B., 1967. Introduction to Geochemistry. McGraw-Hill, New York, N.Y., 721 pp.

Krishnaswami, S. and Lal, D., 1972. Manganese nodules and budget of trace solubles in oceans. Proc. Nobel Symp., 20th, pp. 307—320.

Krishnaswami, S. and Moore, W. S., 1973. Accretion rates of freshwater manganese deposits. Nature Phys. Sci., 243: 114—116.

Krishnaswami, S., Somayajulu, B. L. K. and Moore, W. S., 1972. Dating of manganese nodules using beryllium-10. In: D. R. Horn (Editor), Ferromanganese Deposits on the Ocean Floor. National Science Foundation, Washington, D.C., pp. 117—122.

Kröll, V., 1955. Radium in manganese crusts. Medd. Oceanogr. Inst., Göteborg, 24: 1—10.

Krueger, R. B., 1968. The development and administration of the outer continental shelf lands of the United States. Rocky Mount. Miner. Law Inst. Annual, 14: 643—721.

Krueger, R. B., 1970. The background of the doctrine of the Continental Shelf and the Outer Continental Shelf Lands Act. Nat. Resour. J., 10: 442—514.

Krumbein, W. E., 1971. Manganese-oxidising fungi and bacteria in Recent shelf sediments of the Bay of Biscay and the North Sea. Naturwissenschaften, 58: 56—57.

Krupyanskii, Y. F. and Suzdalev, I. P., 1973. Magnetic properties of ultrafine iron oxide particles. Zh. Eksp. Teor. Fiz., 65: 1715—1725. (Cf. also C.A. 80: 8465c, 1974.)

Kruyt, H. R. (Editor), 1952. Colloid Science, Vol. 1. Irreversible Systems. Elsevier, Amsterdam, 389 pp.

Ku, T. L., 1965. An evaluation of the U^{234}/U^{238} method as a tool for dating pelagic sediments. J. Geophys. Res., 70: 3457—3474.

Ku, T. L. and Broecker, W. S., 1967. Uranium, thorium and protactinium in a manganese nodule. Earth Planet. Sci. Lett., 2: 317—320.

Ku, T. L. and Broecker, W. S., 1969. Radiochemical studies on manganese nodules of deep-sea origin. Deep-Sea Res., 16: 625—637.

Ku, T. L. and Glasby, G. P., 1972. Radiometric evidence for the rapid growth rate of shallow-water, continental margin manganese nodules. Geochim. Cosmochim. Acta, 36: 699—703.

Ku, T. L., Broecker, W. S. and Opdyke, N., 1968. Comparison of sedimentation rates measured by paleomagnetic and the ionium methods of age determination. Earth Planet. Sci. Lett., 4: 1—16.

Ku, T. L., Knauss, K. G. and Lin, M. C., 1975. An evaluation of dating nodules by the uranium-series isotopes. EOS Trans. Am. Geophys. Union, 56(12): 999 (Abstr.).

Kuenen, P. H., 1950. Marine Geology. Wiley, New York, N.Y., 551 pp.

Kuenzler, E. J., 1965. Zooplankton Distribution and Isotope Turnover During Operation Swordfish. U.S.A.E.C. Document NYO-3145-1, New York Operations Office, New York.

Kunda, W., Veltman, H. and Evans, D. J. I., 1970. Production of copper from the ammine carbonate system. In: R. P. Ehrlich (Editor), Copper Metallurgy. Proc. Symp. Copper Metall., Extractive Metall. Div., Am. Inst. Min. Metall. Petrol. Eng., Denver, pp. 27—69.

Kundig, W., Bommel, H., Constabaris, G. and Lundquist, R. H., 1966. Some properties of supported small α-Fe_2O_3 particles determined with the Mössbauer effect. Phys. Rev., 142: 327—333.

Kurbatov, L. M., 1937. On the radioactivity of bottom sediments. Am. J. Sci., 33: 147—153.

Kurbatov, M. H. and Wood, G. B., 1952. Rate of adsorption of cobalt ions on hydrous ferric oxide. J. Phys. Chem., 56: 698—701.

Kurbatov, M. H., Wood, G. B. and Kurbatov, J. D., 1951. Isothermal adsorption of cobalt from dilute solutions. J. Phys. Colloid Chem., 55: 1170—1182.

Lakin, H. W., Thompson, C. E. and Davidson, D. F., 1963. Tellurium content of marine manganese oxides and other manganese oxides. Science, 142: 1568—1569.

Lal, D. and Lerman, A., 1973. Dissolution and behavior of particulate biogenic matter in the ocean: some theoretical considerations. J. Geophys. Res., 78: 7100—7111.

Lalou, C. and Brichet, E., 1972. Signification des mesures radiochimiques dans l'évaluation de la vitesse de croissance des nodules de manganèse. C.R. Acad. Sci., Paris, 275D: 815—818.

Lalou, C., Brichet, E. and Ranque, D., 1973. Certains nodules de manganèse trouvés en surface des sédiments sont-ils des formation contemporaines de la sédimentation? C.R. Acad. Sci., Paris, 276D: 1661—1664.

La Motte, C., 1970. Deepsea Ventures' pilot run is successful. Ocean Ind., 5(10): 7—9, 11.

Landergren, S., 1964. On the geochemistry of deep-sea sediments. Rep. Swed. Deep-Sea Exped., 10: 57—154.

Langmuir, D. and Whittemore, D. O., 1971. Variations in the stability of precipitated ferric oxyhydroxides. Adv. Chem. Ser., 106: 209—236.

Larson, L. T., 1962. Zinc-bearing todorokite from Philipsburg, Montana. Am. Miner., 47: 59—66.

Latimer, W. M., 1952. Oxidation Potentials. Prentice-Hall, New York, 352 pp.

Laughton, A. S., 1967. Underwater photography of the Carlsberg Ridge. In: J. B. Hersey (Editor), Deep-Sea Photography. The Johns Hopkins Oceanographic Studies, 3: 191—206.

Leichhardt, F. W. L., 1847. Journal of an overland expedition in Australia from Moreton Bay to Port Essington, a distance of upwards of 3,000 miles, during the years 1844 to 1845. Boone, London.

Lemoine, M., 1953. Remarques sur les caractères et l'évolution de la paléogéographie de la zone briançonnaise au Secondaire et au Tertiare. Bull. Soc. Geol. Fr., Sér. 6, 3: 105—120.

Lengweiler, H., Buser, W. and Feitknecht, W., 1961. Die Ermittlung der Löslichkeit von Eisen(III)-Hydroxiden mit [59]Fe(II). Helv. Chim. Acta, 44: 796—811.

Levinson, A. A., 1960. Second occurrence of todorokite. Am. Miner., 45: 802—807.

Levinson, A. A., 1962. Birnessite from Mexico. Am. Miner., 47: 790—791.

Lewis, G. J. and Goldberg, E. D., 1954. Iron in marine waters. J. Mar. Res., 13: 183—187.

Li, H. C., 1958. Adsorption of Inorganic Ions on Quartz. Thesis, Massachusetts Institute of Technology, Cambridge, Mass.

Li, Y.-H., Bischoff, J. and Mathieu, G., 1969. The migration of manganese in the Arctic Basin sediment. Earth Planet. Sci. Lett., 7: 265—270.

Li, Y.-H., Takahashi, T. and Broecker, W. S., 1969. Degree of saturation of $CaCO_3$ in the oceans. J. Geophys. Res., 74: 5507—5525.

Lippert, K. K., Pietsch, H. B., Roeder, A. and Walden, H. W., 1969. Recovery of non-ferrous metals impurities from iron ore pellets by chlorination (CV or LDK process). Trans. Inst. Min. Metall., 78C: 98—107.

Lister, C. R. C., 1972. On the thermal balance of a mid-ocean ridge. Geophys. J. R. Astr. Soc., 26: 515—535.

Litherland, M. and Malan, S. P., 1943. Manganiferous stromatolites from the Precambrian of Botswana. J. Geol. Soc., Lond., 129: 543—544.

Ljunggren, P., 1953. Some data concerning the formation of manganiferous and ferrifer-ous bog ores. Geol. För. Stockh. Förh., 75: 277—297.

Ljunggren, P., 1955a. Chemistry and radioactivity of some Mn and Fe bog ores. Geol. För. Stockh. Förh., 77: 33—44.

Ljunggren, P., 1955b. Differential thermal analysis and X-ray examination of Fe and Mn bog ores. Geol. För. Stockh. Förh., 77: 135—147.

Lockyer, J. N., 1888. Notes on meteorites, III. Identity of origin of meteorites, luminous meteors, and falling stars. Nature, 38: 530—533.

Loganathan, P. and Burau, R. G., 1973. Sorption of heavy metal ions by a hydrous man-ganese oxide. Geochim. Cosmochim. Acta, 37: 1277—1293.

Lohmann, G. P., 1973. Stratigraphy and sedimentation of deep-sea oceanic formation on Barbados, West Indies. Bull. Am. Assoc. Petrol. Geol., 57: 791 (Abstr.).

Lonsdale, P., Normark, W. R. and Newman, W. A., 1972. Sedimentation and erosion on Horizon Guyot. Bull. Geol. Soc. Am., 83: 289—316.

Lonsdale, P. F., 1974. Abyssal Geomorphology of a Depositional Environment at the Exit of the Samoan Passage. Ph.D. Thesis, Univ. of California, San Diego, Calif., 105 pp. (unpublished).

Loosanoff, V. L. and Tommers, F. D., 1948. Effect of suspended silt and other sub-stances on the rate of feeding of oysters. Science, 107: 69—70.

Lowman, F. C., Rice, T. R. and Richards, F. A., 1971. Accumulation and redistribution of radionuclides by marine organisms. In: Radioactivity in the Marine Environment. National Academy of Sciences, Washington, D.C., pp. 161—199.

Lukmanova, T. L., Savinkova, E. A. and Vilnyanskii, Y. E., 1965. Chlorination of hydro-lized products from molten carnallite. Izv. Vyssh. Ucheb. Zaved. Tsvet. Metall., 8(6): 63—68.

Lynn, D. C. and Bonatti, E., 1965. Mobility of manganese in diagenesis of deep-sea sedi-ments. Mar. Geol., 3: 457—474.

Lyttle, N. A. and Clarke, D. B., 1975. New analyses of Eocene basalt from the Olympic Peninsula, Washington. Bull. Geol. Soc. Am., 86: 421—427.

MacDonald, R. D. and Murray, J. W., 1969. Marine geology of the Upper Jervis Inlet, British Columbia. Geol. Surv. Can. Pap., 69—1A: 5—8.

MacIsaac, J. J. and Dugdale, R. C., 1969. The kinetics of nitrate and ammonia uptake by natural populations of marine phytoplankton. Deep-Sea Res., 16: 45—57.

MacKenzie, R. C., Follett, E. A. C. and Meldau, R., 1971. The oxides of iron, aluminium, and manganese. In: J. A. Gard (Editor), The Electron-Optical Investigation of Clays. Miner. Soc. Monograph, pp. 315—344.

Mackiw, V. N., Benz, T. W. and Evans, D. J. I., 1962. Hydrometallurgie unter Anwendung von Druck. Chemie-Ing.-Tech., 34: 441.

MacPherson, L. B., Sinclair, N. R. and Hayes, F. R., 1958. Lake water and sediment, III. The effect of pH on the partition of inorganic phosphate between water and oxidised mud and its ash. Limnol. Oceanogr., 3: 318—326.

Malone, T. C., 1971a. The relative importance of netplankton and nanoplankton as pri-mary producers in neritic and oceanic tropical waters. Limnol. Oceanogr., 16: 633—639.

Malone, T. C., 1971b. The relative importance of nanoplankton and netplankton as primary producers in the California Current System. Fish. Bull., 69: 799—820.

Malone, T. C., Garside, C., Paul, A. Z. and Roels, O. A., 1973a. Potential environmental impact of manganese nodule mining in the deep sea. Offshore Technol. Conf., Houston, Texas, April 30—May 2, 1973, Vol. 1, pp. 129—139 (preprint).

Malone, T. C., Garside, C., Anderson, O. R. and Roels, O. A., 1973b. The possible occurrence of photosynthetic microorganisms in deep-sea sediments of the North Atlantic. J. Phycol., 9: 482—488.

Mancke, E. B., 1956. Processes for Treating Nickel Bearing Iron Ores. U.S. Patent 2,775,517.

Manheim, F. T., 1961a. A geochemical profile in the Baltic Sea. Geochim. Cosmochim. Acta, 25: 52—71.

Manheim, F. T., 1961b. In situ measurements of pH and Eh in natural waters and sediments. Stockh. Contr. Geol., 8: 27—36.

Manheim, F. T., 1965. Manganese-iron accumulations in the shallow marine environment. In: D. R. Schink and J. T. Corless (Editors), Symposium on Marine Geochemistry. Occas. Publ. Narragansett Mar. Lab., Univ. Rhode Island, 3: 217—276.

Manheim, F. T., 1972. Composition and origin of manganese-iron nodules and pavements on the Blake Plateau. In: D. R. Horn (Editor), Ferromanganese Deposits on the Ocean Floor. National Science Foundation, Washington, D.C., p. 105 (Abstr.).

Manheim, F. T. and Chan, K. M., 1974. Interstitial waters of Black Sea sediments: new data and review. Mem. Am. Assoc. Petrol. Geol., 20: 155—180.

Margolis, S. V., 1975. Manganese deposits encountered during Deep Sea Drilling Project, Leg 29. In: Initial Reports of the Deep Sea Drilling Project, Vol. 29. U.S. Govt. Printing Office, Washington, D.C., pp. 1083—1091.

Margolis, S. V. and Burns, R. G., 1976. Pacific deep-sea manganese nodules: Their distribution, composition, and origin. Ann. Rev. Earth Planet. Sci., 4.

Margolis, S. V. and Glasby, G. P., 1973. Microlaminations in marine manganese nodules as revealed by scanning electron microscopy. Bull. Geol. Soc. Am., 84: 3601—3610.

Marti, W., 1944. Über die Oxydation von Manganhydroxyd und über höherwertige Oxyde und Oxyhydrate des Mangans. Ph.D. Thesis, Univ. of Bern, 125 pp. (unpublished).

Martinez, E., 1967. The copper segregation process studies by thermoanalysis. Trans. Am. Inst. Min. Metall. Eng., 238: 172—179.

Mason, B. and Berry, L. G., 1968. Elements of Mineralogy. Freeman, San Francisco, Calif., 550 pp.

Masuda, Y., Cruickshank, M. J. and Mero, J. L., 1971. Continuous bucket-line dredging at 12,000 feet. Offshore Technol. Conf., 1971, 1: 1837—1858 (preprint).

Matijevic, E., 1967. Charge reversal of lyophobic colloids. In: S. D. Faust and J. V. Hunter (Editors), Principles and Applications of Water Chemistry. Wiley, New York, N.Y., 328—369.

Matthews, D. H., 1962. Altered lavas from the floor of the eastern North Atlantic. Nature, 194: 368—369.

Mattson, S. and Pugh, A. J., 1934. The laws of soil colloidal behavior, XIV. The electrokinetics of hydrous oxides and their ionic exchange. Soil Sci., 38: 299—305.

McBirney, A. R. and Gass, I. G., 1967. Relations of oceanic volcanic rocks to midoceanic rises and heat flow. Earth Planet. Sci. Lett., 2: 265—276.

McDougal, M. S. and Burke, W. T., 1962. The Public Order of the Oceans: A Contemporary International Law of the Sea. Yale University Press, New Haven, Conn., 1226 pp.

McKeague, J. A. and Cline, M. G., 1963a. Silica in soil solutions, I. The form and concentration of dissolved silica in aqueous extracts of some soils. Can. J. Soil Sci., 43: 70—82.

McKeague, J. A. and Cline, M. G., 1963b. Silica in soil solutions, II. The adsorption of monosilicic acid by soil and other substances. Can. J. Soil Sci., 43: 83—96.

McKelvey, W. E. and Wang, F. F. H., 1969. World subsea mineral resources. Dept. Interior, U.S. Geol. Surv., Washington, D.C.

McKenzie, R. M., 1970. The reaction of cobalt with manganese dioxide minerals. Aust. J. Soil Res., 8: 97—106.

McKenzie, R. M., 1971. The synthesis of birnessite, cryptomelane, and some other odixes and hydroxides of manganese. Miner. Mag., 28: 493—502.

McKenzie, R. M., 1972. The sorption of some heavy metals by the lower oxides of manganese. Geoderma, 8: 29—35.

McKenzie, R. M., 1975. An electron microprobe study of the relationships between heavy metals and manganese and iron in soils and ocean floor nodules. Aust. J. Soil Res., 13: 177—188.

McManus, D. A. et al., 1970. Initial Reports of the Deep-Sea Drilling Project, 5. U.S. Govt. Printing Office, Washington, D.C., 827 pp.

McMurdie, H. F., 1944. Microscopic and diffraction studies on dry cells and their raw materials. Trans. Electrochem. Soc., 86: 313—326.

McMurdie, H. F. and Golovato, E., 1948. Study of the modifications of manganese dioxide. J. Res. Nat. Bur. Stand., 41: 589—600.

Megaw, A., 1934. The crystal structure of hydrargillite Al(OH)$_3$. Z. Kristallogr. Kristallgeom., 87: 185—204.

Meiser, H. J. and Müller, E., 1973. Manganese nodules: A further resource to meet mineral requirements? In: M. Morgenstein (Editor), Papers on the Origin and Distribution of Manganese Nodules in the Pacific and Prospects for Exploration. Hawaii Inst. of Geophysics, Honolulu, pp. 115—124.

Meldau, R., Newesely, H. and Strunz, H., 1973. Zur Kristallchemie von Feitknechtit, β-MnOOH. Naturwissenschaften, 60: 387.

Melson, W. G. and Thompson, G., 1972. Electron probe petrology of glassy basalt and its alteration products from the sea floor near St. Paul's Rocks, equatorial Atlantic. Mem. Geol. Soc. Am., in press.

Menard, H. W., 1960. Consolidated slabs on the floor of the eastern Pacific. Deep-Sea Res., 7: 35—41.

Menard, H. W., 1964. Marine Geology of the Pacific. McGraw-Hill, New York, N.Y., 271 pp.

Menard, H. W., Goldberg, E. D. and Hawkes, H. E., 1964. Composition of Pacific Sea-Floor Manganese Nodules. Scripps Inst. Oceanogr., unpubl. Rep.

Menzel, D. W. and Ryther, J. H., 1960. The annual cycle of primary production in the Sargasso Sea off Bermuda. Deep-Sea Res., 6: 351—367.

Menzel, D. W., Hulburt, E. M. and Ryther, J. H., 1963. The effects of enriching Sargasso Sea water on the production and species composition of the phytoplankton. Deep-Sea Res., 10: 209—219.

Menzies, R. J., George, R. Y. and Rowe, G. T., 1973. Abyssal Environment and Ecology of the World Oceans. Wiley, New York, N.Y., 488 pp.

Mero, J. L., 1952. Manganese. N. Dakota Eng., 27: 28—32.

Mero, J. L., 1959. The Mining and Processing of Deep-Sea Manganese Nodules. Inst. of Marine Research, Univ. of California, Berkeley, Calif., 96 pp.

Mero, J. L., 1962. Ocean-floor manganese nodules. Econ. Geol., 57: 747—767.

Mero, J. L., 1965a. The Mineral Resources of the Sea. Elsevier, Amsterdam, 312 pp.

Mero, J. L., 1965b. Process for Separation of Nickel from Cobalt in Ocean Floor Manganiferous Ore Deposits. U.S. Patent, 3,169,856.

Mero, J. L., 1972. Potential economic value of ocean-floor manganese nodule deposits. In: D. R. Horn (Editor), Ferromanganese Deposits on the Ocean Floor. National Science Foundation, Washington, D.C., pp. 191—203.

Merrill, J. R., Lyden, E. F., Honda, M. and Arnold, J. R., 1960. The sedimentary geochemistry of the beryllium isotopes. Geochim. Cosmochim. Acta, 18: 108—129.

MESA (Marine Eco-Systems Analysis), 1975. Project Development Plan, Deep Ocean Mining Environmental Study. Phase I, Marine Environmental Assessment. U.S. Dept. Comm., NOAA/PMEL, Seattle, Wash., 151 pp. (unpubl. manuscript).

Metallgesellschaft AG, 1975. Manganese nodules: metals from the sea. Metallgesellschaft AG, Frankfurt, Rev. Activ., 18: 87 pp.

Meyer, K., 1973. Surface sediment and manganese nodule facies, encountered on R/V Valdivia cruises 1971/73. In: M. Morgenstein (Editor), Papers on the Origin and Distribution of Manganese Nodules in the Pacific and Prospects for Exploration. Hawaii Inst. of Geophysics, Honolulu, pp. 125—130.

Meylan, M. A., 1968. The mineralogy and geochemistry of manganese nodules from the Southern Ocean. Contr. Sediment. Res. Lab. Fla. St. Univ., 22: 177 pp.

Meylan, M. A., 1974. Field description and classification of manganese nodules. Hawaii Inst. Geophys. Rep., HIG-74-9: 158—168.

Meylan, M. A. and Craig, J. D., 1975. Manganese nodules of the northeastern equatorial Pacific Ocean — descriptive characterization. EOS Trans. Am. Geophys. Union, 56(12): 999—1000 (Abstr.).

Michard, G., 1971. Theoretical model for manganese distribution in calcareous sediment cores. J. Geophys. Res., 76: 2179—2186.

Mikheev, V. I., 1957. X-Ray Determination of Minerals. Gosudarstvennoe Nauchno-Tekhnicheskoe Izdatel'stvo Literatury po Geologii i Okhrane Nedr. (in Russian).

Miller, A. R., Densmore, C. D., Degens, E. T., Hathaway, J. C., Manheim, F. T., McFalin, P. F., Poklington, R. and Jokela, A., 1966. Hot brines and recent iron deposits in deeps of the Red Sea. Geochim. Cosmochim. Acta, 30: 341—359.

Miller, L. P., 1950. Formation of metal sulphides through the activities of sulfate-reducing bacteria. Contr. Boyce Thompson Inst., 16: 85—89.

Miller, R. W., 1967. Soluble silica in soil. Proc. Soil Sci. Soc. Am., 31: 46—50.

Minguzzi, C. and Talluri, A., 1951. Indagine e considerazione sulla presenze e sulla distribuzione dei constituenti minori pelle pririti. Memorie Soc. Tosc. Sci. Nat., Ser. A, 58: 89—120.

Mintz, Y. and Dean, G., 1952. The observed mean field of motion of the atmosphere. Geophys. Res. Pap. U.S., 17: 37—42.

Mohr, P. A., 1959. The distribution of some minor elements between sulphide and silicate phases of sediments. Contrib. Geophys. Obs. Univ. Coll. Addis Ababa, pp. 1—18.

Mohr, P. A., 1964. Genesis of the Cambrian manganese carbonate rocks of North Wales. J. Sediment. Petrol., 34: 819—829.

Molengraaf, G. A. F., 1909. On oceanic deep-sea deposits of Central Borneo. Proc. R. Acad. Sci. Amst., 12: 141—147.

Molengraaf, G. A. F., 1915. On the occurrence of nodules of manganese in Mesozoic deep-sea deposits from Borneo, Timor, and Rotti, their significance and mode of formation. Proc. R. Acad. Sci. Amst., 18: 415—430.

Molengraaf, G. A. F., 1922. On manganese nodules in Mesozoic deep-sea deposits of Dutch Timor. Proc. R. Acad. Sci. Amst., 23: 997—1012.

Montgomery, R. B. and Stroup, E. D., 1962. Equatorial waters and currents at 150°W in July—August 1952. The Johns Hopkins Oceanographic Studies, 1: 68 pp.

Monty, C., 1973. Les nodules de manganèse sont des stromatolithes océaniques. C.R. Acad. Sci., Paris, 276D: 3285—3288.

Moore, E. J., 1910. The occurrence and origin of some bog iron deposits in the district of Thunder Bay, Ontario. Econ. Geol., 5: 528—537.

Moore, J. G., 1966. Rate of palagonization of submarine basalt adjacent to Hawaii. Prof. Pap. U.S. Geol. Surv., 550D: 163—171.

Moore, J. G. and Calk, L., 1971. Sulfide spherules in vesicles of dredged pillow basalt. Am. Miner., 56: 576—588.

Moore, J. R., Meyer, R. P. and Morgan, C. L., 1973. Investigation of the sediments and potential nodule resources of Green Bay, Wisconsin. Univ. Wisconsin Tech. Rep., WIS-SG-73-218: 144 pp.

Moore, T. C., 1970. Abyssal hills in the central equatorial Pacific: Sedimentation and stratigraphy. Deep-Sea Res., 17: 573—593.

Moore, T. C. and Heath, G. R., 1966. Manganese nodules, topography and thickness of Quaternary sediments in the central Pacific. Nature, 212: 983—985.

Moore, W. S., 1973. Accumulation rates of manganese crusts on rocks exposed on the sea floor. In: Inter-University Program of Research on Ferromanganese Deposits of the Ocean Floor — Phase I Report. National Science Foundation, Washington, D.C., pp. 93—97.

Moore, W. S. and Vogt, P. R., 1976. Hydrothermal manganese crusts from two sites near the Galápagos spreading axis. Earth Planet. Sci. Lett., 29: 349—356.

Morgan, J. J., 1964. Chemistry of Aqueous Manganese II and IV. Ph.D. Thesis, Harvard Univ., Harvard (unpublished).

Morgan, J. J., 1967. Chemical equilibria and kinetic properties of manganese in natural waters. In: S. D. Faust and J. V. Hunter (Editors), Principles and Applications of Water Chemistry. Wiley, New York, N.Y., pp. 561—624.

Morgan, J. J. and Stumm, W., 1964. Colloid-chemical properties of manganese dioxide. J. Colloid Sci., 19: 347—359.

Morgan, J. J. and Stumm, W., 1965. The role of multivalent metal oxides in limnologi-cal transformations as exemplified by iron and manganese. In: O. Jaag (Editor), Advances in Water Pollution Research. Proc. 2nd Int. Conf., Tokyo, vol. I. Pergamon Press, London. pp. 103—118.

Morgenstein, M., 1969. Composition and development of palagonite in deep-sea sediments from the Atlantic and Pacific Oceans. M.Sc. Thesis, Syracuse Univ., N.Y., 137 pp. (unpublished).

Morgenstein, M., 1972. Manganese accretion at the sediment—water interface at 400 to 2400 meters depth, Hawaiian Archipelago. In: D. R. Horn (Editor), Ferromanganese Deposits on the Ocean Floor. National Science Foundation, Washington, D.C., pp. 131—138.

Morgenstein, M., 1973a. Sedimentary diagenesis and rates of manganese accretion on the Waho Shelf, Kauai Channel, Hawaii. In: Inter-University Program of Research on Ferromanganese Deposits of the Ocean Floor — Phase I Report. National Science Foundation, Washington, D.C., pp. 121—135.

Morgenstein, M. (Editor), 1973b. Papers on the Origin and Distribution of Manganese Nodules in the Pacific and Prospects for Exploration. Hawaii Inst. of Geophysics, Honolulu, 175 pp.

Morgenstein, M. and Riley, J. J., 1975. Hydration-rind and dating of basaltic glass: a new method for archaeological chronologies. Asian Perspectives, 17: 145—159.

Mortimer, C. H., 1941. The exchange of dissolved substances between mud and water in lakes. J. Ecol., 29: 280—329.

Mortimer, C. H., 1942. The exchange of dissolved substances between mud and water in lakes. J. Ecol., 30: 147—201.

Mortimer, C. H., 1971. Chemical exchanges between sediments and water in the Great Lakes — speculations on probable regulatory mechanisms. Limnol. Oceanogr., 16: 387—404.

Mothersill, J. S. and Shegelski, R. J., 1973. The formation of iron and manganese rich layers in the Holocene sediments of Thunder Bay, Lake Superior. Can. J. Earth Sci., 10: 571—576.

Müller, G. and Förstner, U., 1973. Recent iron ore formation in Lake Malawi, Africa. Miner. Deposita, 8: 278—290.

Mullin, M. N., 1963. Some factors affecting the feeding of marine copepods of the genus *Calanus*. Limnol. Oceanogr., 8: 239—250.

Munk, W., Snodgrass, F. and Wimbush, M., 1970. Tides offshore: transition from California coastal to deep-sea waters. Geophys. Fluid Dynam., 1: 161—235.

Murata, K. J. and Erd, R. C., 1964. Composition of sediments from the experimental Mohole Project (Guadalupe site). J. Sediment. Petrol., 34: 633—655.

Muromtsev, A. M., 1963. The principal hydrological features of the Pacific Ocean [trans. from Russian]. Israel Program for Scientific Translations, Jerusalem. [Gidrometeorol. Izdat., 1958, Leningrad, 417 pp.]

Murphy, R. C., 1936. Oceanic Birds of South America. Macmillan, New York, N.Y., 1245 pp.

Murray, D. J., Healy, T. W. and Fuerstenau, D. W., 1968. The adsorption of aqueous metal on colloidal hydrous manganese oxide. Adv. Chem. Ser., 79: 74—81.

Murray, J., 1876. Preliminary report on specimens of the sea bottom. Proc. R. Soc. Lond., 24: 471—532.

Murray, J., 1895. A summary of the scientific results obtained at the sounding, dredging, and trawling stations of H.M.S. Challenger, II. Rep. Sci. Results Explor. Voyage Challenger, 1608 pp.

Murray, J., 1900. On the deposits of the Black Sea. Scott. Geogr. Mag., 16: 673—702.

Murray, J. and Irvine, R., 1893. On the chemical changes which take place in the composition of the sea-water associated with blue muds on the floor of the ocean. Trans. R. Soc. Edinb., 37: 481—507.

Murray, J. and Irvine, R., 1894. On the manganese oxides and manganese nodules in marine deposits. Trans. R. Soc. Edinb., 37: 721—742.

Murray, J. and Lee, G. V., 1909. The depth and marine deposits of the Pacific. Mem. Mus. Comp. Zool. Harv. Coll., 38: 7—169.

Murray, J. and Renard, A. F., 1884. On the microscopic characters of volcanic ashes and cosmic dust, and their distribution in the deep sea deposits. Proc. R. Soc. Edinb., 12: 474—495.

Murray, J. and Renard, A. F., 1891. Deep-sea deposits. Rep. Sci. Results Explor. Voyage Challenger, 525 pp.

Murray, J. W., 1974. The surface chemistry of hydrous manganese dioxide. J. Colloid Interface Sci., 46: 357—371.

Murray, J. W., 1975a. The interaction of metal ions at the manganese dioxide—solution interface. Geochim. Cosmochim. Acta, 39: 505—519.

Murray, J. W., 1975b. The interaction of cobalt with hydrous manganese dioxide. Geochim. Cosmochim. Acta, 39: 635—647.

Namby, M., Okada, K. and Tanida, K., 1964. Chemical composition of todorokite. Jap. J. Assoc. Miner. Petrogr. Econ. Geol., 51: 30—38.

National Oceanic and Atmospheric Administration, U.S. Department of Commerce, 1974. North Pacific tropical cyclones. Obverse of Pilot Chart of the North Atlantic Ocean, 16: October 1974, Defence Department Mapping Agency Hydrographic, Washington, D.C.

National Petroleum Council, 1969. Petroleum resources under the ocean floor. National Petroleum Council, Washington, 107 pp.

National Petroleum Council, 1971. Petroleum resources under the ocean floor: A supplementary report. National Petroleum Council, Washington, 57 pp.

National Research Council, 1975. Mining in the outer continental shelf and in the deep ocean. National Academy of Sciences, Washington, D.C. 152 pp.

Naumann, E., 1922. Södra och mellestra Sveriges Sjö — och Myrmalmer deras bildningshistoria, utbredning och praktiska betydelse. Sver. Geol. Unders. Afh, Ser. C, 297: 194 pp.

Naumann, E., 1930. Einführung in die Bodenkunde der Seen. Die Binengewässer, vol. 9. Schwerzer Verlag, Stuttgart, 126 pp.

Neihof, R. A. and Loeb, G. I., 1972. The surface charge of particulate matter in seawater. Limnol. Oceanogr., 17: 7—16.

Neihof, R. A. and Loeb, G. I., 1974. Dissolved organic matter in seawater and the electric charge of immersed surfaces. J. Mar. Res., 32: 5—12.

Nemeth, R. and Matijevic, E., 1968. Interaction of silver halides with gelatin of like charge. Kolloid- Z.Z. Polym., 225: 155.

Neumann, G. and Pierson, W. J., 1966. Principles of Physical Oceanography. Prentice-Hall, Englewood Cliffs, N.J., 545 pp.

Nicholls, G. D., Curl, H. and Bowen, V. T., 1959. Spectrographic analyses of marine plankton. Limnol. Oceanogr., 4: 472—478.

Niino, H., 1955. On a manganese nodule and *Perotrochus* dredged from the Banks near the Izu Islands, Japan. Rec. Oceanogr. Works Jap., 2(2): 120—126.

Nikolayev, P. S. and Yefimova, E. I., 1963. On the age of iron-manganese concretions from the Indian and Pacific Oceans. Geochem. Int., 7: 703—714.

Nohara, M., 1972. Manganese minerals in ferromanganese nodules dredged from the sea mounts in the Pacific Ocean. Chishitsugaku Zasshi, 78: 699—701.

Nordenskiöld, A. E., 1881. The Voyage of the Vega round Asia and Europe, vol. 1. Macmillan, London, 524 pp.

Office of the Geographer, 1971. Oceania and miscellaneous insular areas: Civil divisions. Department of State, Washington, D.C., 47 pp.

Ohle, W., 1953. Phosphor als Initialfaktor der Gewässer-Eutrophierung. Vom Wass., 20: 11—23.

Okada, A. and Shima, M., 1969. Study on the manganese nodule (II). Comparison of a manganese nodule collected from the surface of the sea-floor with that collected from a 3-metre deep core. J. Jap. Assoc. Miner. Petrol. Econ. Geol., 61: 41—49.

Okada, A. and Shima, M., 1970. Study on the manganese nodule. J. Oceanogr. Soc. Jap., 26(3): 151—158 (in Japanese, English Abstr.).

Okada, A., Minakuchi, T. and Shima, M., 1972a. Study on the manganese nodule (V). Thermal studies of the iron-manganese phase. J. Oceanogr. Soc. Jap., 28: 39—47.

Okada, A., Okada, T. and Shima, M., 1972b. Study on the manganese nodule (VI). Some aspects of the chemical form of iron in the manganese nodule. J. Jap. Assoc. Miner. Petrol. Econ. Geol., 66: 178—183.

Okada, A., Okada, T. and Shima, M., 1973. Study on the manganese nodule (VI). Magnetic properties and Mössbauer effect on the manganese nodules. J. Jap. Assoc. Miner. Petrol. Econ. Geol., 68: 199—203.

Okamoto, S., 1968. Structure of δ-FeOOH. J. Am. Ceram. Soc., 51: 594—599.

Okamoto, S., Sekizawa, H. and Okamoto, S. I., 1972. Characterization and phase transformation of amorphous ferric hydroxide. In: J. S. Anderson, M. W. Roberts and F. S. Stone (Editors), Reactivity of Solids. Proc. I.S.R.S., 7th, Bristol, pp. 341—350.

Okubo, A., 1971. Oceanic diffusion diagrams. Deep-Sea Res., 18: 789—802.

Onoda, G. Y. and De Bruyn, P. L., 1966. Proton adsorption at the ferric oxide/aqueous solution interface, I. A kinetic study of adsorption. Surface Sci., 4: 48—63.

Opdyke, N. D. and Foster, J. H., 1970. Paleomagnetism of cores from the North Pacific. Mem. Geol. Soc. Am., 126: 83—119.

Öpik, E. J., 1956. Interplanetary dust and terrestrial accretion of meteoric matter. Irish Astr. J., 4: 84—135.

Orgel, L. E., 1966. An Introduction to Transition-Metal Chemistry. Wiley, New York, N.Y., 186 pp.

Ostwald, J. and Frazer, F. W., 1973. Chemical and mineralogical investigations on deep sea manganese nodules from the Southern Ocean. Miner. Deposita, 8: 303—311.

Oswald, H. R. and Wampetich, M. J., 1967. Die Kristallstrukturen von Mn_5O_8 und $Cd_2Mn_3O_8$. Helv. Chim. Acta, 50: 2023—2034.

Otgonsuren, O., Perelygin, V. P. and Flerov, G. N., 1969. The search for remote transuranium elements in iron-manganese concretions. Dokl. Akad. Nauk SSSR, 189: 1200—1203.

Othmer, D. F. and Roels, O. A., 1973. Power, fresh water and food from cold, deep-sea water. Science, 182: 121—125.

Owen, D. M., Sanders, H. L. and Hessler, R. R., 1967. Bottom photography as a tool for estimating benthic populations. In: J. B. Hersey (Editor), Deep-Sea Photography. The Johns Hopkins Oceanographic Studies, 3: 229—234.

Owen, R. W. and Zeitzschel, B., 1970. Phytoplankton production: seasonal change in the oceanic eastern tropical Pacific. Mar. Biol., 7: 32—36.

Ozima, M., 1967. Magnetic properties of manganese nodules associated with dredged submarine basalts. J. Geomag. Geoelect., 19: 253—255.

Palache, C., Berman, H. and Frondel, C., 1944. The System of Mineralogy, Vol. I. Wiley, New York, N.Y., 7th ed., 834 pp.

Park, C. F., 1946. Spilite and manganese problems of the Olympic Peninsula, Washington. Am. J. Sci., 244: 305—323.

Park, P. K., 1968. Seawater hydrogen-ion concentrations: vertical profile. Science, 162: 357—358.

Parks, G. A., 1965. The isoelectric points of solid oxides, solid hydroxides and aqueous hydroxo complex system. Chem. Rev., 65: 177—198.

Parks, G. A., 1967. Aqueous surface chemistry of oxides and complex oxide minerals. Adv. Chem. Ser., 67: 121—160.

Parks, G. A. and De Bruyn, P. L., 1962. The zero point of charge of oxides. J. Phys. Chem., 66: 967—973.

Parsons, T. R., 1975. Particulate organic carbon in the sea. In: J. P. Riley and G. Skirrow (Editors), Chemical Oceanography, Vol. 2. Academic Press, London, 2nd ed., pp. 365—383.

Parsons, T. R. and LeBrasseur, R. J., 1970. The availability of food to different trophic levels in the marine food chain. In: J. H. Steele (Editor), Marine Food Chains. University of California Press, pp. 325—343.

Pauling, L., 1960. The Nature of the Chemical Bond. Cornell University Press, Ithaca, N.Y., 644 pp.

Payne, R. R. and Conolly, J. R., 1972. Pleistocene manganese pavement production: its relationship to the origin of manganese in the Tasman Sea. In: D. R. Horn (Editor), Ferromanganese Deposits on the Ocean Floor. National Science Foundation, Washington, D.C., pp. 81—92.

Pearsall, W. A. and Mortimer, C. H., 1939. Oxidation-reduction potentials in water-logged soils, natural waters and mud. J. Ecol., 27: 483—501.

Perfil'ev, B. V., Gabe, D. R., Gal'pernia, A. M., Rabinovich, V. A., Sapotnitskii, A. A., Sherman, E. E. and Troshanov, E. P., 1965. Applied Capillary Microscopy. Consultants Bureau, New York, N.Y., 122 pp.

Pernet, J., Chenavas, J. and Joubert, J. C., 1973. Caractérization et étude par effet Mössbauer d'une nouvelle variété haute pression de FeOOH. Solid State Comm., 13: 1147—1154.

Perseil, E.-A., 1968. Caractères minéralogiques de quelques types de gisements manganésifères de la France méridionale. Bull. Soc. Geol. Fr., Sér. 7, 10: 408—412.

Pettersson, H., 1943. Manganese nodules and the chronology of the sea floor. Medd. Oceanogr. Inst. Goteborg, 6: 1—39.

Pettersson, H., 1945. Iron and manganese on the ocean floor. Medd. Oceanogr. Inst. Goteborg, Ser. B, 3: 1—37.

Pettersson, H., 1955. Manganese nodules and oceanic radium. Deep-Sea Res., 3 (Suppl.): 335—345.

Pettersson, H., 1959. Manganese and nickel on the ocean floor. Geochim. Cosmochim. Acta, 17: 209—213.

Pettersson, H. and Rotschi, H., 1952. The nickel content of deep sea deposits. Geochim. Cosmochim. Acta, 2: 81—90.

Piggott, C. S., 1933. Radium content of ocean-bottom sediments. Am. J. Sci., 25: 229—238.

Piggott, C. S., 1944. Scientific results of cruise VII of the "Carnegie". Publ. Carnegie Inst., 556: 185—193.

Pilipchuk, M. F. and Volkov, I. I., 1974. Behavior of molybdenum in processes of sediment formation and diagenesis in Black Sea. Mem. Am. Assoc. Petrol. Geol., 20: 542—553.

Piper, D. Z., 1972. Rare earth elements in manganese nodules from the Pacific Ocean. In: D. R. Horn (Editor), Ferromanganese Deposits on the Ocean Floor. National Science Foundation, Washington, D.C., pp. 123—130.

Piper, D. Z., 1974. Rare earth elements in ferromanganese nodules and other marine phases. Geochim. Cosmochim. Acta, 38: 1007—1022.

Pitzl, D., 1974. SEM Photographs of Non-Opaque Minerals in Manganese Nodules. Dept. of Geology, Washington State University, Pullman, Wash. (unpubl. data).

Posselt, H. S., Anderson, I. J. and Weber, W. J., 1968a. Cation sorption on colloidal hydrous manganese dioxide. Environ. Sci. Technol., 2: 1087—1093.

Posselt, H. S., Reidies, A. H. and Weber, W. J., 1968b. Coagulation of colloidal hydrous manganese dioxide. J. Am. Water Works Assoc., 60: 48—68.

Postma, H., 1964. The exchange of oxygen and carbon dioxide between the ocean and the atmosphere. Neth. J. Sea Res., 2: 258—283.

Pratt, R. M. and McFarlin, P. F., 1966. Manganese pavements on the Blake Plateau. Science, 151: 1080—1082.

Present, E. W., 1971. Geochemistry of iron, manganese, lead, copper, zinc, arsenic, antimony, silver, tin and cadmium in the soils of the Bathurst area, New Brunswick. Geol. Surv. Can. Bull., 174: 93 pp.

Presley, B. J., Kolodny, Y., Nissembaum, A. and Kaplan, I. R., 1972. Early diagenesis in a reducing fjord, Saanich Inlet, British Columbia, II. Trace element distribution in interstitial water and sediment. Geochim. Cosmochim. Acta, 36: 1073—1090.

Presley, B. J., Brooks, R. R. and Kaplan, I. R., 1967. Manganese and related elements in the interstitial water of marine sediments. Science, 158: 906—910.

Price, N. B., 1967. Some geochemical observations on manganese-iron oxide nodules from different depth environments. Mar. Geol., 5: 511—538.

Price, N. B., 1976. Diagenesis in marine sediments. In: J. P. Riley and R. Chester (Editors), Chemical Oceanography, Vol. 6. Academic Press, London, pp. 1—58.

Price, N. B. and Calvert, S. E., 1970. Compositional variation in Pacific Ocean ferromanganese nodules and its relationship to sediment accumulation rates. Mar. Geol., 9: 145—171.

Queneau, P., 1961. Extractive Metallurgy of Copper, Nickel and Cobalt. Interscience, New York, N.Y., 647 pp.

Raab, W., 1972. Physical and chemical features of Pacific deep sea manganese nodules and their implications to the genesis of nodules. In: D. R. Horn (Editor), Ferromanganese Deposits on the Ocean Floor. National Science Foundation, Washington, D.C., pp. 31—49.

Raab, W. J. and Norton, D. L., 1973. Trace and Major Element Composition of DSDP Holes No. 37, 40, and 42, Leg 5, With Implications Regarding Distribution of Cu, Ni, Fe, Mn and Co (unpubl. manuscript).

Rampacek, C. and McKinney, W. A., 1960. The copper segregation process. Ann. Mtg. Am. Ins. Min. Metall. Petrol., February 1960.

Rampacek, C., McKinney, W. A. and Waddleton, P. T., 1959. Treating oxidized and mixed oxide—sulfide copper ores by the segregation process. U.S. Bur. Min. Rep. Invest., 5501: 28 pp.

Rancitelli, T. A. and Perkins, R. W., 1973. Major and minor elemental composition of manganese nodules. In: Inter-University Program of Research on Ferromanganese Deposits of the Ocean Floor — Phase I Report. National Science Foundation, Washington, D.C., pp. 1—5.

Rankama, K. and Sahama, Th. G., 1950. Geochemistry. University of Chicago Press, Chicago, Ill., 911 pp.

Rashid, M. A. and Leonard, J. D., 1973. Modifications in the solubility and precipitation behavior of various metals as a result of their interaction with sedimentary humic acid. Chem. Geol., 11: 89—97.

Redfield, A. C., 1942. The processes determining the concentration of oxygen, phosphate and other organic derivatives within the depths of the Atlantic Ocean. Pap. Phys. Oceanogr. Meteorol., 9(2): 1—22.

Redman, M. J., 1972. Gewinnung von Metallen aus zusammengesetzten Erzen. German Patent, 2,135,734.

Redman, M. J., 1973. Extraction of Metal Values from Complex Ores. U.S. Patent, 3,734,715.

Reid, J. L., 1962. On circulation, phosphate-phosphorus content, and zooplankton volumes in the upper part of the Pacific Ocean. Limnol. Oceanogr., 7: 237—306.

Reid, J. L., 1965. Intermediate waters of the Pacific Ocean. The Johns Hopkins Oceanographic Studies, 2: 85 pp.

Reid, J. L., 1969. Preliminary results of measurements of deep currents in the Pacific Ocean. Nature, 221: 848.

Reid, J. L. and Lynn, R. S., 1971. On the influence of the Norwegian—Greenland and Weddell Seas upon the bottom waters of the Indian and Pacific Oceans. Deep-Sea Res., 18: 1063—1088.

Reisenauer, H. M., Tabikh, A. A. and Stout, P. R., 1962. Molybdenum reactions with soils and the hydrous oxides of iron, aluminum and titanium. Proc. Soil Sci. Soc. Am., 26: 23—27.

Reiswig, H. M., 1972. Spectrum of particulate organic matter of shallow-bottom boundary waters of Jamaica. Limnol. Oceanogr., 17: 341—348.

Renzoni, L. S., 1945. Purification of cobalt precipitates containing iron and other impurities. U.S. Patent, 2,367,239.

Revelle, R. R., 1944. Scientific results of the cruise VII of the "Carnegie". Publ. Carnegie Inst., 556: 1—180.

Revelle, R., Bramlette, M., Arrhenius, G. and Goldberg, E. D., 1955. Pelagic sediments of the Pacific. Spec. Pap. Geol. Soc. Am., 62: 221—236.

Rey, M., 1936. The segregation process for low-grade copper oxide ores. Rev. Metall., Paris, 33: 295—302 (in French).

Rickard, D. T., 1969. The chemistry of iron sulphide formation at low temperatures. Stockh. Contr. Geol., 20: 67—95.

Riley, J. P. and Chester, R., 1971. Introduction to Marine Chemistry. Academic Press, London, 465 pp.

Riley, J. P. and Sinhaseni, P., 1958. Chemical composition of three manganese nodules from the Pacific Ocean. J. Mar. Res., 17: 466—482.

Robbins, J. A. and Callender, E., 1975. Diagenesis of manganese in Lake Michigan sediments. Am. J. Sci., 275: 512—533.

Roberts, W. M. B., Walker, A. L. and Buchanan, A. S., 1969. The chemistry of pyrite formation in aqueous solution and its relation to the depositional environments. Miner. Deposita, 4: 18—29.

Robertson, D. E., 1968. Adsorption of trace elements in sea water on various container surfaces. Anal. Chim. Acta, 42: 533—537.

Robertson, D. E., 1970. The distribution of cobalt in oceanic waters. Geochim. Cosmochim. Acta, 34: 553—567.

Robertson, D. E. and Rancitelli, L. A., 1973. Trace element additions to seawater resulting from contact with ferromanganese nodule particles. In: Inter-University Program of Research on Ferromanganese Deposits of the Ocean Floor — Phase I Report. National Science Foundation, IDOE, Washington, D.C., pp. 273—277.

Robideau, R. F., 1972. The discharge of submerged buoyant jets into water of finite depth. Rep. No. U440-72-121, Gen. Dynam. Elect. Boat Div., 57 pp.

Robinson, M., Pask, J. A. and Fuerstenau, D. W., 1964. Surface charge of alumina and magnesia in aqueous media. J. Am. Ceram. Soc., 47: 516—520.

Robinson, R. A. and Stokes, R. H., 1959. Electrolyte Solutions. Butterworths, London, 559 pp.

Roels, O. A., Laurence, S. and Amos, A. F., 1975. Potential mariculture yield of floating sea-thermal power plants. EOS Trans. Am. Geophys. Union, 56(12): 1002 (Abstr.).

Roels, O. A., Amos, A. F., Anderson, O. R., Garside, C., Haines, K. C., Malone, T. C., Paul, A. Z. and Rice, G. E., 1973. The environmental impact of deep-sea mining: Progress report. NOAA Tech. Rep. ERL 290-OD 11, National Oceanographic and Atmospheric Administration, Washington, D.C.

Roels, O. A., 1974. A suggested procedure to ensure the safe development of deep-sea mining. Proc. Ann. Conf. Mar. Technol. Soc., 10th, Washington, D.C., pp. 163—168.

Rolf, R. F., 1969. Process for the selective recovery of manganese and iron from ores. U.S. Patent, 3,471,285.

Ronov, A. B., Balashov, Yu. A. and Migdisov, A. A., 1967. Geochemistry of the rare earths in the sedimentary cycle. Geochim. Int., 4: 1—17.

Rossmann, R., 1973. Lake Michigan Ferromanganese Nodules. Ph.D. Thesis, Univ. of Michigan, Ann Arbor, Mich., 151 pp. (unpublished).

Rossmann, R. and Callender, E., 1968. Manganese nodules in Lake Michigan. Science, 162: 1123—1124.

Rossmann, R. and Callender, E., 1969. Geochemistry of Lake Michigan manganese nodules. Proc. Conf. Great Lakes Res., 12th, 12: 306—316.

Rossmann, R., Callender, E. and Bowser, C. J., 1972. Inter-element geochemistry of Lake Michigan ferromanganese nodules. Proc. Int. Geol. Congr., 24th, Sect. 10: 336—341.

Rothstein, A. J. and Kaufman, R., 1973. The approaching maturity of deep ocean mining — the pace quickens. Proc. Offshore Technol. Conf., 1973, pp. 1323—1344.

Rothstein, A. J. and Kaufman, R., 1974. The approaching maturity of deep ocean mining — the pace quickens. Min. Eng., 26(4): 31—36.

Rowe, G. T., 1971. Benthic biomass and surface productivity. In: J. D. Costlow (Editor), Fertility of the Sea, Vol. 2. Gordon and Breach, London, pp. 441—454.

Rubey, W. W., 1951. Geologic history of sea water. Bull. Geol. Soc. Am., 62: 1111—1148.

Rubin, A. P., 1969. Some legal implications of the Pueblo incident. Int. Comp. Law Q., 18: 961—970.

Ryther, J. H., 1969. Photosynthesis and fish production in the sea. Science, 166: 72—76.

Samoilov, Y. V. and Gorshkova, T. I., 1924. The deposits of the Barents and Kara Seas. Tr. Plov. Morsk. Nauch. Inst., 14: 40 pp.

Samoilov, Y. V. and Titov, A. G., 1922. Iron-manganese rich nodules of the Black, Baltic and Barents Seas. Tr. Geol. Miner. Muz., 3: 25—112.

Sano, M. and Matsubara, H., 1970. Some aspects of the element distribution of manganese nodules and its relationship to mineral composition. Suiyokwai-Shi, 17: 111—114.

Saunders, W. M. H., 1965. Phosphate retention by New Zealand soils and its relationship to free sesquioxides, organic matter and other soil properties. N.Z. J. Agric. Res., 8: 30—57.

Sayles, F. L. and Bischoff, J. L., 1973. Ferromanganoan sediments in the equatorial East Pacific. Earth Planet. Sci. Lett., 19: 330—336.

Schaefer, M. B., 1969. Freedom of scientific research and exploration in the sea. Stanford J. Int. Stud., 4: 46—70.

Schatz, C., 1971. Observation of sampling and occurrence of manganese nodules. Prepr. 1971 Offshore Technol. Conf., 1: 1389—1396.

Schaufelberger, F. A., 1956. Precipitation of metal from salt solution by reduction with hydrogen. J. Metall., 8: 695—704.

Schaufelberger, F. A. and Roy, T. K., 1955. Separation of copper, nickel and cobalt by selective reduction from aqueous solution. Trans. Inst. Min. Metall., 64: 375—393.

Schindler, P. W. and Gamsjäger, H., 1972. Acid-base reactions of the TiO_2 (anatase) — water interface and the point of zero charge of TiO_2 suspensions. Kolloid-Z.Z. Polym., 250: 759—763.

Schnitzer, M. and Skinner, S. I. M., 1966. Organo-metallic interactions in soils, 5. Stability constants of Cu^{+2}, Fe^{+2}, and Zn^{+2} fulvic acid complexes. Soil Sci., 102: 361—365.

Schoettle, M. and Friedman, G. H., 1971. Fresh water iron-manganese nodules in Lake George, New York. Bull. Geol. Soc. Am., 82: 101—110.

Schultz-Westrum, H. H., 1973. The station and cruise pattern of the R/V Valdivia in relation to the variability of manganese nodule occurrences. In: M. Morgenstein (Editor), Papers on the Origin and Distribution of Manganese Nodules in the Pacific and Prospects for Exploration. Hawaii Inst. of Geophysics, Honolulu, pp. 145—149.

Schutz, D. F. and Turekian, K. K., 1965. The investigation of the geographical and vertical distribution of several trace elements in seawater using neutron activation analysis. Geochim. Cosmochim. Acta, 29: 259—313.

Schuylenborgh, J. V. and Arens, P. L., 1950. The electrokinetic behavior of freshly prepared γ- and α-FeOOH. Recl. Trav. Chim., 69: 1557—1565.

Schweisfurth, R., 1971. Manganknollen im Meer. Naturwissenschaften, 58: 344—347.

Scott, M. R., Scott, R. B., Rona, P. A., Butler, L. W. and Nalwalk, A. J., 1974. Rapidly accumulating manganese deposit from the median valley of the Mid-Atlantic Ridge. Geophys. Res. Lett., 1: 355—358.

Scott, R. B., Rona, P. A., Butler, L. W., Nalwalk, A. J. and Scott, M. R., 1972a. Manganese crusts of the Atlantis Fracture Zone. Nature Phys. Sci., 239: 77—79.

Scott, R. B., Rona, P. A., Butler, L. W., Scott, M. R., Kennett, J. P., Nalwalk, A. J., Warme, J. E. and McCrevey, J. A., 1972b. Mn crusts of the Atlantis Fracture Zone. EOS Trans. Am. Geophys. Union, 53(4): 529 (Abstr.).

Semina, H. J., 1968. Water movements and the size of phytoplankton cells. Sarsia, 34: 267—272.

Sevast'yanov, V. F., 1967. Redistribution of arsenic during formation of iron-manganese concretions in Black Sea sediment. Dokl. Akad. Nauk SSSR, 176: 191—193 (English transl. 180—182).

Sevast'yanov, V. F. and Volkov, I. I., 1966. The chemical composition of the ferromanganese concretions of the Black Sea. Dokl. Akad. Nauk SSSR, 166: 701—704 (English transl. 174—176).

Sevast'yanov, V. F. and Volkov, I. I., 1967a. Redistribution of chemical elements during the oxidation process proceeding in the bottom deposits of the oxygen zone of the Black Sea. Tr. Inst. Okeanol., 83: 115—134.

Sevast'yanov, V. F. and Volkov, I. I., 1967b. Redistribution of chemical elements in the oxidised layers of the Black Sea sediments and the formation of iron-manganese nodules. Tr. Inst. Okeanol., 83: 135—152.

Shannon, R. D. and Prewitt, C. T., 1969. Effective ionic radii in oxides and fluorides. Acta Crystallogr., 25B: 925—946.

Shapiro, J., 1957. Chemical and biological studies on the yellow organic acids of lake water. Limnol. Oceanogr., 2: 161—179.

Shapiro, N. I., 1965. The chemical composition of deposits formed by *Metallogenium* and *Siderococcus*. In: Perfil'ev, B. V. et al., Applied Capillary Microscopy. Consultants Bureau, New York, N.Y., pp. 82—87.

Sheldon, R. W. and Parsons, T. R., 1967. A continuous size spectrum for particulate matter in the sea. J. Fish. Res. Bd. Can., 24: 909—915.

Shepard, F. P., 1973. Submarine Geology. Harper and Rowe, New York, N.Y., 3rd ed., 517 pp.

Shima, M. and Okada, A., 1968. Study on the manganese nodule (I). Manganese nodules collected from a long deep-sea core on the mid-Pacific ocean floor. Proc. Soc. Petrol. Miner. Miner. Deposits, 60: 47—56 (in Japanese).

Shimakage, K., Sakata, K., Yamo, H. and Morioka, S., 1968. On the ammonia pressure leaching of nickelferrite. J. Jap. Inst. Metals, 32: 584—589 (in Japanese).

Shimakage, K., Okuyama, H. and Morioka, S., 1969. On the denickelizing of laterite ores by ammonia pressure leaching. J. Min. Metall., Japan, 85: 91—96 (in Japanese).

Shinjo, T., 1966. Mössbauer effect in antiferromagnetic fine particles. J. Phys. Soc. Jap., 21: 917—922.

Shipek, C. J., 1960. Photographic study of some deep-sea floor environments in the eastern Pacific. Bull. Geol. Soc. Am., 71: 1067—1074.

Shterenberg, L. Y., 1971. Some aspects of the genesis of iron-manganese concretions in the Gulf of Riga. Dokl. Akad. Sci. SSSR, 201: 252—255.

Shterenberg, L. Y., Brazilevskaya, Y. S. and Chigireva, T. A., 1966. Manganese and iron carbonates in bottom deposits of Lake Pinnus-Yarvi. Dokl. Akad. Nauk SSSR, 170: 691—694 (English transl. 205—209).

Shterenberg, L. Y., Gorshkova, T. I. and Naktinas, E. M., 1968. Manganese carbonates in ferromanganese nodules in the Gulf of Riga. Litol i Polez. Iskop., No. 4: 63—69 (English transl. 438—443).

Sillén, L. G., 1961. The physical chemistry of sea water. In: M. Sears (Editor), Oceanography. Am. Assoc. Adv. Sci. Publ., 67: 549—581.

Sillén, L. G. and Martell, A. E., 1964. Stability constants of metal-ion complexes. Chem. Soc. Spec. Publ., Lond., 17: 754 pp.

Simons, C. S., 1964. Hydrogen sulfide as a hydrometallurgical reagent. In: M. E. Wadsworth and F. T. Davis (Editors), Unit Process in Hydrometallurgy, Gordon and Breach, New York, N.Y., pp. 592—616.

Singer, P. C. and Stumm, W., 1970. Acidic mine drainage: the rate-determining step. Science, 167: 1121—1123.

Sisselman, R., 1975. Ocean miners take soundings on legal problems, development alternatives. Eng. Min. J., 176(4): 75—86.

Sissons, J. B., 1967. The Evolution of Scotland's Scenery. Oliver and Boyd, Edinburgh, 259 pp.

Skadowsky, S., 1923. Hydrophysiologische und hydrobiologische Beobachtungen über die Bedeutung der Reaction des Mediums für Süsswasser Organismen. Verh. Int. Ver. Limnol., 1: 341—358.

Skarbo, R. R., 1973a. Extraction of Copper and Nickel from Manganese Nodules. U.S. Patent, 3,723,095.

Skarbo, R. R., 1973b. Extraction of Metal Values from Manganese Deep Sea Nodules. U.S. Patent, 3,728,105.

Skirrow, G., 1975. The dissolved gases — carbon dioxide. In: J. P. Riley and G. Skirrow (Editors), Chemical Oceanography, Vol. 2. Academic Press, London, 2nd ed., pp. 1—192.

Skornyakova, N. S., 1965. Dispersed iron and manganese in Pacific Ocean sediments. Int. Geol. Rev., 7: 2161—2174.

Skornyakova, N. S. and Andrushchenko, P. F., 1970. Ferro-manganese in the Pacific Ocean. In: P. L. Bezrukov (Editor), Sedimentation in the Pacific Ocean. Nauka, Moscow, pp. 203—268 (in Russian). (Translation in: Int. Geol. Rev., 16: 863—919.)

Skornyakova, N. S. and Zenkevitch, N. L., 1961. Distribution of iron-manganese nodules in the top layers of the Pacific Ocean deposits. Okeanologiya, 1: 86—94 (in Russian).

Skornyakova, N. S., Andrushchenko, P. F. and Fomina, L. S., 1962. The chemical composition of iron-manganese nodules in the Pacific Ocean. Okeanologiya, 2: 264—277. (Translation in: Deep-Sea Res., 11: 93—104.)

Sly, P. G. and Thomas, R. L., 1974. Review of geological research as it relates to an understanding of Great Lakes limnology. J. Fish. Res. Bd Can., 31: 795—825.

Smales, A. A. and Wiseman, J. D. H., 1955. Origin of nickel in deep-sea sediments. Nature, 175: 464—465.

Smith, B. H. and Leeper, G. W., 1969. The fate of applied molybdate in acidic soils. J. Soil Sci., 20: 247—253.

Smith, J. D. and Burton, J. D., 1972. The occurrence and distribution of tin with particular reference to marine environments. Geochim. Cosmochim. Acta, 36: 621—626.

Smith, R. E., Gassaway, J. D. and Giles, H. N., 1968. Iron-manganese nodules from Nares Abyssal Plain: geochemistry and mineralogy. Science, 161: 780—781.

Sokolova, M. N., 1972. Trophic structure of deep-sea macrobenthos. Mar. Biol., 16: 1—12.

Sokolova, T. A. and Polteva, R. N., 1968. The study of iron-manganese concretions from a strongly podzolic soil profile. Trans. Int. Congr. Soil Sci., 9th, 4: 459—466.

Sokolow, N., 1901. Die Manganerzlager in den tertiären Ablagerungen des Gouvernements Jekaterinoslaw. Mem. Comit. Geol. St. Petersb., 18: 61—82.

Somayajulu, B. L. K., 1967. Beryllium-10 in manganese nodule. Science, 156: 1219—1220.

Somayajulu, B. L. K., Heath, G. R., Moore, T. C. and Cronan, D. S., 1971. Rates of accumulation of manganese nodules and associated sediment from the equatorial Pacific. Geochim. Cosmochim. Acta, 35: 621—624.

Sorem, R. K., 1967. Manganese nodules: nature and significance of internal structure. Econ. Geol., 62: 141—147.

Sorem, R. K., 1972. Mineral recognition and nomenclature in marine manganese nodules. Acta Miner. Petrogr., Szeged, 20: 383—384 (Abstr.).

Sorem, R. K., 1973. Manganese nodules as indicators of long-term variations in sea floor environment. In: M. Morgenstein (Editor), Papers on the Origin and Distribution of Manganese Nodules in the Pacific and Prospects for Exploration. Hawaii Inst. of Geophysics, Honolulu, pp. 151—164.

Sorem, R. K. and Cameron, E. N., 1960. Manganese oxides and associated minerals of the Nsuta manganese deposits, Ghana, West Africa. Econ. Geol., 55: 278—310.

Sorem, R. K. and Foster, A. R., 1969. Growth history of manganese nodules west of Baja California, Mexico. Spec. Pap. Geol. Soc. Am., 121: 287 (Abstr.).

Sorem, R. K. and Foster, A. R., 1972a. Internal structure of manganese nodules and implications in beneficiation. In: D. R. Horn (Editor), Ferromanganese Deposits on the Ocean Floor. National Science Foundation, Washington, D.C., pp. 167—181.

Sorem, R. K. and Foster, A. R., 1972b. Macroprobe X-ray analysis and specimen distance effect. Norelco Rep., 19: 21—26.

Sorem, R. K. and Foster, A. R., 1972c. Marine manganese nodules: importance of structural analysis. Int. Geol. Congr., 24th, Canada, Sect. 8, Mar. Geol. Geophys., pp. 192—200.

Sorem, R. K. and Foster, A. R., 1973. Mineralogical, chemical, and optical properties and standards for study of growth features and economic potential of manganese nodules. In: Inter-University Program of Research on Ferromanganese Deposits of the Ocean Floor — Phase I Report. National Science Foundation, IDOE, Washington, D.C., pp. 23—38.

Sorem, R. K. and Gunn, D. W., 1967. Mineralogy of manganese deposits, Olympic Penin-
    sula, Washington. Econ. Geol., 62: 22—56.
Sorokin, Y. I., 1972. Role of biological factors in the sedimentation of iron, manganese,
    and cobalt and in the formation of nodules. Oceanology, 12: 1—11.
Spencer, D. W. and Brewer, P. G., 1969. The distribution of copper, zinc, and nickel in
    sea water of the Gulf of Maine and the Sargasso Sea. Geochim. Cosmochim. Acta,
    33: 325—339.
Spencer, D. W. and Brewer, P. G., 1970. Analytical methods in oceanography, 1. Inorga-
    nic methods. CRC Critical Rev. Solid St. Sci., 1: 409—478.
Spencer, D. W. and Brewer, P. G., 1971. Vertical advection diffusion and redox potentials
    as controls on the distribution of manganese and other trace metals dissolved in
    waters of the Black Sea. J. Geophys. Res., 76: 5877—5892.
Spencer, D. W., Robertson, D. E., Turekian, K. K. and Folsom, T. R., 1970. Trace ele-
    ment calibrations and profiles at the GEOSECS test station in the northeast Pacific
    Ocean. J. Geophys. Res., 75: 7688—7696.
Spencer, D. W., Brewer, P. G. and Sachs, P. L., 1972. Aspects of the distribution and
    trace element composition of suspended matter in the Black Sea. Geochim. Cosmo-
    chim. Acta, 36: 71—86.
Spooner, E. T. C. and Fyfe, W. S., 1973. Sub-sea-floor metamorphism, heat and mass
    transfer. Contrib. Miner. Petrol., 42: 287—304.
Stang, D. P., 1968. Wet land: the unavailable resource of the outer continental shelf. J.
    Law Econ. Dev., 2: 153—189.
Steele, J. H. and Yentsch, C. S., 1960. The vertical distribution of chlorophyll. J. Mar.
    Biol. Assoc. U.K., 39: 217—226.
Stefánsson, U. and Richards, F. A., 1963. Processes contributing to the nutrient distribu-
    tion off the Columbia River and Strait of Juan de Fuca. Limnol. Oceanogr., 8:
    394—410.
Stephens, E. A., 1956. The manganese deposits of North Borneo. In: J. G. Reyna (Editor),
    Symposium sobre yacimientos de manganeso, 4. Asia y Oceania, Int. Geol. Congr.,
    20th, Mexico, pp. 297—312.
Stevenson, J. R., 1972. Foreign policy issues and the Law of the Sea Conference. In:
    Hearing before the Subcommittee on Oceans and Atmosphere of the Committee on
    Commerce, United States Senate, pp. 65—70.
Stevenson, J. S. and Stevenson, L. S., 1970. Manganese nodules from the Challenger
    Expedition at Redpath Museum. Can. Miner., 10: 599—615.
Straczek, J. A., Horen, A., Ross, M. and Warshaw, C. M., 1960. Studies of the manganese
    oxides, IV. Todorokite. Am. Miner., 45: 1174—1184.
Strakhov, N. M., 1966. Types of manganese accumulation in present day basins: their
    significance in understanding of manganese mineralisation. Int. Geol. Rev., 8:
    1172—1196.
Strunz, H., 1970. Mineralogische Tabellen. Akad. Verlagsgesellschaft, Leipzig, 5th ed.,
    527 pp.
Stumm, W. and Brauner, P. A., 1975. Chemical speciation. In: J. P. Riley and R. Chester
    (Editors), Chemical Oceanography, Vol. 1. Academic Press, London, pp. 173—
    239.
Stumm, W. and Lee, G. F., 1961. Oxygenation of ferrous iron. Ind. Eng. Chem., 53:
    143—146.
Stumm, W. and Morgan, J. J., 1970. Aquatic Chemistry. Wiley-Interscience, New York,
    N.Y., 583 pp.
Stumm, W. and O'Melia, C. R., 1968. Stoichiometry of coagulation. J. Am. Water Works
    Assoc., 60: 514—539.
Stumm, W., Huang, C. P. and Jenkins, S. R., 1970. Specific chemical interaction affecting
    the stability of dispersed systems. Croatica Chem. Acta, 42: 223—245.

Sugawara, K., Koyama, T. and Terada, K., 1958. Coprecipitation of iodide ions by some metallic hydrated oxides with special reference to iodide accumulation in bottom water layers and in interstitial water of muds in some Japanese lakes. J. Earth Sci., Nagoya Univ., 6: 52—61.

Sugawara, K., Tanaka, M. and Okabe, S., 1959. Separation and determination of microgram quantities of molybdenum in natural waters. Bull. Chem. Soc. Jap., 32: 221—222.

Summerhayes, C. P., 1967. Manganese nodules from the south-western Pacific. N.Z. J. Geol. Geophys., 10: 1372—1381.

Sumner, M. E. and Reeve, N. G., 1966. The effect of iron oxide impurities on the positive and negative adsorption of chloride by kaolinites. J. Soil Sci., 17: 274—279.

Suzdalev, I. P., 1969. Superparamagnetism of antiferromagnetic particles of ultramicro size. Proc. Conf. Application Mössbauer Effect, Tihany, pp. 193—196.

Sverdrup, H. U., Johnson, M. W. and Fleming, R. H., 1946. The Oceans, their Physics, Chemistry and General Biology. Prentice-Hall, Englewood Cliffs, N.J., 1087 pp.

Swaine, D. J. and Mitchell, R. L., 1960. Trace-element distribution in soil profiles. J. Soil Sci., 11: 347—368.

Swanson, V. E., Frost, I. C., Rader, L. F. and Huffman, C., 1966. Metal sorption by northwest Florida humate. U.S. Geol. Surv. Prof. Pap., 550C: 174—177.

Tadros, T. F. and Lyklema, J., 1968. Adsorption of potential-determining ions at the silica aqueous electrolyte interface and the role of some cations. J. Electroanal. Chem., 17: 267—275.

Takahashi, M., Satake, K. and Nakamoto, N., 1972. Chlorophyll distribution and photosynthetic activity in the north and equatorial Pacific Ocean along 155°W. J. Oceanogr. Soc. Jap., 28(1): 27—36.

Takahashi, T., Weiss, R. F., Culberson, C. H., Edmond, J. M., Hammond, D. E., Wong, C. S., Li, Y.-H. and Bainbridge, A. E., 1970. A carbonate chemistry profile at the 1969 GEOSECS intercalibration station in the eastern Pacific Ocean. J. Geophys. Res., 75: 7648—7666.

Taniguchi, K., Nakajima, M., Yoshida, S. and Tarama, K., 1970. The exchange of the surface protons in silica gel with some kinds of metal ions. Nippon Kagaku Zasshi, 91: 525—529.

Taylor, R. M. and McKenzie, R. M., 1966. The association of trace elements with manganese minerals in Australian soils. Aust. J. Soil Sci., 4: 29—39.

Taylor, R. M., McKenzie, R. M. and Norrish, K., 1964. The mineralogy and chemistry of manganese in some Australian soils. Aust. J. Soil Res., 2: 235—248.

Tegger Kildow, J. A., 1973. Nature of the present restrictions on oceanic research. In: W. S. Wooster (Editor), Freedom of Oceanic Research. Crane Russak, New York, N.Y., 255 pp.

Terasmae, J., 1971. Notes on lacustrine manganese-iron concretions. Geol. Surv. Can. Pap., 70—69: 13 pp.

Tewari, P. G., Campbell, A. B. and Lee, W., 1972. Adsorption of Co^{2+} by oxides from aqueous solution. Can. J. Chem., 50: 1642—1648.

Theis, T. L. and Singer, P. C., 1974. Complexation of iron (II) by organic matter and its effect on iron (II) oxygenation. Environ. Sci. Technol., 8: 569—573.

Thiel, G. A., 1925. Manganese precipitated by microorganisms. Econ. Geol., 20: 301—310.

Thode, H. G., Kleerekoper, H. and McElcheran, D., 1951. Isotopic fractionation in the bacterial reduction of sulphate. Research, 4: 581—582.

Thomas, D. W. and Blumer, M., 1964. Pyrene and fluoranthene in manganese nodules. Science, 143: 39.

Thomas, W. H., 1966. Surface nitrogenous nutrients and phytoplankton in the northeastern tropical Pacific Ocean. Limnol. Oceanogr., 11: 293—400.

Thomas, W. H., 1969. Phytoplankton nutrient enrichment experiments off Baja Califor-
    nia and in the eastern equatorial Pacific Ocean. J. Fish. Res. Bd Can., 26: 1133—
    1145.
Thomas, W. H., 1970a. On nitrogen deficiency in tropical Pacific oceanic phytoplankton:
    photosynthetic parameters in poor and rich water. Limnol. Oceanogr., 15: 380—385.
Thomas, W. H., 1970b. Effect of ammonium and nitrate concentration on chlorophyll
    increases in natural tropical Pacific phytoplankton populations. Limnol. Oceanogr.,
    15: 387—394.
Thomas, W. H. and Owen, R. W., 1971. Estimating phytoplankton production from
    ammonium and chlorophyll concentrations in nutrient-poor water of the eastern
    tropical Pacific Ocean. Fish. Bull., 69: 87—92.
Thomas, W. H., Renger, E. H. and Dodson, A. N., 1971. Near-surface organic nitrogen
    in the eastern tropical Pacific Ocean. Deep-Sea Res., 18: 65—71.
Thomson, C. W., 1873. Notes from the "Challenger" II. Nature, 8: 51—53.
Thorndike, E., 1975. A deep-sea, photographic nephelometer. Ocean Eng., 3: 1—15.
Thorndike, E. and Ewing, M., 1966. Light-scattering in the sea. In: Underwater Photo-
    Optics Seminar. Proc. Soc. Photo-Optical Instrum. Eng., pp. 1—7 (AIV).
Tizard, T. H., Moseley, H. N., Buchanan, J. Y. and Murray, J., 1885. Narrative of the
    cruise of H.M.S. Challenger with a general account of the scientific results of the
    expedition, Vol. 1, First part. Rep. Sci. Results Explor. Voyage Challenger, 509 pp.
Towe, I. M. and Bradley, W. F., 1967. Mineralogical constitution of colloidal "hydrous
    ferric oxides". J. Colloid Interface Sci., 24: 384—392.
Trimble, R. B. and Ehrlich, H. L., 1968. Bacteriology of manganese nodules, III. Reduc-
    tion of MnO_2 by two strains of nodule bacteria. Appl. Microbiol., 16: 695—702.
Trimble, R. B. and Ehrlich, H. L., 1970. Bacteriology of manganese nodules, IV. Induction
    of an MnO_2-reductase system in a marine bacillus. Appl. Microbiol., 19: 966—972.
Trofimov, A. V., 1939. Oxidising activity and pH of brown sediments of the Barents Sea.
    C.R. Acad. Sci., URSS, 23: 925—928.
Tromp, S. W., 1948. Shallow-water origin of radiolarites in southern Turkey. J. Geol.,
    56: 492—494.
Tucker, M. E., 1971. Devonian manganese nodules from France. Nature Phys. Sci., 230:
    116—117.
Tucker, M. E., 1973a. Ferromanganese nodules from the Devonian of the Montagne Noire
    (S. France) and West Germany. Geol. Rundsch., 62: 137—153.
Tucker, M. E., 1973b. Sedimentology and diagenesis of Devonian pelagic limestones
    (Cephalopodenkalk) and associated sediments of the Rhenohercynian Geosyncline,
    West Germany. Neues Jahrb. Geol. Paläontol. Abh., 142: 320—350.
Tucker, M. E., 1974. Sedimentology of Palaeozoic pelagic limestones: the Devonian
    Griotte (Southern France) and Cephalopodenkalk (Germany). In: K. J. Hsü and
    H. C. Jenkyns (Editors), Pelagic Sediments: on Land and under the Sea. Spec. Publ.
    Int. Assoc. Sedimentol., 1: 71—92.
Turekian, K. K., 1965. Some aspects of the geochemistry of marine sediments. In: J. P.
    Riley and G. Skirrow (Editors), Chemical Oceanography, Vol. 2. Academic Press,
    London, pp. 81—126.
Turekian, K. K., 1967. Estimates of the average Pacific deep-sea clay accumulation rate
    from material balance calculations. In: M. Sears (Editor), Progress in Oceanography,
    4: 227—244.
Turekian, K. K. and Imbrie, J., 1966. The distribution of trace elements in deep-sea
    sediments of the Atlantic Ocean. Earth Planet. Sci. Lett., 1: 161—168.
Turekian, K. K. and Wedepohl, K. H., 1961. Distribution of the elements in some major
    units of the earth's crust. Bull. Geol. Soc. Am., 72: 175—192.
Turekian, K. K., Katz, A. and Chan, L., 1973. Trace element trapping in pteropod tests.
    Limnol. Oceanogr., 18: 240—249.

Turekian, K. K., Cochran, J. K., Kharkar, D. P., Cerrato, R. M., Vaisnys, J. R., Sanders, H. L., Grassle, J. F. and Allen, J. A., 1975. The slow growth rate of a deep-sea clam determined by ^{228}Ra chronology. Proc. Nat. Acad. Sci. U.S.A., 72(7): 2829—2832.

Turner, R. P., 1971. The significance of color banding in the upper layers of Kara Sea sediments. U.S. Coastal Guard Oceanogr. Rep. No. 36 (CG 373-36), 36 pp.

Twenhofel, W. H., McKelvey, V. E. and Feray, D. E., 1945. Sediments of Trout Lake, Wisconsin. Bull. Geol. Soc. Am., 56: 1099—1142.

Ulrich, K. H., Scheffler, U. and Meixner, M. J., 1973. Aufarbeitung von Manganknollen durch Saure Laugung. In: C. Kruppa (Editor), Interocean '73. Seehafen-Verlag Erik Blumenfeld, Hamburg, 1: 445—457.

U.S. Congress, 1975. Ocean Manganese Nodules. Senate, Comm. Int. Insul. Affairs, 94th Congr., 1st Session, June 1975. U.S. Govt. Printing Office, Washington, D.C.

Van Bemmelen, R. W., 1949. The Geology of Indonesia, IA, General Geology of Indonesia and Adjacent Archipelagoes. Government Printing Office, The Hague, 732 pp.

Van Der Giessen, A. A., 1966. The structure of iron (III) oxide-hydrate gels. J. Inorg. Nucl. Chem., 28: 2155—2159.

Van Der Weijden, C. H., 1975. Sorption experiments relevant to the geochemistry of manganese nodules. Ph.D. Thesis, Rijks Univ. Utrecht, Utrecht, 154 pp. (unpublished).

Van Der Zeeuw, A. J., 1972. Copper, nickel, cobalt and iron separation process. U.S. Patent, 3,701,650.

Van Hecke, M. C. and Bartlett, R. W., 1973. Kinetics of sulfation of Atlantic Ocean manganese nodules. Metall. Trans., 4: 941—947.

Van Hemmelrijck, L., 1973. Artificial upwelling pipelines: optimization and thermal losses. Proc. Solar Sea Power Plant Conf. Workshop, Pittsburgh, Pa., June 27—28, pp. 221—239.

Van Schuylenborgh, J. and Arens, P. L., 1950. The electrokinetic behaviour of freshly prepared γ- and α-FeOOH. Recl. Trav. Chim. Pays-Bas Belg., 69: 1557—1565.

Varentsov, I. M., 1964. Sedimentary Manganese Ores. Elsevier, Amsterdam, 119 pp.

Varentsov, I. M., 1972a. Geochemical studies on the formation of iron-manganese nodules and crusts in Recent basins, I. Eningi-Lampi Lake, Central Karelia. Acta Miner. Petrogr., Szeged., 10: 363—381.

Varentsov, I. M., 1972b. On the main aspects of formation of ferromanganese ores in Recent basins. Int. Geol. Congr., 24th, Sect. 4, pp. 395—403.

Varentsov, I. M., 1973. Geochemical aspects of formation of ferromanganese ores in shelf regions of Recent seas. Acta Miner. Petrogr. Szeged., 21: 141—153.

Varentsov, I. M., 1974. Ferromanganese nodules from the Gulf of Finland. Acta Miner. Petrogr., Szeged, 21: 303—304 (Abstr.).

Varentsov, I. M. and Pronina, N. V., 1973. On the study of mechanisms of iron-manganese ore formation in Recent basins: the experimental data on nickel and cobalt. Miner. Deposita, 8: 161—178.

Vasilchikov, N. V., Shirer, G. B., Matespon, Yu. A., Krasnykh, I. F. and Grishankova, E. A., 1968. Iron-manganese nodules from the ocean floor — raw materials for the production of cobalt, nickel, manganese and copper. Tsvet. Metall., 41(1): 40—42 (in Russian).

Veeh, H. H., 1967. Deposition of uranium from the ocean. Earth Planet. Sci. Lett., 3: 145—150.

Verwey, E. J. W. and Overbeek, J. Th. G., 1948. Theory of the Stability of Lyophobic Colloids. Elsevier, Amsterdam.

Vine, F. J. and Hess, H. H., 1970. Sea-floor spreading. In: A. E. Maxwell (Editor), The Sea, Part 2. Wiley, New York, N.Y., pp. 587—622.

Vinogradov, A. P., 1953. The elementary chemical composition of marine organisms. Mem. Sears Found. Mar. Res., 2: 647 pp.

Vogt, J. H. L., 1906. Über mangan Wiesernerz und Über das Verhältnis zwischen Eisen und Mangan in der See- und Wiesenerzen. Z. Prakt. Geol., 14: 217—233.

Vogt, J. H. L., 1915. On manganrik sjømalm i Stors jøen, Nordre Odalen. Norg. Geol. Unders. Aarbok, 75, IV: 3—43.

Von Buttlar, H. and Houtermans, G., 1950. Photographische Bestimmung der Aktivitätverteilung in einer Manganknolle der Tiefsee. Naturwissenschaften, 37: 400—401.

Von Heimendahl, M., Hubred, G. L., Fuerstenau, D. W. and Thomas, G., 1976. A transmission electron microscope study of deep-sea manganese nodules. Deep-Sea Res., 23: 69—79.

Von Mackiw, V. N., Benz, T. W. and Evans, D. J. I., 1962. Hydrometallurgie unter Anwendung von Druck. Chemie-Ing.-Tech., 34: 441—444.

Wadsley, A. D., 1950a. A hydrous manganese oxide with exchange properties. J. Am. Chem. Soc., 72: 1782—1784.

Wadsley, A. D., 1950b. Synthesis of some hydrated manganese minerals. Am. Miner., 35: 485—499.

Wadsley, A. D., 1952. The structure of lithiophorite (Al, Li) MnO_2 $(OH)_2$. Acta Crystallogr., 5: 676—680.

Wadsley, A. D., 1953. The crystal structure of psilomelane, $(Ba, H_2O)_2 Mn_5 O_{10}$. Acta Crystallogr., 6: 433—438.

Wadsley, A. D., 1955. The crystal structure of chalcophanite, $ZnMn_3 O_7 \cdot 3H_2 O$. Acta Crystallogr., 8: 165—172.

Wadsley, A. D., 1964. Inorganic non-stoichiometric compounds. In: L. Mandelcorn (Editor), Non-Stoichiometric Compounds. Academic Press, London, pp. 98—209.

Walter, M. R., Bauld, J. and Brock, T. D., 1972. Siliceous algal and bacterial stromatolites in hot spring and geyser effluents of Yellowstone National Park. Science, 178: 402—405.

Warner, J. P., 1956. Nickel recovery at Fort Saskatchewan. Ind. Chem., 32: 359—363.

Watkins, N. D. and Kennett, J. P., 1971. Antarctic bottom water: major change in velocity during the Late Cenozoic between Australia and Antarctica. Science, 173: 813—818.

Watkins, N. D. and Kennett, J. P., 1972. Regional sedimentary disconformities and upper Cenozoic changes in bottom water velocities between Australasia and Antarctica. Antarctic Res. Ser., 19: 273—293.

Watson, J. A. and Angino, E. E., 1969. Iron rich layers in sediments from the Gulf of Mexico. J. Sediment. Petrol., 39: 1412—1419.

Wauschkuhn, A., 1973. Rezente Sulfiabildung in vulkanischen Seen auf Hokkaido (Japan), I. Geochemie des Ojunuma und des Okunoku. Geol. Rundsch., 62: 774—785.

Wedepohl, K. H., 1960. Spurenanalytische Untersuchungen an Tiefseetonen aus dem Atlantik. Geochim. Cosmochim. Acta, 18: 200—231.

Wehmiller, J. and Hare, P. E., 1971. Racemization of amino acids in marine sediments. Science, 173: 907—911.

Weisz, P. B., 1968. Deep-sea manganese nodules as oxidation catalysts. J. Catalysis, 10: 407—412.

Weisz, P. B. and Silvestri, A. J., 1973. Demetalation of Hydrocarbon Charge Stocks. U.S. Patent, 3,716,479.

Welling, C. G., 1972. Some environmental factors associated with deep-ocean mining. 8th Ann. Mtg Mar. Technol. Soc., Washington, D.C.

Wendt, J., 1963. Stratigraphisch-paläontologische Untersuchungen im Dogger Westsiziliens. Boll. Soc. Paleont. Ital., 2: 57—145.

Wendt, J., 1969a. Stratigraphie und Paläogeographie des Roten Jurakalks im Sonnwendgebirge (Tirol, Österreich). Neues Jahrb. Geol. Paläontol. Abh., 132: 219—238.

Wendt, J., 1969b. Foraminiferen "Riffe" in karnischen Hallstätter Kalk des Feuerkogels (Steiermark, Österreich). Paläontol. Z., 43: 177—193.

Wendt, J., 1970. Stratigraphische Kondensation in triadischen und jurassischen Cephalopodenkalk der Tethys. Neues Jahrb. Geol. Paläontol. Monatsh., 1970: 433—448.

Wendt, J., 1973. Cephalopod accumulations in the Middle Triassic Hallstatt-Limestone of Jugoslavia and Greece. Neues Jahrb. Geol. Paläontol. Monatsh., 1973: 624—640.

Wendt, J., 1974. Encrusting organisms in deep-sea manganese nodules. Spec. Publ. Int. Assoc. Sedimentol., 1: 437—447.

Weyl, P. K., 1970. Oceanography: An Introduction to the Marine Environment. Wiley, New York, N.Y., 535 pp.

Wezel, F. C., 1970a. Numidian flysch: an Oligocene—Early Miocene continental rise deposit off the African platform. Nature, 228: 275—276.

Wezel, F. C., 1970b. Geologia del Flysch Numidico della Sicilia nordorientale. Mem. Soc. Geol. Ital., 9: 225—280.

White, D. E., 1968. Environments of generation of some base-metal ore deposits. Econ. Geol., 63: 301—335.

White, D. E., Hem, J. D. and Waring, G. A., 1963. Chemical composition of subsurface waters. U.S. Geol. Surv. Prof. Pap., 440F: 67 pp.

Wilber, C. G., 1971. Turbidity — general introduction. In: O. Kinne (Editor), Marine Ecology, Vol. 1. Wiley-Interscience, London, pp. 1157—1165.

Wilder, T. C., 1972. Verfahren zur Gewinnung von Nickel und Kupfer aus komplexen Erzen. German Patent, 2,135,731.

Wilder, T. C., 1973. Two Stage Selective Leaching of Copper and Nickel from Complex Ore. U.S. Patent, 3,736,125.

Williams, J. D. H., Syers, J. K., Shukla, S. S., Harris, R. F. and Armstrong, D. E., 1971. Levels of inorganic and total phosphorus in lake sediments as related to other sediment parameters. Environ. Sci. Technol., 5: 1113—1120.

Willis, J. P., 1970. Investigations on the Composition of Manganese Nodules with Particular Reference to Certain Trace Elements. M.Sc. Thesis, Univ. of Cape Town, Cape Town (unpublished).

Willis, J. P. and Ahrens, L. H., 1962. Some investigations on the composition of manganese nodules, with particular reference to certain trace elements. Geochim. Cosmochim. Acta, 26: 751—764.

Wimbush, M., 1972. Tidal movements of the deep sea. Underwater J., 4(6): 239—248.

Winterhalter, B., 1966. Iron-manganese concretions from the Gulf of Bothnia and the Gulf of Finland. Geotekn. Julk, 69: 1—77.

Winterhalter, B. and Siivola, J., 1967. An electron microprobe study of the distribution of iron manganese and phosphorus in concretions from the Gulf of Bothnia, Northern Baltic Sea. C.R. Soc. Geol., Finl., 39: 161—172.

Winters, E., 1938. Ferromanganiferous concretions from some podzolic soils. Soil Sci., 46: 33—40.

Wiseman, J. D. H., 1937. Geological and mineralogical investigations, 1. Basalts from the Carlsberg Ridge, Indian Ocean. Scient. Rep. John Murray Exped., 1933—34, 3: 1—30.

Woo, C. C., 1973. Scanning electron micrographs of marine manganese micronodules, marine pebble-sized nodules, and fresh water manganese nodules. In: M. Morgenstein (Editor), Papers on the Origin and Distribution of Manganese Nodules in the Pacific and Prospects for Exploration. Hawaii Inst. of Geophysics, Honolulu, pp. 165—171.

Wooster, W. S. and Volkman, G. H., 1960. Indications of deep Pacific circulation from the distribution of properties at five kilometers. J. Geophys. Res., 65: 1239—1248.

Wyman, W. F. and Ravitz, S. F., 1947. Sulfur dioxide leaching tests on various western manganese ores. U.S. Bur. Min. Rep. Invest., 4077: 12 pp.

Wyrtki, K., 1963. The horizontal and vertical field of motion in the Peru Current. Bull. Scripps Inst. Oceanogr., 8: 313—346.

Wyrtki, K., 1967. Circulation and water masses in the eastern equatorial Pacific Ocean. Int. J. Oceanol. Limnol., 1: 117—147.

Yochelson, E. L., 1968. Biostratigraphy of the Phosphoria, Park City, and Skedhorn Formations. Prof. Pap. U.S. Geol. Surv., 313D: 571—660.

Yoshimura, T., 1934. Todorokite, a new manganese mineral from the Todoroki Mine, Hokkaido, Japan. J. Fac. Sci. Hokkaido Univ., Sapporo, Ser. 4, 2: 289—297.

Zelenov, K. K., 1964. Iron and manganese in exhalations of the submarine Banu Wahu volcano (Indonesia). Dokl. Akad. Sci. USSR, Earth Sci. Sect., 155: 94—96.

Zimmerly, S. R., 1967. Use of Deep-Sea Nodules for Removing Sulfur Compounds from Gases. U.S. Patent, 3,330,096.

Zingg, Th., 1935. Beiträge zur Schotteranalyse. Schweiz. Min. Petrogr. Mitt., 15: 39—140.

Zumberge, J. H., 1952. The lakes of Minnesota, their origin and classification. Bull. Minn. Geol. Surv., 35: 1—90.

Zwicker, W. K., Groeneveld Meijer, W. O. J. and Jaffe, H. W., 1962. Nsutite — a widespread manganese oxide mineral. Am. Miner., 47: 246—266.

SUBJECT INDEX